Advanced Structured Materials

Volume 85

Series editors
Andreas Öchsner, Esslingen, Germany
Lucas F. M. da Silva, Porto, Portugal
Holm Altenbach, Magdeburg, Germany

More information about this series at http://www.springer.com/series/8611

Andreas Öchsner
Editor

Engineering Applications for New Materials and Technologies

 Springer

Editor
Andreas Öchsner
Faculty of Mechanical Engineering
Esslingen University of Applied Sciences
Esslingen
Germany

ISSN 1869-8433   ISSN 1869-8441 (electronic)
Advanced Structured Materials
ISBN 978-3-319-72696-0   ISBN 978-3-319-72697-7 (eBook)
https://doi.org/10.1007/978-3-319-72697-7

Library of Congress Control Number: 2017961094

© Springer International Publishing AG 2018
This work is subject to copyright. All rights are reserved by the Publisher, whether the whole or part of the material is concerned, specifically the rights of translation, reprinting, reuse of illustrations, recitation, broadcasting, reproduction on microfilms or in any other physical way, and transmission or information storage and retrieval, electronic adaptation, computer software, or by similar or dissimilar methodology now known or hereafter developed.
The use of general descriptive names, registered names, trademarks, service marks, etc. in this publication does not imply, even in the absence of a specific statement, that such names are exempt from the relevant protective laws and regulations and therefore free for general use.
The publisher, the authors and the editors are safe to assume that the advice and information in this book are believed to be true and accurate at the date of publication. Neither the publisher nor the authors or the editors give a warranty, express or implied, with respect to the material contained herein or for any errors or omissions that may have been made. The publisher remains neutral with regard to jurisdictional claims in published maps and institutional affiliations.

Printed on acid-free paper

This Springer imprint is published by Springer Nature
The registered company is Springer International Publishing AG
The registered company address is: Gewerbestrasse 11, 6330 Cham, Switzerland

Preface

Engineering knowledge, skills, and development procedures play an important role for the development of new materials and technologies. As an example, modern joining technologies can be mentioned to fabricate new compound or composite materials, even composed of dissimilar materials. Such materials are many times exposed to harsh environments and must reveal specific properties. Technologies in this context are mainly related to the transportation technologies in their wider sense. This means automotive, marine technology, i.e., ships, amphibious vehicles, docks, offshore structures, and even robots. To achieve a certain performance, computer-based engineering tools are widely used and related to the simulation, evaluation of data and design processes. Typically, finite element and finite volume methods are used in the context of engineering simulations and this volume contains many state-of-the-art applications and developments to highlight their importance.

This volume contains selected and reviewed manuscripts from the International Conference on Marine & Automotive Technology (ICMAT 2017), 'Propelling & Driving Innovative Ideas,' which was held from March 15 to 17, 2017, at the Penang Island, Malaysia. This conference was organized by the Universiti Kuala Lumpur, Malaysian Institute of Marine Engineering Technology, Dataran Industri Teknologi Kejuruteraan Marin, Bandar Teknologi Maritim, Jalan Pantai Remis, 32200 Lumut, Perak, Malaysia. The committee members can be listed as follows:

Patron: Assoc. Prof. Zainorin Mohamad (Head of Campus/Dean). *Advisor:* Assoc. Prof. Dr. Mohd Yuzri Mohd Yusop (Deputy Dean, Academic & Technology), Dr. Puteri Zarina Megat Khalid (Deputy Dean, Student Development and Campus Lifestyle). *Chairman:* Dr. Wardiah Mohd Dahalan. *Co-Chairman:* Ahmad Makarimi Abdullah. *Chief of Publication:* Azman Ismail. *Publication:* Fauziah Ab Rahman, Amirrudin Yaacob, Sarah Nadiah Rashidi, Mohamad Rafiuddin Che Othman. *Secretary:* Noorazlina Mohamid Salih, Muhamad Fadli Ghani. *Protocol:* Norfadhlina Khalid, Fatin Zawani Zainal Azaim. *Audit/Jury:*

Azman Ismail, Atzroulnizam Abu, Dr. Norshakila Haris. *Finance:* Neirul Nisa Abd Latib. *Registration:* Siti Nur Azreen Ismail. Promotion: Mohd Saidi Hanaffi, Azila Ayub. *Logistics:* Daud Saari. *IT/Photography:* Yusmizhar Serad. *Prizes/Certificates:* Nordiana Jamil. *Sponsorship:* Mohd Shahrizam Saharudin. *IIIP:* Siti Noor Kamariah Yaakop.

Esslingen, Germany   Andreas Öchsner

Contents

A Study on the Effect of Parameters on the Tensile Strength of Friction Stir Welded AA6061 1.5 mm Thin Plate Butt-Joints 1
Nurfadilah Sukara, Azman Ismail, Darulihsan Abdul Hamid, Fauziah Ab Rahman, Bakhtiar Ariff Baharudin and Puteri Zarina Megat Khalid

Physical and Chemical Properties of Perak River Sand for Greensand Casting Molds 13
Muhammad Arieef Bin Hussain, Azhar Bin Abdullah and Rusmay Bin Abdullah

Potential Use of Cellulose Fibre Composites in Marine Environment—A Review 25
Muhamad Firdaus Muhamad Yang, Hisham Hamid and Ahmad Makarimi Abdullah

Development of a Batik Fiberglass Composite for Marine Applications Based on Water Absorption Testing 57
Muhammad Abdul Mun'aim Bin Mohd Idrus, Siti Amelia Binti Amiruddin, Siti Afiza Binti Johari Jamaludin and Shamsul Effendy Bin Abdul Hamid

Flame Spread Behavior over Kenaf Fabric, Polyester Fabric, and Kenaf/Polyester Combined Fabric 67
Mohd Azahari Razali, Azwan Sapit, Akmal Nizam Mohammed, Mohd Faisal Hushim, Azmahani Sadikin, Md Norrizam Mohmad Ja'at, Hazahir Peraman and Mirnah Suardi

Design and Analysis of an Automotive Oil Filter Gripper Socket Special Tool 77
Chee Guan Choong, Chun Lin Saw and Yoke Yin Chiang

The Development of Hovercraft Design with a Horizontal Propulsion System 91
Khairul Anuar Mat Saad, Kogulan Murugan
and Mohd Amin Hakim Ramli

A Rule Based Method to Auto-recognize Fillet Features of B-Rep Mill Parts.. 105
Pramodkumar Siddappa Kataraki and Mohd Salman Abu Mansor

Experimental Study of Direct Injected Marine Auxiliary Diesel Engine Performance, Emission and Cylinder Pressure Using Biodiesel Fuels Derived from Jatropha Curcas Oil............................. 115
Ridwan Saputra Nursal, Zakiman Zali, Ismail Zainol
and Mohd Nazri Mohd Sabri

Integrated Full Electric Propulsion System for Tanker Ships with Combined Diesel and Hydro Generator Drive 137
Wilfredo Yutuc, Hamzah Jamal and Bahktiar Afandi

A Stress Analysis and Design Improvement of a Car Door Hinge for Compact Cars 151
Tajul Adli Abdul Razak, Muhammad Najib Abdul Hamid,
Ahmadiamri Mohd Ghazali, Shahril Nizam Mohamed Soid
and Khairul Shahril Shafee

Development of a Global Warming Impact Prediction Analysis Tool for Mobile Vehicles 163
Chun Lin Saw, Chee Guan Choong, Yoke Yen Chiang and Azmi Naroh

Surface Recognition and Volume Generation for Symmetrical Parts Using a Mirror Approach 171
Ahmad Faiz Zubair and Mohd Salman Abu Mansor

Influence of Passenger Car Air Conditioner System Thermostat Level Setting to Fuel Consumption and Thermal Comfort 183
Rawaida Muhammad, Muhammad Khiril Kamaruddin and Yeoh Poh See

Investigation of the Piston Bowl Shape Effect on the Diesel Spray Development... 197
Azwan Sapit, Mohd Azahari Razali, Akmal Nizam Mohammad,
Mohd Faisal Hushim, Azmahani Binti Sadikin
and Md Norrizam Bin Mohmad Ja'at

Improvement of the Switching of Behaviours Using a Fuzzy Inference System for Powered Wheelchair Controllers 205
Jaya Bhanu Rao and Ahmad Zakaria

Mesh Filtering Algorithm for Virtualisation of Rapid Prototype Models Based on Digitised Data 219
Nur Ilham Aminullah Abdulqawi and Mohd Salman Abu Mansor

The Development of a Mobile Campus Information Sharing Android Application 231
Shareen Adlina Shamsuddin, Mohammad Arif Kader, Norazlina Abdul Nasir, Nurul Akmal Radzi and Fatimah Abdul Hamid

Service Restoration Based on Simultaneous Network Reconfiguration and Distributed Generation Sizing for Loss Minimization Using a Modified Genetic Algorithm 239
W. M. Dahalan, H. Mokhlis, M. K. P. Zarina, A. G. Othman and N. M. Salih

Improved Design of the UniKL Amphibious Research Crawler II for Underwater Exploration 259
Ahmad Makarimi Abdullah, Khairul Arieff Abu Jalil, Hisyam Hamid, Nursyahida Izzati Zakaria, Norhafizah Othman, AshmanYusoff Iqmal Mohd Haidhir and Mohd Iqram Mohd Kamro

Preliminary Design and Analysis Study of Propeller for Autonomous Underwater Vehicle (AUV) 269
Mohd Amin Hakim Ramli, Mohd Iqram Bin Mohd Kamro, Muhamad Fadli Ghani, Ahmad Makarimi Abdullah, Norhafizah Othman, Ashman Yusoff Iqhmal Mohd Haidhir, Nursyahida Izzati Zakaria, Khairul Arieff Bin Abu Jalil, Muhammad Haziq Bin Noor Zaharil Ehsan, Muhammad Farhan Bin Mohamad Noor and Fatin Zawani Binti Zainal Azain

Preliminary Study on the Development of Two Degree of Freedom Robotic Arms for Underwater Applications 279
Noorazlina Mohamid Salih, Ashman Yusoff Iqmal Mohamad Haidhir, Muhamad Fahezal Ismail, Ahmad Makarimi Abdullah, Khairul Arieff Abu Jalil, Norhafizah Othman, Muhamad Fadli Ghani, Muhamad Faizal Muhamad Isa and Mohd Iqram Mohd Kamro

Design and Analyses of a Ship Floating Dry-Dock 291
Roslin Ramli and Anwar Faris Sobri

3D Design of a Ship Floating Dry-Dock by Using Simulation Software 299
Roslin Ramli, Nur'Aqilah Mohd Sabri and Nurul Asima Zainon

Hybrid Combination Product Between Aluminum Can with Reinforcement Fiberglass for Autonomous Underwater Vehicles 313
Mohd Amin Hakim Ramli, Syajaratunnur Yaaakup,
Fatin Zawani Zainal Azaim, Khairul Anuar Mat Saad,
Mohd Faizal Abdul Razak, Muhammad Fadli Ghani, Naqib Idris
and Ahmad Shaharibudin Dato Abd. Aziz

The Comparison of Impact Energy and Three Point Bending Properties on Coconut Fiber Composite for Marine Application 319
Amirrudin Yaacob, Jaswar Koto and Mohd Yazid Bin Yahya

Tensile and Hardness Analysis of Dissimilar Friction Stir Welding Between AA6061 with AA5083 and Mild Steel 335
Wan Mohd Syafiq Wan Sulong, Mohd Afendi Rojan
and Mohd Noor Mazlee

Analysis of Drum Brake System for Improvement of Braking Performance 345
Siti Nor Nadirah Baba, Muhammad Najib Abdul Hamid,
Shahril Nizam Mohamed Soid, Mohd Nurhidayat Zahelem
and Mohd Suyerdi Omar

Ultrasonic Based Technique to Measure Residual Stresses in Offshore Structures 359
Ramesh Ramasamy and Zainah Ibrahim

Influence of Tool Plunge Depth on the Joint Strength and Hardness of Friction Stir Welded AA6061 and Mild Steel 373
Wan Mohd Syafiq Wan Sulong, Mohd Afendi Rojan
and Mohd Noor Mazlee

Analysis of Production Layout Model to Improve Production Efficiency 385
Ahmad Razlee AB Kadir, Baizura Zubir, C. A. Mohd Norzaimi,
M. Sabri, M. Zaki, W. Faradiana and Noor Helinahani

Effect of Vibration on Occupant Driving Performances: Measurement by Simulated Driving 401
Mohd Amzar Azizan

Simulation Studies of a New Magnetorheological Brake with Difference Gap Size Using Combination of Shear and Squeeze Mode 413
Lailatul Hamidah Hamdan, Saiful Amri Mazlan, Fitrian Imaduddin,
Shamsul Sarip and Ashadi Yusop

Optimization of Air-Fuel Ratio and Compression Ratio to Increase the Performance of Hydrogen Port Fuel Injection Engines 425
Mohd Fazri Shaari, Shahril Nizam Mohamed, Surenthar Magalinggam, Muhammad Najib Abdul Hamid, Mohd Nurhidayat Zahelem, Zahelem and Mohd Farid Muhammad Said

Evaluation of the Hardness Distribution and Fracture Location in Friction Stir Welded AA6063 Pipe Butt Joints 439
Azman Ismail, Mokhtar Awang, Fauziah Ab Rahman, Bakhtiar Ariff Baharudin, Puteri Zarina Megat Khalid and Darulihsan Abdul Hamid

Study on Gas Emission of Saline Water from a Hydrogen System 445
Zakiman Zali, Norsheila Buyamin, Nazihah Mohd Noor, Rawaida Muhammad, Normah Ishak and Rohimi Yusof

Analysis of Human Behavior During Braking for Autonomous Electric Vehicles 453
Khairul Ikram, Wan Khairunizam, A. B. Shahriman, D. Hazry, Zuradzman M. Razlan, Hasri Haris, Hafiz Halin and Chin S. Zhe

The Mapping of Full Weld Cycle Heat Profile for Friction Stir Welding Pipe Butt Joints 461
Azman Ismail, Mokhtar Awang, Fauziah Ab Rahman, Bakhtiar Ariff Baharudin, Puteri Zarina Megat Khalid, Darulihsan Abdul Hamid and Kamal Ahmad

Assessment of Thermal Comfort in a Car Cabin Under Sun Radiation Exposure 469
Mohamad Asyraf Othoman, Mohd Sahril Mohd Fouzi and Adzuieen Nordin

Defects of Post Weld Heat Treatment on A36 Carbon Steel Welded by Shielded Metal Arc Welding 481
Norfadhlina Khalid, Zaherrudin Yusof and M. A. Mun'aim Mohd Idrus

Automatic Tug Assistance 491
Yasseen Adnan Ahmed

Excitation Force Between Two Ship Models in Waves 505
Faizul Amri Adnan

A Simplified Computational Fluid Dynamics Approach for a Self-propelled Ship Using the Actuator Disc Theory 523
Iwan Mustaffa Kamal, Muhamad Sufri Shamsuddin and Jonathan Binns

Study of MSI300 Propeller Characteristics Using Computational Fluid Dynamics Analysis 541
Mohamad Sabri Mohamad Sidik, Ruziah Bolhassan,
Mohd Nurhidayat Zahelem and Muhamad-Husaini Abu-Bakar

Mathematical Model of the Manoeuvring Motion of a Ship 551
Yassen Adnan Ahmed

A Trim Tank Control System for an Autonomous Underwater Vehicle (AUV) 567
Md Salim Kamil, Noorazlina Mohamid Salih, Atzroulnizam Abu,
Muhammad Muzhafar Abdullah, Norshakila Abd Rasid
and Mohd Shahrizan Mohd Said

The Use of Backscatter Classification and Bathymetry Derivatives from Multibeam Data for Seabed Sediment Characterization 579
Razak Zakariya, Mohd Azhafiz Abdullah, Rozaimi Che Hasan
and Idham Khalil

A Review of Piezoelectric Design in MEMS Scanner 593
Nur Azirah Abdul Rahim, Ishak Abdul Azid and Loke Kean Koay

Review of the Control System for an Unmanned Underwater Remotely Operated Vehicle 609
Ahmad Makarimi Abdullah, Nursyahida Izzati Zakaria,
Khairul Arief Abu Jalil, Norhafizah Othman,
Wardiah Mohd Dahalan, Hisham Hamid, Muhamad Fadli Ghani,
Ashman Yussof Iqmal bin Mohd Haidir and Mohd Iqram Mohd Kamro

Maneuvering and Submerged Control System for a Modular Autonomous Underwater Vehicle 633
Mohamid Salih Noorazlina, Md Salim Kamil,
Mohammad Amiruddin Hashim, Nordiana Jamil and Hanisah Johor

Development of an Electric Turbo Generator for Automotive Application 645
Khairul Shahril Shaffee, Muhammad Idham Hassan Azmi,
Shahril Nizam Mohamed Soid, Muhammad Najib Abdul Hamid
and Tajul Adli Abdul Razak

A Study on the Effect of Parameters on the Tensile Strength of Friction Stir Welded AA6061 1.5 mm Thin Plate Butt-Joints

Nurfadilah Sukara, Azman Ismail, Darulihsan Abdul Hamid,
Fauziah Ab Rahman, Bakhtiar Ariff Baharudin
and Puteri Zarina Megat Khalid

Abstract Although fusion welding of aluminium alloys is the most common method used for welding in the industry, it is always accompanied by defects like porosity, solidification, and cracking. Friction stir welding, which is a unique welding method that does not undergo a melting process has been invented and promotes a defect-free welded joint. Nevertheless, some hindrances like severe softening in the heat affected zone (HAZ), and defects like pin-hole and crack formation will potentially occur due to an improper plastic flow of the materials. These defects can lead to a decrease in the tensile properties of the welded materials. Since the welding parameters play a major role in deciding the weld quality, this experiment considers the effects of using different rotational speeds and tool traverse speeds on the strength of the welded joint. Furthermore, defects that may occur during the welding process of aluminium thin plates also were determined.

N. Sukara · D. Abdul Hamid
Universiti Kuala Lumpur Institute of Product Design and Manufacturing,
No. 9, Jalan Perdana 7/3, Taman Shamelin Perkasa, 56100 Kuala Lumpur, Selangor,
Malaysia
e-mail: fadilah.sukara@gmail.com

D. Abdul Hamid
e-mail: darulihsan@unikl.edu.my

A. Ismail (✉) · F. Ab Rahman · B. Ariff Baharudin · P. Z. Megat Khalid
Universiti Kuala Lumpur Malaysian Institute of Marine Engineering Technology, Jalan
Pantai Remis, 32200 Lumut, Perak, Malaysia
e-mail: azman@unikl.edu.my

F. Ab Rahman
e-mail: fauziahabra@unikl.edu.my

B. Ariff Baharudin
e-mail: bakhtiarab@unikl.edu.my

P. Z. Megat Khalid
e-mail: puterizarina@unikl.edu.my

© Springer International Publishing AG 2018
A. Öchsner (ed.), *Engineering Applications for New Materials and Technologies*,
Advanced Structured Materials 85, https://doi.org/10.1007/978-3-319-72697-7_1

Keywords Friction stir welding · Thin plate · Tensile strength
Butt joint

1 Introduction

Welding, brazing, reverting, or soldering, are some of the joining methods that are applicable to aluminium. In the 1940s, the inert gas welding processes that were introduced became revolutionary for aluminium as a structural metal. This invention created a way to make high strength welds without corrosive fluxes at high speeds in all positions [1].

The inert gas welding process is categorized as a type of fusion welding, which is one of the most common processes to join the base metal in industrial production. This process is where two or more metal workpieces are joined through melting (fusing) and solidifying [1]. A major drawback is, that the properties of the base metal will be undermined by the welding heat that is produced during the welding process. Sometimes, the addition of a filler metal during fusion welding lowers the properties of the weld metal that possibly deteriorates the tensile strength of the welded joint [1].

Therefore, the development of friction stir welding (FSW) in December 1991 by The Welding Institute (TWI), UK, prevents these potential drawbacks by utilizing a low heat input [2]. FSW is an appropriate solid state welding technique to effectively join aluminium alloys, even for dissimilar alloys [3]. It utilizes the heat generated by friction to join the sheet and plate material [3]. Therefore, an attempt is made here to investigate the friction stir welding of the 1.5 mm AA6061 aluminum alloy using the MILKO 37 milling machine.

Even though FSW is free from those type of defects, the wrong selection of welding parameters can cause the materials to be heated significantly and thus be improperly welded, which may eventually cause a loss in mechanical properties [4], such as a decrease of the tensile properties and ductility strength of the welded materials [4]. Thus, a proper consideration of FSW parameters is vital to yield a superior weld joint quality.

2 Experimental Setup

2.1 Experiment Preparation

To study the effects on the quality of the aluminium weld joint, the experiment will test different parameters on the weld quality. Each parameter being used during the project will lead to different joining conditions in terms of quality, microstructural and strength. For the materials, the aluminium alloy 6061 plate and H13 tool steel were cut as has been determined and required by the American Welding Society (AWS) [5]. Based on the literature concerning tool geometry, the pin length was at 80–95% of the plate's thickness, and the pin probe diameter was about 1.5 mm with flat design [6]. The aluminium alloy 6061 was cut with a shear cutting machine to the dimensions of 1.5 mm × 356 mm × 210 mm for 30 pieces. H13 steel was used for making the tools, and manufactured with a pin diameter of 5, 1.25 mm pin length, and fabricated by a lathe machine.

2.2 Experiment Commencement

The Milko 37 is a universal milling machine that has been used to perform the welding process in this study. It can operate at variable speeds of rotation and traverse speeds. To run the FSW process on the specimens, the first step is to remove the oxide film on the specimen's surfaces by using the power brush or wire brush, and polishing it until clean. This is to ensure that the material mixture becomes viscous, strong, and does not mix with the oxide film that could lead to failure of the mechanical properties of the specimen.

Every set of welding trials was conducted after the function of the milling machine was checked, such as the traverse speed control button and speed option, rotation speed control button and speed option, and the vertical elevation movement. Before the actual welding process was performed, the tools were tested by running the milling machine with the selected parameters. During this process, the welding parameters that have been defined were checked by visual inspection to ensure the welding specimens do not have any defect. The tilt angle of the tool was set up at 2° as a constant variable while the tool's rotational speed and tool's traverse speed were chosen based on the best parameters from previous research concerning thin plates [7]. Selected welding parameters of tool rotational speed and tool traverse are set as manipulated variables, as seen in Table 1.

Table 1 Welding parameters of tool rotational and tool traverse speed

Sample No.	Tool rotational speed (RPM)	Tool traverse speed (mm/min)
1	180	340
2	240	340
3	352	340
4	352	495
5	352	695

2.3 Tensile Test

A tensile test has been conducted to measure the resistance of the material to a static or slowly applied force. For this, a machined specimen is placed in the testing machine and load is applied. A strain gauge or extensometer is used to measure elongation. In a stress versus strain diagram, the ultimate tensile strength (UTS) is the stress attained at the highest applied force while the stress at which a specified amount of plastic deformation is produced is the yield strength. Elongation in the tensile test describes the extent to which the specimen stretched before fracture [8]. All the information concerning the strength, stiffness, and ductility of a material can be obtained from a tensile test. Sample in this experiment was in accordance to ASTM E8.

The tensile testing specimens were taken from four parts of the weld specimen. It is because each section has its own strength because of different levels of heat, the tensile test specimens were taken from near at the start (2), and near the end of weld joint (2). The applied maximum tensile load was set up to 100 kN with the speed of the testing machine at 0.05 and 0.5 mm/min. The parameter of the speed of testing when determining tensile strength for materials of elongation of more than 5% was in accordance to the ASTM E8 standard. Then, the extensions were recorded during the test for the calculations of stress and strain (Figs. 1 and 2).

2.4 Macrostructure Analysis

The specimens for macrostructure testing were taken from two parts which were from the start and another from the end of plate. To allow the metallographic structures of the sample to be measured under an optical microscope, the mounting

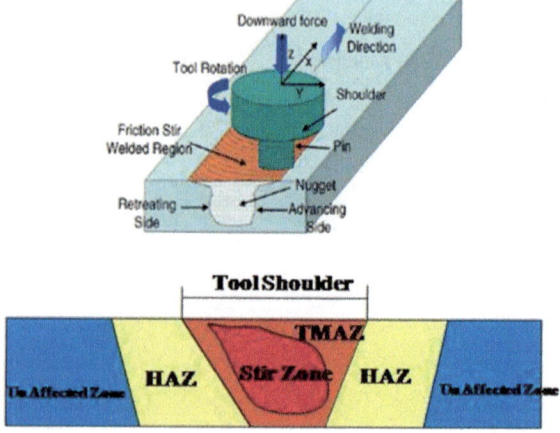

Fig. 1 Schematic of friction stir welded joint main microstructural zones [9]

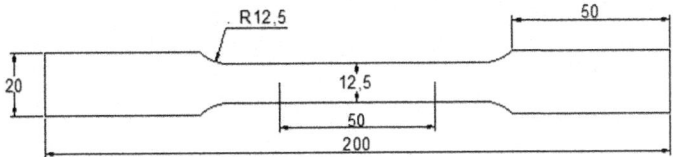

Fig. 2 Dimensions of rectangular tension test specimen by ASTM E8 [10]

press operation had to be prepared for the sample to be grinded and polished. The grinding process is needed to make sure that the surface is free from all scratches and smooth for macrostructure observation. The specimens are manually grinded, alternating the longitudinal and latitudinal direction in increasing order. After the grinding process, the polishing process takes place to polish the specimens for a smooth surface of specimen. The sequence for polishing plates is determined by the different types of diamond paste. Then the etching process takes around 30 min, by which the exposed sections of the specimens are removed. A metallographic etching process was conducted to reveal microstructural details that would otherwise not be evident on the polished specimens. The solution that had been used is Keller's reagent for the aluminium alloy. Then the image of the microstructure was taken by optical microscope with capture scale set to 10 µm.

3 Results and Discussion

3.1 Surface Appearance

The welded specimen surfaces for each selected parameter of tool rotational speed and tool traverse speed were captured and evaluated. All of the specimens were evaluated visually based on their appearance. By visually inspect the welded specimen surfaces, researchers can determine the welded features and defects that occurred. This action is important to decide whether the sample can undergo tensile testing to determine its tensile strength. The macrostructure appearance can be seen for 5 samples in Figs. 3, 4, 5, 6 and 7.

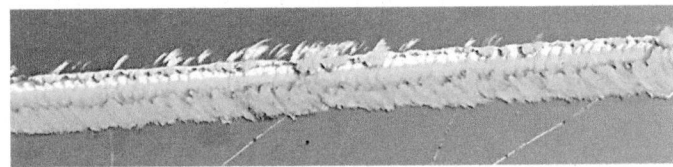

Fig. 3 Welded specimen surface of sample number 1. Poor welded joint surface; severe flash at the side of welded joint; material removal to the side of welded joint; produce wormhole defect along welded joint

Fig. 4 Welded specimen surface of sample number 2. Poor welded joint surface; severe flash at the side of welded joint; material removal to the side of welded joint; produce wormhole defect along welded joint

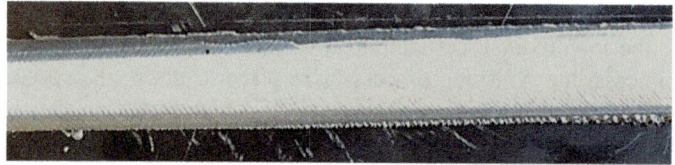

Fig. 5 Welded specimen surface of sample number 3. Smooth and fine welded joint surface; produce hole at the end of welded joint; no defects in welded joint (visual inspection); good welded penetration; slight flash produced

Fig. 6 Welded specimen surface of sample number 4. Smooth welded joint surface; no defects in welded joint (visual inspection); good welded penetration; no flash produced

Fig. 7 Welded specimen surface of sample number 5. Smooth welded joint surface; no defects in welded joint (visual inspection); good welded penetration; no flash produced

3.2 Tensile Test

In this experiment, only three samples underwent tensile testing, which were selected based on visual observations of the welded joint surface and penetration. The results show the characteristics of the specimen after the friction stir welding process using different traverse and rotational speeds. Sample number 1 and 2 did not pass the testing requirement because of welding defects such as severe wormhole were identified by visual inspection before the tensile tests were conducted.

The Galdabini universal testing machine automatically provides detailed data on load, stress, tensile strength, and elongation. The speed used to perform the testing is 0.05 mm/mm/min. The machine also provides a plotted graph of load versus elongation, from which the stress and strain can be obtained. The formulas used to determine stress and strain data are shown in (1) and (2).

$$Stress, \sigma \left(\frac{N}{m^2}\right) = \frac{Load\ (N)}{Crossectional\ Area\ (m^2)} \quad (1)$$

$$Strain, \varepsilon \left(\frac{mm}{mm}\right) = \frac{\%\ Elongation}{100\%} \quad (2)$$

After all the graphs were generated, the specimens were compared in order to determined which has the best result in term of its maximum force, tensile strength, and Young's Modulus. All the results were compared with the standard specifications for aluminium and aluminium-alloy sheets and plates, from the ASTM B209 standard.

In terms of maximum load strength result (see Fig. 8), the maximum load strength recorded is 1794 N obtained for sample 3, welded at 352 rpm with a traverse speed of 340 mm/min. Therefore, this specimen can withstand high loads before fracture occurs.

Based on the results in Fig. 9, different values of strength have been determined in terms of elongation and ultimate tensile strength. The results have been determined based on the tensile test data in the elongation part, the longest elongation

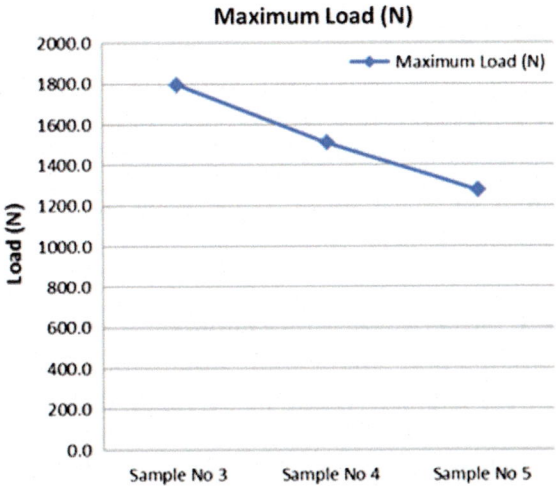

Fig. 8 Comparison of specimen's maximum load

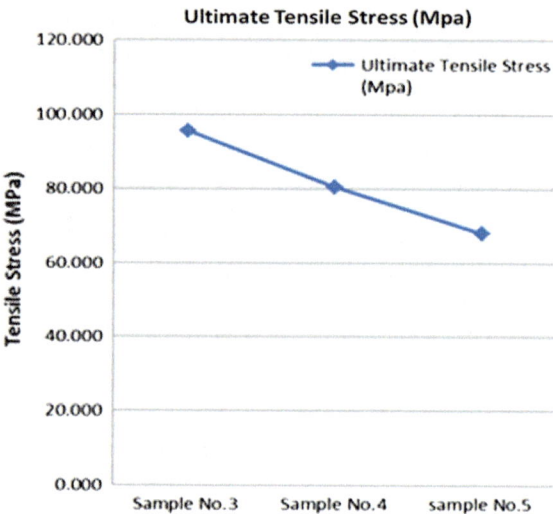

Fig. 9 Comparison of specimen's ultimate tensile strength

observed in sample number 3, which underwent welding at 352 rpm and a traverse speed of 340 mm/min. Sample number 3 sustained the highest tensile stress at 95.68 MPa with an elongation of 0.022 mm/mm. It shows that the specimen has a good ductility with the selected parameters of tool rotation and traverse speed.

The results for the Young's modulus are shown in Fig. 10, indicating that the joint was elastic. For sample 30,024 welded at 352 RPM with a traverse speed of 495 mm/min, the elastic limit is 39,268.9 MPa. It shows, that using a specimen parameter of 352 rpm at traverse speed of 495 mm/min results in the specimen with the highest elasticity.

The graph in Fig. 11 shows that, even though the tool rotational speed of 352 rpm with traverse speed 340 mm/min gives the highest value of tensile strength, this value is lower than tensile strength value of base material. To determine if the result is acceptable or not, the welded join efficiency factor should be considered.

Friction stir welding can produce satisfactory butt welds of AA6061 grade with a joint efficiency of around 60–90%. However, here the highest tensile strength obtained is 95.68 MPa, which gives a joint efficiency factor of 0.63, thus around 63.08%. Thus, it can be said the result is satisfactory as the joint efficiency contribution is in the range between 60 and 90%.

3.3 Macrostructure Analysis

Welded specimen surfaces for each selected parameter of tool rotational speed and tool traverse speed were captured and evaluated. All the specimens were evaluated visually based on their appearance. By visually inspecting the welded specimen

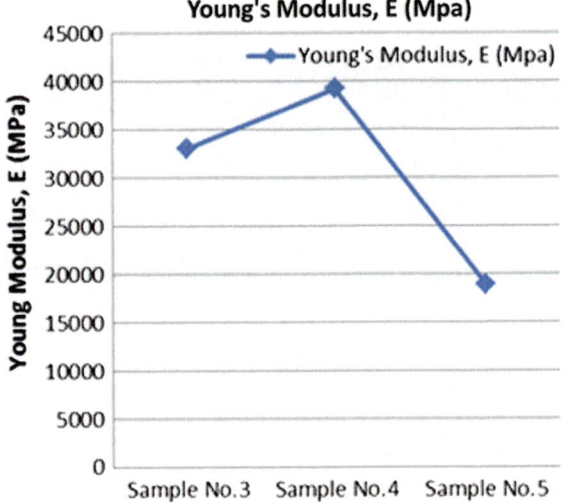

Fig. 10 Comparison of specimen's Young Modulus

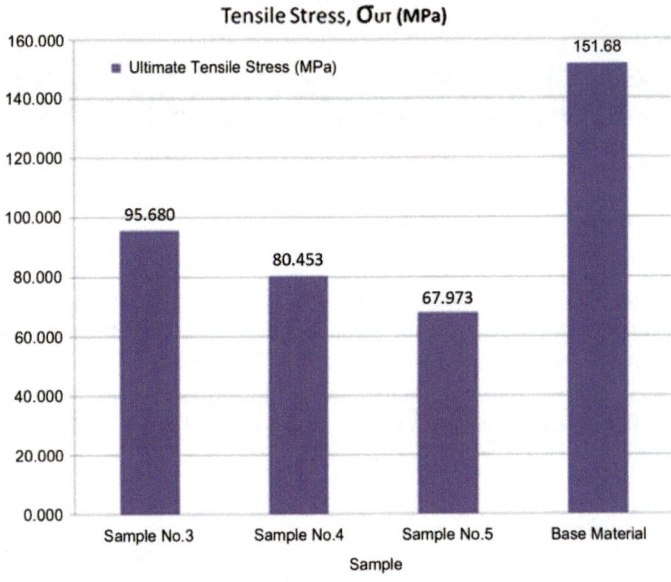

Fig. 11 Comparison of specimens and base material tensile strength

surfaces, researchers can determine the welded features and defects that occurred. This action is importance to decide if the sample can undergo tensile testing to determine its tensile strength. Results for macro testing were captured and noted for the welded specimen's cross sectional surfaces in accordance to the five different

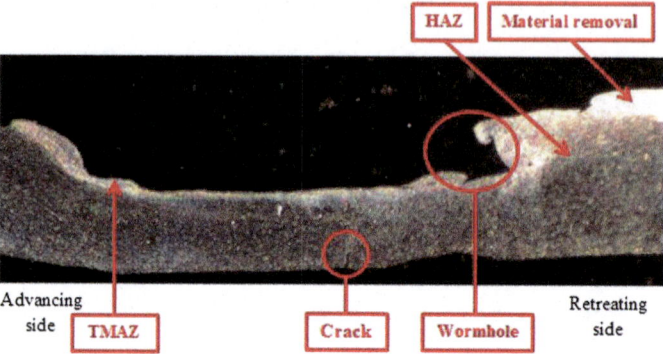

Fig. 12 Cross section of sample number 1

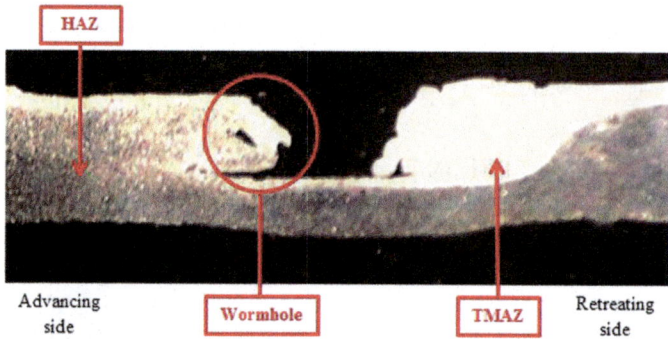

Fig. 13 Cross section of sample number 2

selected parameters. By conducting macro testing, researchers can determine the macro structural zones of friction stir welded joint and the defects that occurred. The microstructure of the parameter's tool rotational and traverse speeds of 352 rpm with 340, 495, and 695 mm/min on the other hand shows a free defect welded joint. The results after undergoing macro testing are shown in Figs. 12, 13, 14, 15 and 16.

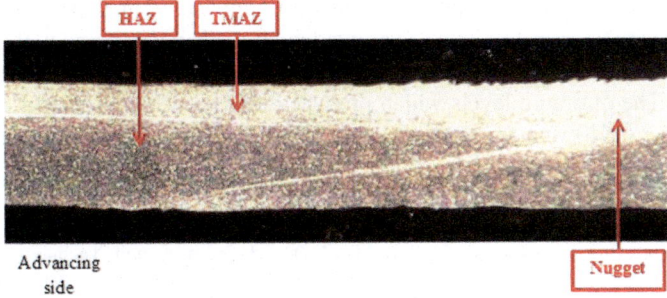

Fig. 14 Cross section of sample number 3

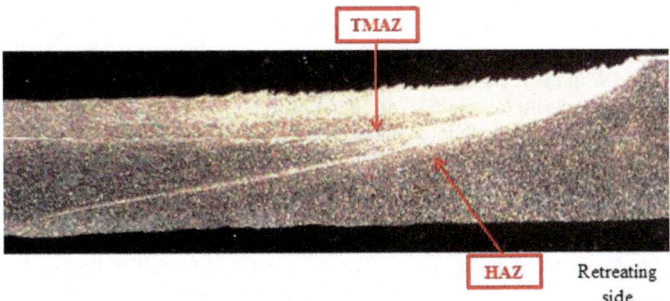

Fig. 15 Cross section of sample number 4

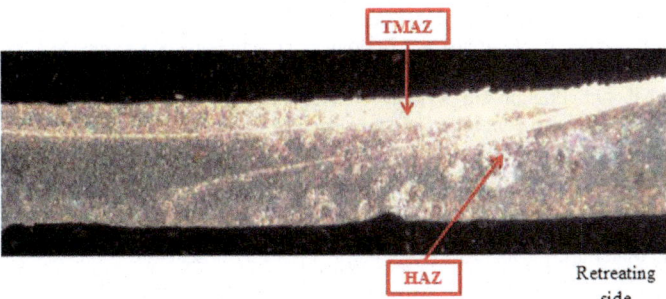

Fig. 16 Cross section of sample number 5

4 Conclusion

According to the hypothesis, the translational and transverse forces follow the general relationship of increased rotational speed. However, as the rotational speed is increased, a new relationship becomes apparent. This project concludes that the parameters of 495 and 695 mm/min, both with a tool rotational speed of 352 rpm, can generate the best result in terms of weld quality when compared to the other set of parameters, but in term of its tensile strength, it is reduced as the traverse speed increases. The optimum levels which produced the best tensile strength among the samples, were 340 rpm for rotational speed, 352 mm/min for transverse speed and 2° of tilt angle.

Acknowledgements The authors thanked the University Kuala Lumpur Malaysian Institute of Marine Engineering Technology (UniKL MIMET) and Sciences and Technology Research Institute for Defence (STRIDE), Lumut for research facilities and financial assistance provided for this project.

References

1. Welding and Joining Technologies, pp. 3–13. Tokyo (2010)
2. Hattingh, D., Blignault, C., van Niekerk, T., James, M.: Characterization of the influences of FSW tool geometry on welding forces and weld tensile strength using an instrumented tool. J. Mater. Process. Technol. **203**(1–3), 46–57 (2008)
3. Palanivel, R., Koshy Mathews, P., Murugan, N., Dinaharan, I.: Effect of tool rotational speed and pin profile on microstructure and tensile strength of dissimilar friction stir welded AA5083-H111 and AA6351-T6 aluminum alloys. Mater. Des. **40**, 7–16 (2012)
4. Jayaraman, M., Sivasubramanian, R., Balasubramanian, V.: Effect of Process Parameters on Tensile Strength of Friction Stir Welded Cast LM6 Aluminium Alloy Joints, vol. 25, issue 5, pp. 655–663 (2009)
5. American Welding Society: Specification for Friction Stir Welding of Aluminium alloys for Aerospace Applications, pp. 14–20 (2009)
6. Ismail, Azman, Awang, Mokhtar, Rojan, Mohd Afendi: An effective jig for friction stir welding of pipe butt joint. Appl. Mech. Mater. **752–753**, 491–495 (2015)
7. Galvao, I., Leitao, C., Loureiro, A., Rodrigues, D.: Friction stir welding of very thin plates. Soldagem Inspeçao **17**(1), 02–10 (2012)
8. Twi-global.com: Friction stir welding of aluminium alloys. Retrieved 22 Oct 2015, from http://www.twi-global.com/technical-knowledge/published-papers/friction-stir-welding-of-aluminium-alloys/ (2015)
9. Mohan, D.: Friction stir welding tools and overview. Int. J. IT, Eng. Appl. Sci. Res. **3**(4), 11–15 (2014)
10. An American National Standard. ASTM E8: Standard Test Methods for Tension Testing of Metallic Materials, pp. 1–27 (2010)

Physical and Chemical Properties of Perak River Sand for Greensand Casting Molds

Muhammad Arieef Bin Hussain, Azhar Bin Abdullah and Rusmay Bin Abdullah

Abstract Silica sand is the main mineral used as foundry sand and considered as one of the important industrial minerals in Malaysia. In Malaysia, silica sand is obtained from sand mining (natural sand) and ex-tin mining (tailing sand). Malaysia has an estimated resource of 640 Mt of silica sand, of which 148 Mt are natural silica and 492 Mt tailing sand. The objective of this study is to determine the physical and chemical properties of the Perak River sand as a molding mineral for making greensand casting molds. Six samples were taken randomly at the marked identified locations. Clay content, grain size distribution and chemical composition of the samples were tested and analyzed. These samples were compared with requirements as stated in the Foseco Non-Ferrous Handbook. A grain size above 425 μm was eliminated in order to comply the foundry sand size and tested for grain size distribution and clay content. The results from the analysis show that the clay content is ranging from 8.88 to 11.30%. The chemical compositions showed that the silica content for all samples was in the range of 86–91%, which is considered as natural molding sand. As a conclusion, this research discovered that Perak River sand is physically and chemically suitable for making greensand casting molds for non-ferrous castings. Further study on the mechanical properties is highly suggested.

Keywords Silica sand · Grain size · Grain shape · Clay content

---

M. A. B. Hussain (✉) · A. B. Abdullah · R. B. Abdullah
Mechanical Engineering Department, Ungku Omar Polytechnic, Ipoh, Malaysia
e-mail: arieef@puo.edu.my

A. B. Abdullah
e-mail: azhar@puo.edu.my

R. B. Abdullah
e-mail: rusmay@puo.edu.my

© Springer International Publishing AG 2018
A. Öchsner (ed.), *Engineering Applications for New Materials and Technologies*,
Advanced Structured Materials 85, https://doi.org/10.1007/978-3-319-72697-7_2

1 Introduction

Casting is one of the main and adaptable processes for manufacturing engineering components where sand casting is one of the specific processes. It is a common process for manufacturing automotive and marine components such as propellers, pumps impellers, pump casings, couplings, valves, and sub-surface platforms [2]. Sand casting is a manufacturing process by which a liquid metal at high temperature is poured into a cavity of a mold (which is made by mixture of sand, binder and other additives) of the required shape, which solely depends on the design of the pattern, and then is allowed to solidify. The mold is broken out to complete the process. One of the processes in sand casting is called greensand casting which is composed of high quality silica sand (85–95%) blended together with bentonite clay (4–10%) as a binder and water (2–5%) [18].

The advantage of greensand casting is that many components for various applications can be molded in different sizes, shapes and cheaper in cost [7]. Sand casting remains the most popular method of casting, despite that it has a poor finish and tolerance, and accounts for more than 90% of the cast metal production [18].

Sand is the most commonly mineral used for making sand molds in foundry industry. Sand is found in many different locations and can have wide variations in surface, physical, and chemical characteristics due to environmental, ecological, climatic and geological factors. Different sands have different foundry properties and thus, necessary standard laboratory tests are properly carried out on it [16].

There are many types of sand used as molding mineral such as silica, zircon, olivine and chromite. Silica sand is the most popular due to the availability and cost. Silica sand grains are of paramount importance in molding sand because they impart refractoriness, chemical resistivity and permeability to the sand [15]. They are specified according to their average size and shape. Zircon, olivine and chromite sand are used for special applications where the price of these is expensive. Silica sand can be broadly grouped into two types, depending essentially upon their chemical composition, which are clay-free sand (synthetic sand) or washed sand and clay-bearing sand (naturally-bonded sand) [5, 14, 15].

In Malaysia, synthetic sand is available from tin mine tailing sand where during the extraction of tin, the soil was washed out by using gravel pumps. One of the residues from tin extraction is tailing sand, which has a silica content of 94–99.5%. Malaysia has an estimated resource of 640 Mt of silica sand, of which 148 Mt are natural silica and 492 Mt tailing sand [12]. Previous research on the tin mine tailing sand as a molding mineral found that it is suitable for making greensand casting molds [1].

The examples of naturally-bonded sand silica are river sand, shore sand and lake sand where the sand is produced due to the geology history such as by water, wind or glacier [11]. Synthetic sand is the sand that has low clay content which is 0–1.5% with high silica content is 94–99% while natural-bonded sand has high clay content which is 5–20% with low silica content which is 80–90% [14]. The ideal purity for silica sand as a foundry sand is that the material should have a silica content of

95–96% although 98–99% is often preferred [14]. The higher the content of silica the more refractory the sand. The presence of impurities in sand such as Fe_2O_3, K_2O, CaO, LOI and true clay reduced the silica content and lowered the ability of refractoriness and avoiding fusing in the casting process. Impurities such as K_2O, Na_2O contain high acid [14, 15]. Thus, this will affect the amount of binder especially the consumption of resin for chemical bonded sand casting mold. The same will apply to alkaline sand.

Loss on ignition (LOI) methods have long been used by soil testing laboratories to measure soil organic matter and soil organic carbon [9]. It is also a common and widely used method to estimate sediment properties (i.e., water content, organic matter, inorganic carbon and minerogenic residue), because it is the quickest and cheapest among all the methods employed for determining some of those parameters. It is pointed out that the method is useful for analyses of a large number of samples [4, 6, 13]. Most sediments are composed of a mixture of clastic silicates and oxides (sand, silt and clay fractions), organic material, carbonates and water. Quantitative determinations of such sediment parameters by means of loss-on-ignition are based on the sequential heating of the samples (other techniques involve gas collecting) and measuring the amount either volumetrically or chromatically in a muffle furnace [6, 10].

After oven drying of the sediment to constant weight (or dry matter) and cooling to room temperature, the organic matter is combusted in a first reaction to ash and carbon dioxide at temperatures between 500 and 5500 °C. The amount of organic matter is the weight difference between the dry sediment and the 5500 °C ash [6, 10, 13]. True clay includes fine particles of silt or dead clay, which do not function as effective clay or new clay and do not contribute to bonding properties. Thus, the test of clay grade is expressed as the percentage of true (natural) clay particles in sand. The percentage of these impurities should be controlled because they are affecting the quality of the mold and also the quality of casting. Table 1 shows the limitation of the impurities in the sand mold.

Grain size distribution is a major determinant of the mechanical properties of the molding such as green compression strength and permeability besides surface finish of the casting [3, 5, 8, 14]. Table 2 shows the typical grading for sand suitable for greensand systems. Finer sand will give better surface finish but this must be negotiated against permeability which is better with coarser grains and prevents

Table 1 Control of chemical composition in sand mold [5]

Compound	Limitation	Result
Silica, SiO_2	95–96% min	The higher the silica the more refractory the sand
Loss on ignition, LOI	0.5% max	Represents organic impurities
Fe_2O_3	0.3% max	Iron oxide reduces the refractoriness
CaO	0.2% max	Raises the acid demand value
K_2O, Na_2O	0.5% max	Reduces refractoriness
Acid demand value to pH4	6 ml max	High acid demand adversely affects acid catalysed binders

Table 2 Foseco typical grading of sand suitable for greensand system [5]

Sieve Mesh size in micrometer (μm)	%
710	0.14
500	1.42
355	5.44
250	22.88
212	19.64
150	26.74
106	9.70
75	2.84
Pan	1.10

casting defects. Finer grains appear to be more easily fused than coarser ones [5]. Also for finer grains, the more intimate will be the contact and lower the permeability. However, fine grains tend to fortify the mold and lessen the tendency to get distorted [15]. Foundry sands usually fall into the range 150–400 μm, with 220–250 μm being the most commonly used [5]. In general, a size of 200–300 μm or AFS 75–50 is preferred with a grain size distribution spread of 95% over four to five consecutive sieves [8].

The grain size is determined by grading on the standard sieves to separate selected fractions for weighing [5, 11, 14, 15, 17]. The easier way is by comparing the graphs rather than rows of figures of percentages of sand retained on each sieve, the simplest and clearest method of illustrating how the sand is divided into the different sieve sizes, known as grain distribution, is to plot a curve graph of the cumulative percentage against each sieve number [5]. Figure 1 shows the curves of

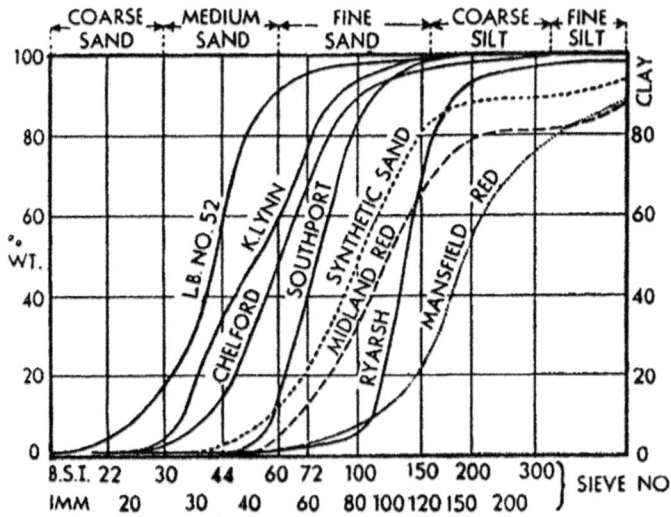

Fig. 1 Sand's grading and classifications [11]

foundry sands from various locations in the UK. The 'S' shape of synthetic sand (dotted line) represents the well sorted of the sand [11]. The more vertical the curve such as Southport and Ryarsh, the more uniform is the grading of the sand, i.e. a large proportion of the sand grains rest on the two or three adjacent sieves in the stack. On the other hand, a wide distribution of the grains, with a small percentage of the grains on each of the sieves, produces a curve that begins in the bottom left-hand corner and slopes at an angle of 45° to the upper right-hand corner, for instances Midland Red sand and Mansfield Red sand [11].

This preliminary research is conducted to identify the physical and chemical composition of Perak River sand as a molding mineral for making greensand casting molds focusing on non-ferrous metals such as aluminium based and copper based. The study on the physical and chemical composition is crucial because if the result would discover that the sand does not comply with the requirement from the Foseco Non-Ferrous Foundryman's Handbook or common practices in foundry industry, the study on the molding mechanical properties could not proceed. Six samples were taken randomly at the marked identified locations. The grain size distribution and average grain size fall under physical properties while the clay content and chemical composition of the samples were tested and analyzed.

2 Materials and Methods

A total of six samples representing deposit from Perak River sand located at Kampung Raja, Parit, weighing approximately 40 kg were obtained for testing the clay grade, chemical composition, and grain sand distribution. All lab testing, method and equipment are conducted in the Department of Mineral and Geoscience Malaysia, Perak Branch and strictly followed the lab standard procedure.

2.1 Clay Grade

To determine the clay grade, a 100 g sample from a dried sand specimen was mixed with 500 ml distilled water and 25 ml ammonia (to maintain the clay in suspension during the test). The sand/solution mixture was then stirred with a glass rod and then allowed to settle during a period of exactly 10 min. The solution was then removed and water was added for the repeated process. The procedures were repeated until a 5-min settling period gave perfectly clear conditions. The remaining sand and water were dried at 110 °C for an hour and the weight of clay washed dried sand was taken and the percentage of the weight of dry sand taken and calculated as below:

$$\% \text{ Clay Grade} = \frac{100\,\text{g} - \text{Weigh of Sand After Clay Wash}}{100\,\text{g}} \times 100\% \qquad (1)$$

2.2 Chemical Composition

The chemical composition of a sand sample was determined by weighting up 10 g washed sand. The sand powder then was produced by grinding the sample. A 1 g sample was taken for LOI and the rest for other chemical compositions which were SiO_2, Fe_2O_3, CaO and K_2O. LOI was identified by a combustion chamber by using Thermconcept while for SiO_2 it was by LAC. The standard used to identify the content of Fe_2O_3, CaO and K_2O was according to ASTM (146-80). The analysis was done by the principle of balancing Mettler AE 163.

2.3 Grain Sand Distribution

The sand specimen is graded by using an Andelcott sieve shaker with an aperture size of 710, 500, 355, 250, 212, 150, 90, 63 µm and pan. The sieve grading was as follows:

a. Weigh 100 g of dry sand.
b. Place the sample into the top sieve of a nest of ISO sieves on a vibrator. Vibrate for 15 min.
c. Remove the sieves and, beginning with the top sieve, weigh the quantity of sand remaining on each sieve.

2.4 Average Grain Size

The adoption of the ISO metric sieves means that the old AFS grain fineness number can no longer be calculated. Instead, the average grain size, expressed in micrometres (µm) is now used. Table 3 shows an example of calculating the average grain size.

Table 3 Example of an average grain size calculation

Aperture size (μm)	Multiplier A	Sand retained (%) B	Product C A × B
710	1180	0.000	0.00
500	600	0.017	12.00
355	425	8.397	3570.00
250	300	25.267	7581.00
212	212	12.793	2711.48
150	150	19.937	2991.00
90	106	20.117	2132.72
63	75	8.873	665.25
Pan	38	4.500	171.00
Total		99.900	19,834.45

Average grain size = Total of C/Total of B = 19,834.35/99.9 = 199 μm

3 Results and Discussion

3.1 Clay Grade

The clay grade resulted from the analysis in the range from 8.88 to 11.30% as shown in Table 4. The results show that the content of true clay is high thus Perak River sand can be classified as naturally-bonded sand.

3.2 Chemical Composition

Table 5 shows that the SiO_2 content for all samples was in the range of 86–91%. The silica content is found to be low and can be grouped as naturally-bonded sand. The reason that can be explained is due to the high true clay content that was found from the results of the clay grade test. The content of impurities, which are Fe_2O_3 (0.35–0.60%), K_2O (2.71–3.07%), CaO (0.05–0.11%), and LOI (0.69–0.93%), were discovered to be above the maximum limit. These impurities and true clay

Table 4 Percentage of true clay in a various Perak River sand samples

Sample	% True clay
Sample 1	11.30
Sample 2	10.93
Sample 3	10.30
Sample 4	9.90
Sample 5	8.88
Sample 6	9.84

Table 5 Results for chemical composition

Specimen	Percentage (%)				
	SiO_2	Fe_2O_3	K_2O	CaO	LOI
Specimen 1	88.6	0.53	2.93	0.07	0.84
Specimen 2	87.6	0.50	2.86	0.07	0.87
Specimen 3	88.4	0.55	2.99	0.11	0.93
Specimen 4	90.9	0.37	1.12	0.05	0.69
Specimen 5	91.1	0.37	1.10	0.06	0.78
Specimen 6	91.0	0.36	1.09	0.05	0.73

content will affect the quality of mold sand and thus should be controlled. It is suggested that Perak River sand should undergo a washing process to minimize the impurities to achieve the requirement of chemical composition for molding sand.

3.3 Grain Sand Distribution

Figure 2 shows the grain size distribution for all samples. Sample 4 and 6 have been discovered with a well sorted grain distribution where they have a 'S' curve with a 45° slope and the curves are similarly closed to Foseco's 'S' curve. Even though other samples (sample 1, 2, 3 and 5) also produced almost to look like a 'S' curve, their distribution mostly falls more to left-hand of the graph. Thus, it can be concluded that most of the grain samples have coarser grains. Coarser grain will

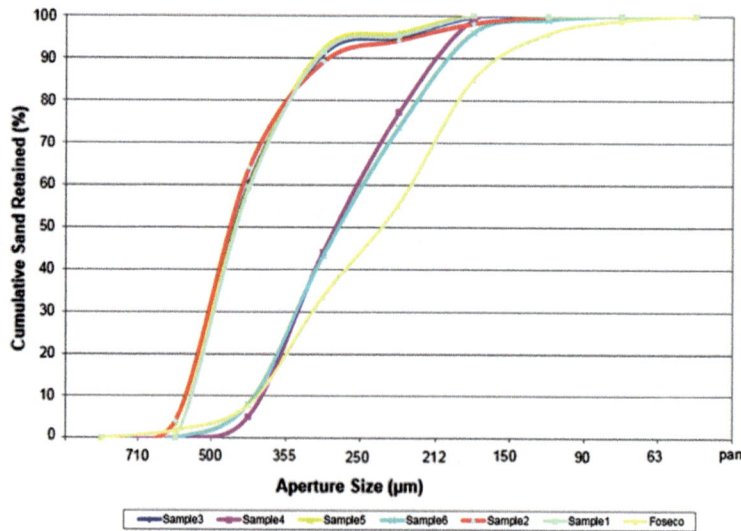

Fig. 2 The grain size distribution for all samples

produce coarser surface at the end product but allow a smooth gas flow through the particles but resulting in a lower strength. Meanwhile for samples 4 and 6, it showed that they had a large proportion of the sand grains resting on the two or three adjacent sieves with the acceptable size (150–400 μm). Samples 4 and 6 comprise suitable grains with the preferred size suggested by Foseco (220–250 μm). Too fine grains produce a smooth surface and good strength but are hard for gas to flow smoothly between the particles. This will be countered by the mixtures of clay consumption in order to produce the preferred surface and strength.

3.4 Average Grain Size

Table 6 shows the calculation of the average grain size for Sample 1 and it was found that the average grain size is 364 μm and it is within the requirement for foundry sand grain size. Table 7 shows the calculation of the average grain size for Sample 2 and it was found that the average grain size is 373 μm and it is within the requirement for foundry sand grain size.

Table 8 shows the calculation of the average grain size for Sample 3 and it was found that the average grain size is 371 μm and it is within the requirement for foundry sand grain size.

Table 9 shows the calculation of the average grain size for Sample 4 and it was found that the average grain size is 242 μm and it is within the requirement for foundry sand grain size.

Table 10 shows the calculation of the average grain size for Sample 5 and it was found that the average grain size is 371 μm and it is within the requirement for foundry sand grain size.

Table 6 Calculation of average grain size for sample 1

Aperture size (μm)	Sand retained (%)	Multiplier	Product
710	0.00	1180	0.00
500	0.00	600	0.00
355	59.20	425	25,160.00
250	32.20	300	9690.00
212	3.80	212	805.60
150	4.70	150	705.00
90	0.00	106	0.00
63	0.00	75	0.00
Pan	0.00	38	0.00
Total	100		36,360.60

Average grain size = 36,360.60/100 = 364 μm

Table 7 Calculation of average grain size for sample 2

Aperture size (μm)	Sand retained (%)	Multiplier	Product
710	0.00	1180	0.0
500	4.00	600	2400.00
355	59.90	425	25,457.5
250	25.10	300	7530.0
212	5.00	212	1060.00
150	4.00	150	600.00
90	1.70	106	180.20
63	0.20	75	150.00
Pan	0.00	38	0.00
Total	99.90		37,242.70

Average grain size = 37,242.70/99.9 = 373 μm

Table 8 Calculation of average grain size for sample 3

Aperture size (μm)	Sand retained (%)	Multiplier	Product
710	0.00	1180	0.00
500	3.70	600	2220.00
355	56.90	425	24,182.50
250	30.70	300	9210.00
212	3.90	212	826.80
150	4.80	150	720.00
90	0.00	106	0.00
63	0.20	75	15.00
Pan	0.00	38	0.00
Total	100.2		37,174.30

Average grain size = 37,174.30/100.2 = 371 μm

Table 9 Calculation of average grain size for sample 4

Aperture size (μm)	Sand retained (%)	Multiplier	Product
710	0.00	1180	0.00
500	0.00	600	0.00
355	5.00	425	2125.00
250	39.00	300	11,700.00
212	33.00	212	6996.00
150	22.00	150	3300.00
90	0.80	106	84.80
63	0.10	75	7.50
Pan	0.00	38	0.00
Total	99.9		24,213.30

Average grain size = 24,213.30/99.9 = 242 μm

Table 10 Calculation of average grain size for sample 5

Aperture size (μm)	Sand retained (%)	Multiplier	Product
710	0.00	1180	0.00
500	3.50	600	2100.00
355	56.40	425	23,970.00
250	32.40	300	9720.00
212	3.70	212	784.40
150	4.20	150	630.00
90	0.00	106	0.00
63	0.00	75	0.00
Pan	0.00	38	0.00
Total	100.0		37,204.40

Average grain size = 37,204.40/100.0 = 371 μm

Table 11 Calculation of average grain size for sample 6

Aperture size (μm)	Sand retained (%)	Multiplier	Product
710	0.00	1180	0.00
500	0.00	600	0.00
355	8.00	425	3400.00
250	35.00	300	10,500.00
212	30.10	212	6381.20
150	22.30	150	3345.00
90	3.00	106	318.00
63	1.00	75	75.00
Pan	0.00	38	0.00
Total	99.4		24,019.20

Average grain size = 24,019.20/99.4 = 242 μm

Table 11 shows the calculation of the average grain size for Sample 6 and it was found that the average grain size is 242 μm and it is within the requirement for foundry sand grain size.

4 Conclusions

The physical property investigated in this research was the grain size distribution and the average grain size will determine the mechanical properties. Whereas the chemical properties were the true clay content and the chemical composition and they will determine the refractoriness of Perak river sand. Samples 4 and 6 were identified to have the most suitable grain size distribution and average grain size among the investigated samples. Furthermore, they have a low true clay content and the silica content is over 90%. The major finding from this research is that Perak river sand, physically and chemically is suggested to be used as molding sand for making non-ferrous casting molds. Further research on the mechanical properties such as green compression strength, shatter index and permeability number is suggested to be conducted.

References

1. Abdullah, A., Sulaiman, S., Baharudin, B.T.H.T., Ariffin, M.K.A., Vijayaram, T.R., Hamid, N.H.A.: Mechanical moulding properties of tailing sand-clay mixture from Batu Gajah, Perak, Malaysia for making greensand casting mould. In: Materials Science Forum, vols. 773–774, pp. 211–217 (2014)
2. Azhar, A., Rusmay, A., Redzuan, A.: Tin mine tailing sand for making greensand casting mould in copper based marine applications, ARPN J. Eng. Appl. Sci. **12**(6) (2017)
3. Beeley, P.: The moulding material: properties, preparation and testing. Foundry Technology, 2nd edn, pp. 178–238. Butterworth-Heinemann, Oxford (2001)
4. Bengtsson, L., Enell, M.: Chemical analysis. In: Berglund, B.E. (ed.) Handbook of Holocene Palaeoecology and Palaeohydrology, pp. 423–445. Wiley, Chichester (1986)
5. Brown, J.R., John, R.B.: Sands and Green Sand, Foseco Non-Ferrous Foundryman's Handbook, 11th edn, pp. 149–163. Butterworth-Heinemann, Oxford (1999)
6. Dean, W.E.: Determination of carbonate and organic matter in calcareous sediments and sedimentary rocks by loss on ignition: comparison with other methods. J. Sed. Petrol. **44**, 242–548 (1974)
7. Degul, Y.M., Sridhar, K.: Marine propeller manufacturing—a new approach. Am. J. Eng. Res. (AJER). **03**(05), 207–211 (2014)
8. Griffith, J.: Minerals in Foundry Casting: Investment in the Future. Industrial Minerals, pp. 39–51 (1990)
9. Konare, H., Yost, R.S., Doumbia, M., McCarty, G.W., Jarju, A., Kablan, R.: Loss on ignition: measuring soil organic carbon in soils of the Sahel, West Africa. Afr. J. Agric. Res. **5**(22), 3088–3095 (2010)
10. Heiri, O., Lotter, A.F., Lemcke, G.: Loss-on-ignition as a method for estimating organic and carbonate content in sediments: reproducibility and comparability of results. J. Paleolimnol. **25**, 101–110 (2010)
11. Howard, E.D.: Modern Foundry Practice, 3rd edn. Odhams Press Limited, London (1958)
12. Mackay, I.M.C., Schnellmann, Final Report of Mineral Processing Consultancy Silica Sand. International Mining Consultants Limited, Kuala Lumpur (2000)
13. Maher, L.J.Jr.: Automating the dreary measurements for loss on ignition. INQUA Sub-Commission on Data-Handling Methods. Newsletter **18** (1998)
14. Webster, P.D.: Fundamentals of Foundry Technology. Portcullis Press, Redhill, Surrey (1980)
15. Jain, P.L.: Principles of Foundry Technologies, 4th edn. Tata McGraw-Hill Publishing Company Limited, New Delhi (2003)
16. Patel, D., Deshpande, V., Jha, E., Patel, V., Desai, S., Patel, J.: The Proceeding of 4th International Conference on Progress in Production, Mechanical and Automobile Engineering (ICPMAE-2015) (2015)
17. Salmon, W.H., Simons, E.N.: Foundry Practice. Sir Isaac Pitman & Sons Ltd., London (1966)
18. Siddique, R., Schutter, Gd, Noumowec, A.: Effect of used-foundry sand on the mechanical properties of concrete. Constr. Build. Mater. **23**, 976–980 (2008)

Potential Use of Cellulose Fibre Composites in Marine Environment—A Review

Muhamad Firdaus Muhamad Yang, Hisham Hamid and Ahmad Makarimi Abdullah

Abstract Sustainability, renewability, eco-friendly, and green chemistry are guiding towards the new development of the next generation of materials in marine applications. More recently, natural fibre reinforced plastic (NFRPs) have provided an alternative solution for boat builders with enhanced bio reinforcement, however, they suffer from questionable mechanical properties and their performance. The combination of bio-reinforcement such as Kenaf, bamboo, banana, and pineapple leaf fibre with polymer matrixes from both nonrenewable and renewable resources to produce composites materials that are comparable with synthetic composites requires special attention, i.e., bio-reinforcement, interphase optimization and fabrication techniques. This review aims to provide an overview of the factors that are affecting the mechanical properties, performance of NFRPs and potential cellulose based polymer for marine application.

Keywords Green chemistry · Cellulose fibres · Fabrication techniques
Bio-composites · Marine application

M. F. M. Yang (✉) · A. M. Abdullah
Malaysian Institute of Marine Engineering Technology, Universiti Kuala Lumpur,
Jalan Pantai Remis, 32200 Lumut, Perak, Malaysia
e-mail: mfirdaus365@gmail.com

A. M. Abdullah
e-mail: makarimi@unikl.edu.my

H. Hamid
Malaysia France Institute, Universiti Kuala Lumpur, Section 14, Jalan Teras
Jernang, 43650 Bandar Baru Bangi, Selangor, Malaysia
e-mail: hishamhamid@unikl.edu.my

© Springer International Publishing AG 2018
A. Öchsner (ed.), *Engineering Applications for New Materials and Technologies*,
Advanced Structured Materials 85, https://doi.org/10.1007/978-3-319-72697-7_3

1 Introduction

In the past few decades, it was found that polymers have replaced many of the conventional metals or synthetic materials in various applications worldwide. This is possible because the advantages of polymers offer great value compared to conventional materials. The most important advantages of using polymers are the ease of processing and cost reduction. In most of these applications, the properties of polymers are modified using fibres and matrixes to suit with the industrial requirement or other usage. One of modified fibre such as NFRP offer advantages over other conventional materials, comparing in certain specific properties. These composites are finding applications in diverse fields including marine [1].

The earliest attempts to fabricate boat hulls with fibre reinforced polymer composites (FRPC) were in 1947 when twelve small surf boats were made for the United States Navy. As in record refers to FRPC material was used in the construction of boats shortly after World War 2. It shows that the usage of FRPC in marine application increased drastically instead of timber manufactured boats. The use of FRPC because of timber were becoming increasingly deficient and expensive. The timber was losing favor with many boat builders and owners due to degradation by seawater and barnacles. Therefore it is required ongoing maintenance and repairs which can be expensive. Most maritime boats are built using glass reinforced polyester (GRP) composites; meanwhile sandwiched composites and advanced FRPC materials containing carbon are aramid fibre's with vinyl ester or epoxy matrixes, which are commonly used for high performance structural applications [2]. In details, FRPC materials are consisting of high strength fibres or reinforcement embedded in polymeric matrixes. Fibres in these materials are the load-carrying elements which provide strength and rigidity. While the polymer matrixes maintain the fibres alignment, position, orientation, protection against the environment, and possible damage. A polymer is necessary to have essential mechanical properties for application in a specific field. The reinforcement by high strength fibres provides the polymer substantially enhanced mechanical properties and makes the fiber reinforced polymer composites (FRPC) suitable for a large number of diverse applications.

The FRPC are developed primarily using synthetic fibres such as glass, carbon, aramid, Kevlar etc. Synthetic fibre reinforced composites (SFRC) have unique advantages. Besides their high strength and stiffness, these composites have long fatigue life and adaptability to the desired function of the structure. Additional improvements can also be realized in the SFRC in regards to corrosion resistance, wear resistance, temperature-dependent behavior, environmental stability, thermal insulation, conductivity, and finished product [3]. Specific properties of these materials such as high stiffness, high strength, and low density as compared to metals make the material highly desirable also in military and civilian aircraft structures. These SFRC are also used in various forms in the transportation industry. Ship structures incorporate SFRC in various forms. Although the SFRC possess exclusive mechanical strength, they have some serious drawbacks such as high cost, high density, high embodied energy, poor recyclability and non-biodegradable properties as compared to natural composites. For these reasons, over the last few years NFRP are increasingly gaining attention as a viable alternative to SFRC [4, 5].

Other than that, the main limitations of SFRC is the use of two different synthetic components, which makes the reuse and recycling quiet difficult, to such an extent that it is often preferred to perform the direct disposal in a dump [6, 7]. Another drawback related to SFRC is that, these composites are often disposed in unsatisfactory ways such as an incineration which causes air pollution to the environment [8–10]. It is imperative to note that most of these polymers are made from petroleum-based non-renewable resources. Environmental problems in recent decades have urged the necessity to look for new alternatives, which could replace the conventional SFRC with lower environment impact materials [11–14]. Furthermore, it is worsened by the fact that plastics production requires a remarkable consumption of oil-based resources, which are notoriously non-renewable.

This created a renewed interest in natural materials which could be used as reinforcements or fillers in the composites and are thus referred to as NFRP [15]. They are also called as "bio-composites" [16]. Several researchers came up with ideas of reinforcing different natural fibres in polymers to produce eco-friendly composites for several applications which do not require excellent mechanical properties, such as secondary or tertiary building structures, car panels, packaging, etc. [13]. The use of natural fibres as reinforcements in composites has been growing since then and has replaced several SFRC in several fields such as automotive, construction, aerospace, and marine application [11–14, 17]. This is mainly due to extensive research undertaken with its biodegradability characteristic, low cost, and interesting physical and mechanical properties [14, 16]. In chronological order, the very first attempts focused on replacing the synthetic fibres by natural fibres in petroleum-based matrixes to form composites [13]. The production and characterization of these composites were studied and the mechanical properties were improved substantially [18–24]. These composites gained support due to the environment-friendly aspect, considerable reduction of non-renewable resources used in composites, and the replacement of mineral-inorganic materials with natural-organic ones [13]. The use of natural fibres in composite industry created a new alternative for the manufactures. The possibilities of using recycled plastics like polyolefin succeeded as it reduced the consumption of non-biodegradable polymers [25–29]. Furthermore, the biodegradable polymers were derived from renewable resources, i.e., polylactic acid thermoplastic polymer from corn starch, soy bean based thermoset matrix, etc., and the composites were made completely from renewable resources, and were thus referred to as bio-composites [30–37].

In certain composite applications NFRP have shown to be competitive in relation to GRP [38]. NFRP with thermoplastic matrixes have successfully proven their qualities in various fields of application. The growing interest in using natural plant fibres as reinforcement of polymer-based composites is mainly due to their availability from renewable natural resources, high specific strength and stiffness, light weight, low cost and biodegradability [39, 40]. The biodegradability of the natural plant fibres may provide an improve environment quality while the low costs and good performance of these fibres are able to fulfill the economic interest of the industry [41]. Furthermore, in the 21 century, researchers and technologies have moved toward renewable raw materials and processes which are more environmental friendly and sustainable [42]. The use of natural fibres as reinforcement in polymer composites for making engineering materials has generated much interest in recent years [43]. These fibres multiple the eco-design potential of new composite materials for designers and industrial engineering, give new impetus to the research and development of new process techniques.

Natural fibres are grouped into several types: seed hair, bast fibres, leaf fibres, and etc. depending upon the source. Some examples are cotton (seed hairs), ramie, jute, and a flax (bast fibres), and sisal and abaca (leaf fibres). Of these fibres, kenaf, jute, ramie, flax, bamboo and sisal are the most commonly used fibres for polymer composites. Another type of natural fibres in the form of wood flour has also been often used for preparation of NFRP [1].

Although the mechanical properties of natural fibres, kenaf included, are lower than those of conventionally used fibres such as glass, carbon and Kevlar, due to their low density the properties of these natural fibres are comparable to those of the conventional synthetic fibres. A single fiber of kenaf can have a tensile strength and modulus as high as 11.9 and 60 GPa, respectively [44]. Another study showed that bamboo fibre, which is a stronger and stiffer plant fibre, has been used to produce bio-composites structures including boats and surfboards [45]. However, another concerning factor of polymer composites is degradation. For example, polyanhydrides and polyesters both degrade through hydrolysis but at significantly different rates; 0.1 h and 3.3 years respectively [46], whilst polyether's are non-biodegradable since they do not contain a hydrolysable bond. By making copolymers from polymers with different degradation rates, this property can be tailored to the application [47, 48]. Another example of biodegradation is pure PLA which will degrade to carbon dioxide, water and methane, within 2 years, whereas petroleum-based plastics require hundreds of years [49]. Biodegradation is a desired quality since it prevents stack of solid waste, which is a major consideration for composite materials in general and hinders their use in products with a limited service life [50, 51]. Bio-composites could enter new markets as their biodegradability offers a serious advantage over synthetic composites.

2 Fibre Selection

A great number of fibres can be found in nature. Figure 1 provides an overview of some of the more important 'natural fibres' [52]. The first main difference to be seen is that the term 'natural fibres' includes both organic and inorganic fibres. Both can be used as composite reinforcement; however researchers are only concerned with the most industrially significant organic cellulose fibres such as kenaf, bamboo, flax, hemp, jute, ramie, coir, abaca and sisal, with particular focus on kenaf and bamboo [52].

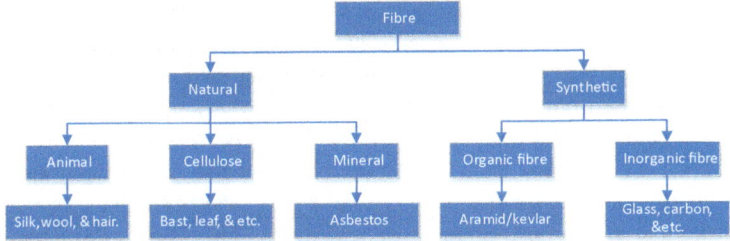

Fig. 1 Classification of fibres [101]

There is an enormous variety of cellulose fibres, their properties and morphology depend upon both the location of the fibres in the plant and their function. Cellulose fibres are found in different positions in the stems of monocotyledonous (plants with one cotyledon; abbreviated to monocot) and dicotyledonous (plants with two embryonic leaves; abbreviated dicot) plants as well as dicotyledonous and gymnosperm trees. Of particular importance as composite reinforcement are the so-called bast (cellulose) fibres such as Kenaf, flax, hemp, jute, and ramie. These fibres are to be found in the outer portion, the bark, of dicotyledonous plants and are of special interest because of their excellent specific strength and stiffness properties [52].

2.1 Cellulose Fibres

Bast fibres are obtained from the stems of various dicotyledonous plants and are also referred to as 'soft' fibres to distinguish them from leaf fibres. They are composed of elongated thick-walled ultimate cells joined together both end to end and side by side and arranged in bundles along the length of the stem. Bast fibre bundles are removed from the parent material by the decorticating process, which consists of removing from the stem, the 'cortex' comprising the bast and outer barks. The separated fibres are then washed in water and dried [52].

Kenaf has been cultivated for 4000 years but has only recently been exploited as a source of fibre. It is grown in tropical and sub-tropical areas. In tropical countries it is cultivated throughout the year. The bark fibres of kenaf plant are long and stringy but the inner core is much like balsa wood. Kenaf can produce about 2.47 kg of pulp per season, which is the same amount that a pine tree can produce after 20 years of growth [53]. Application for which kenaf fibres find markets are very similar to those of other bast fibres, namely paper making, pulp and more importantly is its used in making construction materials or automotive interior parts [54].

Bamboo has a hollow stem called the culm with the cellulose fibres aligned along the length of the culm carrying nutrients between the leaves and roots. It has a light coloured lignin. Bamboo can be grown in both the tropics and in temperate climates. There are over 1250 species of bamboo and over 10,000 tonnes are produced annually [55]. It is one of the main building materials used in developing countries, surpassed only by wood [54].

Another potential cellulose fibres are the leaf type fibres, also referred to as 'hard' fibres, are obtained from the leaves or leaf stalks of various monocotyledonous plants. Monocotyledons have parallel veined leaves and have one seed leaf [56]. Pineapple is the best example of leaf fibres. The fibres are in bundles of individual cells obtained from the leaf sheaths. They are removed from the strands by boiling in an alkali solution and are smooth with uniform diameter. The fibre lumen is large in relation to the cell wall [54].

Pineapple plants are largely grown in tropical America, in the Far-East Asia countries and Africa. It is in the Philippines, Taiwan, and India where the pineapple plant is largely used as a source of fibre. The pineapple fibre bundles are separated from the pineapple leaf by hand and sometimes by machines. The hand separation involves the stripping off of the fibre from the retted leaf. This method is considered to be laborious and costly and tends to lose a lot of weaker fibres. The use of machines to separate the fibre bundles is slower than the hand method of extraction but facilitates production processes. The yield of hand separated pineapple fibre bundles is in the range 2–3% dry fibre from about 1 tonne of pineapple leaf, that is, 20–27 kg of dry fibre [57].

In addition, banana fibres are also preferable as reinforment. Banana fibre Musa paradiisiaca L. var Sapientum or Musa ulugurensis Warb. is the most cultivated banana plant [58]. There are about 300 species of banana and about 20 are used for consumption. In order to obtain the best fibre the plants are cut when they are almost at the flowering stage, before any fruit has formed. The separation process is done manually and it involves cutting pieces of banana from the stem and passing them through a mangle to remove excess moisture (water), combing and drying at ambient temperature. The fibre obtained is usually of low quality because the separation of the fibre bundles is done either after the fruits have just developed or when they have matured ready for food purposes [54]. Finally, Table 1 shows a previous report of potential cellulose fibre for composites manufacturing.

The mechanical properties of the composites are depended on several factors such as type of fibres (refer Table 2), length, volume content and orientation embedded in the matrix. When a load is applied to the matrix, stress transfer occurs by shear at the interface along the fibre length and ends of the fibres. The extent of load transfer is a function of the critical fibre length (aspect ratio), the direction and orientation of fibre and the compatibility between fibre–matrix interphase. Depending on the fibre orientation at the matrix, three types of composite are prepared. Firstly, longitudinally aligned fibre composites generally have higher tensile strength but lower compressive strength (due to fibre buckling). Secondly, transversely directed fibres undergo very low tensile strength, which is lower than the matrix strength. Finally, randomly orientated short fibre composites have different mechanical properties. This is due to the complexities of load distribution at different direction along the interphase, consistent mechanical properties of these

Table 1 Reported work on cellulose fibre composite

Fibre	Orientation	Treatment	Method	Matrix	Summary	Sources
1. Pine apple leave fibres (PALF) 2. Kenaf fibres (KF)	–	6% NaOH + 2% Silane	–	Phenolic	1. Effects of alkali, silane and combined alkali and silane treatments on the mechanical (tensile), morphological, and structural properties PALF and KF	[130]
1. Kenaf	Non-woven mat	–	RTM	Bisphenol A epoxy matrix CP 210DF part A + amine hardener CP 210DF part B	1. The effect of kenaf fibre volume fraction on the composites' tensile properties and Poisson's ratio was investigated 2. The performance of the Tsai–Pagano model in predicting the composites' tensile modulus and Poisson's ratio was compared with the Manera and Cox-Krenchel model	[131]
1. Kenaf	–	Xylene	Hot pressed	Polypropylene	1. The mechanical properties values of rPP/DVB/PP-g-AA/Hall is as follows: TS is 32.32 MPa; FS is 26.07 MPa; and IT is 12.8 kJ/m^2. The effect of mechanical properties by adding 20% (w/w) second reinforcement KF of rPP/DVB/PP-g-AA/KF/Hall composites is as follows: the TS value increased by 18% (38.28 MPa); FS increased by 28% (33.48 MPa) and IT increased by 27% (15.2 kJ/m^2), compared to that of composites without KF	[132]
1. Kenaf 2. Oil palm	Non-woven mat	–	Hand layup	Epoxy (DER 331) + JOIN TMINE 905-3S	1. Dispersing nano filler (nano OPEFB, MMT, and OMMT) at 3% loading 2. Alternative constructional materials respect steel, brick, and cement for Malaysia	[133]

(continued)

Table 1 (continued)

Fibre	Orientation	Treatment	Method	Matrix	Summary	Sources
1. Kenaf	Roving form	–	Injection molding	PP matrix	1. This research is focused on partially eco-friendly hybrid long fiber reinforced thermo plastics with natural kenaf fiber to enhance the desired mechanical properties for car bumper beams as automotive structural components 2. A specimen without any modifier is tested and compared with a typical bumper beam material called LFRT	[134]
1. Kenaf	Roving form	Alkaline and silane	Hot melt press	Biodegradable aliphatic–aromatic copolyester	1. Study the morphology and properties of a biodegradable aliphatic–aromatic copolyester mixed with kenaf fiber were investigated 2. This composite showed improved thermal, thermomechanical, and mechanical properties	[135]
1. Kenaf 2. Banana	Woven and random	Alkaline and sodium lauryl sulphate	–	Polyester	1. The maximum tensile and flexural strength was found at 40% volume fraction of woven hybrid fiber composites 2. The sodium lauryl sulfate treatment improves the mechanical properties in both random mix and woven mix hybrid composites	[136]
1. Bamboo 2. Glass fibre	Random	MAPP	–	PP matrix	1. The percentage improvement of tensile strength after treated with MAPP fibres composites was 5.7% compared to untreated fibres composites 2. Similarly the tensile modulus, flexural strength and modulus have been increased to 8.3, 23.5 and 32.3%, respectively	[137]

(continued)

Table 1 (continued)

Fibre	Orientation	Treatment	Method	Matrix	Summary	Sources
					3. Carlo Santulli40 investigated the impact properties of bamboo/GFs-reinforced unsaturated polyester composites with bamboo fiber of 6.2% and GF of 18.8% weight content of 25% total fiber content. The maximum impact strength of 32 kJ/m^2 was obtained	
1. Kenaf	Nano-fibre	Maleic anhydride	Extrusion	PP matrix	1. The addition of MA-g-PP into nano-kenaf filled PP composite improved T.S., elongation%, and I. S. while it showed decreased M.I. compared to without MA-g-PP composite 2. Morphology observation showed that this was due to increased interfacial adhesion between nano-kenaf surface and PP by formation of a compatible interface. Regardless of MA-g-PP, nano-kenaf fibres showed better adhesion with PP compared to micron-kenaf fibres 3. The addition of MA-g-PP further improved wettability between kenaf surface and PP matrix regardless of kenaf fiber diameter	[116]

Table 2 Mechanical properties of natural fibres [101]

Properties	Tensile strength (MPa)	Specific tensile strength (MPa)	Young's modulus	Failure strain (%)
Abaca	12	–	41	3–4
Banana	529–914	392–677	27–32	1–3
Pineapple	413–1627	287–1130	60–82	0–1.6
Sisal	80–840	55–580	9–22	2–14
Bamboo	575	383	27	–
Flax	500–900	345–620	50–70	1.3–3.3
Hemp	310–750	210–510	30–60	2–4
Jute	200–450	140–320	20–55	2–3
Kenaf	295–1191	–	22–60	–
Ramie	915	590	23	3.7

composites are far more difficult. By controlling factors such as the aspect ratio, the dispersion and orientation of fibres, considerable improvements in the properties can be achieved [59–61].

In common, with all engineering applications for NFRP, fibre alignment and placement should be optimised to suit the application [62]. The authors considered stiffness index of fibres as a function of their geometry, for examples, aligned unidirectional (UD), woven, cross-laminated (orthogonal), and randomly oriented. The stiffness index is the product of the fibre volume fraction and orientation factor [63]. There are clear advantages in maximizing the elastic properties of natural fibres in boat hulls by packing them closely in a unidirectional configuration and using untwisted yarns [64].

However, the handling of natural fibres arrays is more easily achieved in the form of a woven fabric. The authors describe the microstructure and properties of jute are co-mingled with thermoplastic PLA filaments, derived from corn starch, which are then woven to produce Biotex fabrics [65]. The production of natural fibre composite components can be achieved by hot pressing Biotex fabrics to produce net shape components. In a second venture, Composites Evolution has supplied Biotex flax fabric to Amber Composites to produce a prepreg impregnated with Multipreg 8020 epoxy matrix [64].

3 Moisture Uptake of Natural Fibres

The cellulose fibres are hydrophilic and tend to absorb moisture. Many hydrogen bonds (hydroxyl groups OH) are present between the macromolecules in the fibre cell wall. When moisture from the atmosphere comes in contact with the fibres, the hydrogen bond breaks and hydroxyl groups form new hydrogen bonds with water molecules. The cross section of the fibre becomes the main access of water penetration. The interaction between hydrophilic fibres and hydrophobic matrixes causes fibre swelling within the matrix. This results are weakening the bonding strength at the interface, which leads to dimensional instability, matrix cracking and poor mechanical properties of the composites [66]. Therefore, the removal of moisture from fibres is an essential step for the preparation of composites. The moisture absorption of fibres can be reduced by eliminating hydrophilic hydroxyl groups from the fibre structure through different chemical treatments [67].

4 Modification of Cellulose Fibres

The reaction zone is a diffusion zone between fibre–matrix also called as interphase or boundary, in which two phases are chemically and/or mechanically combined [22]. If there is poor adhesion between the fibre-matrix boundary, then a weak distribution of force will occur which results in poor mechanical performance [68]. In order to optimize the interphase bonding, modification of cellulose fibres through chemical treatment are necessaries. A number of studies showed that chemically treated cellulose fibres have improved (1) surface roughness, (2) cleanliness (remove unnecessary element such as lignin, wax, oil, and dirts) [69]. This can be seen in Fig. 2a, b where the differences between non-treated and treated cellulose fibres are shown. As indicated in Fig. 2a, non-treated cellulose fibres have shown a smooth fibre surface and plenty amount of voids. According to Fig. 2b, treated cellulose fibres have shown a rough fibre surface and a reduced amount of voids. The author's showed the differences before and after chemical treatment which then relatively improved the mechanical performance of the CFRP [69].

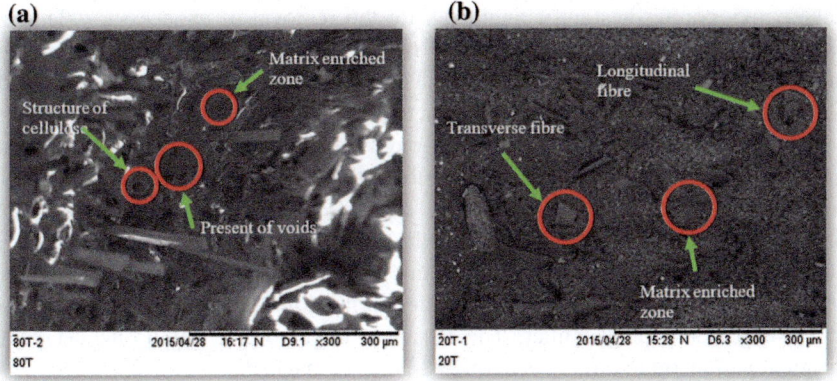

Fig. 2 a Non-treated CFRP [69]. **b** Treated CFRP [69]

4.1 Alkaline Treatment

Alkaline treatment or also called as mercerization is one of the most used chemical treatments of cellulose fibres when used to reinforce thermoplastics and thermosets. The important modification done by alkaline treatment is the disruption of hydrogen bonding in the network structure, thereby increasing surface roughness. This treatment removes a certain amount of lignin, wax and oils covering the external surface of the fiber cell wall, depolymerizes cellulose and exposes the short length crystallites [70]. Soaking cellulose fibres into sodium hydroxide (NaOH) in liquid form promotes the ionization of the hydroxyl group to the alkoxide [71]:

$$\text{Fiber}-\text{OH} + \text{NaOH} \rightarrow \text{Fiber}-\text{O}-\text{Na} + \text{H}_2\text{O} \quad (1)$$

Thus, alkaline processing directly influences the cellulosic fibril, the degree of polymerization and the extraction of lignin and hemicellulosic compounds [72]. In alkaline treatment, fibres are immersed in NaOH solution for a given period of time. Several researcher's treated jute and sisal fibres with 5% aqueous NaOH solution for 2 h up to 72 h at room temperature [73, 74], similar to treat flax fiber [75]. The author's reported that 2% alkali solution for 90 s at 200 °C and 1.5 MPa pressure was suitable for removal of dirt up to a single fibres [76]. These researchers observed that alkali led to an increase in amorphous cellulose content at the expense of crystalline cellulose. It is reported that alkaline treatment has two effects on the fiber: (1) it increases surface roughness resulting in better mechanical interlocking; and (2) it increases the amount of cellulose exposed on the fiber surface, thus increasing the number of possible reaction sites [77]. Consequently, alkaline treatment has a lasting effect on the mechanical behavior of flax fibres, focusing on fiber strength and stiffness [72]. The author' reported that alkaline treatment improved up to a 30% the tensile properties (both strength and modulus) for flax fiber–epoxy composites and coincided with the removal of unnecessary elements [78]. It also significantly improved the mechanical, impact fatigue and dynamic mechanical behaviors of CFRP [79–81]. The author's examined the effect of NaOH concentration (0.5, 1, 2, 4 and 10%) for treating sisal fiber-reinforced composites and concluded that maximum tensile strength resulted from the 4% NaOH treatment at room temperature and shows different opinion on optimum alkaline concentration [81]. While Mishra et al. reported that 5% NaOH treated sisal fiber-reinforced polyester composite is the optimum concentration to be used in NaOH treatment [82]. The different opinions occurred due to higher alkali concentration and this may provoke excess delignification of cellulose fibers, resulting in a weaker or damaged fiber. The tensile strength of the composite decreased drastically after optimum NaOH concentration is reached.

4.2 Silane Treatment

Silane is a chemical compound (SiH_4). It was used as coupling agent to let glass fibres adhere to a polymer matrix and stabilizing the composite material. Silane coupling agents may reduce the number of cellulose hydroxyl groups in the fiber–matrix interphase. In the presence of moisture, hydrolyzable alkoxy group leads to the formation of silanols. The silanol then reacts with the hydroxyl group of the fibres, forming stable covalent bonds to the cell wall that are chemically absorbed onto the fibres surface [71]. Therefore, the hydrocarbon chains provided by the application of silane restrain the swelling of the fibres by creating a crosslinked network due to covalent bonding between the fibres and matrixes. The reaction formulae are given as follows [71]:

$$CH_2CHSi\,(OC_2H_5)_3 \rightarrow CH_2CHSi\,(OH)_3 + 3C_2H_5OH \qquad (2)$$

$$CH_2CHSi(OH)_3 + Fiber-OH \rightarrow CH_2CHSi\,(OH)_2O-Fiber + H_2O \qquad (3)$$

Many researchers applied silane treatment in surface modification of GRP [83–86]. Silane coupling agents were also found to be effective in modifying CFRP matrix interphase and strengthen the interfacial bond. Three aminopropyl trimethoxy silane with concentration of 1% in a solution of acetone and water (50/50 by volume) for 2 h was reportedly used to modify the flax surface [78]. The author's soaked sisal fiber in a solution of 2% aminosilane in 95% alcohol for 5 min at a controlled pH value of 4.5–5.5 followed by 30 min air drying to hydrolyze the coupling agent [87]. Several researcher's conducted silane solutions in a water and ethanol mixture with concentration of 0.033 and 1% respectively to treat henequen fibers and oil palm fibres [71, 88]. It was verified that the results between the silane treatment modified fibres and the matrixes was increased, which led to composites with higher tensile strength compared to alkaline-treated fibres [88]. Thermal stability of the composites was also improved after silane treatment [71].

4.3 Stearic Acid Treatment

Stearic acid (CH3(CH2)16COOH) in ethyl alcohol solution is used to modify the fibre surfaces. The carboxyl group of stearic acid reacts with the hydrophilic hydroxyl groups of the fibre and improves water resistance properties. This treatment removes unnecessary elements covering materials constituents from the fibres structure. As a result, fibre bundles are breaking down and more fibrillation occurs [89, 90]. Fibre alignment embedded into the matrix provides a better bonding at the interphase and improved properties of the composites. The author's used stearic acid treatment on sisal fibre in ethyl alcohol solution to increase the mechanical properties [91]. Other authors used 3% stearic acid treated on sisal fibre reinforced with polyethylene composites and reported 23% higher shear strength ability compared to the non-treated [28]. The reaction between fibre and stearic acid is shown as follows:

$$\text{Fiber}-\text{OH} + \text{CH}_3(\text{CH}_2)16\text{COOH}\,\text{CH}_3(\text{CH}_2)16\text{COO}-\text{O}-\text{Fibre} + \text{H}_2\text{O} \quad (4)$$

4.4 Maleated Treatment

Maleated coupling agents provide efficient synergy with the functional surface of the fibres and matrixes. During grafting, maleic anhydride reacts with the hydroxyl groups (OH) in the amorphous region of cellulose structure and removes it (OH groups) from the fibre cells. This produces brush like long chain polymer coating on the fibre surface and reduces the hydrophilic tendency. Therefore, a maleated coupler forms carbon to carbon bond to the polymer chain with the matrixes [19]. This covalent bonding between the hydroxyl groups of the fibres and the anhydride groups of the maleic anhydride provides balanced properties to improve the interlocking efficiency [92]. Maleic anhydride grafted polypropylene (MAPP) copolymer is activated at a temperature of 1700 °C and then the esterification of the cellulose fibres is carried out. This treatment results in a better wettability of the fibre and enhances the interfacial adhesion [92, 93]. The authors used jute fibre (30% fibre loading, 6 mm on fibre length) treated with 0.5% MAPP concentration in toluene for 5 min and reinforced in PP matrix [94]. The composite showed around 72% higher flexural strength properties compared to the non-treated. Water absorption tendency by the treated fibre composite was also reduced. Similar studies were performed by another author and reported, 1% MAPP treated sisal fibre–PP composite exhibited around 50, 30 and 58% higher tensile, flexural and impact strength properties, respectively [95].

5 Thermal Stability of Cellulose Fibers

Cellulose fiber generally starts degrading at about 240 °C. Structural constituents of the fibre (cellulose, hemicellulose, lignin, etc.) are sensitive with different range of temperatures. It was reported that, lignin starts degrading at a temperature of around 200 °C and hemicelluloses and cellulosic constituents degraded at higher temperatures [61]. The thermal stability of the fibre can be enhanced by removing certain proportion of hemicelluloses and lignin constituents by different chemical treatments. The degradation of natural fibres is an important issue in the development of composites in both manufacturing (curing, extrusion or injection molding) and materials in service [20, 96].

6 Matrix Selection

A matrix is a binder material that is used to hold fibres in position and transfer external loads to internal reinforcements. It is used in conjunction with a reinforcing fibre to form a laminate. In cellulose fibre reinforced polymer composites, both thermoset and thermoplastic matrices such as unsaturated polyesters, epoxies and phenolics, and polypropylenes, polyethylenes and elastomers, respectively are widely used for composites applications [61, 97]. These matrices have different chemical structures and undergo different reactivities with the surface molecules of fibres in composites [22]. The various types of matrixes are described in the following.

6.1 Polyester Matrix

Generally, a polyester matrix can be made by a dibasic organic acid and a dihydric alcohol. They can be classified as saturated polyester, such as polyethylene terephthalate, and unsaturated polyester. To form the network of the composite matrix, the unsaturated group or double bond needs to exist in a portion of 8 the dibasic acid. By varying the acid and alcohol, a range of polyester matrices can be made. Orthophthalic polyesters are made by phthalic anhydride with either maleic anhydride or fumaric acid. Isophthalic polyesters, however, are made from isophthalic acid or terephthalic acid. The polyester matrix is usually dissolved in monomer (styrene is the most widely used), which will copolymerize with it and contribute to the final properties of the cured matrix. The addition of catalyst will cause the matrix to cure. The most frequently used catalyst is methyl ethyl ketone peroxide (MEKP) or benzoyl peroxide (BPO) and the amount varies from 1–2%. The catalyst will decompose in the presence of the polyester matrix to form free radicals, which will attack the unsaturated groups (like C=C) to initiate the polymerization. The processing temperature and the amount of the catalyst can control the rate of polymerization, the higher temperature or the more the catalyst, the faster the reaction. After the matrix turned from liquid to brittle solid, post cure at higher temperature may need to be done. The purpose of the post cure is to increase Tg of the matrix by completing the cross-linking. The properties of the polyester matrix are affected by the type and amount of reactant, catalyst and monomers as well as the curing temperature. The higher the molecular weight of polyester and the more points of unsaturation in molecules, the higher is the strength of the cured matrix. Orthophthalic polyesters are environmentally sensitive and have limited mechanical properties. They have been replaced in some applications by isophthalic polyesters due to the excellent environment resistance and improved mechanical properties of the latter [98].

The unsaturated polyester amounts to about 75% of all polyester matrices used in the world. It is produced by the condensation polymerization of dicarboxylic acids and dihydric alcohols. The formulation contains an unsaturated material such as maleic anhydride or fumaric acid which is a part of the dicarboxylic acid component. The formulation affects the viscosity, reactivity, resiliency and heat deflection temperature (HDT). The viscosity controls the speed and degree of wet-out (saturation) of the fibres. The reactivity affects the cure time and peak exotherm (heat generation) temperatures. High exotherm is needed for a thin section curing at room temperature and low exotherm for a thick section. Resiliency or flexible grade composites have a higher elongation, lower modulus and HDT. The HDT is a short term thermal property which measures the thermal sensitivity and stability of the matrices. The advantages of unsaturated polyester are its dimensional stability and affordable cost. Other advantages include ease in handling, processing, and fabricating. Some of the special formulations are high corrosion resistant and fire retardants. This matrix is probably the best value for a balance between performance and structural capabilities [99, 100].

6.2 Epoxy Matrix

The epoxies used in composites are mainly the glycidyl ethers and amines. The material properties and cure rates can be formulated to meet the required performance. Epoxies are generally found in diverse applications. The high viscosity in epoxy matrices limits its use to certain processes such as molding, filament winding, and hand lay-up. The right curing agent should be carefully selected because it will affect the type of chemical reaction, pot life and finish material properties. Although epoxies can be expensive, it may be worth the cost when high performance is required [98].

6.3 Vinyl-Ester Matrix

The vinyl ester matrixes were developed to take both advantage of the epoxy matrixes (high performance) and the fast curing of the polyesters. The vinyl ester has higher physical properties than polyesters but costs less than epoxies. The acrylic esters are dissolved in a styrene monomer to produce vinyl ester matrixes, which are cured with organic peroxides. A composite product containing a vinyl ester matrix can withstand high toughness demand [100].

They have properties similar to an epoxy matrix, being hard and strong with the addition of excellent corrosion resistance. They are formed by combining of vinyl groups and ester linkages to an epoxy based matrix [101].

7 Gel Coat

The issue of gel-coats for boat hulls and other marine structures requiring a quality surface finish was addressed in a paper by authors from the University of Plymouth, UK (Summerscales and Hoppins), PERA Innovation (Anstice, Brooks, Wiggers and Yahathugoda), Magnum Venus Plastech Ltd (Harper), W Ball and Sons Ltd (Wood) and Scott Bader Company Ltd (Cooper). Normal methods for applying gel-coats result in volatile organic compound (VOC) emissions and the paper discussed an improved application method suitable for RTM and other closed mold processes [102].

The patented in-mold gel-coating process involves the use of an impermeable separator layer between the laminate and the gel-coat. Laminates with acceptable surface finish have been produced using the method, which is still under development. It was suggested that, by reducing or obviating harmful emissions, the new process can improve workplace safety and reduce environmental impact [102].

A special formulated polyester matrix used for the outer surface of the craft is known as gel coat (styrene monomer 30–40%). Gel coat is designed to be highly waterproof, scratch resistant and to cure rapidly. To provide a decorative finish gel coat comes in a range of colors (pigment). It is also available as a clear liquid. A thin layer of gel-coat is usually sprayed onto the mold prior to the lay-up of the reinforcing material and the matrix [102].

8 Cellulose Nanofillers

In order to define nanometer scale items (10^{-9} m) the term nano is used [103]. Nanomaterials are categorized into three groups; nanotubes, nanoparticles, and nanolayers [104, 105]. Nano-particles are regarded as important potential filler materials for the enhancement of physical and mechanical properties of fibre reinforced plastics [106]. The unique nanometric size, capable of producing large specific surface areas, even more than 1000 m^2/g, along with their other unique properties currently draws great attention among researchers in the field of advanced materials [107]. Nanofillers possess the tendency to improve or to alter the variable properties of the materials into which they are integrated, such as physical, mechanical, thermal, and even the materials ability to withstand heat or fire retardant. Nanofillers are integrated in fibre reinforced plastics at rates from 1 to 10% (in mass) [108]. There are various types of nanofillers being used in fibre reinforced plastic such as nanoclays, nano-oxides, carbon nanotubes, and organic nanofillers. However, nanofillers can be distributed as classified in ISO/TS 27687 [109]. A nanocomposite is where a nano-object (particle) is distributed into a matrix [108].

The nanocomposites exhibit unique characteristics and comparably better properties than conventional composites such as fibreglass composites [110]. The nanofiller in fibre reinforced plastics are the main components and can be constituted of inorganic or organic sources. Fibre reinforced plastics are polymers (thermoplastics, thermosets, or elastomers) that have been reinforced with small quantities (less than 5% by weight) of nano-sized particles having high aspect ratios (L/h > 300) [111]. The reinforcement of polymeric matrix materials (thermosets) with nano-sized particles to form nanocomposites is considered as an attractive area of research. Polymer/layered nanocomposites, in general, can be classified into three different types, namely (i) Intercalated nanocomposites, (ii) flocculated nanocomposites, and (iii) exfoliated nanocomposites [112, 113]. Considerably larger interfacial matrix material surface (interphase) are presented by nanocomposites, depicting properties quite dissimilar from the bulk polymer caused by high specific surface area of the nanofiller [114, 115]. Thus, when the dimensions of polymer fiber materials are shrunken or reduced from micrometers to nanometers, numerous unique characteristics, such as flexibility in surface functionalities, and greater surface area to volume ratio with superior mechanical performance (such as stiffness and tensile strength) can be achieved, compared with conventional fibre reinforced plastic products [106]. Recognizing the morphological and mechanical benefits of nanofillers, some researchers produced nanocomposites by using different polymeric matrix and reinforcing by a wide variety of clays, which exhibited improved and better properties. A recent study by Kim [116] reports three main improvements from nanofillers, (1) improve in tensile strength, (2) increase interfacial adhesion between nano-kenaf surface and matrix (morphology), further improved wettability between kenaf surface and matrix regardless of fibre diameter [116]. Over conventional composites, nanocomposites display improvements in physical, mechanical, thermal, and resistant (fire) properties. Furthermore, the nano particles cause a significant reduction in flammability and also preserve the transparency or clearness of the polymer matrix [12, 117]. However, there are two ways of defining nano fibre reinforced plastic, (1) developed from renewable materials such cellulose and petroleum-derived matrix like PP, PE, and epoxies, (2) derived from biopolymers (e.g., PLA and PHA) and inorganic nanofillers (e.g., carbon nanotubes and nanoclay) [118, 119].

9 Presence of Porosity

During the insertion of fibres into the matrix, air or other volatile substances may be trapped inside the composite. After the curing process micro-voids are formed along the individual fibre tows and in the matrix rich regions. This causes sudden failure of the composites and shows poor mechanical properties. The curing and cooling rate of the composites are also responsible for the void formation [61]. High void content (over 20% by volume) is responsible for lower fatigue resistance, greater affinity to water diffusion and increase variation (scatter) in mechanical properties [120, 121]. Composites at higher fibre content display more risk for void formation [121]. Other factors contribute to the structural porosity content is manufacturing techniques. For thermoset composites, the basic fabrication is known as 'hand lay-up'. This is a manual laminating procedure between fibre and matrix at which the byproduct of the composites in terms of fibre to matrix ratio, sample thickness, and porosity content are fully dependent on the workmanship skill. In contrast, in vacuum assisted matrix infusion (VARI), the matrix is under vacuum pressure inside the mould and infuses with the fibres. In this case, the matrix impregnation in the composite is of much higher quality compared to manual fabrication. Thus, porosity can be kept as minimal [101].

10 Manufacturing Techniques for Cellulose Fibres

Thermosetting cellulose fibre composites represent almost one third of the cellulose fibre composite applications [101].

10.1 Hand Lay-Up

The hand lay-up method was developed in the USA 70 years ago. It is normally used for glass fibre fabrics with final volume fraction of 30–50%. The usual thermosetting matrix is catalyzed by an accelerated polyester, although the use of vinylester and phenolics is also possible. The use of hand lay-up can also be extended to cellulose fibres (woven and non-woven fabrics). The principal of the hand lay-up method is shown in Fig. 1 and carried out as follows: First, the mould surface should be cleaned, waxed and gel coated, then the fabric is laid on the desired mould. Matrix impregnation is done manually. Depending on the type of matrix, matrix curing can be heat accelerated. Mould forming details are found in many references elsewhere. Figure 3 shows a schematic for the technique [101].

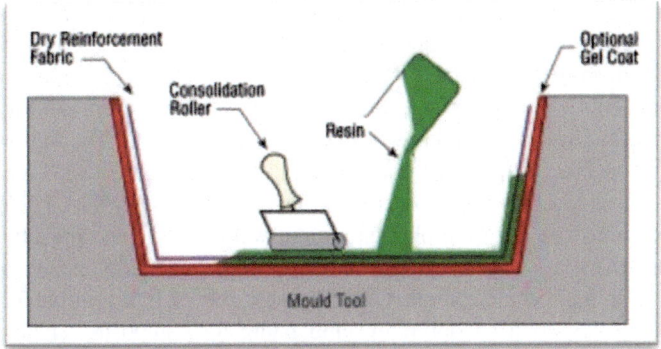

Fig. 3 Schematic for the hand lay-up technique

For low performance structures, spray and hand lamination are often chosen because the raw materials are relatively inexpensive and it is a low cost process. The process can use relatively unskilled labour with simple practice procedure, relatively good surface, easy insertion of accessories and the mould tools do not need to be vacuum tight. However, these open mould processes produce composites with low fibre volume fractions, excessive matrix residue and high levels of voids (porosity). Another disadvantage is that end product of the hand lay-up process also contains other mechanical defects such as porosity, low infusion, and low adhesion between matrix and fibre orientation, thus leads to a low mechanical properties or performance of cellulose fibre reinforced plastic. This also leads to poor product quality (dimensions, wrinkling, accurate fibre volume fraction), which is dependent on human factors. In addition, for certain conditions, labour cost are higher, especially with severe work conditions (matrix fumes and cost of achieving healthy ventilation conditions) [22].

10.2 Compression Molding (CM)

Compression moulding (CM) is generally used for a thermoplastic matrix with loose chopped fibre or mats of short or long fibre either randomly oriented or aligned, but can also be used with a thermoset matrix. The fibres are normally stacked alternately with thermoplastic matrix sheets before pressure and heat are applied. The viscosity of the matrix during pressing and heating needs to be carefully controlled, in particular for thick samples to make sure the matrix is impregnated fully into the space between fibres. Good quality composites can be produced by controlling viscosity, pressure, holding time, temperature taking account of the type of fibre and matrix [122]. Film stacking has been recommended as it limits cellulose fibre degradation due to involvement of only one temperature cycle [123]. Temperature still needs to be carefully controlled as commonly there is little difference in temperature between that at which a particular matrix can be processed and that at which fibre degradation will occur. Reduction of fibre strength has been shown to occur at temperatures as low as 150 and at 200 °C, with strength reducing by 10% in ten minutes. Flexural properties were found to be less dependent on temperature below 150 °C, but reduced significantly at higher temperatures. For composites made from jute yarn and the bacterial copolyester Biopol the optimum compression temperature for a range of mechanical properties was found to be approximately 180 °C. The highest strength was obtained at 200 °C for non-woven mat reinforced PP matrices. Alternatively to film stacking, sheet moulding compounds have been used in CM [124]. Overall, there is a compromise between obtaining good wetting and avoiding fibre degradation that leads to an optimum temperature for a particular composite material or geometry. This has been explored with film stacking of flax reinforced poly(ester amide) composites, with an optimum temperature for composite tensile properties found to be 150 °C [125].

10.3 Resin Transfer Molding (RTM)

In RTM, the liquid thermoset matrix is injected into a mould containing a fibre preform. The main variables with this process are the temperature, injection pressure, matrix viscosity, preform architecture and mould configuration. Advantages compared with other processes include lower temperature requirements and avoidance of thermomechanical degradation. Compaction in this process is affected by the structure of cellulose fibres including the effect of lumen closing and due to lower degrees of fibre alignment, cellulose fibre composites are less compactable than glass fibre composites. Good component strength can be achieved with this process which is suitable for low production runs [126]. While RTM is appropriate for relatively small components, the mould closure forces become excessive as component size increases. However, the process is limited to a specific dimension. The process of producing a complex mould also becomes an issue. Finally, a complicated procedure comes with higher costs because more accessories are required [101].

10.4 Vacuum Assisted Resin Infusion (VARI)

This process is normally applied with fabrics (woven or non-woven). The multiple layers required for the cellulose fibre fabrics are first cut into shape and then placed over the mould. Fibres and mould are then vacuum–bagged. The matrix is allowed to infuse into the fabrics smoothly and slowly for wet out, assisted by the vacuum. After complete matrix infusion, the cellulose fibre composites are left to cure under vacuum for 24 h at room temperature, after which they are unpacked and taken for inspection. Figure 4 shows a schematic for a simplified VARI process [101].

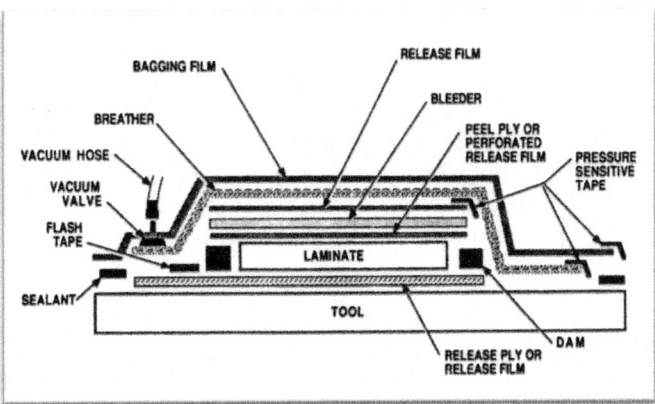

Fig. 4 Schematic for the VARI technique

VARI provides a better adhesion between matrix and fibre orientation which provides a higher degree of integration. Other advantages of VARI is the possibility of producing large complex form/structures compared to previous processes mention before. However, the probability of leakage may lead to total loss of the product. Besides, the preparation takes a long period of time due to the absence of automation and also because the product must be demoulded. After a complete curing, all the accessories need to be removed. Finally, a complicated procedure comes with higher costs because more accessories are required (vacuum pump, connections, matrix distribution spirals, sealant material like tacky tapes, etc.) [101].

11 Marine Qualification Issue

A feature of composites that is both a strength and a weakness is their chameleon-like adaptability to a wide range of applications on the one hand, and their variability on the other. It consists of many products and process variations as there are individual structures, a fact that makes it difficult to formulate industry standards and qualification routes [102]. This was acknowledged in a paper by Weitzenboeck, M. et al., of the Norwegian classification society Det Norske Veritas (DNV) and Echtermeyer from the Norwegian University of Science and Technology. Nevertheless, standards are essential to establish minimum requirements before constructions [102]. For instance, manufacturing of wind turbine blades. Development of the DNV offshore Standard for the Design and Manufacture of Wind Turbine Blades (DNV-OS-J102) provides a comfort zone by offering a contractual reference and guidelines for manufacturers [102]. Other similar cases, DNV's Offshore Standard for Concrete Structures was enlisted in qualifying a new type of fibre reinforced plastic (FRP) reinforcement bar (rebar) for concrete. This novel product incorporated basalt fibres as the major composite reinforcement, and a new manufacturing process [102]. Another area crying out for standards is the business of composites for marine high speed and light crafts. The requirements in this section are applied to fibre reinforced plastic (FRP) sandwich constructions for structural design (hull, structures, and etc.) [127]. Maritime certification authorities like DNV are well known when it comes to marine vessels. The platforms have several decades of experience handling composite materials. Combining existing rules for ships and high speed crafts with guidelines for composites usage provides a working basis for manufacturers [102].

12 Marine Application

A canoe comprising cellulose fibre on a marine plywood and solid pine frame was manufatured. A weave cellulose fibre reinforced a matrix of UV-cured EcoComp UV-L matrix containing 95% vegetable oil manufactured by Sustainable

Composites. While, a successful collaboration between Movevirgo Ltd and Sustainable Composites have led to the manufacturing of biodegradable surfboards and boats cured from EcoComp UV-L matrix [128]. Huntman Advance Material were responsible for providing the epoxy matrix for the manufacture of a lightweight Mini Transat 6.5 composite racing boat prototype named the 'Araldite', which contained a fibre content of 50% cellulose fibre and 50% carbon fibre. The natural fibre were supplied by Lineo, who pre-impregnate natural fibres with proxy matrix to prevent moisture absorption [64]. Lineo's FlaxPly and FlaxPreg (epoxy flax prepreg) products have been used to manufacture various marine products such as canoes and yachts. As well as boosting the eco-profile of the 'Araldite', the cellulose fibres endow the composite structure with good damping characteristics. The external hull was constructed from sandwich laminates comprising cellulose fibre fabrics and structural foam. The sandwich laminates were impregnated with epoxy matrix by vacuum infusion and cured at 50 °C. Internal framing and reinforcement were fabricated from carbon fibre composite. Niels Haarbosch from Holland has also constructed a natural fibre epoxy catamaran based on an Aero Skimmer design [64]. In a study of the performance of cellulose fibre composites versus synthetic materials for the fabrication of surfboards, evaluated hemp fibre as a substitute for glass fibre with matrixes of polyester or epoxy matrix [129]. Other authors concluded that cellulose composites were inferior to industry standard. However, bamboo fibre, which is stronger and stiffer compared to other cellulose fibre, has been used to produce biocomposite structures [45].

12.1 Superstructure

They are two main requirements from boat manufacturers, (1) lowering the weight of the boat, (2) obtaining parts that are very easy to manoeuver. Superstructure fittings are the key parts. Incorporating cellulose fibres is a solution that provides lightweight yet high strength fittings. Currently industries had manufactured an environment friendly pulley for which the flanges are cellulose fibre reinforced. These parts are obtained by injecting 15% by weight of cellulose fibre aggregate into a 100% bio-matrix PA11 (castor bean oil) [101].

12.2 Racing Sailboats

There are even more prototypes for sailboats made of CFRPs, since they are highly appreciated for their lightness, maneuverability, glidability, and environment friendliness [101]. The 'pioneer' turns out to be one of the ten most high-performance models in the 6.50 yacht racing sector. Its main added value is the low weight: 50% of the total reinforcement fibre weight consists of cellulose fibres. For the hull and deck, about 80% of the fibres used are cellulose bases, making the boat one of the lightest in its category at 750 kg [101]. The hull and deck include composite parts that are 100% CFRP, processed in the form of unidirectional prepreg tapes (180 g/m^2) that are embedded with the epoxy matrixes. The cellulose composite parts were assembled using UD or 2D carbon tapes laminated with an epoxy matrix system [101]. The infusion-moulded composite materials are sandwich structures with a Styrene-Acrylonitrile (SAN) foam core (density: 80 kg/m3) and skins with a 35% cellulose fibre volume content. This racing boat has already participated in numerous races and regattas in which it has proved itself to be reliable [101].

13 Conclusion

This review intends to determine the potential candidate of cellulose fibres for marine application. Thus, lead to the usage of CFRPs in future marine composites. The use of polymer composites reinforced with cellulose fibre, in replacement of synthetic materials, is of great interest in order to reduce the usage of petroleum-based or nonrenewable resources. These CFRPs with good sound absorption ability, density around 0.9 g/cm^3, stiffness around 3000 MPa, and impact strength of 25 kJ/m^2 which can be improved further respectively looks promising [1]. Meanwhile, for a long term overview, it is a more intelligent utilization of financial and environmental resources [13]. These "biocomposites" can find several industrial applications, although some limitations occurred regarding mainly process ability and mechanical performance. Worldwide researchers have been spending much effort in order to develop efficient solutions through chemical modification of the reinforcement, use of adhesion promoters or additives (coupling agents). However, a full biodegradability, and thus a really improved environmental impact, can be obtained only by replacing synthetic composites (coming from non-renewable resources) with biodegradable materials. In these cases, however, new limitations arise and current scientific investigations have been focusing on the selection of the most suitable biodegradable fibre/matrix and the optimization of all of the initial preparation and manufacturing parameters. As regards, in the commercial perspective, it can be stated that the market is still in an opening phase, therefore much can still be done in order to find new possible applications, improving the mechanical properties, and the marketability of these materials especially in marine industry. All of these issues continue to require significant research efforts in finding an effective formulations (synthetic or biodegradable composites; types and final product quality), characterize them and to be applied for the right applications, including to refine processing techniques. As soon as the market for these composites increases, reduction of costs and improvement of the quality will be achieved.

References

1. Saheb, D.N., Jyoti, P.J.: Natural fiber polymer composites: a review. Adv. Polym. Technol. **18**(4), 351–363 (1999)
2. Selvaraju, S., Ilaiyavel, S.: Applications of composites in marine industry. J. Eng. Res. Stud. **II**, 89–91 (2011)
3. Daniel, I.M., Ishai, O., Daniel, I.M., Daniel, I.: Engineering Mechanics of Composite Materials, vol. 3. Oxford University Press, New York (1994)
4. Coutinho, F.M.B., Thais H.S.C.: Performance of polypropylene–wood fiber composites. Polym. Test. **18**(8), 581–587 (1999)
5. Bettini, S.H.P., Uliana, A.T., Holzschuh, D.: Effect of process parameters and composition on mechanical, thermal, and morphological properties of polypropylene/sawdust composites. J. Appl. Polym. Sci. **108**(4), 2233–2241 (2008)
6. Netravali, A.N., Chabba, S.: Composites get greener. Mater. Today **6**(4), 22–29 (2003)
7. Rozman, H.D., Lee, M.H., Kumar, R.N., Abusamah, A., Mohd Ishak, Z.A.: The effect of chemical modification of rice husk with glycidyl methacrylate on the mechanical and physical properties of rice husk-polystyrene composites. J. Wood Chem. Technol. **20**(1), 93–109 (2000)
8. Henshaw, J.M., Han, W., Owens, A.D.: An overview of recycling issues for composite materials. J. Thermoplast. Compos. Mater. **9**(1), 4–20 (1996)
9. Pickering, S.J.: Recycling technologies for thermoset composite materials—current status. Compos. Part A Appl. Sci. Manuf. **37**(8), 1206–1215 (2006)
10. Conroy, A., Halliwell, S., Reynolds, T.: Composite recycling in the construction industry. Compos. A Appl. Sci. Manuf. **37**(8), 1216–1222 (2006)
11. Bledzki, A.K., Gassan, J.: Composites reinforced with cellulose based fibres. Prog. Polym. Sci. **24**(2), 221–274 (1999)
12. Faruk, O., Bledzki, A.K., Fink, H.-P., Sain, M.: Biocomposites reinforced with natural fibers: 2000–2010. Prog. Polym. Sci. **37**(11), 1552–1596 (2012)
13. La Mantia, F.P., Morreale, M.: Green composites: a brief review. Compos. Part A Appl. Sci. Manuf. **42**(6), 579–588 (2011)
14. John, M.J., Thomas, S.: Biofibres and biocomposites. Carbohydr. Polym. **71**(3), 343–364 (2008)
15. Bogoeva-Gaceva, G., Avella, M., Malinconico, M., Buzarovska, A., Grozdanov, A., Gentile, G., Errico, M.E.: Natural fiber eco-composites. Polym. Compos. **28**(1), 98–107 (2007)
16. Mohanty, A.K., Misra, M., Hinrichsen, G.: Biofibres, biodegradable polymers and biocomposites: an overview. Macromol. Mater. Eng. **276**(1), 1–24 (2000)
17. Koronis, G., Silva, A., Fontul, M.: Green composites: a review of adequate materials for automotive applications. Compos. B Eng. **44**(1), 120–127 (2013)
18. Ku, H., Wang, H., Pattarachaiyakoop, N., Trada, M.: A review on the tensile properties of natural fiber reinforced polymer composites. Compos. Part B Eng. **42**(4), 856–873 (2011)
19. George, J., Sreekala, M.S., Thomas, S.: A review on interface modification and characterization of natural fiber reinforced plastic composites. Polym. Eng. Sci. **41**(9), 1471–1485 (2001)
20. Sgriccia, N., Hawley, M.C., Misra, M.: Characterization of natural fiber surfaces and natural fiber composites. Compos. A Appl. Sci. Manuf. **39**(10), 1632–1637 (2008)
21. Cheung, H., Ho, M., Lau, K., Cardona, F., Hui, D.: Natural fibre-reinforced composites for bioengineering and environmental engineering applications. Compos. B Eng. **40**(7), 655–663 (2009)
22. Kabir, M.M., Wang, H., Lau, K.T., Cardona, F.: Chemical treatments on plant-based natural fibre reinforced polymer composites: an overview. Compos. Part B Eng. **43**(7), 2883–2892 (2012)

23. Xie, Y., Hill, C.A.S., Xiao, Z., Militz, H., Mai, C.: Silane coupling agents used for natural fiber/polymer composites: a review. Compos. Part A Appl. Sci. Manuf. **41**(7), 806–819 (2010)
24. Dittenber, D.B., GangaRao, H.V.S.: Critical review of recent publications on use of natural composites in infrastructure. Compos. A Appl. Sci. Manuf. **43**(8), 1419–1429 (2012)
25. Malkapuram, R., Kumar, V., Negi, Y.S.: Recent development in natural fiber reinforced polypropylene composites. J. Reinf. Plast. Compos. **28**(10), 1169–1189 (2009)
26. Shubhra, Q.T.H., Alam, A.K.M.M., Quaiyyum, M.A.: Mechanical properties of polypropylene composites: a review. J. Thermoplast. Compos. Mater. **26**(3), 362–391 (2013)
27. Yuan, Q., Wu, D., Gotama, J., Bateman, S.: Wood fiber reinforced polyethylene and polypropylene composites with high modulus and impact strength. J. Thermoplast. Compos. Mater. **21**(3), 195–208 (2008)
28. Torres, F.G., Cubillas, M.L.: Study of the interfacial properties of natural fibre reinforced polyethylene. Polym. Test. **24**(6), 694–698 (2005)
29. Sobczak, L., Brüggemann, O., Putz, R.F.: Polyolefin composites with natural fibers and wood-modification of the fiber/filler–matrix interaction. J. Appl. Polym. Sci. **127**(1), 1–17 (2013)
30. Plackett, D., Andersen, T.L., Pedersen, W.B., Nielsen, L.: Biodegradable composites based on L-polylactide and jute fibres. Compos. Sci. Technol. **63**(9), 1287–1296 (2003)
31. Siracusa, V., Rocculi, P., Romani, S., Dalla Rosa, M.: Biodegradable polymers for food packaging: a review. Trends Food Sci. Technol. **19**(12), 634–643 (2008)
32. Bax, B., Müssig, J.: Impact and tensile properties of PLA/Cordenka and PLA/flax composites. Compos. Sci. Technol. **68**(7), 1601–1607 (2008)
33. Meier, M.A.R., Metzger, J.O., Schubert, U.S.: Plant oil renewable resources as green alternatives in polymer science. Chem. Soc. Rev. **36**(11), 1788–1802 (2007)
34. Galià, M., de Espinosa, L.M., Ronda, J.C., Lligadas, G., Cádiz, V.: Vegetable oil-based thermosetting polymers. Eur. J. Lipid Sci. Technol. **112**(1), 87–96 (2010)
35. de Espinosa, L.M., Meier, M.A.R.: Plant oils: the perfect renewable resource for polymer science?! Eur. Polym. J. **47**(5), 837–852 (2011)
36. Sharma, V., Kundu, P.P.: Addition polymers from natural oils—a review. Prog. Polym. Sci. **31**(11), 983–1008 (2006)
37. Güner, F.S., Yağcı, Y., Erciyes, A.T.: Polymers from triglyceride oils. Prog. Polym. Sci. **31**(7), 633–670 (2006)
38. Mohanty, A.K., Misra, M., Drzal, L.T. (eds.) Natural Fibers, Biopolymers, and Biocomposites. CRC Press, USA (2005)
39. Begum, K., Islam, M.: Natural fiber as a substitute to synthetic fiber in polymer composites: a review. Res. J. Eng. Sci. ISSN 2278, 9472 (2013)
40. Baley, C.: Analysis of the flax fibres tensile behaviour and analysis of the tensile stiffness increase. Compos. A Appl. Sci. Manuf. **33**(7), 939–948 (2002)
41. Karmaker, A.C., Hoffmann, A., Hinrichsen, G.: Influence of water uptake on the mechanical properties of jute fiber-reinforced polypropylene. J. Appl. Polym. Sci. **54**(12), 1803–1807 (1994)
42. Cai, J., Zhang, L., Zhou, J., Qi, H., Chen, H., Kondo, T., Chen, X., Chu, B.: Multifilament fibers based on dissolution of cellulose in NaOH/urea aqueous solution: structure and properties. Adv. Mater. **19**(6), 821–825 (2007)
43. Kalia, S., Kaith, B.S., Kaur, I.: Pretreatments of natural fibers and their application as reinforcing material in polymer composites—a review. Polym. Eng. Sci. **49**(7), 1253–1272 (2009)
44. Karnani, R., Krishnan, M., Narayan, R.: Biofiber-reinforced polypropylene composites. Polym. Eng. Sci. **37**(2), 476–483 (1997)
45. Khalil, H.P.S.A., Bhat, I.U.H., Jawaid, M., Zaidon, A., Hermawan, D., Hadi, Y.S.: Bamboo fibre reinforced biocomposites: a review. Mater. Des. **42**, 353–368 (2012)
46. Göpferich, A.: Mechanisms of polymer degradation and erosion. Biomaterials **17**(2), 103–114 (1996)

47. Guarino, V., Lewandowska, M., Bil, M., Polak, B., Ambrosio, L.: Morphology and degradation properties of PCL/HYAFF11® composite scaffolds with multi-scale degradation rate. Compos. Sci. Technol. **70**(13), 1826–1837 (2010)
48. Baker, B.M., Gee, A.O., Metter, R.B., Nathan, A.S., Marklein, R.A., Burdick, J.A., Mauck, R.L.: The potential to improve cell infiltration in composite fiber-aligned electrospun scaffolds by the selective removal of sacrificial fibers. Biomaterials **29**(15), 2348–2358 (2008)
49. Hu, R., Lim, J.-K.: Fabrication and mechanical properties of completely biodegradable hemp fiber reinforced polylactic acid composites. J. Compos. Mater. **41**(13), 1655–1669 (2007)
50. Satyanarayana, K.G., Arizaga, G.G.C., Wypych, F.: Biodegradable composites based on lignocellulosic fibers—an overview. Prog. Polym. Sci. **34**(9), 982–1021 (2009)
51. Klemchuk, P.P.: Degradable plastics: a critical review. Polym. Degrad. Stab. **27**(2), 183–202 (1990)
52. Verpoest, I., et al. Flax and Hemp Fibres: A Natural Solution for the Composite Industry. JEC Composites. JEC, Paris (2012)
53. Anon: Kenaf—an alternative fibre crop. In: Natural Life Magazine, International Kenaf Association, p. 42 (2001)
54. Mwaikambo, L.Y.: Review of the history, properties and application of plant fibres. Afr. J. Sci. Technol. **7**(2), 120–133 (2006)
55. Robson, D.: Survey of Natural Materials for use in Structural Composites as Reinforcement and Matrices. Biocomposites Centre, University of Wales (1993)
56. Perry, D.R. (conv.): Identification of Textile Materials. The Textile Institute, Manara Printing Services, London (1975)
57. Kirby, R.H.: Vegetable Fibers. Interscience Publishers Inc., New-York, 473 p (1963)
58. Al-Qureshi, H.A.: The use of banana fibre reinforced composites for the development of a truck body. In: Second International Wood and Natural Fibre Composites Symposium, Kassel/Germany, pp. 1–8 (1999)
59. Fakirov, S., Bhattacharyya, D. (eds.) Engineering Biopolymers: Homopolymers, Blends, and Composites. Carl Hanser Verlag GmbH Co KG (2015)
60. Mwaikambo, L.Y., Ansell, M.P.: The effect of chemical treatment on the properties of hemp, sisal, jute and kapok fibres for composite reinforcement. Die angewandte makromolekulare Chemie **272**, 108–116 (1999)
61. Joseph, P.V., Joseph, K., Thomas, S., Pillai, C.K.S., Prasad, V.S., Groeninckx, G., Sarkissova, M.: The thermal and crystallisation studies of short sisal fibre reinforced polypropylene composites. Compos. Part A Appl. Sci. Manuf. **34**(3), 253–266 (2003)
62. SHAZ, IZ.: Optimising the properties of green composites. Green Compos. Polym. Compos. Environ. 154 (2004)
63. Bader, M.G.: Polymer composites in 2000: structure, performance, cost and compromise. J. Microsc. **201**(2), 110–121 (2001)
64. Ansell, M.P.: Natural fibre composites in a marine environment. Nat. Fibre Compos. Mater. Processes Prop. 365 (2014)
65. Weager, B.: Growing for gold. Mater. World **21**(1), 24–25 (2013)
66. Zakaria, S., Kok Poh, L.: Polystyrene-benzoylated EFB reinforced composites. Polym. Plast. Technol. Eng. **41**(5), 951–962 (2002)
67. Wang, B., Panigrahi, S., Tabil, L., Crerar, W.: Pre-treatment of flax fibers for use in rotationally molded biocomposites. J. Reinf. Plast. Compos. **26**(5), 447–463 (2007)
68. Alavudeen, A., Rajini, N., Karthikeyan, S., Thiruchitrambalam, M., Venkateshwaren, N.: Mechanical properties of banana/kenaf fiber-reinforced hybrid polyester composites: effect of woven fabric and random orientation. Mater. Design (1980–2015) **66**, 246–257 (2015)
69. Mohd Idrus, M.A.M, Firdaus, M., Yang, M., Ismail, S.B., Abdul Hamid, S.E., Syed Abdullah, S.F.A.: Development of Treated Kapok/Fiberglass Hybrid Composite for Marine Application. Applied Science and Advance Tech. Section, Universiti Kuala Lumpur Malaysian Institute of Marine Engineering Technology, Perak, Malaysia (2015)

70. Mohanty, A.K., Misra, M., Drzal, L.T.: Surface modifications of natural fibers and performance of the resulting biocomposites: an overview. Compos. Interfaces **8**(5), 313–343 (2001)
71. Agrawal, R., Saxena, N. S., Sharma, K.B., Thomas, S., Sreekala, M.S.: Activation energy and crystallization kinetics of untreated and treated oil palm fibre reinforced phenol formaldehyde composites. Mater. Sci. Eng. A **277**(1), 77–82 (2000)
72. Jähn, A., Schröder, M.W., Füting, M., Schenzel, K., Diepenbrock, W.: Characterization of alkali treated flax fibres by means of FT Raman spectroscopy and environmental scanning electron microscopy. Spectrochim. Acta Part A Mol. Biomol. Spectro. **58**(10), 2271–2279 (2002)
73. Ray, D., Sarkar, B.K., Rana, A.K., Bose, N.R.: Effect of alkali treated jute fibres on composite properties. Bull. Mater. Sci. **24**(2), 129–135 (2001)
74. Mishra, S., Misra, M., Tripathy, S.S., Nayak, S.K., Mohanty, A.K.: Graft copolymerization of acrylonitrile on chemically modified sisal fibers. Macromol. Mater. Eng. **286**(2), 107–113 (2001)
75. Morrison Iii, W.H., Archibald, D.D., Sharma, H.S.S., Akin, D.E.: Chemical and physical characterization of water-and dew-retted flax fibers. Ind. Crops Prod. **12**(1), 39–46 (2000)
76. Dupeyre, D., Vignon, M.R.: Fibres from semi-retted hemp bundles by steam explosion treatment. Biomass Bioenerg. **14**(3), 251–260 (1998)
77. Valadez-Gonzalez, A., Cervantes-Uc, J.M., Olayo, R.J.I.P., Herrera-Franco, P.J.: Effect of fiber surface treatment on the fiber–matrix bond strength of natural fiber reinforced composites. Compos. B Eng. **30**(3), 309–320 (1999)
78. Van de Weyenberg, I., Ivens, J., De Coster, A., Kino, B., Baetens, E., Verpoest, I.: Influence of processing and chemical treatment of flax fibres on their composites. Compos. Sci. Technol. **63**(9), 1241–1246 (2003)
79. Joseph, K., Thomas, S., Pavithran, C.: Effect of chemical treatment on the tensile properties of short sisal fibre-reinforced polyethylene composites. Polymer **37**(23), 5139–5149 (1996)
80. Sarkar, B.K., Ray, D.: Effect of the defect concentration on the impact fatigue endurance of untreated and alkali treated jute–vinylester composites under normal and liquid nitrogen atmosphere. Compos. Sci. Technol. **64**(13), 2213–2219 (2004)
81. Jacob, M., Thomas, S., Varughese, K.T.: Mechanical properties of sisal/oil palm hybrid fiber reinforced natural rubber composites. Compos. Sci. Technol. **64**(7), 955–965 (2004)
82. Mishra, S., Mohanty, A.K., Drzal, L.T., Misra, M., Parija, S., Nayak, S.K., Tripathy, S.S.: Studies on mechanical performance of biofibre/glass reinforced polyester hybrid composites. Compos. Sci. Technol. **63**(10), 1377–1385 (2003)
83. Kim, J.-K., Sham, M.-L., Wu, J.: Nanoscale characterisation of interphase in silane treated glass fibre composites. Compos. Part A Appl. Sci. Manuf. **32**(5), 607–618 (2001)
84. Ishak, Z.A.M., Ariffin, A., Senawi, R.: Effects of hygrothermal aging and a silane coupling agent on the tensile properties of injection molded short glass fiber reinforced poly (butylene terephthalate) composites. Eur. Polym. J. **37**(8), 1635–1647 (2001)
85. Lee, G.-W., Lee, N.-J. Jang, J., Lee, K.-J., Nam, J.-D.: Effects of surface modification on the resin-transfer moulding (RTM) of glass-fibre/unsaturated-polyester composites. Compos. Sci. Technol. **62**(1), 9–16 (2002)
86. Debnath, S., Wunder, S.L., McCool, J.I., Baran, G.R.: Silane treatment effects on glass/resin interfacial shear strengths. Dent. Mater. **19**(5), 441–448 (2003)
87. Rong, M.Z., Zhang, M.Q., Liu, Y., Yang, G.C., Zeng, H.M.: The effect of fiber treatment on the mechanical properties of unidirectional sisal-reinforced epoxy composites. Compos. Sci. Technol. **61**(10), 1437–1447 (2001)
88. Valadez-Gonzalez, A., Cervantes-Uc, J.M., Olayo, R., Herrera-Franco, P.J.: Chemical modification of henequen fibers with an organosilane coupling agent. Compos. B Eng. **30**(3), 321–331 (1999)
89. Paul, S.A., Oommen, C., Joseph, K., Mathew, G., Thomas, S.: The role of interface modification on thermal degradation and crystallization behavior of composites from

commingled polypropylene fiber and banana fiber. Polym. Compos. **31**(6), 1113–1123 (2010)
90. Paul, S.A., Joseph, K., Mathew, G.D.G., Pothen, L.A., Thomas, S.: Influence of polarity parameters on the mechanical properties of composites from polypropylene fiber and short banana fiber. Compos. Part A Appl. Sci. Manuf. **41**(10), 1380–1387 (2010)
91. Kalaprasad, G., Francis, B., Thomas, S., Kumar, C.R., Pavithran, C., Groeninckx, G., Thomas, S.: Effect of fibre length and chemical modifications on the tensile properties of intimately mixed short sisal/glass hybrid fibre reinforced low density polyethylene composites. Polym. Int. **53**(11), 1624–1638 (2004)
92. Keener, T.J., Stuart, R.K., Brown, T.K.: Maleated coupling agents for natural fibre composites. Compos. A Appl. Sci. Manuf. **35**(3), 357–362 (2004)
93. Li, X., Tabil, L.G., Panigrahi, S.: Chemical treatments of natural fiber for use in natural fiber-reinforced composites: a review. J. Polym. Environ. **15**(1), 25–33 (2007)
94. Mohanty, S., Nayak, S.K., Verma, S.K., Tripathy, S.S.: Effect of MAPP as a coupling agent on the performance of jute–PP composites. J. Reinf. Plast. Compos. **23**(6), 625–637 (2004)
95. Mohanty, S., Verma, S.K., Nayak, S.K., Tripathy, S.S.: Influence of fiber treatment on the performance of sisal–polypropylene composites. J. Appl. Polym. Sci. **94**(3), 1336–1345 (2004)
96. Taj, S., Munawar, M.A., Khan, S.: Natural fiber-reinforced polymer composites. Proc. Pak. Acad. Sci. **44**(2), 129 (2007)
97. Pickering, K.L., Li, Y., Farrell, R.L., Lay, M.: Interfacial modification of hemp fiber reinforced composites using fungal and alkali treatment. J. Biobased Mater. Bioenergy **1**(1), 109–117 (2007)
98. Bagherpour, S.: In: Hosam El-Din M.S. (ed.) Fibre Reinforced Polyester Composites. p. 167 (2012)
99. Mathews, F., Rawlings, R.: Polymer Matrix Composite. Composite Materials: Engineering and Sciences, pp. 168–200 (1994)
100. Tuttle, M.: Introduction. In: Structural analysis of Polymeric Composite Materials. University of Washington, pp. 1–40 (2004)
101. Ell, R.: Natural fibres: types and properties. Prop. Perform. Nat. Fibre Compos. 1 (2008)
102. Marsh, G.: Marine composites—drawbacks and successes. Reinf. Plast. **54**(4), 18–22 (2010)
103. Kamel, S.: Nanotechnology and its applications in lignocellulosic composites, a mini review. Express Polym. Lett. **1**(9), 546–575 (2007)
104. Alexandre, M., Dubois, P.: Polymer-layered silicate nanocomposites: preparation, properties and uses of a new class of materials. Mater. Sci. Eng. R Rep. **28**(1), 1–63 (2000)
105. Kumar, A.P., Depan, D., Tomer, N.S., Singh, R.P.: Nanoscale particles for polymer degradation and stabilization—trends and future perspectives. Prog. Poly. Sci. **34**(6), 479–515 (2009)
106. Njuguna, J., Pielichowski, K., Desai, S.: Nanofiller-reinforced polymer nanocomposites. Polym. Adv. Technol. **19**(8), 947–959 (2008)
107. Njuguna, J., Pielichowski, K., Alcock, J.R.: Epoxy-based fibre reinforced nanocomposites. Adv. Eng. Mater. **9**(10), 835–847 (2007)
108. Marquis, D.M., Chivas-Joly, C., Guillaume, E.: Properties of Nanofillers in Polymer. INTECH Open Access Publisher (2011)
109. Nanotechnologies—Terminology and Definitions for Nano-Objects—Nanoparticle, Nanofibre and Nanoplate. ISO/TS 27687; International Organization for Standardization (ISO), Geneva, Switzerland (2008)
110. Biswas, M., Ray, S.S.: Recent progress in synthesis and evaluation of polymer-montmorillonite nanocomposites. In: New Polymerization Techniques and Synthetic Methodologies. Springer, Berlin, pp. 167–221 (2001)
111. Denault, J., Labrecque, B.: Technology group on polymer nanocomposites–PNC-Tech. Ind. Mater. Inst. Nat. Res. Counc. Can. **75** (2004)

112. Wypych, F., Satyanarayana, K.G.: Functionalization of single layers and nanofibers: a new strategy to produce polymer nanocomposites with optimized properties. J. Colloid Interface Sci. **285**(2), 532–543 (2005)
113. Ray, S.S., Okamoto, M.: Polymer/layered silicate nanocomposites: a review from preparation to processing. Prog. Polym. Sci. **28**(11), 1539–1641 (2003)
114. De Azeredo, H.M.C.: Nanocomposites for food packaging applications. Food Res. Int. **42**(9), 1240–1253 (2009)
115. Schadler, L.S., Brinson, L.C., Sawyer, W.G.: Polymer nanocomposites: a small part of the story. JOM J. Miner. Metals Mater. Soc. **59**(3), 53–60 (2007)
116. Kim, K.-J.: Modification of nano-kenaf surface with maleic anhydride grafted polypropylene upon improved mechanical properties of polypropylene composite. Compos. Interfaces **22**(6), 433–445 (2015)
117. Gacitua, W., Ballerini, A., Zhang, J.: Polymer nanocomposites: synthetic and natural fillers a review. Maderas. Ciencia y tecnología **7**(3), 159–178 (2005)
118. Oksman, K., Mathew, A.P., Bondeson, D., Kvien, I.: Manufacturing process of cellulose whiskers/polylactic acid nanocomposites. Compos. Sci. Technol. **66**(15), 2776–2784 (2006)
119. Darder, M., Aranda, P., Ruiz-Hitzky, E.: Bionanocomposites: a new concept of ecological, bioinspired, and functional hybrid materials. Adv. Mater. **19**(10), 1309–1319 (2007)
120. Bowles, K.J., Frimpong, S.: Void effects on the interlaminar shear strength of unidirectional graphite-fiber-reinforced composites. J. Compos. Mater. **26**(10), 1487–1509 (1992)
121. Vaxman, A., Narkis, M., Siegmann, A., Kenig, S.: Void formation in short-fiber thermoplastic composites. Polym. Compos. **10**(6), 449–453 (1989)
122. Ho, M., Wang, H., Lee, J.-H., Ho, C., Lau, K., Leng, J., Hui, D.: Critical factors on manufacturing processes of natural fibre composites. Compos. B Eng. **43**(8), 3549–3562 (2012)
123. Bodros, E., Pillin, I., Montrelay, N., Baley, C.: Could biopolymers reinforced by randomly scattered flax fibre be used in structural applications? Compos. Sci. Technol. **67**(3), 462–470 (2007)
124. Herrmann, A.S., Nickel, J., Riedel, U.: Construction materials based upon biologically renewable resources—from components to finished parts. Polym. Degrad. Stab. **59**(1–3), 251–261 (1998)
125. Jiang, L., Hinrichsen, G.: Flax and cotton fiber reinforced biodegradable polyester amide composites, 2. Characterization of biodegradation. Die Angewandte Makromolekulare Chemie **268**(1), 18–21 (1999)
126. Pickering, K.L., Efendy, M.G.A., Le, T.M.: A review of recent developments in natural fibre composites and their mechanical performance. Compos. Part A Appl. Sci. Manuf. **83**, 98–112 (2016)
127. Veritas, D.N.: STANDARD FOR CERTIFICATION No. 2.21. (2010)
128. Stewart, R.: Better boat building-trend to closed–mould processing continues. Reinf. Plast. **55**(6), 30–36 (2011)
129. Johnstone, J.: Flexural testing of sustainable and alternative materials for surfboard construction, in comparison to current industry standard materials. Plymouth Student Sci. **4**(1), 109–142 (2011)
130. Asim, M., Jawaid, M., Abdan, K., Ishak, M.R.: Effect of alkali and silane treatments on mechanical and fibre-matrix bond strength of kenaf and pineapple leaf fibres. J. Bionic Eng. **13**(3), 426–435 (2016)
131. Andre, N.G., Ariawan, D., Mohd Ishak, Z.A.: Elastic anisotropy of kenaf fibre and micromechanical modeling of nonwoven kenaf fibre/epoxy composites. J. Reinf. Plast. Compos. **35**(19), 1424–1433 (2016)
132. Suharty, N.S., Ismail, H., Diharjo, K., Handayani, D.S., Firdaus, M.: Effect of kenaf fiber as a reinforcement on the tensile, flexural strength and impact toughness properties of recycled polypropylene/halloysite composites. Procedia Chem. **19**, 253–258 (2016)

133. Saba, N., Paridah, M.T., Abdan, K., Ibrahim, N.A.: Effect of oil palm nano filler on mechanical and morphological properties of kenaf reinforced epoxy composites. Constr. Build. Mater. **123**, 15–26 (2016)
134. Jeyanthi, S., Janci Rani, J.: Improving mechanical properties by kenaf natural long fiber reinforced composite for automotive structures. 淡江理工學刊 **15**(3), 275–280 (2012)
135. Mokhothu, T.H., Guduri, B.R., Luyt, A.S.: Kenaf fiber-reinforced copolyester biocomposites. Polym. Compos. **32**(12), 2001–2009 (2011)
136. Thiruchitrambalam, M., Alavudeen, A., Athijayamani, A., Venkateshwaran, N., Elaya Perumal, A.: Improving mechanical properties of banana/kenaf polyester hybrid composites using sodium lauryl sulfate treatment. Mater. Phys. Mech. **8**, 165–173 (2009)
137. Thwe, M.M., Liao, K.: Effects of environmental aging on the mechanical properties of bamboo–glass fiber reinforced polymer matrix hybrid composites. Compos. Part A Appl. Sci. Manuf. **33**(1), 43–52 (2002)

Development of a Batik Fiberglass Composite for Marine Applications Based on Water Absorption Testing

Muhammad Abdul Mun'aim Bin Mohd Idrus,
Siti Amelia Binti Amiruddin, Siti Afiza Binti Johari Jamaludin
and Shamsul Effendy Bin Abdul Hamid

Abstract This research presents the investigation of composite materials lamination using a combination of synthetic fiber and natural fiber materials in the structure of lamination. The study proposed alternative method to replace synthetic fiber chopped mat strands (CSM) by using batik fabric, which is cotton, for GRP product making and to comply with the standards. The project followed the Dot Net Veritas (DNV) Regulations, American Society of Testing Materials (ASTM) standards and International Organization of Standardization (ISO) 62. The composite lamination has been produced using the hand lay-up technique. The effect of batik fabric was tested using tensile strength, flexural strength and water absorption test to determine the most optimum strength of composite materials. An implication of this study showed and proven with the difference of the lamination structure system of composite materials, it presented a difference in the behavior of composite materials. The result showed that different lamination of batik fabric will increase the tensile strength but not achieve the minimum standard. The bending stress showed an increase in strength and achieved the standard required for marine applications. In addition, the water absorption increased with batik fabric loading but at acceptable level for marine standards.

Keywords GRP product · Synthetic fiber · Natural fiber · Batik fabric
Water absorption

M. A. M. B. M. Idrus (✉) · S. A. B. Amiruddin · S. A. B. J. Jamaludin · S. E. B. A. Hamid
Universiti Kuala Lumpur, Malaysian Institute of Marine Engineering Technology, Jalan Pantai Remis, 32200 Lumut, Perak, Malaysia
e-mail: mamunaim@unikl.edu.my

S. A. B. Amiruddin
e-mail: saa.ameliaamiruddin@gmail.com

S. A. B. J. Jamaludin
e-mail: fezapple91@gmail.com

S. E. B. A. Hamid
e-mail: seffendy@unikl.edu.my

© Springer International Publishing AG 2018
A. Öchsner (ed.), *Engineering Applications for New Materials and Technologies*,
Advanced Structured Materials 85, https://doi.org/10.1007/978-3-319-72697-7_4

1 Introduction

The global fiberglass market is dominated by the developed markets [1]. In naval applications, the navy requires GRP composites with high strength-to-weight ratio capability and inherent resistance to weather and the corrosive effects of salt air and sea [2]. Composite materials have been created to improve mechanical characteristics, such as stiffness and toughness by the combination of materials, called reinforcement [3]. In marine industry, cost is the main important issues that need to be considered in making a GRP product. Fiberglass, a material produced by using a plastic matrix is used in several applications because of the high quality, lightness and low costs [4]. Using batik fabric, it will reduce the costs of pigmentation because batik has its pattern and color. Natural fibers are one of such proficient materials, which can replace synthetic materials and its related products for some low-weight and energy conversion applications [5]. Natural fibers can be recycled and the fibers come from renewable resources. Their moderate mechanical properties restrain the fibers from using them in high-tech applications, but for many reasons natural fibers can compete with glass fibers [6]. This paper investigated the use of Batik as an alternative material to replace chop strength mat (CSM) in the production of fiberglass.

2 Materials and Methodology

The hand lay-up process was applied in conducting these experiments. The standard and procedure were followed according to the DNV Standards and ensured that the used technique was precise and accurate. Each layer of the fiber was applied with resin and ensured that the fibers were completely wet-out [7]. A roller was used to remove air bubbles during the lamination process.

Note that these materials can only be combined when all dry materials were fully prepared including wax surface preparation. The weights of the gel coat and resin are determined by the mold surface, which is using the surface of mold of $2\text{ m} \times 1\text{ m} = 2\text{ m}^2$, let gelcoat = surface size × 600 g m^{-2}, therefore 2 m^2 × 600 g m^{-2} = 1.2 kg. Next, the next elements were 2% of resin mixed with the catalyst, 14 g (each) of cotton fiber (batik fabric) and 600 g m (each) of woven roving. The thickness of the plate panel was between 0.25 and 0.5 mm. The product specimen for the lamination testing is shown in Table 1.

2.1 Laminating Process

Firstly, the surfaced was as incorporated in the wax procedure. Next, the gel coat was mixed with the catalyst and stirred by using a wooden stick (the color of gel

Table 1 Product specimen testing

Specimen	Number of mat		
Specimen A	Batik fabric	Layer 1	1 pcs
	Woven roving	Layer 2	1 pcs
	Batik fabric	Layer 3	1 pcs
	Woven roving	Layer 4	1 pcs
	Batik fabric	Layer 5	1 pcs
Specimen B	Batik fabric	Layer 1	3 pcs
	Woven roving	Layer 2	1 pcs
	Batik fabric	Layer 3	4 pcs
	Batik fabric	Layer 4	4 pcs
	Batik fabric	Layer 5	3 pcs
Specimen C	Woven roving	Layer 1	1 pcs
	Batik fabric	Layer 2	4 pcs
	Woven roving	Layer 3	1 pcs
	Batik fabric	Layer 4	4 pcs

coat changed a bit from the initial color) and applied the gel coat by using a brush or roller brush. The gel coat was applied evenly and quickly onto the mold surface to avoid the hardening of the gel coat. Then, the gel coat was allowed to rest for 20 min to harden depending on the weather. Meanwhile, the resin was mixed with the catalyst and stirred by using a wooden stick and then applied on the top of the gel coat by using a brush. After that, the Batik fabric was applied piece by piece and layered by the resin. The number of cloths depends on the type of specimen as outlined in Table 1 and rolled using a roller to let air bubbles out. Then, resin was applied again and woven roving was placed on top of it and the air bubbles out rolled out (the number of woven roving refers to the table shown). Steps were continued and step number 4 and 5 were repeated until the specimen requirements were fulfilled (see Table 1 of chapter "Potential Use of Cellulose Fibre Composites in Marine Environment—A Review"). Lastly, the specimens were dried overnight.

2.2 Demolding Flat Panel

After the flat panel was completely dried, it was gently removed from the mold by lifting the masking tape. The edges of the specimen were softened by using a jigsaw and sand paper.

2.2.1 Mechanical Testing

i. Tensile Strength Testing
   The tensile testing was performed using a universal testing machine equipped with a 5 kN load cell and in accordance with ASTM D-3039 standard, see

Fig. 1 Immersion of specimens according to ISO 62

Fig. 1. A constant cross-head speed of 2 mm/min of the testing equipment were used based on ASTM standards for test methods of tensile properties of polymer matrix composite materials [8]. In these tests, the purpose was to identify the ultimate tensile strength of the specimen to be used in the next experiment.

ASTM standards require at least 5 (five) specimens for each sample configuration [9]. The sample of specimen A with dimensions of 250 × 25 × 3 mm was tested as shown in Fig. 1. The geometry of the specimen followed the ASTM D-3039 standard for a rectangular specimen size [10]. Specimen type B had dimensions of 250 × 25 × 6 mm, and specimen type C had dimensions of 250 × 25 × 5 mm and were tested under the same conditions.

The specimens were placed in the grips of the universal testing machine. Then, the force was applied to the test specimen. The displacement-controlled speed was set at 2 mm/min. The higher the applied force onto the specimen, the longer the elongation of the specimen until its breaking point.

ii. Flexural Strength Testing

The test on flexural strength was carried out according to ASTM D-790 for unreinforced and reinforced plastics and electrical insulating materials using a universal testing machine. The aim of this test was to study and investigate the strength of the composite material after the load was applied onto the specimen. The same cross-head speed of 2 mm/min and the load of 5 kN equipment were used based on the ASTM standards as shown in Fig. 2. The ASTM standard requires at least 5 specimens per sample configuration [11]. First the specimen type A with dimensions of 125 × 12.7 × 3 mm was tested.

Fig. 2 Specimen A

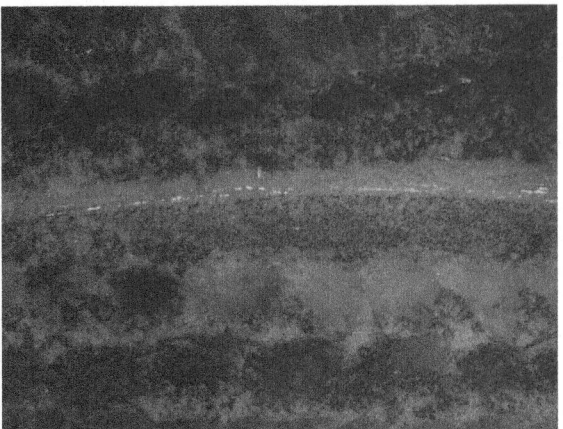

Then, specimen type B with dimensions of 125 × 12.7 × 6 mm and specimen type C with dimensions of 125 × 12.7 × 5 mm were tested [12].

The specimens were placed in the universal testing machine at the span. Then, the load was applied onto the specimens. The higher the applied load was, the more bending deformation of the specimen would occur. The machine control applied the load until the specimen fractured at the breaking point.

iii. Water Absorption Testing

The test of water absorption was performed according to International Organization for Standardization (ISO) 62:1999 English version of DIN EN ISO 62 for plastics. The objective of this experiment was to observe the percentage of the absorbed water after immersion in the liquid. The recommended specimen size is 10 cm by 10 cm (specimen A, B and C) [13]. The immersion time according to ISO 62 was 24 h and the time interval 48 h, 72 h and a total of 3 (three) days. The specimens were placed in the oven for 50 °C for 1 (one) hour to make sure that the samples were completely dried before the immersion process [14]. Each specimen was soaked and fully immersed in a container of 300 ml of distilled water as shown in Fig. 3 [15]. Then, the specimens were weighed and the data noted in the reporting sheet.

iv. Optical Microscopy

The morphology of the batik fiberglass composites is represented in Figs. 2, 3 and 4. It was observed that the surface morphology of all composite specimen types (A, B and C) showed smoothness and less roughness. The different color showed that the layer of batik fabric in the composite was well dispersed in the composites. There were no clear voids seen, meaning that the composite absorbed less water after immersion. This will lead to better mechanical properties of the composite.

Fig. 3 Specimen B

Fig. 4 Specimen C

3 Results and Discussion

i. Tensile Strength Testing

The tensile strength properties of the batik fabric composite at different layer loading are shown in Fig. 5. According to the DNV standards (materials and manufacturing, Sect. 4 FRP structural design c, single skin construction, A 700 mechanical properties laminate, 701), the requirement for a structural laminate for tensile strength is 80 MPa. The result was gained after conducting the tensile testing as shown in Fig. 1. The evaluation revealed that the maximum stress for specimen A was 72.179 MPa, for specimen B it was 31.145 MPa, and for specimen C it was 67.276 MPa. Specimens A and C were the strongest compared to specimen B. The experiment was carried out and the tensile strength according to the DNV standard is 80 MPa.

The evaluated result for specimen A was 72.179 MPa. Thus, specimen type A did not comply with the DNV standards. In addition, during the testing, the specimen broke at the fixation grips of the testing machine due to failure preparation of the specimens.

ii. Flexural Strength Testing

The flexural strength of batik fabric fiberglass composites at different layer loading is shown in Fig. 6. Referring to the DNV standards for flexural strength testing (materials and manufacturing, Sect. 4 FRP structural design c. single skin construction), the required bending strength is 130 MPa. The evaluated result revealed that the maximum stress for specimen A was 166.182 MPa, for specimen B it was 191.567 MPa, and for specimen C it was 265.748 MPa. The result showed that the flexural strength increased with an increase in batik fabric loading. Specimens A and C were the strongest compared to specimen B. The flexural strength gradually increased which was due to the increase in resistance to shearing in the composite structure due to the presence of the fibers. For flexural testing according to the DNV standards, a minimum of 130 MPa is required. The result found that the bending strength was 166.182 MPa for specimen A. As per DNV standards, this result is acceptable.

iii. Water Absorption Testing

According to the DNV standards for water absorption (in part 2, of this chapter composite material 1999 requirement for approval testing Grade 2: resin with normal water resistance, Manufacturer's Specified Minimum Value (MSMV)), the maximum value is 100 mg. After conducted the testing according to ISO 62, each of the specimens absorbed water. The collected data of absorbed water in each of the specimens is shown in Fig. 7.

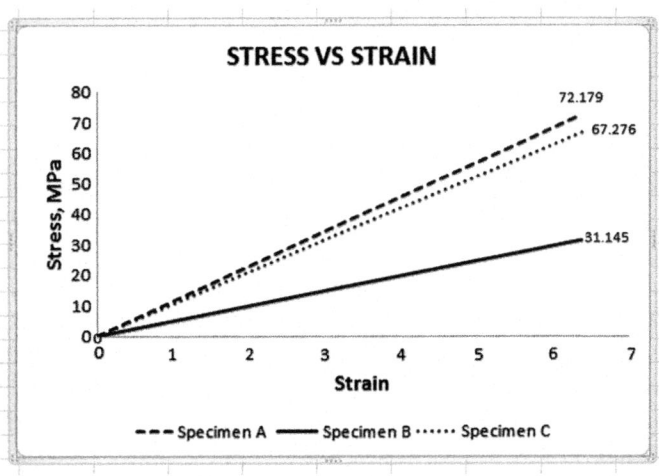

Fig. 5 Tensile strength stress versus strain

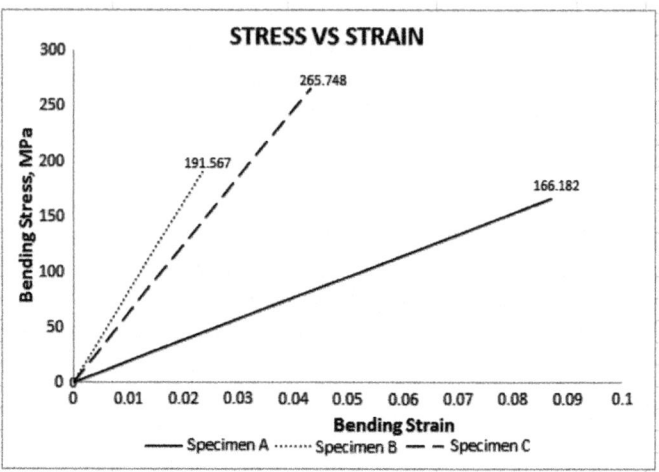

Fig. 6 Flexural strength stress versus strain

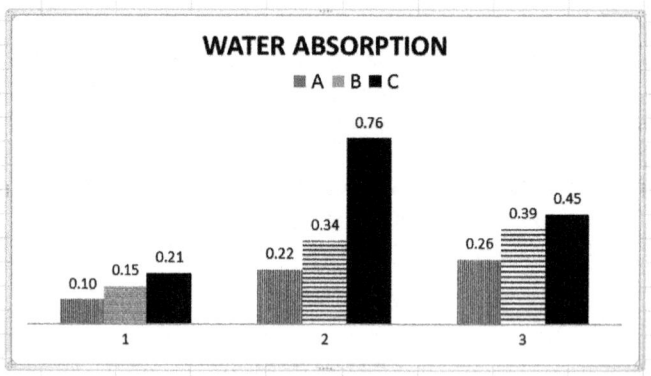

Fig. 7 Water absorption in each specimen

Figure 7 shows the average absorbent of each specimen. Label 1 represents the 24 h immersion time, label 2 stands for 48 h of immersion and label 3 for the 72 h of immersion time. The blue bar represents the specimen of type A, then followed by the read bar which is for the specimen type B and the green bar for the specimen type C. The water absorption varied depending on the fabric loading. As shown in the figure, the lowest water absorption was obtained in the case of specimen A and the highest water absorption in the case of specimen C. This was due to the higher fabric content in the composite, which is responsible that more water can be absorbed. Therefore, specimen A was chosen as the most optimum specimen.

Water absorption testing was conducted, at which each of the specimen was immersed in a container and absorbed a certain amount of water. Refer to the DNV standards, the maximum amount of absorbed water is 100 mg (Grade 2: resin with normal water resistance). This minimum amount of absorbed water, compared to all the specimens, was obtained for the specimen type A. Therefore, the result found to comply with the DNV standards.

4 Conclusion

In this work, a batik fabric/woven roving fiberglass composite was produced to study the tensile strength, flexural strength and water absorption properties. An experimental testing procedure was conducted based on the marine standard. The experiments were carried to reveal the difference in stress level and absorbed water content for different specimen configurations. The comparison between the three types of composite modifications shows a good correlation between each other. It can be concluded that the most optimum strength was obtained for the specimen type A. The result found that this sample type passed the minimum flexural strength testing and water absorption testing.

Acknowledgements The author would like to thank the Universiti Kuala Lumpur Malaysian Institute of Marine Engineering Technology (MIMET) for continuous support that has made this work possible.

References

1. Roving, Yarns, Fiberglass Market (Composites, insulations) and by Regions—Trends and Forecast to 2020, Browse 86 tables and 53 spread through 160 pages and in-depth TOC
2. American Composite Manufacturers Association (Market Overview)
3. Arifin, A.M.T., Abdullah, S., Zulkifli, R., Wahab, D.A.: Investigating the fatigue phenomenon of CSM-woven roving composite materials lamination 2013 International Conference on Technology, Informatics, Management, Engineering & Environment (TIME-E 2013) Bandung Indonesia, 23–26 June 2013
4. ACMA staff composites manufacturing. Fiberglass Still Material of Choice for Most Boatbuilders. 20 July 2020
5. Sanjay, M.R., Arpitha, G.R., Naik, L.L., Gopalakrishana, K., Yogesha, B.: Applications of natural fibers and its composites: an overview. Department of Mechanical Engineering, Malnad College of Engineering, Visvesraya Technological University Belagavi, India, 11 March 2016
6. Abilash, N., Sivapragash, M.: Environmental benefits of eco-friendly natural fiber reinforced polymeric composite materials. Int. J. Application Innovation Eng. Manag. (IJAIEM), 2(1), January 2013
7. DNV Standards, Chapter 3 Materials and manufacturing' Section 3, Fiber reinforcement plastics B-600 Manual Lamination'

8. ASTM D3039 Tensile properties of polymer matrix composite materials section 11, procedure
9. ASTM D3039 Tensile properties of polymer matrix composite materials Section 8, Sampling and Test Specimens
10. ASTM D3039 Tensile properties of polymer matrix composite materials Section 8.2.2.1, Table 1 Tensile Specimen Geometry Requirements
11. ASTM D790-63 Flexural properties of unreinforced and reinforced plastics and electrical insulating materials Section, 8, Number of Test Specimens
12. ASTM D790-63 Flexural properties of unreinforced and reinforced plastics and electrical insulating materials Section, 7.2.1, Materials
13. ISO 62:1999 "Plastics", Determination of Water Absorption, Section 5.3 Test specimens of Reinforced Plastics Affected By Anisotropic Diffusion Effects, part b
14. ISO 62:1999 "Plastics", Determination of Water Absorption, Section 6.2, Method 1: Determination of Water Content Absorbed After Immersion in Water At 23 °C
15. Absorption, Section 6.1, part 6.1.1, General Conditions

Flame Spread Behavior over Kenaf Fabric, Polyester Fabric, and Kenaf/Polyester Combined Fabric

Mohd Azahari Razali, Azwan Sapit, Akmal Nizam Mohammed,
Mohd Faisal Hushim, Azmahani Sadikin,
Md Norrizam Mohmad Ja'at, Hazahir Peraman and Mirnah Suardi

Abstract Flame spread behavior is one of the important topics related to fire safety engineering. It is essential to examine factors, which influence the flame spread behavior over fabrics. It is known that natural fibers exhibit a different flame spread behavior than the one of synthetic fibers. This difference may influence the flame spread behavior over combined fabrics. The purpose of this research is to study the effect of materials on the flame spread behavior over kenaf/polyester fabrics. Before analyzing this effect, it is important also to know the flame spread behavior over 100% kenaf fabric and 100% polyester fabric. Thus, several experiments have been conducted for different materials of fabric made up of 100% kenaf, 100% polyester, and combined fabric of kenaf/polyester. For the combined fabric, experiments have been done for different weft thread angle of $\theta = 0°$ and $\theta = 90°$. A burner is used for igniting the fabric at a point on its top edge. The data collected is recorded via videos and captured images for measuring the flame spread rate and detail observation of characteristics during the burning process. From the results obtained, it is seen that the material and thread angle influence on the flame spread behavior over fabrics. The flame spread rate on kenaf is lower than the flame spread rate on combined fabrics of kenaf/polyester while the flame spread rate on polyester is undetermined. The flame spread velocity also changes when the weft thread angle change from $\theta = 0°$ to $\theta = 90°$.

Keywords Kenaf · Flame spread behavior

M. A. Razali (✉) · A. Sapit · A. N. Mohammed · M. F. Hushim · A. Sadikin
M. N. M. Ja'at · H. Peraman · M. Suardi
Faculty of Mechanical and Manufacturing Engineering, Centre for Energy
and Industrial Environment Studies (CEIES), Universiti Tun Hussein Onn
Malaysia, 86400 Pt. Raja, Batu Pahat, Johor, Malaysia
e-mail: azahari@uthm.edu.my

1 Introduction

Fire safety engineering is an application of science to improve the safety from the destructive effects of fire. There are many researches on the fire safety engineering done in the previous years and it is beginning to focus the research on the technical capability for fire preventions in complex circumstances. In order to improve the technical capability for fire preventions, the fundamental approach is essential which can be applied at the design stage. Such an approach requires a detailed understanding of the combustion process from an engineering viewpoint.

Flame spread over a combustible solid is a basic problem in the field of fire safety engineering. Currently, there are a lot of fire losses caused by fabric. It is essential to study extensively the flame spread behavior over fabric, such as curtains, bed sheets, clothes, table clothes or carpets etc., in order to reduce the destruction caused by fire. It is noted that almost all the fabrics as stated above are made from combined fabrics. The combined fabric is fabricated by using two different types of thread; natural/natural threads, natural/synthetic threads or synthetic/synthetic threads. An example of natural thread is cotton and the one of synthetic thread is polyester. The natural fiber has a different flame spread characteristic than the one of synthetic thread. This different characteristic may influence the flame spread characteristic of a combined fabric.

The flame spread rate is the speed of a flame that creeps forward towards the fuel. A fire hazard can be indicated by the flame spread rate or/and the flame shape. The non-uniformity of the fuel sample and the heat loss from the side is caused by the variability of the spread rate. Several researches have been done in relation to the flame spread rate. Bhattacharjee et al. [1] have designed an apparatus to study the flame spread rate. The spread rate can be measured by tracking the position of the visible edge of the flame through video recording. There are several factors, which influence the flame spread rate. Fernandez-Pello et al. [2] found that the flame spread rate can be affected by ambient and fuel conditions. Ding et al. [3] simulated the fire spread. Results show that spread characteristics have major relation with factors such as the material of the solid, width, thickness, placed angle and atmospheric pressure. It was also discovered that the increment in oxygen concentration will cause the external radiant flux required for flame spread to decrease. Flame spread characteristics of many thin materials will be varying depending on the material thickness, external heat flux, oxygen concentration, pressure and forced flow velocity [4]. Higher heat flux results in higher heat release rate and peak mass loss rate and the fabrics will burn more violently. The density of the structure and moisture content also has effects on the burning behavior of fabrics. The arrangement order of types of fabrics in increasing fire risk are as follows: wollen fabrics > sponge fabrics > cotton fabrics > linen fabrics [5]. It is stated that cellulosic fabrics burn with a yellow flame, light smoke, and have glowing embers. Synthetic fabrics catch fire quickly or shrink from the flame initially. Nylon, lastol, olefin, polyester, and spandex burn slowly and melting. It also can melt and pull away from small flames without igniting. The residue is a hot molten mass and difficult to dispose. The burning of these fabrics can be self-extinguishing [6]. For the testing of the flame spread over fabrics, it is more effective to use a downward flame as the higher flame height of upward flames would interrupt the observation while burning the fabrics [7]. Several experiments have been conducted to explored the flame spread behavior over cotton/polyester fabric [8]. It is found that a significant difference is seen in the flame spread behavior between $\theta = 0°$ and $\theta = 90°$.

Recently, there are a lot of industrial applications which use combined fabrics of kenaf as the natural thread and polyester as the synthetic thread. It is also known that kenaf fibers have lower thermal resistance than synthetic fibers [9]. Higher cellulose content in natural fibers also can cause the flammability to be higher while higher lignin content increases the formation of char with lower degradation temperature [10]. However, the data of flame spread behavior over combined fabrics of kenaf/polyester is still insufficient to describe the behavior in detail. Thus, in this study, the flame spread behavior over this combined fabric is explored.

2 Experimental Setup

Fabric samples are made by means of a weaving machine. Plain weave is chosen as the structure of the fabric. The plain weave has warp threads perpendicular to weft threads. The fabric composed of different materials is referred to as a combined fabric. For the combined fabric, kenaf is used as the warp thread and polyester as the weft thread. Table 1 shows the characteristic values of these threads.

Table 2 shows the characteristic values of these fabrics. The thread distance of all samples has been controlled to be approximately 3.0 mm. For the combined fabric, samples are cut with different weft thread angle of 0° and 90°. The weft thread angle is referred to the angle between the polyester thread and the horizontal line as shown in Fig. 1.

Figure 2 shows the experimental setup. Before the burning test, the fabric sample is dried in a desiccator for more than 48 h and the humidity in the desiccator is controlled to be below 40%. The fabric sample is clamped to the holder and its area of 100 mm × 200 mm is exposed to burn. A burner is used for igniting the fabric at a point on its top edge. The surface of the sample is videotaped to observe the flame spread behavior.

Table 1 Characteristic values of threads

	Diameter (mm)
Kenaf thread	0.7
Polyester thread	0.5

Table 2 Characteristic values of fabrics

	Number of threads per unit length (cm^{-1})	
	Kenaf thread	Polyester thread
Kenaf	4	–
Polyester	–	4
Kenaf/polyester	4	4

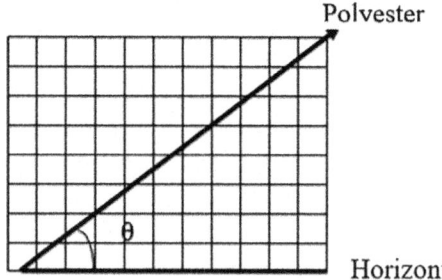

Fig. 1 Illustration of sample

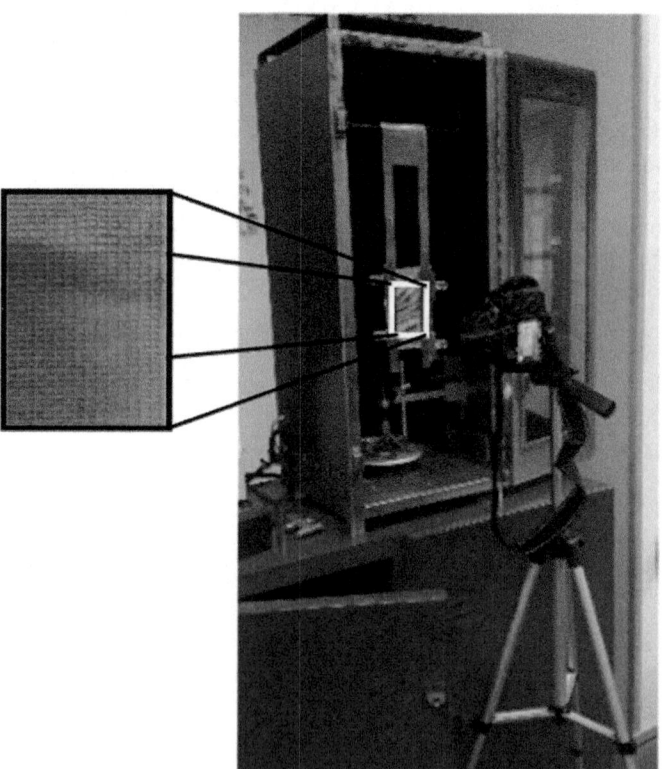

Fig. 2 Experimental setup

3 Results and Discussion

a. Flame spread behavior of different materials

Figure 3 shows the shape of the flame front for different types of fabrics. The flame front is determined at the front edge of discoloration when it decomposes. As seen in Fig. 3a, the kenaf fabric burns and the flame front is in flat shape. On the other hand, the flame is distinguished when the polyester fabric is burned as shown in Fig. 3b. For the combined fabric of kenaf/polyester, a 'V' shape is observed during the flame spreading as shown in Fig. 3c.

Figure 4 shows some details of the observation of the flame front shape. The flame front is determined at the front edge of discoloration when it decomposes. The

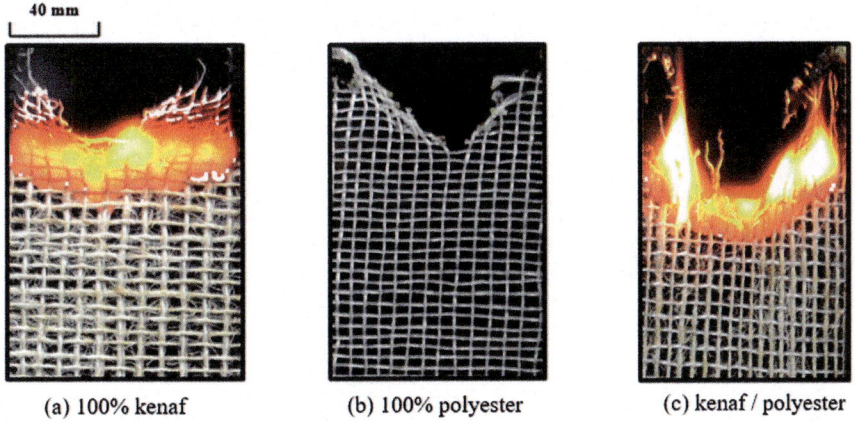

(a) 100% kenaf   (b) 100% polyester   (c) kenaf / polyester

Fig. 3 Shape of flame front of different materials

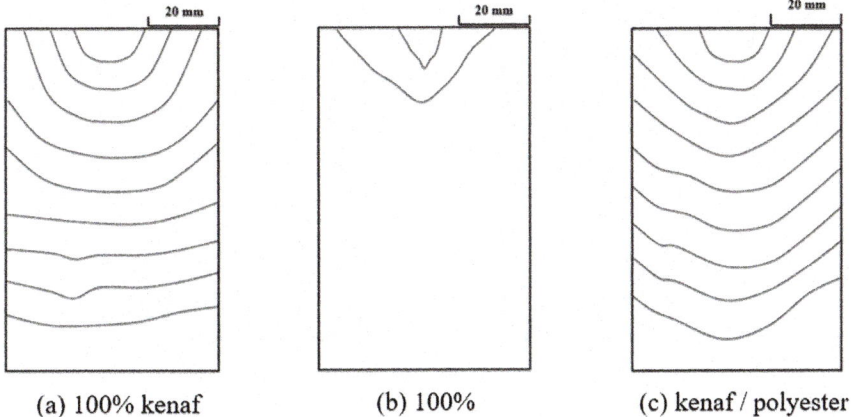

(a) 100% kenaf   (b) 100%   (c) kenaf / polyester

Fig. 4 Shape of flame front for every 5 s

observation is done by capturing images for every 5 s. Figure 4a shows the shape of the flame front for the 100% kenaf fabric. After the ignition, the flame spreads outwards in all directions from the ignition point. After the burning front expands over the whole width of the sample, the shape turns to a flate shape. Figure 4b shows some details of the observation of the shape of the flame front of the 100% polyester fabric. It is seen that the flame was distinguished just after the ignition point. The flame did not spread due to vaporization of the polyester thread during the combustion process. It can be simplified that the 100% polyester fabric does not burn when it is ignited. Figure 4c shows some details of the observation of the shape of the flame front for the combined fabric of kenaf/polyester. It is seen that the flame spreads to the outward after the ignition. The 'V' shape is seen after the flame spreads over the whole area. The shape remains during the combustion process.

By comparing Fig. 4a and c, it can be concluded that the shape of the burning front of 100% kenaf is flatter than the shape of the burning front of the combined fabric, which is in the 'V' shape. This behavior may be resulting from the presence of the polyester thread in the combined fabric. It is noted that the polyester thread melts and vaporizes when exposed to the heat. This phenomenon affects the thread movement at the leading point of the flame spread, which can cause the shape of the flame front.

b. Flame spread behavior of combined fabric with different angles

Figure 5 shows the shape of the flame front of the combined fabric of kenaf/polyester at $\theta = 0°$ and $\theta = 90°$. Results shows that a different flame shape is seen when the weft thread angle changes between $\theta = 0°$ and $\theta = 90°$. When the polyester thread is in the horizontal direction, the flame shape is in 'V' shape.

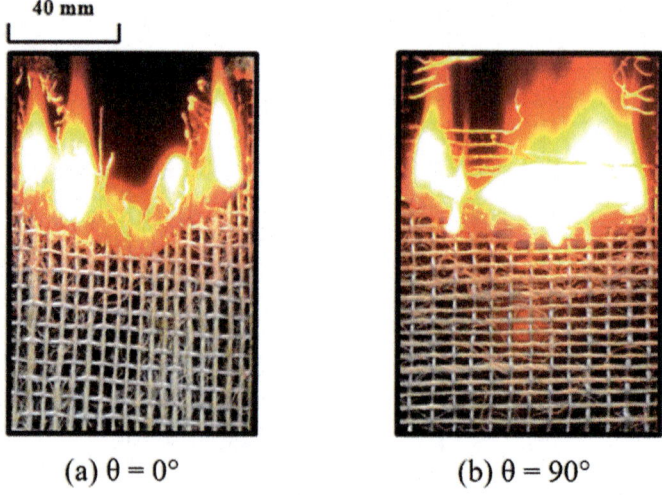

Fig. 5 Shape of flame front of combined fabric for different angles

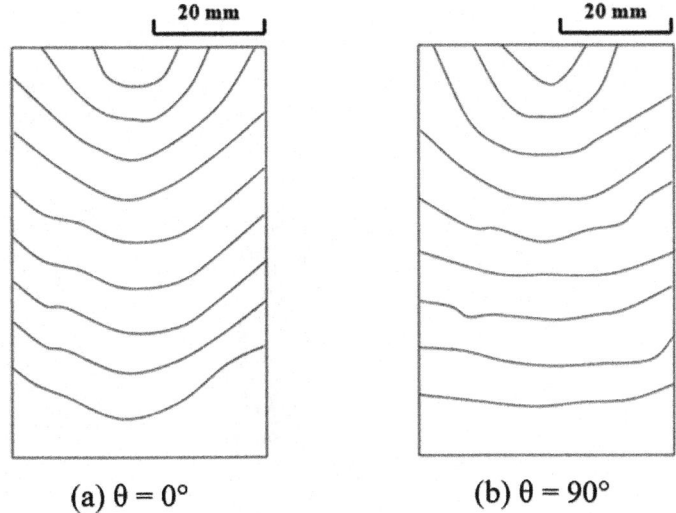

Fig. 6 Flame front shape of combined fabric for every 5 s

However, the shape becomes flat when the polyester thread is perpendicular to the horizontal direction.

Figure 6 shows the shape of the flame front of the combined fabric for every 5 s of kenaf/polyester at $\theta = 0°$ and $\theta = 90°$. It is seen that the flame spread is different between $\theta = 0°$ and $\theta = 90°$. After the flame spreads over the whole area, the 'V' shape remains during the flame spreading for $\theta = 0°$. On the other hand, the flame turns to a flat shape when $\theta = 90°$. The results infer that the weft thread angle may have some influence on the flame spread behavior. However, further investigation is needed to examine this factor.

c. Flame spread rate

Figure 7 shows the flame spread rate between different types of fabrics. Flame spread rates are obtained from Figs. 4, 5 and 6 by measuring the position of the most preceding point of the burning front at each time. In order to avoid the influence of the initial transition process after the ignition, the position is measured from $x = 20$ mm (to 60 mm), where x is the vertical distance from the top edge of the fabric in downward direction.

Results show that the flame spread rate for the kenaf fabric is 0.931 mm/s. On the other hand, it is seen that the polyester fabric is unburned. The value for pure kenaf fabric is found to be less than any values for the combined kenaf/polyester fabric. The values of the flame spread rate for the combined fabric at $\theta = 0°$ and $\theta = 90°$ are 1.4702 and 1.4122 mm/s, respectively. Results show that the materials of the fabric have some influence on both the flame spread shape and rate.

Results also show that the flame spread rate decreases when the weft thread angle changes between $\theta = 0°$ and $\theta = 90°$. This discrepancy is caused by the

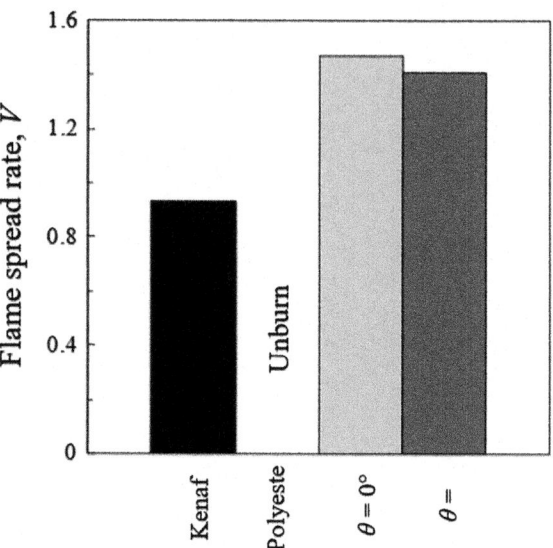

Fig. 7 Flame spread rate between different types of fabric

difference in flame front shape between the thread angle, as seen in Fig. 6. This phenomenon may be caused by the flame spread behavior at the leading point of the flame spreading. From these results, it also concluded that weft thread angle on not only affects the flame spread shape, but also the flame spread behavior. However, further investigation is needed to examine the behavior at the leading point of the flame spreading and the influence of the weft thread angle on the flame spread behavior.

4 Conclusion

In this study, flame spread over kenaf fabric, polyester fabric, and kenaf/polyester fabrics is examined and the following results are obtained:

a. Shapes of the flame front differ between fabrics. Kenaf fabric burns and the flame spreads in a flat shape. However, the flame is distinguished as soon as polyester fabric is ignited. For the combined fabric, the fabric burns and the shape turns to be in 'V' shape during the burning process.
b. The flame spread rate for kenaf is lower than the values of the combined fabric at $\theta = 0°$ and $\theta = 90°$. The value of the spread rate for polyester cannot be determined since the flame is distinguished after the ignition.
c. The flame spread shape and the flame spread rate are differed for the combined fabric at $\theta = 0°$ and $\theta = 90°$. For $\theta = 0°$, the shape turns to be in 'V' shape but the flame is in flat shape when $\theta = 90°$. The flame spread rate at $\theta = 0°$ is higher than the one at $\theta = 90°$. It is indicated the possibility that the weft thread angle may have some influence on the flame spread shape and the flame spread rate.

Acknowledgements The authors would like to thank Universiti Tun Hussein Onn Malaysia (UTHM) and Ministry of Higher Education Malaysia (MOHE) for their financial support of the present work through Fundamental Research Grant Scheme (FRGS-1465).

References

1. Bhattacharjee, S., Bundy, M., Paolini, C., Patel, G., Tran, W.: A novel apparatus for flame spread study. In: Proceedings of the Combustion Institute, pp. 2513–2521. San Diego State University, San Diego, USA (2012)
2. Fernandez-Pello, A.C., Ray, S.R., Glassman, I.: Proceedings of Combustion Institute, pp. 579–589 (1981)
3. Ding, Y., Wang, C., Lu, S.: Large eddy simulation of fire spread. In: Procedia Engineering, pp. 537–543. University of Science and Technology of China, Heifei, China (2014)
4. Osorio, A.F., Fernandez-Pello, C., Urban, D.L., Ruff, G.A.: Limiting conditions for flame spread in fire resistant fabrics. In: Proceedings of the Combustion Institute, pp. 2691–2697. University of California, Berkeley, USA (2012)
5. Bei, P., Liwei, C., Chang, Lu.: An experimental study on the burning behavior of fabric used indoor. In: Proceedings of the Combustion Institute, pp. 2513–2521. San Diego State University, San Diego, USA (2012)
6. North Central Regional Extension Publication: Facts about Fabric Flammability. USA (2003)
7. Quintiere, J.G.: The effects of angular orientation on flame spread over thin materials. Fire Saf. J. **36**, 291–312 (2001)
8. Mohd Azahari, B.R., Azwan, B.S., Mohd Faisal, B.H., Amir, B.K., Suzuki, M., Masuda, W.: Flame spread behavior over combined fabric of cotton/polyester. ARPN J. Eng. Appl. Sci. **11**(6), 7553–7557 (2016)
9. Azwa, Z.N., Yousif, B.F.: Thermal degradation study of kenaf fiber/epoxy composites using thermo gravimetric analysis. Malays. Postgrad. Conf. **2013**, 256–264 (2013)
10. Dittenber, D.B., Gangrao, H.V.S.: Critical review of recent publications on use of natural composites in infrastructure. Compos. A Appl. Sci. Manuf. **43**, 1419–1429 (2012)

Design and Analysis of an Automotive Oil Filter Gripper Socket Special Tool

Chee Guan Choong, Chun Lin Saw and Yoke Yin Chiang

Abstract In repair and car maintenance, it is often necessary to change oil filters. The oil filter is an important part of the car engine operation system. In this study, a new proposed oil filter gripper socket is designed and analysed for the purpose of opening and installing oil filter of cars. The existing oil filter grip socket has encountered problems such as slippage, ruptured and cannot accommodate various sizes of oil filters. The new design is made by using concepts of gear mechanism or to be more precise gear mechanism where forces are transmitted to the gears during the gripping process. These concepts are applied to ensure the socket to hold the oil filter efficiently while performing the operation of tightening or loosening oil filters. In order to overcome the problems of slippage or broken oil filters, a combination of spur gear and spring is used to achieve the highest performance of gripping force. Analysis using the finite element method is carried out to determine the maximum allowable stress on the gripper and deformation of the gripper in a 3D modelling approach.

Keywords Oil filter · Socket · Gear mechanism · Gripping force 3D modelling

C. G. Choong (✉) · C. L. Saw
Department of Mechanical Engineering, Politeknik Ungku Omar,
31400 Ipoh, Perak, Malaysia
e-mail: cgchoong@gmail.com

C. L. Saw
e-mail: sawchunlin@puo.edu.my

Y. Y. Chiang
Department of Mathematics, Science and Computer,
Politeknik Ungku Omar, 31400 Ipoh, Perak, Malaysia
e-mail: yenchiang@puo.edu.my

1 Introduction

The concept of the proposed oil filter gripper socket is based on the current situation at the workshops especially in automotive industries. Automotive maintenance is the act of inspection or testing the condition of car sub-systems and servicing or replacing parts and fluids. Thus, regular maintenance is critical to ensure the car safety, reliability, drivability, comfort and durability.

The aim of the project is to design a new tool that can help to facilitate the installation and removal of oil filters from cars. The designed prototype could be used to achieve two objectives such as:

(i) To install and remove the oil filter with ease;
(ii) To increase safety and handling performance of craftsmen.

An oil filter is designed to remove contaminants from engine oil, transmission oil, lubricating oil, or hydraulic oil. Oil filters are used in different types of hydraulic machineries. The major use of oil filter is in internal combustion engines of motor vehicles, light aircrafts, and naval vessels. Other usage of vehicle hydraulic systems, are such as those in automatic transmissions and power steering, which is also often equipped with an oil filter. The problem arose when the craftsmen want to install or remove the oil filters from cars. Sometimes the oil filter might be stuck and could not be dismantled with bare hands. Hence, the oil filter wrench is used to remove the oil filters. Sometimes changing the oil filter using this wrench might have slippage due to the condition of oily oil filters. Moreover, it is hard to find a suitable wrench that can be used to open different sizes of oil filters. Certain cars have different sizes of oil filter, which might be big or small in sizes [1]. Therefore, the oil filter gripper socket was designed to overcome these problems.

The purpose for this paper is to outline the new concept of the oil filter opener/installer. This special tool can be used as a ratcheting socket wrench. The goal of this invention is to design an oil filter gripper socket tool that can be used as a function to remove or install oil filters in a vehicle more efficiently and safely. This paper will elaborate design criteria for similar tools and their errors of functionality, initial design idea's, the process of choosing the design, and the project build.

2 Literature Review

Existing designs of oil filter openers are widely available in the market. The oil filter wrench is a type of opener where the cap wrench is operated by turning the drive in the middle. It is commonly used to open or tighten oil filters. Another type of opener is the fixed grip type for certain size of oil filters. It is used to open up to 74 mm outer diameter oil filters. It can support heavy load, but it is not suitable for any other sizes of oil filters. This design is one of the simplest kinds of ways to open an oil filter which is widely used [2].

The operation is basically just to put the wrench on the end of any oil filter and it will fit onto the outer body of the oil filter. By using either ratchet or spanner to move the wrench, this oil filter opener can only fit to one size thus, when it comes to different sizes of oil filters, a different size of cap need to be used.

Another commonly used tool is the adjustable oil filter wrench designed to make the removal easy. It is designed with three covered jaws, which were made from heat-treated carbon steel, finished with durable black oxide. This tool is compact, and the design allows gripping in narrow areas of confinement. Besides, the tools are engineered from durable steel to be hard-wear [3].

3 Methodologies

This project was inspired by using only mechanical mechanisms. By using this special tool, it can install and uninstall the oil filter from the engine easily without bringing any harm to the user. By using the concept of springs and gears, this project can grip tightly the oil filter without damaging any part of it.

To develop the best design, a number of design criteria had to be established to help to form and evaluate a good design compared to the existing ones. The initial design was to form the oil filter gripper socket itself where a ratchet wrench is attached at the centre of the tool as shown in Fig. 1. The size of the centre attachment must be able to provide ample space for the movement of the ratchet wrench. Thus, this will make it easier for craftsmen to allocate this tool onto the filter.

Another important function is designing the socket criteria. The designed tool must be able to continually to tighten the grips on the filter as the wrench is turned

Fig. 1 Position of the wrench

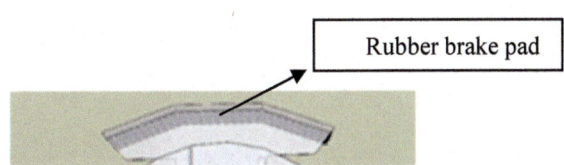

Fig. 2 Bicycle brake pads were added to the jaws

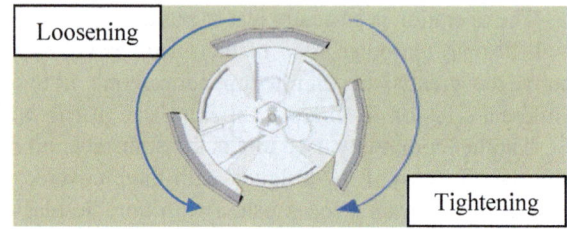

Fig. 3 Movement of the gripping jaws

either right or left. Most of the oil filter wrenches work based on this concept. The friction of the jaw that touches the oil filter surface is added by putting a bicycle rubber brake pad as shown in Fig. 2. As more force is applied to the wrench, the built up force will tighten the grip on the filter as the wrench is turned.

Next, the designed tool is able to work both ways; in clockwise and anti-clockwise direction (refer to Fig. 3). This is to ensure that the tool can be easily switched between the process of tightening and loosening.

The last design criterion is to specify that the designed prototype can be applied on multiple filter sizes. This tool needs to be functioning efficiently on different sizes of oil filters.

Once the design is finalised, finite element analysis on the oil filter gripper socket is carried out using Structural Static. The 3D model of the oil filter gripper is converted into IGES format before imported into Structural Static. The mesh is generated after components and materials are selected and renamed. Next, the boundary conditions as revolution rotational, fixed support, frictionless support, displacement and gripping forces are declared. Hence, analysis of the von Mises stress, deformation of structure, moment on gear and moment on the sleeve are simulated. A flowchart of the simulation process is shown in Fig. 4.

4 Equations and Mathematics

The spur and rack gear is selected as the gear mechanism for the oil filter gripper socket. The spur and rack gear is selected because spur gears are simple, easily manufactured gears and are usually the first choice when exploring gear options. Transmitting power between parallel axes, the teeth project radially on the disc. The sole variance in their identity remains the rack [4].

Therefore, a rack gear is designed to translate rotational into linear motion or vice versa and the rack gear is also durable and ideal for high load applications. The calculation to get gear ratio it outlined in the following:

$$\textbf{Gear ratio} = \textbf{T}_2/\textbf{T}_1 \qquad (1)$$

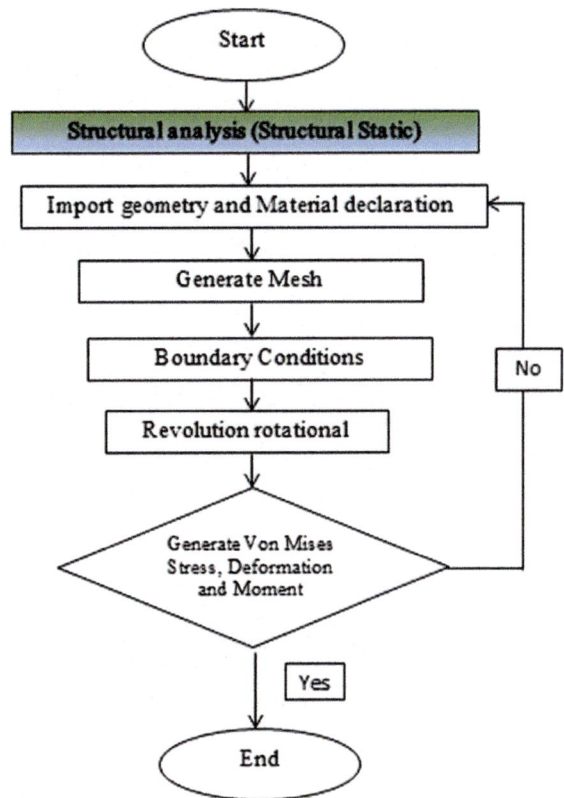

Fig. 4 Simulation using finite element analysis tool

where

T_1 = Number of teeth on first gear
T_2 = Number of teeth on second gear

Power analysis for either summation of torque summed or summation of speed gear arrangements are based on the relationship of:

$$\text{Power} = T\omega \qquad (2)$$

where

T is the torque and ω is the angular velocity. The power can be measured at any gear in the gear train. At the output shaft (arm) of the planetary gear train the total power output would be the output torque multiplied by the angular velocity of the arm. The torque required to dismantle the oil filter is from 1.2 to 1.6 kgfm (where 1 kg force meter = 9.80,665 N).

In any gear train, there will be torque (power) losses due to friction occurring at each gear mesh (approximately a 3% loss for each gear mesh for hardened steel gears) [5]. The basic equation is given as:

$$\text{Efficiency} = \eta = \frac{Power\ Out}{Power\ In} = \frac{Torque\ Out}{Torque\ In} \qquad (3)$$

The force calculated to open the oil filter manually is given as:

(i) Small filter (60–80 mm) = 150 Nm
(ii) Big filter (75–100 mm) = 250 Nm

These results were obtained from researches on existing oil filters [6].

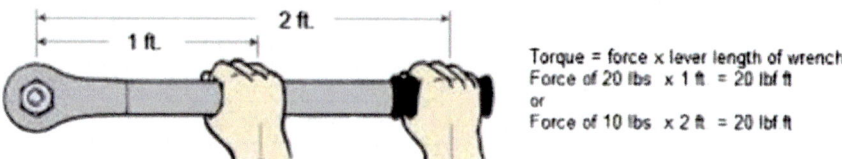

Normal length ratchet = 17 cm /0.17 m
Formula:

Force = Torque(N.m)/Lever length of wrench (m)

The force needed to open the oil filter is estimated from 60 to 100 N and the average force applied is 80.76 N (Tables 1 and 2).

Table 1 Comparison between oil filter openers

No	Types		Conditions
1.	Oil filter wrench		Advantages: • Features a precision-stamped • Deep-reach cap-style for a perfect fit • Heavy-duty construction for strength and durability
			Disadvantages: • Does not work on vehicles with traditional canister filters • Non-adjustable socket • Slippage
2.	Adjustable oil filter wrench		Advantages: • The covered jaws can grip the filter tighter • Compact in size • Allows gripping in tight areas of confinement
			Disadvantages: • Slippage due to improper grip of oil filter • Oil filter leakages due to slippage

Table 2 Average force calculations

Torque (kgfm)	Torque (Nm)	Force (N)
1.2	11.767980	69.22
1.3	12.748645	74.99
1.4	13.729810	80.76
1.5	14.709975	86.53
1.6	15.690640	92.30
Average force		80.76

5 Results and Discussions

Contact connection or constraint of the components in the oil filter gripper assembly design is shown in Figs. 5 and 6. The details on the components constraints are listed in Table 3. Table 4 shows the mechanical properties of steel used for the oil filter gripper.

The total deformation of the oil filter gripper is 0.0396 m maximum as shown in Fig. 7, while the directional deformation in x-axis could reach 0.017,483 m maximum as shown in Fig. 8. There is also compression happening to the gripper as the deformation in the simulation showed a negative value.

The stress simulated on the oil filter gripper assembly reached a maximum von Mises stress of 1.988×10^{13} Pa as depicted in Fig. 9.

Stress and moment were simulated on Gear 1, Gear 2 and Gear 3 as depicted in Figs. 10 and 11, respectively. The value of the moment occurred to retract the gripper with 300 N force are −98,448 Nm (x-axis), 1.736×10^5 Nm (y-axis) and 8.335×10^5 Nm (z-axis). Meanwhile, the stress simulated showed a value of 3.023×10^{13} Pa (x-axis), the same stress value was obtained in the analyses when gears and grippers were analysed together as shown in Fig. 12.

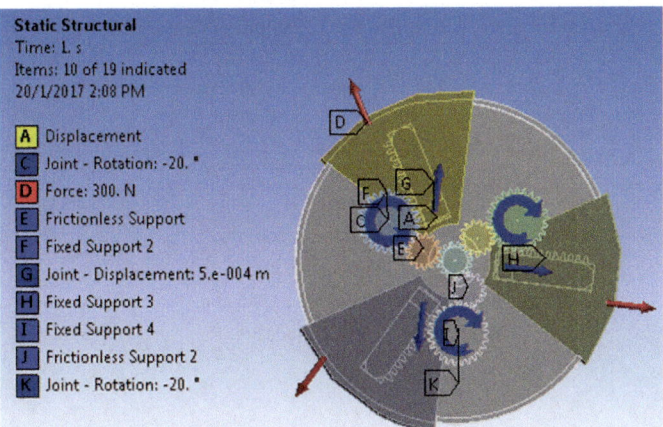

Fig. 5 Constraints of the oil filter gripper bottom view

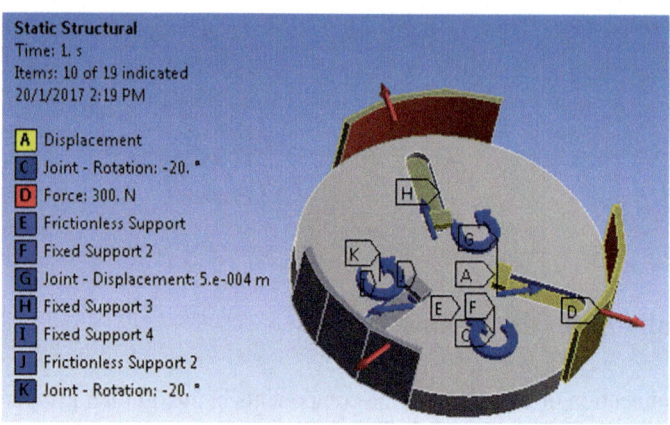

Fig. 6 Constraints of the oil filter gripper plan view

Table 3 Connection of the components constraints

No	Components	Constraint
1.	Gear 1	Fixed support/Revolution rotation
2.	Intermediate gear 1	Frictionless support
3.	Gear 2	Fixed support/Revolution rotation
4.	Intermediate gear 2	Frictionless support
5.	Gear 3	Fixed support/Revolution rotation
6.	Intermediate gear 3	Frictionless support
7.	Main gear	Frictionless support
8.	Upper housing	Frictionless support
9.	Lower housing	Fixed support
10.	Sleeve 1	Revolution displacement
11.	Sleeve 2	Revolution displacement
12.	Sleeve 3	Revolution displacement

Table 4 Mechanical properties of steel

No.	Mechanical properties	Material (steel)
1.	Density	7850 kg/m^3
2.	Young's modulus	2×10^{11} Pa
3.	Poisson's ratio	0.3
4.	Thermal expansion	1.2×10^{-5} (1/ °C)
5.	Shear modulus	7.692×10^{10} Pa

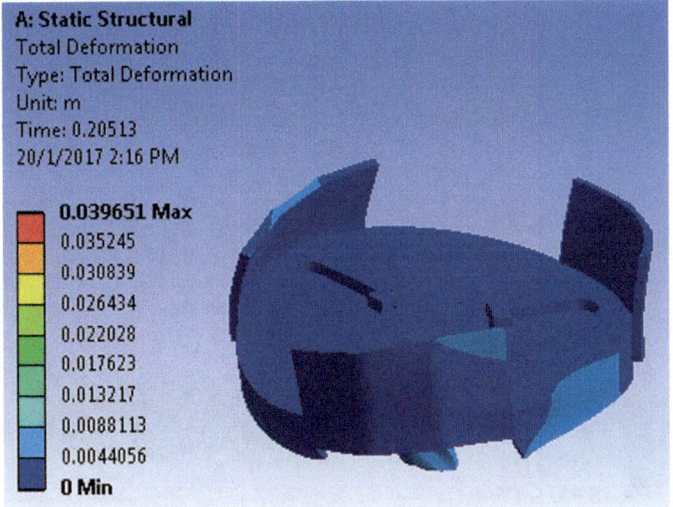

Fig. 7 Total deformation of the oil filter gripper

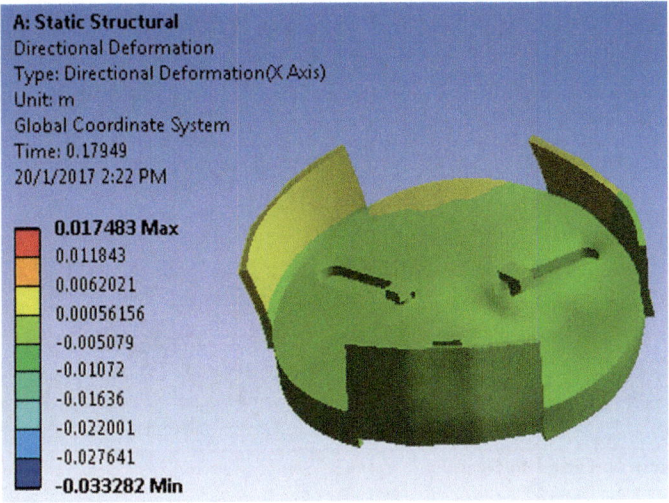

Fig. 8 Directional deformation of the oil filter gripper

This final 3D prototype was fabricated with PLA plastic material, which consists of the essential elements. Each and every part of these elements played an important role. As such, the roles for the elements are described in the Table 5.

The working processes of changing an oil filter is shown in Fig. 13.

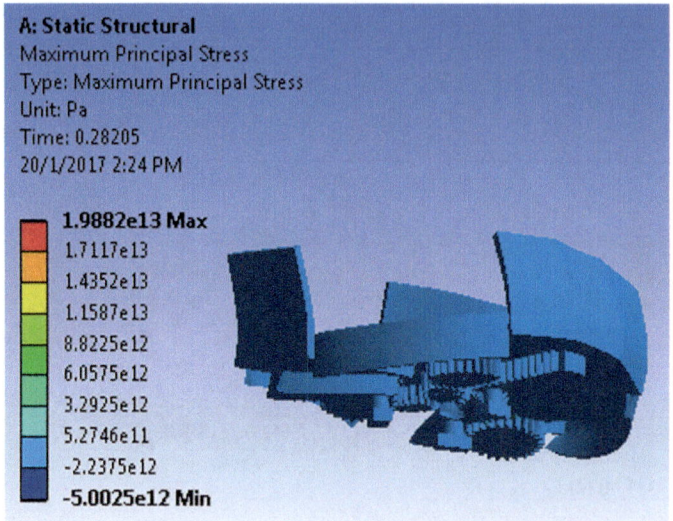

Fig. 9 Stress on oil filter gripper

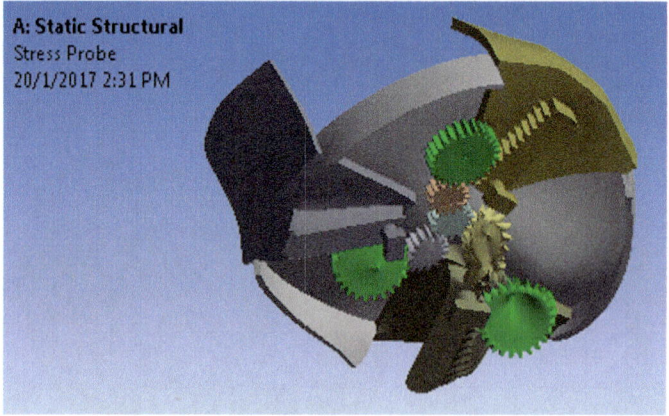

Fig. 10 Stress on Gear 1 to Gear 3

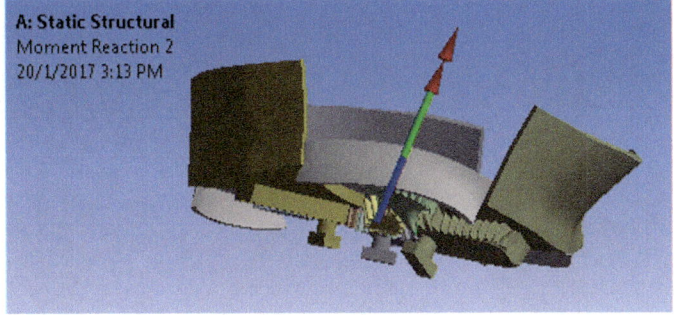

Fig. 11 Moment on Gear 1 to Gear 3

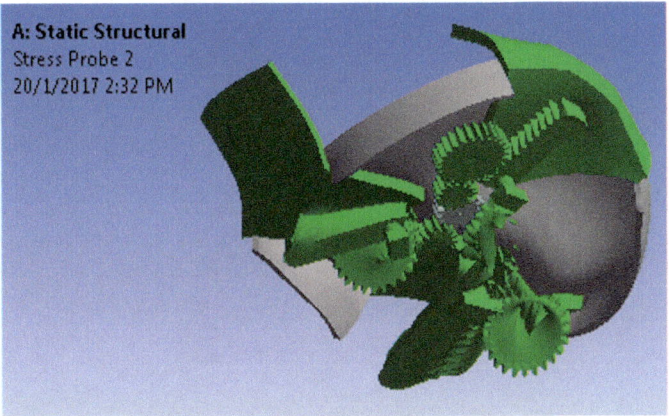

Fig. 12 Stress on gears and grippers

Table 5 Prototype components and function

No	Components	Function
1.	Spur gear	To move the rack gear
2.	3-jaws gripper	To grip the oil filter
3.	Rubber stopper	To add the friction onto the oil filter
4.	Ratchet lock	To lock and hold the ratchet
5.	Rack gear	To move the jaw gripper

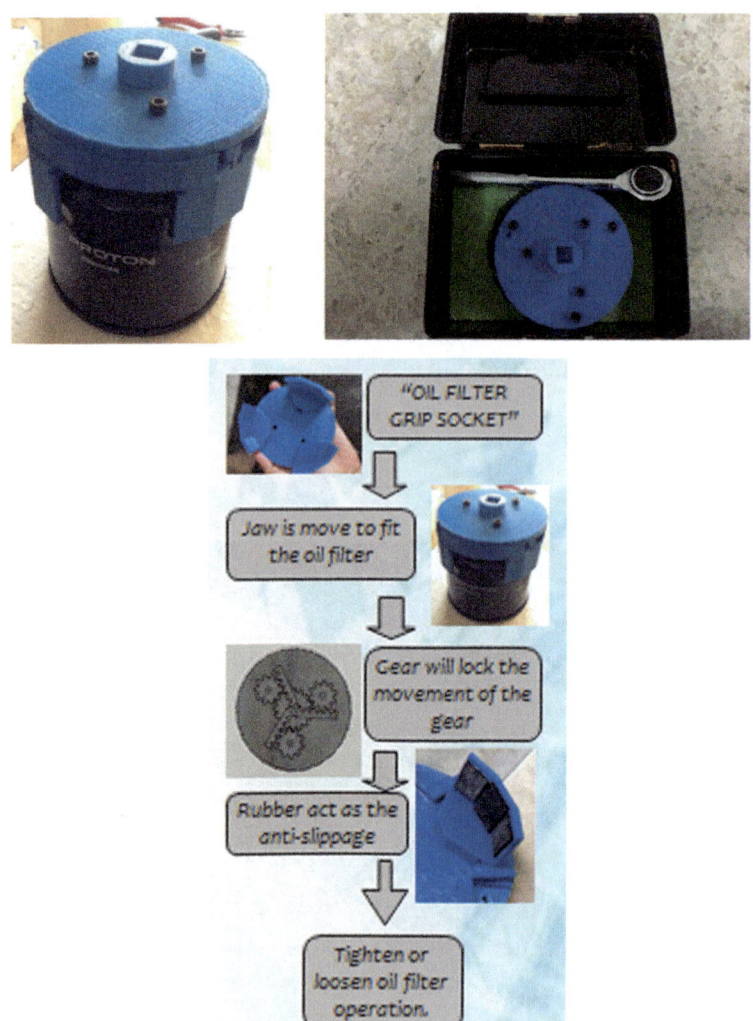

Fig. 13 The process of changing oil filter

6 Conclusions

This prototype design could be definitely continuing to be improved further. A few problems were still found with the latest design iterations. The base part was still free to move with the rotation part and must be held in place with one hand while the other applies force to the wrench. This could be solved by having the lower arms pinch on the oil filter, but that would add more complications to the tool and also decrease the size flexibility. Another issue often seen in the prototype design is that the direction switch can fall off easily. This would likely not be an issue in a production part; however, the simulation on 'Inventor' resources do not offer high precision in the parts to have a tighter fit while still allowing for the required motion.

The overall success of this design will be hard to estimate without producing the actual solid material prototype that is strong enough to be vehicle tested. The designed tool of this prototype would likely perform as well as some of tools available in the market but would be higher in cost since it has more parts and requires more assembly. As the part evolved, the gripping element may be pursued in order to find a precise, more compact solution.

References

1. Jones, F.D., Horton, H.L., Newell, J.A.: Ingenious mechanisms for designers and inventors. Industrial Press Inc, USA (1967). (ISBN: 0831110848)
2. Greenbank, S.J.: Oil filter wrench attachment. Honors Research Projects. (2015). Retrieved from http://ideaexchangeuakron.edu/
3. Crolla, D.A.: Automotive engineering. butterworth heinemann, USA. (2009)
4. Abdo, E.: Power equipment engine technology. Cengage Learning, USA (2010)
5. Serway, R.A., Jewett, J.W.: Physics for scientists and engineers. **1**, Chapters 1–22. Cengage Learning, USA. (2009)
6. Rao, J.S., Dukkipati, R.V.: Mechanism and machine theory. New Age International (P) Limited, India (2007)

The Development of Hovercraft Design with a Horizontal Propulsion System

Khairul Anuar Mat Saad, Kogulan Murugan and Mohd Amin Hakim Ramli

Abstract Over the centuries there were many efforts to reduce the element of drag between moving vehicles. A hovercraft is a relatively new means of transportation. The concept of the hovercraft was born when engineers developed an experimental design to reduce the drag force of a vessel. The revolutionary idea was to use a cushion of air between the boat and the water to reduce the drag force. The aim of this project was to design and model a hovercraft with a horizontal propulsion system. Usually, a hovercraft is designed to be used with a vertical propulsion system to produce lift and thrust forces. This model can perform the basic functions of a hovercraft and travel on water and land surfaces. It produces a reasonable air cushion to lift its body and carry some loads, and at the same time it produces a suitable thrust to move inland operation. The objectives of this project are to develop a basic hovercraft model with a horizontal propulsion system. Moreover, a well-functioning hovercraft model needs to be designed in consideration of the horizontal propulsion system to reduce drag force, minimize the lift force and improve the trust force. The expected outcome for this project is a model of a hovercraft that works by producing lift and thrust.

Keywords Hovercraft · Hull · Design

K. A. Mat Saad (✉) · K. Murugan · M. A. H. Ramli
Universiti Kuala Lumpur Malaysian Institute of Marine Engineering Technology, Jalan Pantai Remis, 32200 Lumut, Perak, Malaysia
e-mail: khairulanuar@unikl.edu.my

K. Murugan
e-mail: mkcjkogu@hotmail.com

M. A. H. Ramli
e-mail: mohdamin@unikl.edu.my

1 Introduction

A hovercraft is a vehicle that flies like a plane but can float like a boat, can drive like a car but will traverse ditches and gullies. A hovercraft is also sometimes called an air cushion vehicle because it can hover over, or move across, land or water surfaces while being held off from the surfaces by a cushion of air. It can travel over all types of surfaces including grass, mud, muskeg, sand, quicksand, water and ice. Hovercrafts prefer a gentle terrain although they are capable of climbing slopes of up to 20%, depending on the surface characteristics. Modern hovercrafts are used for many applications where people and equipment need to travel at high speeds over water and be able load and unload on land. For example, they are used as passenger or freight carriers, as recreational machines and even as warships. Hovercrafts are very exciting to fly and the feeling of effortlessly traveling from land to water and back again is unique.

2 Problem Statement

The aim of this project was to design the shape of a small working model of a hovercraft with only a horizontal propulsion system. Basically, a hovercraft is designed to be used with a vertical and horizontal propulsion system to produce lift and thrust force to operate the hovercraft. The model must be well designed and powered. The working model hovercraft must be able to perform the basic functions of a hovercraft and able to travel on water and land surface. This model must be able to produce enough air cushion to hover its body and be able to lift some loads. At the same time, it also needs to produce enough thrust to move along the ground by itself.

3 Objective

The objectives of this project are to develop a basic hovercraft model with a horizontal propulsion system. So, by designing the basic model hovercraft with a horizontal propulsion system, it will help to consider the usage of a horizontal propulsion system.

4 Methodology

In this chapter, the design process of the RC model of the basic hovercraft unit with a horizontal propulsion system is briefly explained. At the same time, the ideal design and the major component which will be used to construct the RC model will also be explained.

4.1 Design Selection

Concept selection is the process of evaluating concepts with respects to customer needs and other criteria, comparing the relative strengths and weaknesses and selecting some concepts for further investigation, testing or development of the product design process. The concept generation process begins with a set of customer needs and target specifications and results in a set of product concepts, which the product design will make a final specification. To develop a small-scale hovercraft unit, there are 6 main criteria for the product concept: overall simplicity, ease of operation, design simplicity, lightness, ease of manufacturing, and it must be amphibian.

As in Figs. 1, 2 and 3, and Table 1, various designs of the RC hovercraft model with the horizontal propulsion system and the criteria of the design are shown.

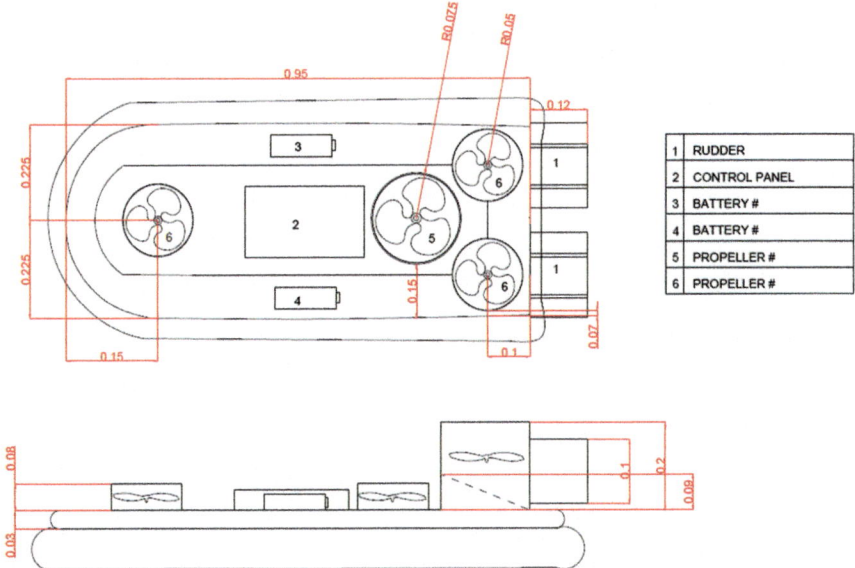

Fig. 1 Design 1

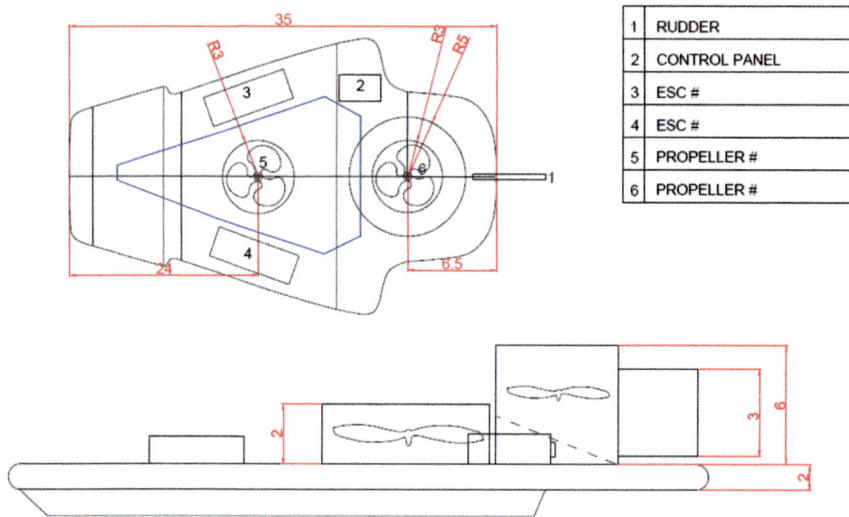

Fig. 2 Design 2

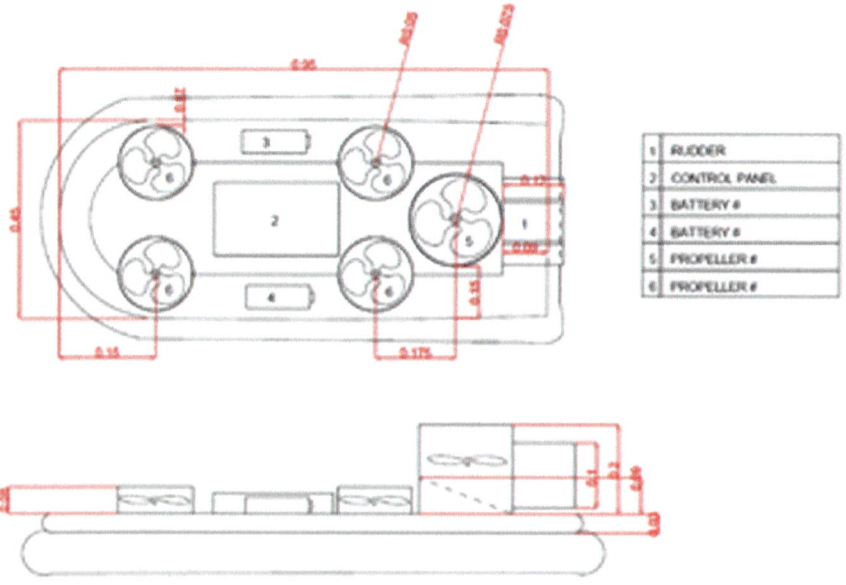

Fig. 3 Design 3

Table 1 Design criterion

Requirement	Design 1	Design 2	Design 3
Simple design	x	x	x
Easy to operate	x	x	x
Lightweight	x		
Simplicity	x	x	x
Easy to build	x		
Amphibious	x	x	x

The design criteria were based on the concept of the existing real hovercrafts and the RC hovercraft model.

Table 1 clearly states the criteria which need to be achieved by the design. Design 1 achieves all criteria except the low cost of building and designs 2 and 3 miss the criteria concerning weight, cost and ease to build.

Referring to design 1, it has two motors which support the lift force and thrust force. The motor used to produce the lift force is bigger than the motor which produces the thrust force. This is because the thrust force must be at a high level to lift the hovercraft model. If the thrust motor produces a higher force than the lifting force, most probably the lifting force will not be enough to lift the hovercraft model. Then, the thrust force would make the hovercraft move, which could wear out the skirt.

In design 2, four small motors at every edge of the RC hovercraft are used to produce the lift force which is more stable than a big size motor at aft of the hovercraft to produce the thrust force which makes the RC hovercraft to move. Because of the larger motor used to produce the thrust force, it misses the criteria of being lightweight, easy to build, and low-cost.

In design 3, there are two similar sized motors at the rear of the hovercraft which produce the thrust force and two different sized motor in the middle and front of the hovercraft which produce the lift force. The size was different because, at the rear of the hovercraft model there are two thrusters which add weight. Therefore, the larger motor is placed at the rear of the hovercraft. In front of the hovercraft there are small sized motors because only small amount of force is needed to lift the hovercraft model. The design 3 was the actual design of a small class hovercraft unit. It misses the criteria of being lightweight, easy to build, and low-cost.

As an ideal design, design 1 has been chose to develop a RC basic hovercraft model unit with a horizontal propulsion system. Costing also been considering in the choice of the design. Design 1 was low-cost when compared with the other designs.

4.2 Material and Components

Hull and Skirt—Polyurethane
Motor—NTM Prop Drive 2315 kV and 3500 kV (56 g)
ESC—HBA 18 A and 25 A
Propeller—Propeller Fan 10 g
Battery—Ni-MH 2/3 A1600 mAh 7.2 V
Servo—H0903 9 g

4.3 Final Design of Hull and Skirt

The design of hull was created in Autodesk AutoCAD 2014 in 2D form. After deciding on the best design concept, construction of the model begun. Before starting the construction, the 2D drawing had to be developed into a 3D drawing to get a clear view of the model. Autodesk Inventor 2013 was used to create a 3D form and to helps analyze the model's estimated weight, center point, momentum, among other parameters. The material to construct was chosen according to weight, having a high buoyancy point, heat resistance, being waterproof and also price. The result of the 3D drawing is shown in Figs. 4 and 5.

4.4 Material to Construct RC Model

a. **Material Selection for Platform (Hull)**

This is the basic structure on which the hovercraft floats when the engine is stopped while moving over water. It supports the whole weight of the craft. As it supports the whole weight and floats on water, the material must have a high buoyancy point,

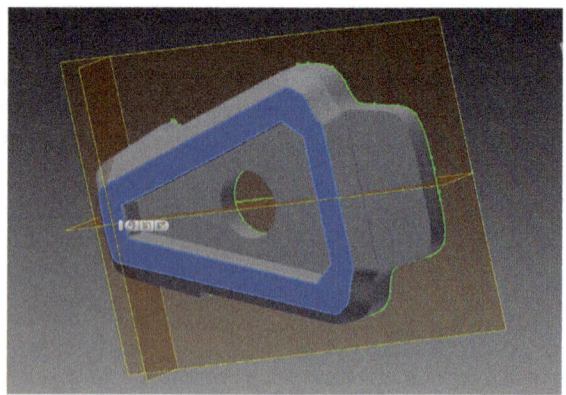

Fig. 4 Underneath view of hull design and skirt

Fig. 5 Final design with material

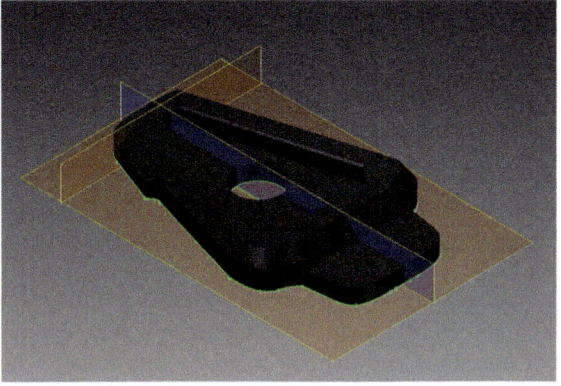

be waterproof and strong enough to support the hovercraft. The chosen material therefore is polyurethane.

b. **Material Selection for Skirt**

They are air bags inflated by air and fitted around the perimeter of the craft holding air under the craft and thus upon a cushion of air. It enables the hovercraft to obtain a greater hover height. As for this RC model, the same material as for the model hull has been chosen, polyurethane. This is because the hull was made in polyurethane which is like a hard sponge, it hard to use a different material for the skirt which later needs to be attached to the hull. There is a high probability that the attachment between the hull and the skirt could not withstand the pressure produced by the lifting fan.

Moreover, in Hovercraft Regulation, in Regulation 8, Topic 2.4 clearly states that the material for the skirt, which is not specified, must have certain characteristic to avoid the damages of the skirt. Therefore, by considering the characteristic of the material and the regulation, polyurethane was chosen for the skirt.

c. **Material Selection for Fan and Rudder**

To produce the lifting and thrusting force, a fan plays the main role as the rudder mostly determines the direction of maneuverings. A drone fan was chosen for the hovercraft model which acts as a propeller in the horizontal axis. The reason for choosing the drone fan is that it produces a high speed of air circulation which makes the drone lift. In the designed hovercraft model, two main propulsions are used for lifting and thrusting. Both propulsions use the same kind of the fan.

For the rudder, the rudder made by alloys is light in weight and strong enough to divert the air produced by the thrusting propulsion. The rudder was designed according to the clearance space between the fan holder and the rudder holder.

4.5 Process of Construction

a. **First Stage**

At first the hull and the skirt were prepared because they are the major part of the hovercraft model. To produce the hull and skirt according to the design, a polystyrene cutter has been used to shape the polyurethane.

b. **Second Stage**

In the second stage, the servo to control the rudder was prepared, and the motor to produce the lift and thrust force was installed. Because of the motor being used was for the drone category, it come up with a motor holder. For the part to produce lifting force, the holder must be attached on the hull according to the designed dimension.

For the motor, which produces thrust, it requires a cover to cover the motor holder to trap the air produce, and force it out in a single hole. This will produce a high pressure inside of it, and produce the thrusting force.

During the preparation of the cover, there will be a small size hole for the shaft at the bottom of the cover to connect the rudder with the servo. As the servo works, by connection of the shaft, it moves the rudder to the left and right. The motor holder is attached on top of the cover to avoid damages to the cover or the motor holder, when the motor produces high speed air.

c. **Third Stage**

At last the electrical speed controller (ESC) is installed. Because the two motors together have one connection with one controller, the ESCs also must be combined. Each motor needs one ESC. The main function of the ESC is to control the rotation speed of the motor by limiting the flow of current. When combining the two ESCs, the wires are connected according to the positive and negative wire, while the switch button connects it with any ESC. This allows turning on and off the power supply to the hovercraft model. Figure 6 shows the complete hovercraft model.

Fig. 6 Complete arrangement of hovercraft model

5 Results

The sea trial of the RC horizontal propulsion hovercraft was performed in accordance to the Hovercraft Regulation and Hovercraft Act of 1968. It explains all sea trial activity which had to be accomplished by the RC model to consider it a real hovercraft. By following the Hovercraft Regulation and Hovercraft Act 1968, we have achieved the expected outcome.

a. Aspect Ratio

To build the concept model of the hovercraft, a real hovercraft has been used as a benchmark and reference. The chosen hovercraft is the Hivus A5 which is a product of the Aerohod Ltd from Russia. Figures 7 and 8 show the Hivus A5 Hovercraft and its specifications.

Fig. 7 Hivus A5 hovercraft

Payload, kg	600
Passenger capacity, prs.	5 (without the driver)
Length (transportation length), m	7,25 (6,8)
Beam (transportation width), m	3,08 (2,5)
Max height (transportation height), m	2,65 (2.2)
Engine type	petrol or diesel engine
Engine power, hp	143-173
Speed on water, km/h	50
Speed on snow, km/h	70

Fig. 8 Specification of Hivus A5 hovercraft

$$\text{Ratio} = \frac{\text{Reference length (m)}}{\text{Model length (m)}}$$
$$\text{Ratio} = \frac{6.8}{0.35}$$
$$\text{Ratio} = 1 : 19.43$$

Hovercraft base area, A

To create a 3D design, Autodesk Inventor 2013 was used. The software could set up the product material and help calculate the product area, volume, mass, momentum, among others. Basic information like mass, area and volume needed for mathematical calculation of hull and skirt which gained from Inventor 2013. Figure 9 shows the result of the analysis of model hovercrafts.

Area: 153,750.071 mm^2
Area: 0.15375 m^2
Volume: 1,496,756.658 mm^3
Volume: 0.001496757 m^3
Mass: 1.497 kg
Cushion Pressure, Pc

$$\text{Pc} = \frac{\text{Total Weight}}{\text{Area of Base}}$$
$$\text{Pc} = \frac{1.497 \text{ kg}}{0.15375 \text{ m}^2}$$
$$\text{Pc} = 9.7366 \text{ kg/m}^2$$

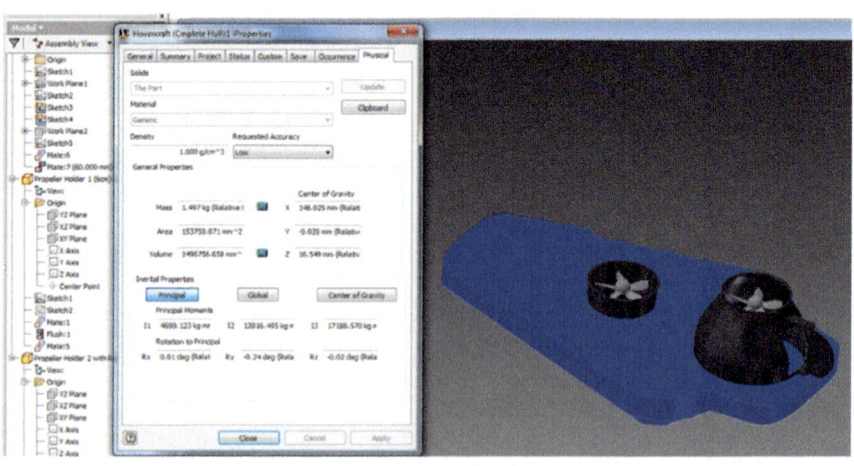

Fig. 9 Analysis software

Escape Velocity, Ve

$$Ve = \sqrt{\frac{2 \times Pc}{\rho}}$$

where, ρ is density of air which is 1.22 kg/m^3

$$Ve = \sqrt{\frac{2 \times 9.7366}{1.22}}$$
$$Ve = 3.995 \text{ m/s}$$

Perimeter of skirt

Skirt Perimeter = 0.23 + 0.23 + 0.06 + 0.17

Skirt Perimeter = 0.69 m

Hover Gap = 2 mm = 0.002 m
Escape Area, Ae

Ae = Skirt Perimeter × Hover Gap
Ae = 0.00138 m^2

Volume of Air Lost

Volume of Air Lost = Ve × Ae
Volume of Air Lost = 0.0055131 m^3/s

Total Airflow Required

This volume of air is required to lift the hovercraft. From the total airflow generated by the turbine fan, 33% is used to lift the hovercraft. This 33% corresponds to the volume of air lost. Therefore, the total volume of air must be generated are three times of this quantity.

$$\text{Total of Airflow Required} = 3.33 \times \text{Volume of Air Lost}$$
$$\text{Total of Airflow Required} = 3.33 \times 0.0055131 \text{ m}^3/\text{s}$$
$$\text{Total of Airflow Required} = 0.01836 \text{ m}^3/\text{s}$$

Hence the fan must provide a cushion pressure of 9.7366 kg/m² and 0.01836 m³/s of total airflow.

b. **Specific Gravity Calculation**

The specific gravity of the hovercraft is determining from the measurement of the hull. The resulting coordinates for specific gravity of the hull are shown in Table 2.

$$X = \frac{\sum XW}{\sum W} = 0.1673 \text{ m @ } 16.732 \text{ cm}$$
$$Y = \frac{\sum YW}{\sum W} = 0.085 \text{ m @ } 8.5 \text{ cm}$$

Table 2 Weight and position of components

Description	Weight (kg)	Weight, W (N)	X (m)	Y (m)	XW (Nm)	YW (Nm)
Hull and Skirt	1.497	14.686	0.175	0.094	2.5701	1.3805
Motor 1 (Lifter)	0.1	0.981	0.155	0.088	0.1521	0.0863
Motor 2 (Thruster)	0.2	1.962	0.277	0.068	0.5435	0.1334
Battery	0.3	2.943	0.06	0.05	0.1766	0.1472
Total		20.572			3.4422	1.7474

6 Discussion

Skirting is one of the major parts of the hovercraft, which has been assumed to be made by plastic or polymer. But, after considering the Hovercraft Regulation and Hovercraft Act of 1968, there is no requirement for the skirt material to be from plastic or a polymer. Instead, it clearly states the characteristics the skirt material must have.

For the thrust and lifting section of this model hovercraft, many types of fan could be used. Theoretically, more propeller blades on the fan produces more pressure due to larger amounts of airflow.

Lastly, the calculation of specific gravity was included in the weight of the hull, skirt, and motor which produce the thrust force, lift force, and battery. These components are the main parts of the model hovercraft which contribute to weight. For easy calculation, ESC components have been included as motor weight.

7 Conclusion

The scope of this research was to develop a basic model of a hovercraft with the horizontal propulsion system only. In addition, research and study must be made to better understand the concept of the hovercraft which operated by a vertical and horizontal propulsion system.

The objectives of this project were to develop a basic hovercraft model with a horizontal propulsion system. Moreover, a well-functioning hovercraft model needed to be designed well per the horizontal propulsion system.

The expected outcome of this project been achieved, that is, a hovercraft model with horizontal propulsion was designed and tested that works by producing lift and thrust forces which make the model move by itself without applying any other external forces.

Acknowledgements The authors would like to thank the UniKL MIMET staff and the technical staff for their assistance during the development and testing phase and the 'RCs community' for their assistance and advice.

Author Biography

Dr. Khairul Anuar Mat Saad graduated a BE degree in Mechanical Engineering at University Science of Malaysia, ME degree in Mechanical (Maritime Technology) at University Technology of Malaysia and a Doctor of Philosophy in Maritime Engineering at Australian Maritime College. He is currently working at Universiti Kuala Lumpur as academic staff and majoring in Ship Construction and Maintenance. His research specialties are in experimental and computational fluid dynamics for the development and application of underwater and surface vehicle.

A Rule Based Method to Auto-recognize Fillet Features of B-Rep Mill Parts

Pramodkumar Siddappa Kataraki and Mohd Salman Abu Mansor

Abstract The transition features such as fillet features provide an aesthetic look for a product and feature recognition of these features is as important as recognition of volumetric and surface based features. Many research works contribute to the recognition of surface based features, volumetric features but a gap still exists in recognition of fillet features. In this paper, an effort is made to develop an algorithm that can auto-recognize fillet features of B-rep parts by a rule based method. The top cover of a cell phone model in .SAT file format is input to test the algorithm and the auto-generated results show the algorithm is able to automatically recognize fillet edges and the faces they belong to.

Keywords B-rep · Recognition · Mill · Fillet

1 Introduction

The computer aided process planning (CAPP) is the bridge between computer aided design (CAD) and computer-aided manufacturing (CAM). However, to achieve an efficient CAPP, the recognition of surfaces and features is necessary. The research works performed so far have been able to recognize regular form volumetric features of simple parts by the volume decomposition method [1]. A rule based method called attribute-based technique was applied to recognize regular form features such as boss, hole, and step. The technique used faces, edges and protrusion or depression attributes of features of a part model to recognize the feature [2]. Decomposed delta volume obtained by subtraction of part model in stock model was used to recognize machining features such as step, open pocket.

P. S. Kataraki · M. S. Abu Mansor (✉)
School of Mechanical Engineering, Engineering Campus, Universiti Sains Malaysia,
Seri Ampangan, 14300 Nibong Tebal, Seberang Perai Selatan, Pulau Pinang, Malaysia
e-mail: mesalman@usm.my

P. S. Kataraki
e-mail: psk13_mec014@student.usm.my

The decomposition time increases with increase in feature intersections [3]. The delta volumes for finishing and roughing processes were auto-generated for surfaces of a regular-freeform part model by auto-recognizing these surfaces. The developed algorithm auto-recognized surfaces of a part model but not its features [4, 5]. The freeform surface based features of sheet metal parts were recognized by a rule based method [6] and also by the type and number of separating curves [7]. The basic deformation features of sheet metal parts such as bend and wall were recognized by the developed algorithm. The work mainly concentrated on deformation features having constant thickness [8]. A review on rule based methods revealed that all the recent feature recognition works mainly dealt with orthogonal features while minimal attention was given to non-orthogonal features [9].

The above literature reviews that express volume decomposition and rule based methods have been successfully applied to recognize features such as volumetric features of regular form parts, surface based features of sheet metal parts, respectively. But the reviews do not reveal anything on the recognition of transition features like fillet on a complex part that has fillets internally and externally. In this paper, an effort is made to develop an algorithm that applies a rule based method to automatically recognize fillet features of complex B-rep parts. The algorithm also automatically generates result tables showing the recognized fillet features, edges and faces they belong to.

2 Methodology

The CAD part model in .SAT file format is input to the algorithm and the faces of the part model are recognized. The recognized faces are then segmented into top, contour, bottom regions and based on the number of loops a face has, categorization into faces with depression and faces without depression is done. Finally, rules are applied to find the fillet features of the part model and the following flow chart shows the steps followed by the algorithm from the input of the part model to the recognition of fillet features (Fig. 1).

2.1 Part Model

The CAD part model input to developed the algorithm is of B-rep (boundary representation) since B-rep models contain faces, loops, edges and vertices [10] and B-rep models are a perfect input for feature recognition techniques [11].

Fig. 1 Flow chart of the algorithm

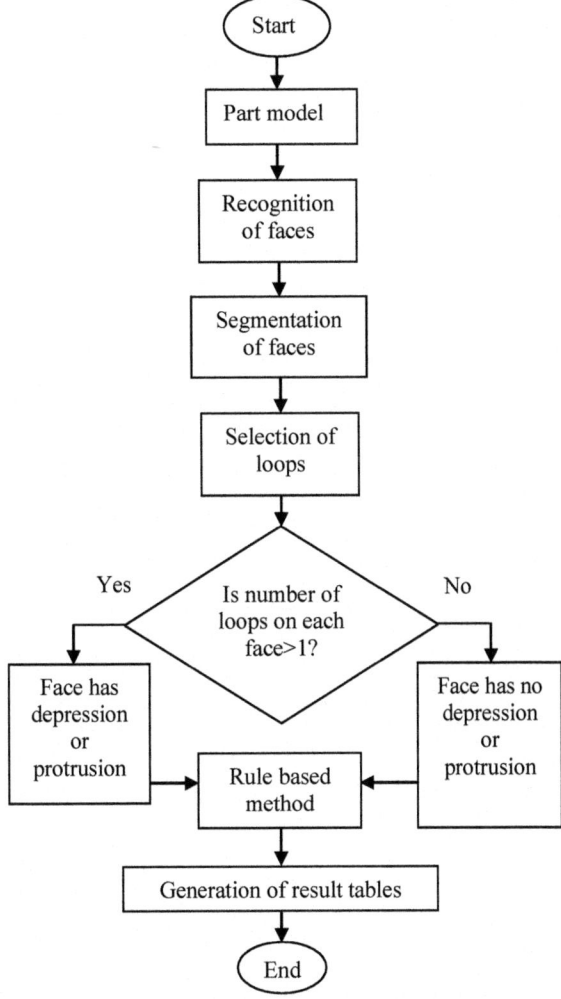

2.2 Recognition of Faces

The algorithm recognizes all the faces of the part model one by one and separates each face into a regular form and freeform categories based on their geometrical shapes such as planar, conical, cylindrical, spline (see Fig. 2).

Fig. 2 a Regular form part model [12]. b Freeform part model

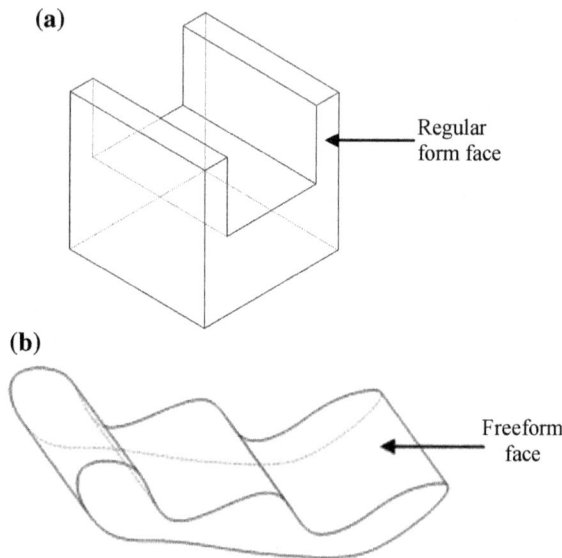

2.3 Segmentation of Faces

The algorithm recognized faces are usually segmented into three different regions (i) top region (ii) contour region and (iii) bottom region. The segmentation is achieved by finding the direction of a face's normal vector at its midpoint and the equation used to determine the midpoint of a face is [1]:

$$\begin{aligned} X_1 &= X_0 + sU_1 + tV_1 \\ Y_1 &= Y_0 + sU_2 + tV_2 \\ Z_1 &= Z_0 + sU_3 + tV_3 \end{aligned} \quad (1)$$

Where X_1, Y_1, Z_1 are coordinates and (U_1, U_2, U_3), (V_1, V_2, V_3) are coordinates of **U**, **V** vectors. Figure 3 shows the face segmentation into the top, contour and bottom region.

2.4 Selection of Loops

The edges together form a loop and in turn a loop binds a face. A face is made of one or more loop(s) and based on the number of loops, a face is categorized into (see Fig. 4) (i) a face without depression or protrusion (FWoD), (ii) a face with depression or protrusion (FWD).

Fig. 3 Regional segmentation of faces of a part model

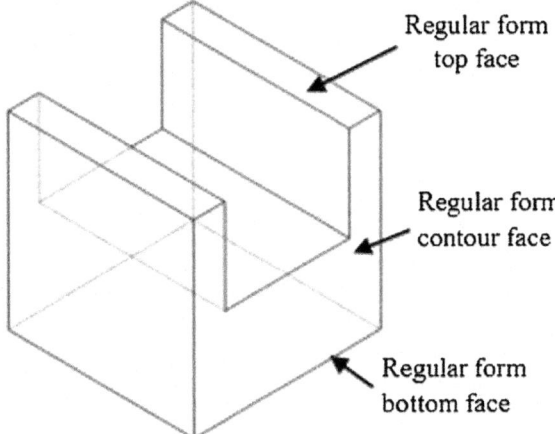

Fig. 4 Faces and their loops of a part model

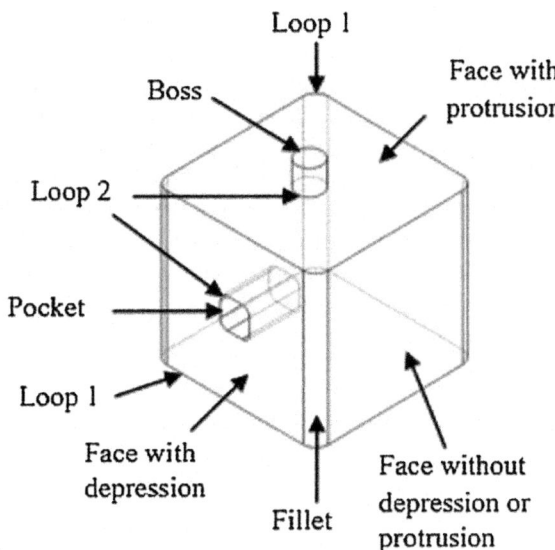

(i) A face without depression or protrusion (FWoD)

$$F \in FWoD$$
$$\text{If } 0 < N < 1;$$

where F = face; N = number of loops;

(ii) A face with depression or protrusion (FWD)

$$F \in \text{FWoD}$$
$$\text{If } 1 < N < \infty;$$

where F = face; N = number of loops;

Figure 4 illustrates a part model having faces with and without depression or protrusion. The loops 1 and 2 indicate a peripheral loop and an internal loop of a face, respectively. The top face has a protrusion in the form of boss, and the contour face has a depression in the form of a pocket. The four contour faces are connected each other by a fillet feature.

2.5 Rule Based Method

The concave face of a part model that connects two adjacent faces is called a fillet face. The fillet face contains two curve edges and the edges are called the fillet edges. A set of new rules is applied by the algorithm to identify a fillet feature wherein the fillet edges of a fillet feature are recognized. A face is said to have a fillet edge (see Fig. 5) if it satisfies the following conditions: (i) The edge is curve and (ii) the co-edges of the curve edge are straight.

The peripheral loop of the contour face of the part model (see Fig. 5) has five edges E_1, E_2, E_3, E_4, and E_5. E_2 is a curve edge and E_1, E_3 are its straight co-edges, so the face is said to have a fillet edge.

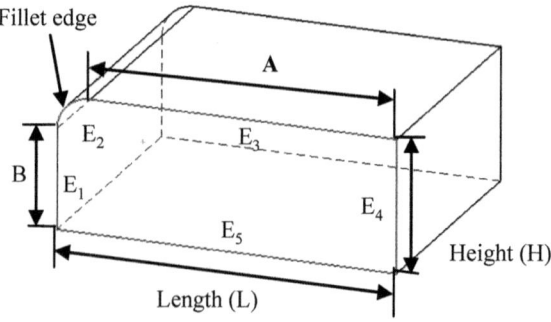

Fig. 5 Part model with a fillet feature

A Rule Based Method to Auto-recognize Fillet Features ...   111

3 Implementation

The top cover of a cell phone having fillet edges is used as test part model to test the developed algorithm (see Fig. 6a).

The algorithm automatically calculates the volume of the test part model and the number of faces, edges, vertices the model has, as shown in Fig. 7.

The algorithm recognizes regular form faces, freeform faces and then segments each face into top, contour and bottom regions in the following ways

(i) If $Z_1 > 0$; then the face is segmented to top region.
(ii) If $Z_1 == 0$; then the face is segmented into contour region.
(iii) If $Z_1 < 0$; then the face is segmented to bottom region.

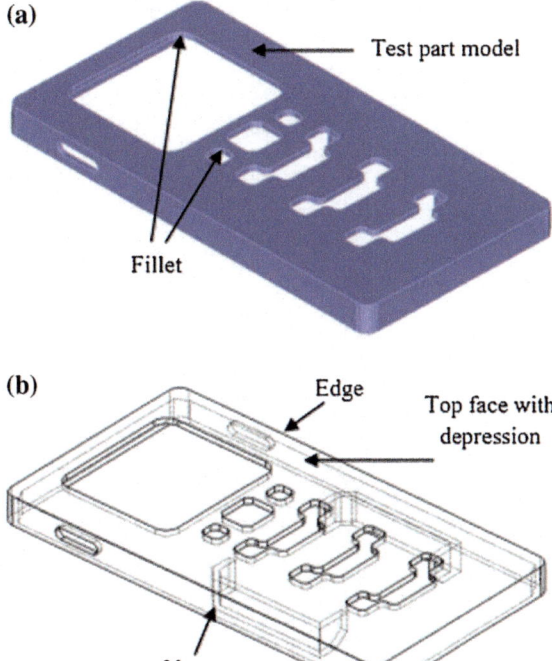

Fig. 6 a Isometric view of test part model. b Test part model and its faces, edges, vertices

Fig. 7 Auto-generated data of test part model

```
Part model volume is 27390.8 mm3
The original model has :
        1 lumps
        1 shells
        0 wires
        159 faces
        462 edges
        308 vertices
```

(a)	Regular form face without depression or protrusion (top region)
	No fillet edges

(c)	Regular form face without depression or protrusion (bottom region)
	No fillet edges

(b)	Regular form face with depression (top region)
	Face [6]
	Edge [1] is a FILLET
	Edge [3] is a FILLET
	Edge [5] is a FILLET
	Edge [7] is a FILLET
	Edge [9] is a FILLET
	Edge [11] is a FILLET
	Edge [13] is a FILLET
	Edge [15] is a FILLET
	Edge [17] is a FILLET
	Edge [19] is a FILLET
	Edge [21] is a FILLET
	Edge [23] is a FILLET
	Edge [25] is a FILLET
	Edge [27] is a FILLET
	Edge [29] is a FILLET
	Edge [31] is a FILLET
	Edge [33] is a FILLET
	Edge [35] is a FILLET
	Edge [37] is a FILLET
	Edge [39] is a FILLET
	Edge [41] is a FILLET
	Edge [43] is a FILLET
	Edge [45] is a FILLET
	Edge [47] is a FILLET
	Edge [49] is a FILLET
	Edge [51] is a FILLET
	Edge [53] is a FILLET
	Edge [55] is a FILLET
	Edge [57] is a FILLET
	Edge [59] is a FILLET
	Edge [61] is a FILLET
	Edge [63] is a FILLET
	Edge [65] is a FILLET
	Edge [67] is a FILLET
	Edge [69] is a FILLET
	Edge [71] is a FILLET
	Edge [73] is a FILLET
	Edge [75] is a FILLET
	Edge [77] is a FILLET
	Edge [79] is a FILLET
	Edge [81] is a FILLET
	Edge [83] is a FILLET
	Edge [85] is a FILLET
	Edge [87] is a FILLET
	Edge [97] is a FILLET
	Edge [99] is a FILLET
	Edge [101] is a FILLET
	Edge [103] is a FILLET
	Edge [105] is a FILLET
	Edge [107] is a FILLET
	Edge [109] is a FILLET
	Edge [111] is a FILLET

(d)	Regular form face with depression (bottom region)
	Face [0]
	Edge [1] is a FILLET
	Edge [3] is a FILLET
	Edge [5] is a FILLET
	Edge [7] is a FILLET
	Edge [9] is a FILLET
	Edge [11] is a FILLET
	Edge [13] is a FILLET
	Edge [15] is a FILLET
	Edge [17] is a FILLET
	Edge [19] is a FILLET
	Edge [21] is a FILLET
	Edge [23] is a FILLET
	Edge [25] is a FILLET
	Edge [27] is a FILLET
	Edge [29] is a FILLET
	Edge [31] is a FILLET
	Edge [33] is a FILLET
	Edge [35] is a FILLET
	Edge [37] is a FILLET
	Edge [39] is a FILLET
	Edge [41] is a FILLET
	Edge [43] is a FILLET
	Edge [45] is a FILLET
	Edge [47] is a FILLET
	Edge [49] is a FILLET
	Edge [51] is a FILLET
	Edge [53] is a FILLET
	Edge [55] is a FILLET
	Edge [57] is a FILLET
	Edge [59] is a FILLET
	Edge [61] is a FILLET
	Edge [63] is a FILLET
	Edge [65] is a FILLET
	Edge [67] is a FILLET
	Edge [69] is a FILLET
	Edge [71] is a FILLET
	Edge [73] is a FILLET
	Edge [75] is a FILLET
	Edge [77] is a FILLET
	Edge [79] is a FILLET
	Edge [81] is a FILLET
	Edge [83] is a FILLET
	Edge [85] is a FILLET
	Edge [87] is a FILLET
	Edge [97] is a FILLET
	Edge [99] is a FILLET
	Edge [101] is a FILLET
	Edge [103] is a FILLET
	Face [1]
	Edge [1] is a FILLET
	Edge [3] is a FILLET
	Edge [5] is a FILLET
	Edge [7] is a FILLET

Fig. 8 Results of auto-recognized faces and their fillet edges

Based on the number of loops, the algorithm categorizes each segmented face into a face with depression or protrusion and a face without depression or protrusion as shown in Sect. 2.4. Next, the rule based method is applied to identify the fillet features and the auto-generated results show that a regular form top region face without depression or protrusion has no fillet edges (see Fig. 8a) while a regular form top region face with depression having face number [6] has 52 fillet edges which means it has 26 fillet faces (see Fig. 8b). Similarly the regular form bottom region face without depression or protrusion has no fillet edges (see Fig. 8c) while a regular form bottom region face with depression having face number [0] has 48 fillet edges which means it has 24 fillet faces and face number [1] has 4 fillet edges which means it has two fillet faces (see Fig. 8d).

4 Discussion

The algorithm calculated the test part model volume and also the number of faces, edges, and vertices present on model was calculated. The faces were categorized into regular form based on their geometrical shapes and were segmented into top, contour, bottom regions. Based on the number of loops, faces were categorized into FWoD and FWD. A rule based method was applied to automatically recognize fillet edges and their faces of the test art model.

5 Conclusion

The developed algorithm is able to generate accurate data of volume and entities of the test part model. The geometrical shapes of all the faces were successfully recognized and faces were categorized into regular form face and freeform face. The algorithm successfully segmented faces into three regions and then categorized faces into FWoD and FWD. The auto-generated data of entities and tables show that the test part model has 159 faces and out of these 159 faces, face numbers [6], [0] and [1] contain 52, 48, 4 fillet edges, respectively.

Acknowledgements This research is supported by Universiti Sains Malaysia under the Research University Grant (No: 814247). The first author also would like to thank the support of National Overseas Scholarship by Ministry of Tribal Affairs, New-Delhi, India.

References

1. Kataraki, P.S., Abu Mansor, M.S.: Auto-recognition and generation of material removal volume for regular form surface and its volumetric features using volume decomposition method. Int. J. Adv. Manuf. Technol. **90**, 1479–506 (2017)

2. Abu, R., Masine, T.: Attribute based feature recognition for machining features. J. Teknol. **46**, 87–103 (2007)
3. Woo, Y., Sakurai, H.: Recognition of maximal features by volume decomposition. Comput. Aided Des. **34**, 195–207 (2002)
4. Bok, A.Y., Mansor, M.S.A.: Computers & industrial engineering generative regular-freeform surface recognition for generating material removal volume from stock model q. Comput. Ind. Eng. **64**, 162–178 (2013)
5. Zubair, A.F., Mansor, M.S.A.: Cylindrical axis detection and part model orientation for generating sub delta volume using feature based method. ARPN J. Eng. Appl. Sci. **11**, 13415–13419 (2016)
6. Sunil, V.B., Pande, S.S.: Automatic recognition of features from freeform surface CAD models. Comput. Aided Des. **40**, 502–517 (2008)
7. Gupta, R.K., Gurumoorthy, B.: Automatic extraction of free-form surface features (FFSFs). Comput. Aided Des. **44**, 99–112 (2012)
8. Gupta, R.K., Gurumoorthy, B.: Classification, representation, and automatic extraction of deformation features in sheet metal parts. Comput. Aided Des. **45**, 1469–1484 (2013)
9. Babic, B., Nesic, N., Miljkovic, Z.: A review of automated feature recognition with rule-based pattern recognition. Comput. Ind. **59**, 321–337 (2008)
10. Wu, M.C.: Analysis on machined feature recognition techniques based on B-rep. Comput. Aided Des. **28** (1996)
11. Subrahmanyam, S., Michael, W.: An overview of automatic feature recognition techniques for CAPP. Comput. Ind. **26**, 1 (1995)
12. Falcidieno, B., Giannini, F.: Incorporating free-form features in aesthetic and engineering product design: State-of-the-art report. Comput. Ind. **59**, 626–637 (2008)

Experimental Study of Direct Injected Marine Auxiliary Diesel Engine Performance, Emission and Cylinder Pressure Using Biodiesel Fuels Derived from Jatropha Curcas Oil

Ridwan Saputra Nursal, Zakiman Zali, Ismail Zainol and Mohd Nazri Mohd Sabri

Abstract Experimental studies have been carried out to investigate the effects of biodiesel fuels on the performance of a 4-stroke direct injected (DI) marine auxiliary diesel engine, brake specific fuel consumption (BSFC), carbon monoxide (CO), carbon dioxide (CO_2), oxides of nitrogen (NOx) and hydrocarbon (HC) emissions, exhaust gas temperature (EGT), and cylinder pressure. Biodiesel fuels were prepared from Jatropha curcas oil and blended with petroleum diesel to specific designated concentration. All the measured parameters for biodiesel fuels are compared with the base diesel, i.e. petroleum diesel for different engine speeds. The biodiesel resulted in a slightly increased of brake power (BP), brake mean effective pressure (BMEP) and exhaust gas temperature (EGT) proportional to the engine speed under all rated loads while slightly reduced the brake thermal efficiency (BTE) throughout engine speeds as load was exerted to the engine. In the other hand, there is no significant change of BSFC by biodiesel fuels as well as diesel corresponding to the range of engine speed and loads. The present work contributes in using biodiesel fuels as alternative fuel for diesel engines without major changes for engines parts. For comparison between

---

R. S. Nursal (✉) · Z. Zali
Department of Marine Engineering, Centre of Technology
in Marine Engineering, Politeknik Ungku Omar, 31400 Ipoh, Perak, Malaysia
e-mail: ridwansaputra@puo.edu.my; r.s.nursal@gmail.com

Z. Zali
e-mail: zakizali@puo.edu.my

I. Zainol
Marine Engineering Technology Section, Universiti Kuala Lumpur,
Malaysian Institute of Marine Engineering Technology, Lumut, Perak, Malaysia
e-mail: ismailz@unikl.edu.my

M. N. M. Sabri
Department of Mechanical Engineering, Centre of Automotive
Engineering Technology, Politeknik Sultan Azlan Shah,
35950 Behrang, Perak, Malaysia
e-mail: mohd.nazri@psas.edu.my

© Springer International Publishing AG 2018
A. Öchsner (ed.), *Engineering Applications for New Materials and Technologies*,
Advanced Structured Materials 85, https://doi.org/10.1007/978-3-319-72697-7_9

biodiesel and diesel fuels, the viscosity is not the main parameter affecting on engine performance and emissions.

Keywords Biodiesel · Marine diesel engine · Performance · Emissions Jatropha curcas

1 Introduction

Scientists around the world have explored several alternative energy resources, which have the potential to quench the ever-increasing energy thirst of today's population. The production of crude oil globally is estimated to have a decreasing trend after the years. However, there will be an increase in the vehicle population every year which will demand an increase in crude oil imports. With this scenario the need for an alternate fuel arises to maintain the economy of countries. Biodiesel has received significant attention both as a possible renewable alternative fuel and as an additive to the existing petroleum-based fuels [1–4]. Several studies revealed the power output and fuel consumption of engines are very close when compared with that of pure diesel. Biodiesels can directly be used in diesel engines, as they have calorific values very close to diesel fuel but high viscosity and low volatility of vegetable oil makes it difficult to atomize the fuel [1].

The local implementation of the B5 program (5% biodiesel blended with 95% diesel) by the government of Malaysia started on February 3, 2009. Yet the government has implemented the B5 mandate to be fulfilled immediately, with the product available throughout Malaysia by the end of 2013. The government of Malaysia was then introducing B7 at the end of 2014 and planned to implement the use of B10 in transportation in 2017 [5, 6]. However, the biodiesel used in the blending is derived from palm oil bases.

Despite of that, the biodiesel produced from Jatropha is still in its nascent state in Malaysia even though considerable interest has been shown by the government and private sectors. Malaysia has adequate land area, and its climate is suitable for Jatropha cultivation. A few initial measures have already been taken to explore Jatropha cultivation in Malaysia through partnerships with government agencies and private sectors. Several private companies are planting Jatropha seeds in land areas of up to 1000 ha [7]. Jatropha oil is obtained from the seeds of Jatropha Curcas. Jatropha is non edible and one of the advantages of Jatropha is that it can be cultivated on waste land and thus does not compete with food crops. Jatropha crops do not require much fertilizer and water, yet lead to the reductions of plantation cost which render the price of biodiesel produced from Jatropha extremely competitive with diesel from fossil fuels [8]. It is also reported that Jatropha biodiesel blend with petroleum diesel could provide optimum performance without any engine modification nor preheating. However, some studies found that Jatropha to contain as much as 34 wt% of saturated fatty acids. Hence, it is expected that Jatropha based biodiesel may exhibit poor operability at low temperatures [9].

The use of biodiesel fuels as an alternative fuel for petroleum diesel fuel is reported to benefit for the reduction of CO_2 emission, because biodiesel is produced from renewable biomass resources. Biodiesel is usually blended with conventional diesel in various proportions. Many researches and studies has been conducted to prove it. A performance and emission characteristics study of a diesel engine fuelled with palm, Jatropha, and moringa oil methyl ester has been carried out by Rahman et al. [10] and remarked the reductions of CO and HC emissions to a large extent by biodiesel blended fuels. Abedin et al. [11] investigated the performance, emissions and heat losses of palm and Jatropha biodiesel blends in a diesel engine under full load condition and reported that there are significant reductions of CO and HC emissions compared to diesel. Khalid et al. [12] reported that blending the petroleum diesel with biodiesel has advantages in reductions of CO_2 and HC emissions relative to diesel. Kannan et al. [13] investigated the influence of ethanol blend addition in diesel and Jatropha methyl ester on compression ignition engine and notified the increase of engine efficiency. Performance, emission and combustion characteristics of Jatropha oil blends have been experimented by Agarwal and Dhar [14] in a direct injection CI engine. Manieniyan and Sivaprakasam [15] have bench tested the diesel engine using biodiesel (methyl ester of Jatropha oil) in various blending ratios for various injection timing and injection pressures. The outcome was a significant reduction of NOx emissions by B20 and they discovered that the advanced injection timing gives a better performance for a higher blend ratio. The experimental results reported by Jaroonjitsathian et al. [16] on heavy-duty common-rail diesel engine when evaluating 5–20% biodiesel blends have shown that NOx emissions did not increase with a higher content of biodiesel associated with the smoke intensity that was less when using higher biodiesel grades. Tan et al. [17] have studied the particle number and size distribution from a diesel engine with Jatropha biodiesel fuel. Kumar et al. [18] investigated Jatropha oil methanol in a dual fuel engine.

This study is to determine the level to which blending can be done with diesel without scarifying much in the performance and emission characteristics of a marine auxiliary diesel engine when fuelled with these blend fuels without any engine modifications.

2 Experimental Procedure

2.1 Production Process of Jatropha Curcas Biodiesel Fuel

The pure biodiesel fuels were produced from Jatropha seeds. The production of the biodiesel was performed in the biodiesel pilot plant located in the Universiti Tun Hussein Onn Malaysia. The biodiesel was produced by the chemical transformation of feedstocks into the corresponding long-chain fatty acid alkyl esters known as transesterification. The transesterification of triglycerides in oil with a lower alcohol in the presence of an acidic or a basic catalyst to reduce the viscosity of vegetable oil. The basic transesterification reaction process is shown in Fig. 1.

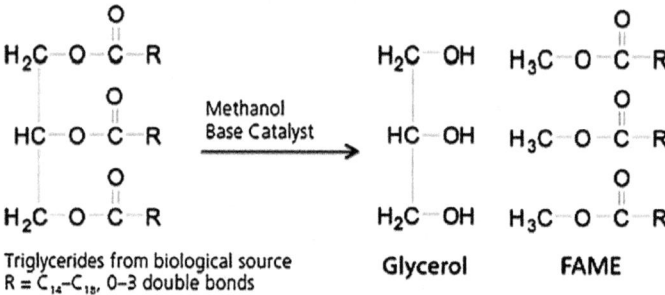

Fig. 1 Reaction scheme for the production of biodiesel

The production of Jatropha based biodiesel began with the first module for crude oil pre-treatment which converts crude Jatropha oil into bleach oil, the second one for refining and separating which converts and processes the bleached oil into the end product of biodiesel that meets standard specifications. In the pre-treatment module, the process began with degumming when the crude Jatropha oil is treated with phosphoric acid to remove natural gums, and followed by bleaching with activated earth under vacuum to remove the colouring matters as well as to adsorb any metal ions content. Other impurities such as dirt and solid particles were removed using a bleaching earth filter. This is to prepare the raw materials quality to be fit for the second module. The product of this module is the refined bleach oil. The subsequent process is esterification and transesterification of the bleached oil with methanol under the presence of acid and base catalyst, respectively. The esterification process converts free fatty acids containing in the bleached oil to methyl esters under mild conditions. These reaction products are then transferred into the transesterfication reactor with methanol to form fatty acid methyl ester (FAME) or biodiesel. The methyl esters are then washed, filtered, and vacuum dried before sending to the storage [19, 20].

2.2 Blending Process of Jatropha Curcas Biodiesel with Petroleum Diesel Fuel

Biodiesel fuels that have been produced in the pilot plant then underwent the blending process with petroleum diesel [21]. The Jatropha curcas biodiesel fuel blends were prepared in the UTHM's automotive laboratory. The purified biodiesel oil from Jatropha curcas feedstocks was blended with petroleum diesel (DF) in various concentrations of 5, 10 and 15% by volume which does mean that 5% biodiesel was mixed and blended with 95% diesel, 10% biodiesel was mixed and blended with 90 and 15% biodiesel was mixed and blended with 85% diesel by volume through a specific procedure. These blended fuels are designated as B5, B10 and B15 for 5, 10 and 15% vol Jatropha curcas biodiesel blends fuel,

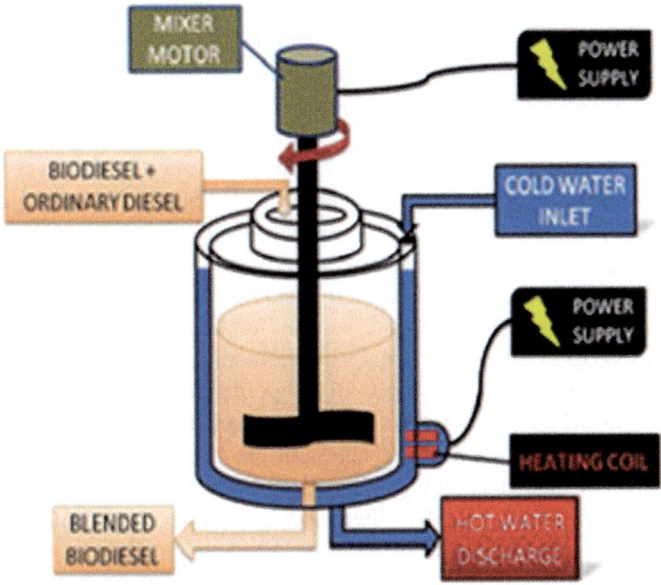

Fig. 2 Illustration of equipment apparatus setup for blending process

respectively. The purified Jatropha crude oil methyl ester and diesel from separate tank were released to the mixing chamber by controlling the mass flow of fluids to attain the preferred ratio. The fluids were premixed before entering the main mixing chamber. During the blending process, the scale blending machine operated under a temperature of around 70 °C and the mixture was stirred at 60 °C for one hour. The mixture was stirred continuously with a rotating blender and the blade speed was set to maintain at 270 rpm. The equipment apparatus setup for the blending process is shown in Fig. 2.

On the other side, the petroleum diesel fuel (DF) with grade IV i.e. Euro 4 diesel rated was tested along with the biodiesel fuel blends and nominated as a standard reference. This petroleum diesel fuel was produced and obtained from Petronas Melaka Refinery [22].

2.3 Measurement of Fuel Properties

Next, the properties of biodiesel blends fuel have been analysed and recorded. In the study, the properties of biodiesel fuels were analysed precisely to comply with the standard specification provided in the Malaysia Standard since they must be comparable to standard diesel properties because it will directly affect the performance and emissions of diesel engines [23, 24]. The properties of biodiesel are depending very much on the nature of its raw material or feedstocks as well as the technology and process utilised for its production. The properties of DF and all types of biodiesel fuel blends have been measured at ambient temperature to be conform with the MS Standard. Table 1 summarizes the fuel properties that have been measured between the DF and biodiesel at particular blending ratios.

The blended biodiesel fuel properties were tested to ensure that they meet the MS Standard. The kinematic viscosity of biodiesel was measured by a Hydromation Viscolite 700 model VL700-T15 viscometer. The kinematic viscosity is denoted as the resistance of a liquid to flow. It is referring to the time taken by a volume of biodiesel sample (liquid form) to flow under gravity through a calibrated glass capillary viscometer. Based on Malaysian Standard the range of kinematic viscosity is 1.9–6.0 mm^2/s. The density of fuel was measured by a Metter Toledo Diamond Scale model JB703-C/AF. The density is the connexion between the mass and volume of fuel and expressed in units of kilograms per metre cubic (kg/m^3). The density of biodiesel adopts in Malaysian Standard is in the range of 860–900 g/cm^3. The density of biodiesel gives an indication of the delay between the injection and combustion of the fuel (ignition quality) and the energy per unit mass (specific energy). The flash points of biodiesel were measured by using a flash point analysis instrument namely the Pensky-Martens model PMA 4. The flash point is the lowest temperature at which fuel emits enough vapour to ignite and is set to a minimum temperature range of 43 °C for petro-diesel and 101 °C for biodiesel as adopted in the Malaysian Standard. The water content and acid value in the biodiesel samples was measured by a Volumetric KF Titrator model v20 and underwent titration process. The term of acid value is expressed as the amount in (milligram) of potassium hydroxide (KOH) required to neutralize 1 g of the biodiesel and is set to a maximum value of 0.8 mg KOH/g as in the Malaysian Standard. A higher acid content can cause severe corrosion in the fuel supply system and internal combustion engine. During acid value measurement, the biodiesel sample is titrated with alcoholic KOH using phenolphthalein as indicator. The standards of water content for biodiesel in the Malaysia Standard is set to a maximum 500 mg/kg or limit 0.05% maximum by volume.

Table 1 Fuel properties

Fuel properties/testing method	Fuel type			
	DF	B5	B10	B15
Density (kg/m^3) ASTM D1298 (88 max)	833.74	846.22	847.87	849.26
Kinematic viscosity (mm^2/s) ASTM D445 (1.9–6.0)	3.01	4.01	4.6	5.3
Flash point (°C) ASTM D93 (43 min)	83	118	118	120
Water content (mg/kg) ASTM E203 (500 max)	79.6	90.1	80.4	80.2
Acid value (mgKOH/g) ASTM D664 (0.8 max)	0.423	0.732	1.389	1.786

2.4 Experiment Apparatus Setup

A marine auxiliary diesel engine also known as marine auxiliary machinery is a marine engine for usage, which is not used to propel a water-borne vessel or ship prime mover. The direct injected (DI) marine auxiliary diesel engine is generally installed on-board medium sized vessels or ships. Whilst the common marine engines used in common are 2-stroke diesel engines, steam turbines and gas turbines. The marine auxiliary diesel engine as well as marine machinery must meet higher standards of reliability and greater demands for weight and volume reduction and access for maintenance. The marine machinery also must be capable of withstanding the marine environment, which tends toward extreme ambient conditions, high humidity, sea-water corrosion, vibration, sea motions, shock, variable demand and fluctuating support services.

Facilities and equipment in this study were set up in the Multifuel Engine Laboratory in the Department of Marine Engineering, Ungku Omar Polytechnic Malaysia. Fuels were tested using a marine auxiliary diesel engine, model TF120-ML with 0.638 l capacity, single cylinder, horizontal type, 4-cycle engine with air-cooled system made by Yanmar Motors. The engine was designed with compact dimensions and in all applications with a power requirement up to 8.8 kW at 1 h due to its low weight. Table 2 summarizes the specification of the test engine. The main engine performance parameters or output parameters measured are temperature, torque and brake power. These parameters were measured by an Eddy current dynamometer system. The Eddy current dynamometer which works as a load controller to the engine or engine brake which is connected to the engine through a coupling shaft.

In conjunction, a specific computer set was integrated and synchronised with the engine particularly the transducer at the engine dynamometer. The function of this transducer is to analyse and measure data from the dynopack and send to the

Table 2 Test engine specification

Item	Specification
Engine model	YANMAR TF120-ML
Type	1 cylinder, horizontal, air and water-cooled, 4 stroke
Combustion system	Direct injection
Aspiration	Natural aspiration
Cylinder bore × stroke	92 mm × 96 mm
Displacement	0.638 l
Max. output power	12 Ps (8.8 kW) at 2400 rpm
Starting system	Manual and auto starting
Lubrication system	Complete enclosed forced lubricating system
Electrical system	Alternator, 12 V–45 W
Fuel tank capacity	11 l
Engine oil capacity	2.8 l

encoder/data acquisition system (DAQ) before being converted/encoded for computer usage. The dashed line shows the link connection from the main particular sensor attached to the engine dynopack to DAQ and connected to a Supervisory Control and Data Acquisition (SCADA) unit, which is computer monitored. The test engine and other testing apparatus set up have been arranged as shown in Fig. 3. The operating parameters such as loads exerted during engine operation and

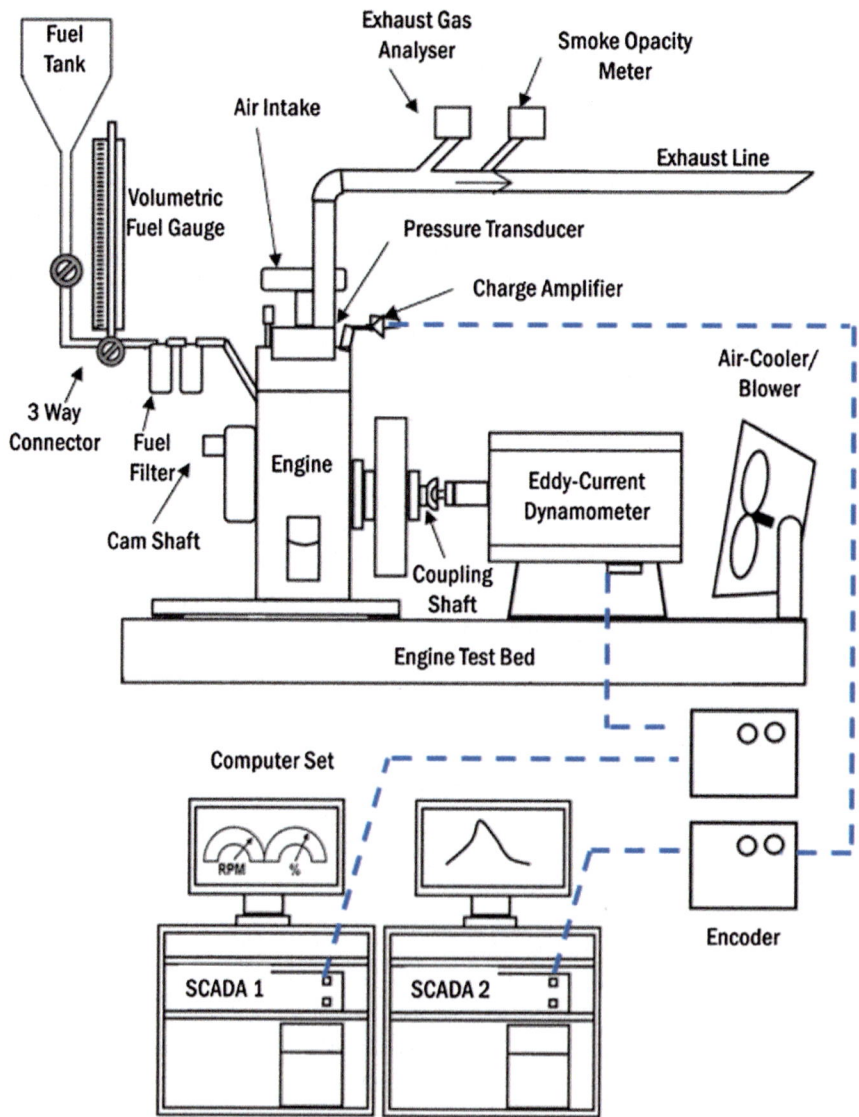

Fig. 3 Schematic of experimental setup

Table 3 Exhaust gases analysis specifications

Parameter	Resolution	Accuracy	Range
CO_2 (Carbon dioxide)	0.01%	±4% rel.	0–10%
CO (Carbon monoxide)	1 ppm	±4% rel.	0–3000 ppm
NOx (Oxide of nitrogen)	1 ppm	±4% rel.	0–5000 ppm
HC (Hydrocarbon)	1 ppm	±2%	0–30000 ppm

engine speeds will be controlled manually, set by user while the overall operation of the engine is automatically monitored by the SCADA unit through the computer. Additionally, the engine is also installed with a combustion analyser probes or pressure transducer which is used to analyse the combustion characteristics as well as ignition behaviour. The DEWESOFT v7.11 software is used to monitor the input parameters and measures the performance of the engine while the DYNO-MAX 2010 software is used to analyse the combustion characteristics.

Separately, the exhaust gas emissions such as carbon monoxide (CO), carbon dioxide (CO_2), hydrocarbon (HC) and oxides of nitrogen (NOx) is analysed and measured using an autocheck 5 channel gas analyser model IMR 2800-A as specified in Table 3. The sampling probes of the gas analyser were mounted at the centre of the exhaust pipe during measurement of the gas emissions.

For this research, the fuels tested of DF, B5, B10 and B15 are used for running in a manner of similar engine operation conditions which means the total 4 different fuels in numbers has been tested. The engine was operated under dynamometer loads at 0, 50 and 90% load conditions. Since the fuel test or engine operating hours taking a longer period of times, the maximum load i.e. 100% was not taken into account to prevent the engine and its component from fatigue and the risk of engine failure as well as safety precaution. Another parameter in the study is the various running speeds of the engine which are simulated at 800 rpm (minimum speed), 1200, 1600 and 2000 rpm (maximum speed).

The parameters measured with respect to engine performance consist of engine torque, brake power (BP), brake specific fuel consumption (BSFC), brake mean effective pressure (BMEP), brake thermal efficiency (BTE), exhaust gas temperature (EGT) and fuel mass flow rate ($\dot{m}_f$). In conjunction, the measurements of exhaust gas emissions comprised of carbon dioxide (CO_2), carbon monoxide (CO), oxides of nitrogen (NOx) and hydrocarbon (HC) using the exhaust gas analyser.

3 Results and Discussions

All types of biodiesel blends that comprise of Jatropha curcas biodiesel fuel oil, in particular B5, B10 and B15, were analysed concurrently by comparing to the standard diesel fuel (DF) results of engine performance, exhaust gas emissions and cylinder. The discussion encompasses the analysis of performance and emission characteristics through engine speeds of 800–2000 rpm at an interval of 400 rpm under 0% (no load), 50% (intermediate load) and 90% (maximum) dynamometer loads condition. The engine performance factors that have been measured consist of the torque (τ), brake power (P_b), brake specific fuel consumption (BSFC), brake mean effective pressure (BMEP), brake thermal efficiency (BTE) and exhaust gas temperature (EGT).

Figures 4, 5 and 6 show the changes in torque, brake power, brake specific fuel consumption (BSFC), brake mean effective pressure (BMEP), brake thermal efficiency (BTE) and exhaust gas temperature (EGT) from marine diesel engine fuelled by B5, B10 and B15 biodiesel blends and DF corresponding to the range of engine speed under 0, 50 and 90% load conditions. According to these figures, it is clearly seen that the torque produced by Jatropha biodiesel blends fuel as well as DF slightly increased as the engine speed increased. However, by referring to Fig. 4, B5 and B10 fuels at 0% load condition indicates the slight decrement of torque corresponding to the increment of engine speeds. Aside, BSFC by all fuels shows about comparable measurement relative to DF except for B15 which shows a bit higher BSFC at the lower speed but approaching the BSFC by DF as speed increased from 1600 to 2000 rpm.

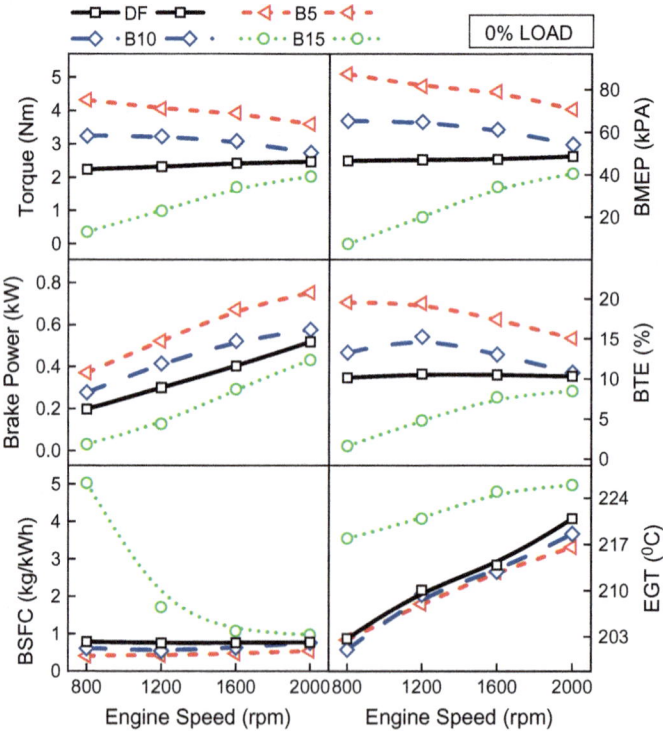

Fig. 4 Effects of engine speed on engine performance by Jatropha BDF blends without load condition

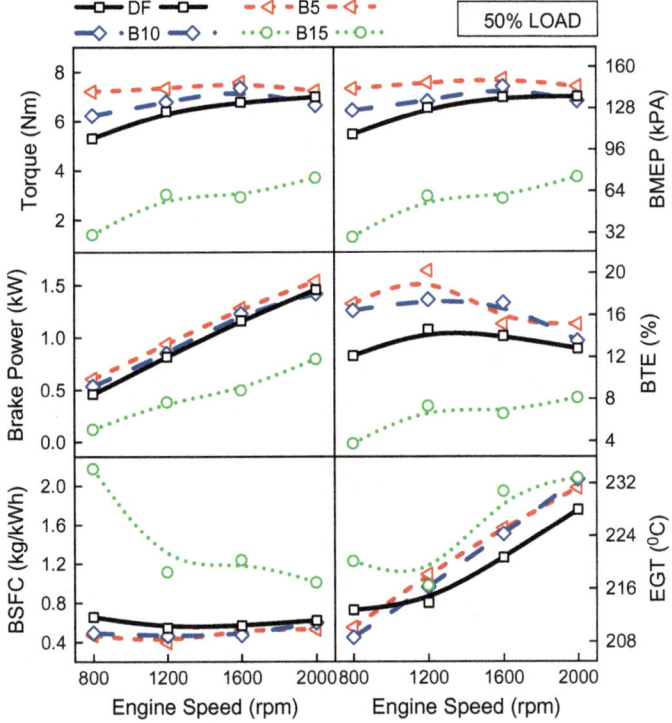

Fig. 5 Effects of engine speed on engine performance by Jatropha BDF blends under 50% load condition

Based on Fig. 5 under intermediate load significance or 50% dynamometer brake loads, it is indicating that the torque shows similar trends of graphs where the measurement for BDF are almost similar to that of DF especially at higher engine speed of 2000 rpm which explained those fuels exhibits the same combustion process except for B15 which demonstrates lower torque compared to DF.

Based on Figs. 4, 5, and 6, it was clearly illustrated that the brake power and EGT are increased proportional to the range of engine speed under 0, 50 and 90% load conditions by Jatropha BDF blends and DF as well. These results proved that the use of B5 and B10 in diesel engine is giving some advantage in terms of power production. With regard to BSFC, it seems that there are small changes of BSFC for B5 and B10 which is plotted lower than DF with respect to the increasing engine speed. Nevertheless, B15 shows a different form of BSFC where it was slightly higher than that of DF especially under a lower engine speed of 800 rpm and then declining contrariwise proportional to the engine speed under all load conditions. With reference to Figs. 5 and 6, the BMEP of Jatropha BDF blends increased over the increment of engine speed with B5 demonstrates the greatest BMEP while B15 for the lowest relative to DF under 50% and 90% loads condition. From Figs. 5 and 6, there are reductions in BTE over engine speed when load is applied due to lower

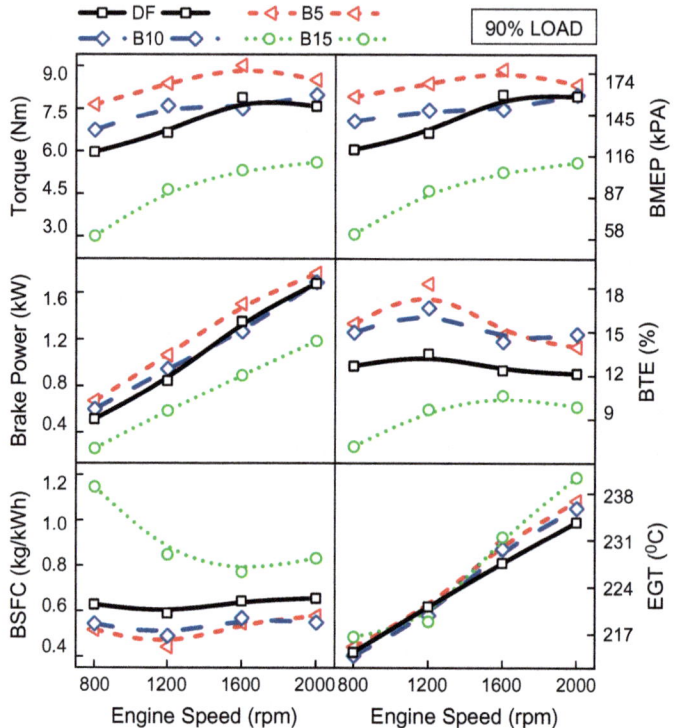

Fig. 6 Effects of engine speed on engine performance by Jatropha BDF blends under 90% load condition

average burning temperature as well as lower heat losses. Based on results demonstrated by Figs. 4, 5 and 6 an early assumption that can be made is that B5 shows the best and comparative performance to that of standard DF when tested in marine auxiliary diesel engine while B15 shows the poorest performance among the tested fuels.

Next, the discussion is now focused on the emissions and its scale produces by the BDF as well as DF fuels. Figures 7, 8 and 9 show the changes in carbon dioxide (CO_2), carbon monoxide (CO), oxides of nitrogen (NOx) and hydrocarbon (HC) emissions from marine auxiliary diesel engine fuelled by biodiesel blends B5, B10 and B15 as well as standard diesel fuel (DF) with respect to the increasing of engine speed from 800 to 2000 rpm under 0, 50 and 90% load conditions. The figures clearly show that DF represents the lowest CO_2 emissions under all load conditions as compared to that of Jatropha BDF blends fuel.

At all load conditions, it is observed that B5 contributes the highest CO_2 emissions around 20% with the similar rates throughout engine speed followed by B10 with inclined trends over the range of speeds. From Figs. 7 and 8 at 50% load and below, it is illustrated that the formation of CO_2 gases by B15 gradually

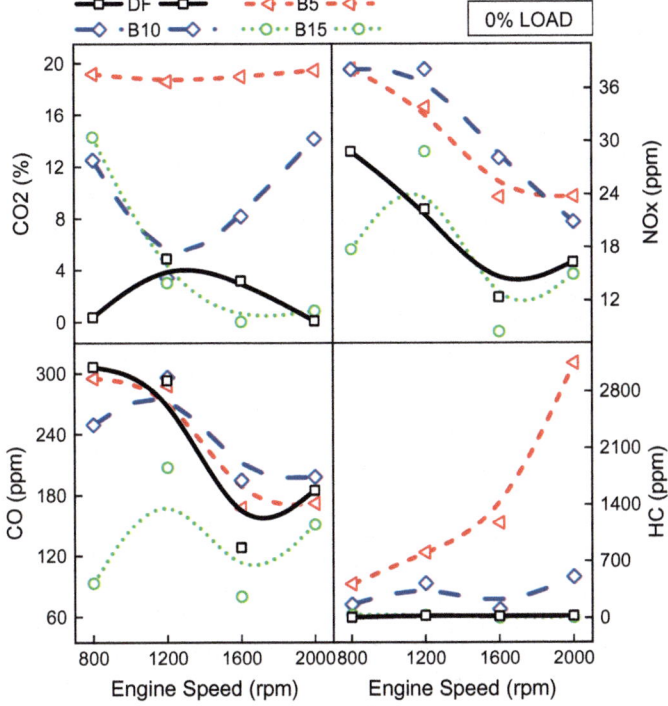

Fig. 7 Effects of engine speed on exhaust gas emissions by Jatropha BDF blends without load condition

decreased when the engine speed increased, associated with a lower CO gas formation over the speed range relative to DF at all rated loads signified a better emission reduction of CO_2 and CO by that of the B15 fuel.

Apart from that, burning of B10 fuel in diesel engine is resulting in a gradually increment of CO_2 emissions corresponding to the increased engine speed from 800 to 2000 rpm at 50 and 90% load conditions as depicted in Figs. 8 and 9. In average, it seems that most of all Jatropha BDF fuel blends exhibit a reduction behaviour of CO gas formation inversely proportional to the increasing engine speed where the parameter values of CO emission during 2000 rpm are almost similar to that of DF. It is also observed in Fig. 7 that the formation of NOx gases by Jatropha BDF blends as well as DF decreased with respect to the increasing engine speed at no loads.

Meanwhile, in Figs. 8 and 9 under 50 and 90% load conditions, higher emissions of NOx are remarked where the main reason for this is considered to be due to the increasing gas temperature in the burning-zone [25]. Again at these rated loads, NOx emissions are then reduced slightly as the engine operated at a higher speed of 2000 rpm. Although, the emissions of the NOx gases are still higher than emitted by DF fuel around 20–30 ppm. On the other hand, HC gas emissions at all rated

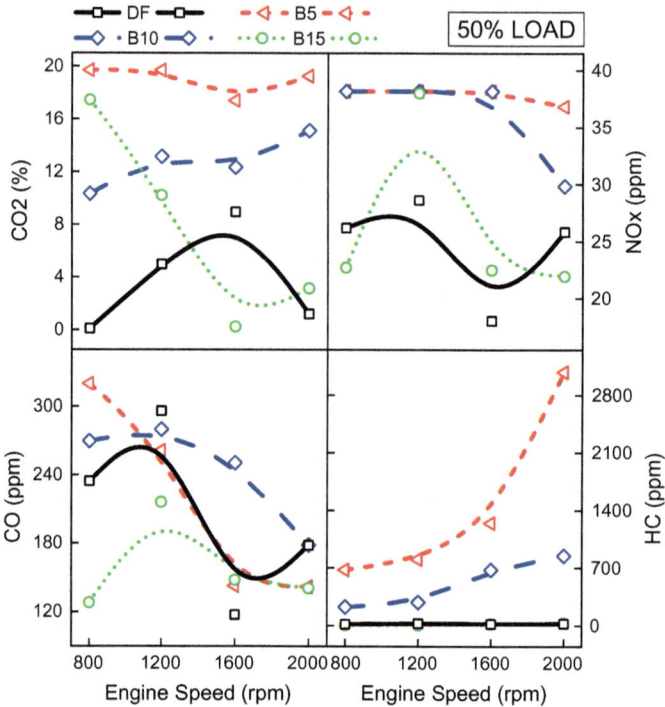

Fig. 8 Effects of engine speed on exhaust gas emissions by Jatropha BDF blends under 50% load condition

loads by B5 gradually increased as the engine speeds increased with a radical increment remarkably between a speed of 600–2000 rpm. While B10 significance the steady increment of HC formation is proportional to the increment of engine speed. Nevertheless, DF and B15 blends fuel contributes to the lowest HC emissions below 50 ppm which are seen from Figs. 7, 8 and 9, associated with uniform emission values that were measured throughout the range of speed. Figures 10, 11, 12, 13, 14 and 15 illustrate the measurements of cylinder pressure of the marine auxiliary diesel engine with respect to the changes of crank angle, which has been recorded through a specific sensor namely the pressure transducer that was mounted onto the top of cylinder block. The discussions give attention to the characteristic of maximum or peak cylinder pressure that occurred by the BDF and DF fuels tested under specific load and engine speed. Figures 10 and 11 indicate the cylinder pressure of the combustion of all tested fuels under 0% rated load and engine speed of 800 and 1200 rpm, respectively.

The peak pressure depends upon the burning rate after the start of ignition during the uncontrolled combustion stage. The major aspect that determines the uncontrolled or diffusion combustion phase is the mixture preparation rate and atomization which is influenced by the viscosity and volatility of the fuel [13].

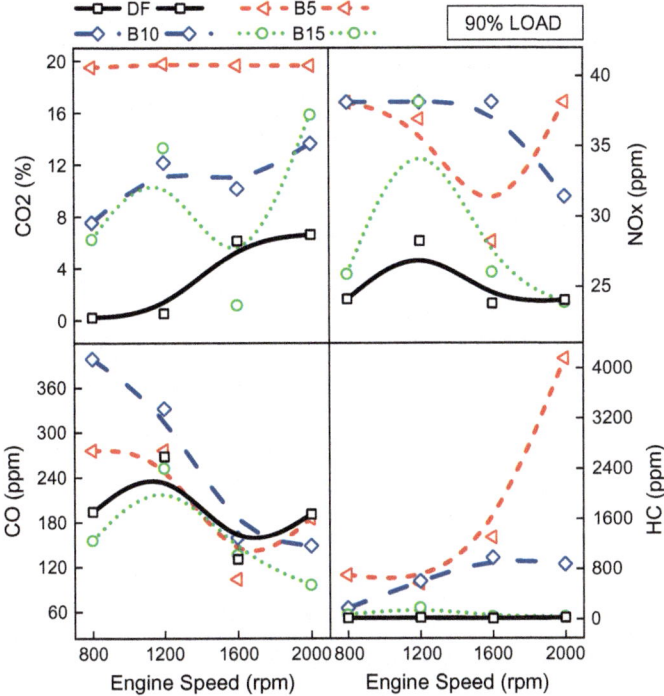

Fig. 9 Effects of engine speed on exhaust gas emissions by Jatropha BDF blends under 90% load condition

Based on Fig. 10, it is observed that the highest peak pressure among the fuels tested during 800 rpm of engine speed at 0% rated load is 50.29 bar at 1°CA before top dead centre (bTDC) which lead by B5 and followed by DF of 50.02 bar at the same crank angle position. As the engine speed increased to 1200 rpm under similar load conditions, once again the combustion of B5 fuel in diesel engine represents the highest cylinder peak pressure of 54.05 bar at TDC i.e. 0°CA followed by 53.34 bar at 1°CA bTDC as depicted in Fig. 11. Peak pressures for all BDF blends are higher than DF due to high oxygen content in the biodiesel molecules leading to an increased rate of combustion, peak temperature and pressure. Aside, from Figs. 10 and 11 it is remarked that B15 demonstrates early combustion and yet exhibits lower cylinder peak pressure from shorter ignition delay, which may be directed to the suppression of NOx formation [26].

On the other hand, Figs. 12 and 13 show the cylinder pressure of BDF and DF combust under 50% load condition through 1200 and 1600 rpm speed, respectively. From these figures, it is remarked that the highest cylinder peak pressure is 58.66 bar at 3°CA aTDC leads by B5 while DF exhibits the lowest cylinder peak

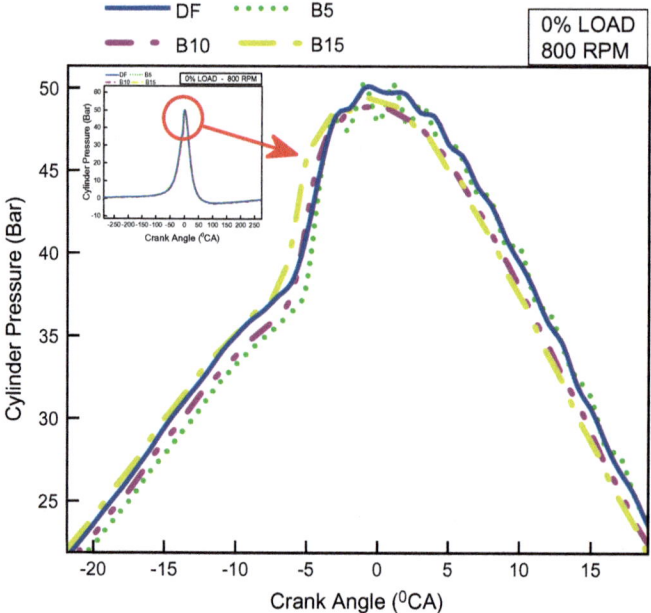

Fig. 10 Cylinder pressure of Jatropha BDF combustion without load through 800 rpm engine speed

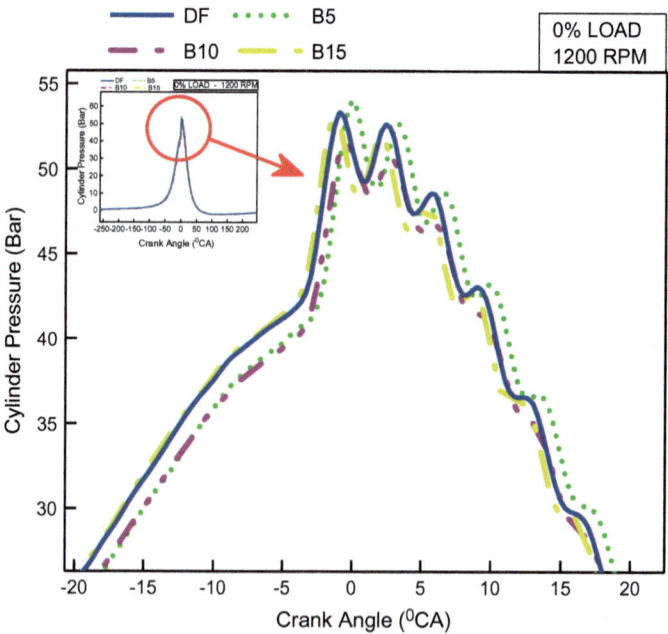

Fig. 11 Cylinder pressure of Jatropha BDF combustion without load through 1200 rpm engine speed

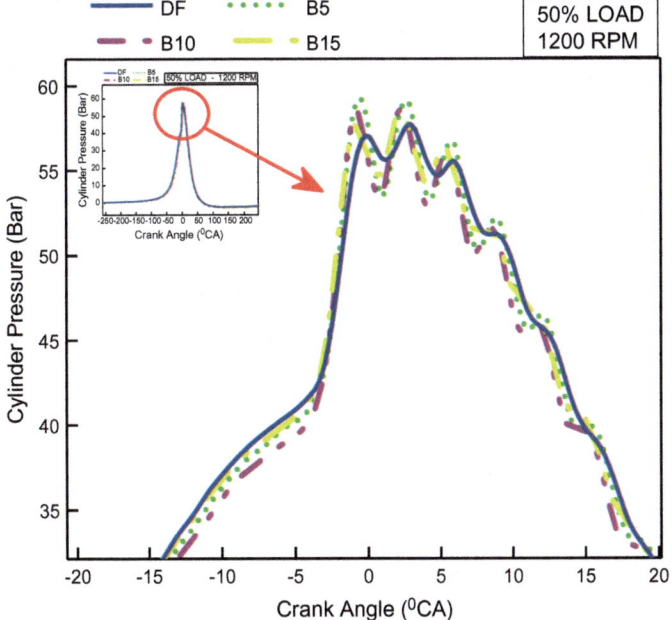

Fig. 12 Cylinder pressure of Jatropha BDF combustion under 50% load through 1200 rpm engine speed

pressure, 57.72 bar at the same crank angle position during 1200 rpm engine speed. When the speed was increased to 1600 rpm, the maximum pressure of 64.41 bar at 1°CA aTDC represents by B15 while the lowest peak pressure is shown by B5 at 60.94 bar at a similar crank angle. Moreover, the combustion of all tested fuels shows the rise of overall peak pressure as the engine speed increased and increased more when loads were applied. Aside, these graph also shows that fluctuations aTDC of cylinder pressure is greater as the engine was operated above 800 rpm.

Despite from that, almost a similar characteristic of cylinder peak pressure to the previous graph was observed by B5, B10, B15 and DF as the engine speed and load were increased as depicted in Figs. 14 and 15. Under 90% dynamometer load exerted and speed of 1600 rpm, the highest cylinder peak pressure was led by B15 fuel of 63.68 bar followed by B10 of 63 bar both are at 1°CA aTDC, while the maximum cylinder pressure of B5 was 62.09 bar at 2°CA aTDC and the lowest was DF of 61.82 bar at 2°CA aTDC as per Fig. 14.

In conjunction, more rise of cylinder peak pressure was recorded when the engine run at a maximum engine speed of 2000 rpm under maximum rated loads as per Fig. 15. From this figure, the B5, B10 and B15 exhibit 65.63 bar at 3°CA aTDC, 66.46 bar at 2°CA aTDC and 69.8 at 2°CA aTDC of cylinder peak pressure, respectively. Apart from that, it is remarked that DF reached 67.27 bar at 4°CA aTDC as shown in Fig. 15.

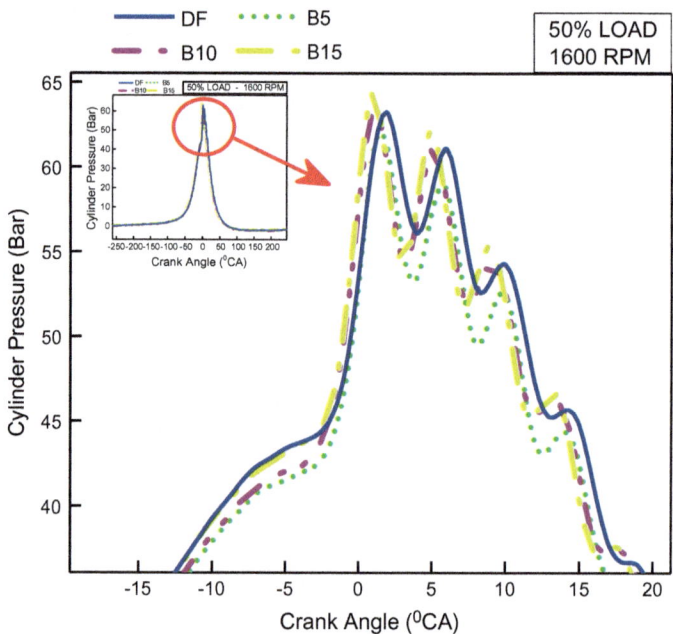

Fig. 13 Cylinder pressure of Jatropha BDF combustion under 50% load through 1500 rpm engine speed

From the graphs of cylinder pressure, the facts that have been acknowledged are that the peak pressure increases with the increase of engine speed from 800 to 2000 rpm. The cylinder peak pressure also increases with the increase of dynamometer loads. The cylinder pressure is mainly dependent on the fraction of fuel burning during the premixed phase and ability of the fuel to mix well with air. The premixed phase is generated by the ignition delay period and by the mixture preparation during the delay period. The peak pressure depends on the rate of combustion in the initial stage, which in turn is influenced by the amount of fuel burning in the premixed phase. The position of the maximum pressure for DF for different speeds are more delayed aTDC than for all BDF due to the lowest cetane number and the longer ignition delay period for diesel fuel. The high viscosity and low volatility of BDF led to poor atomization and mixing with air. Thus, a low burning rate of BDF during the ignition delay period was the reason for the decreasing peak pressure with the increase of speed.

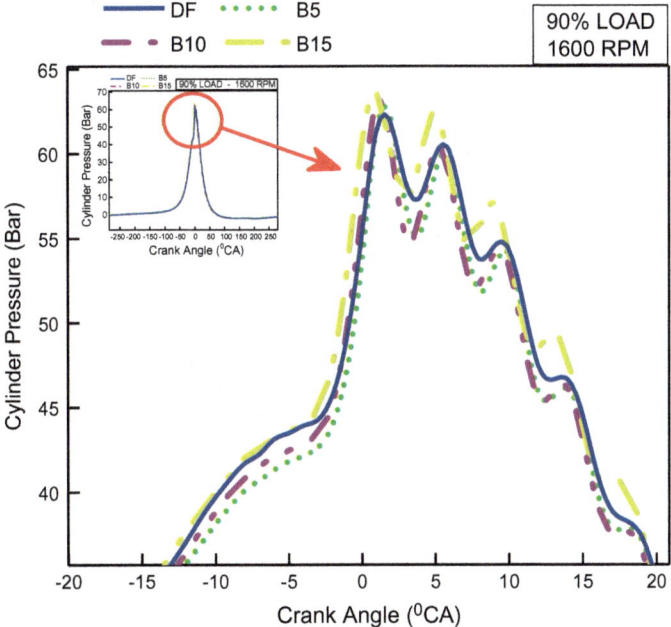

Fig. 14 Cylinder pressure of Jatropha BDF combustion under 90% load through 1600 rpm engine speed

4 Conclusion

The experimental study has been completed using biodiesel fuel (BDF) produced from Jatropha curcas feedstocks that has been blended into 5% (B5), 10% (B10) and 15% (B15) blending concentration along with neat diesel (DF) as standard reference fuel in order to investigate the engine performance, exhaust emissions and cylinder pressure characteristic of direct injected (DI) marine auxiliary diesel engine under 0, 50 and 90% dynamometer loads throughout 800, 1200, 1600 and 2000 rpm engine speeds. The conclusion found from the experimental study were focussed on the characteristics of engine performance, exhaust emissions and cylinder pressure and are drawn as follows:

(1) B5 promotes the greatest results of engine performance among the biofuels tested where significant enhancement on overall engine performance including rises of torque, brake power, BMEP, BTE, lower BSFC and small differences of EGT as compared to that of DF. In conjunction, B5 promotes comparable CO emissions but for rest emissions signified rises of gas emissions relative to that of DF.

(2) B10 promotes a fair and better result on the overall engine performance with slightly lower BSFC under all rated loads condition compared to that of DF. In

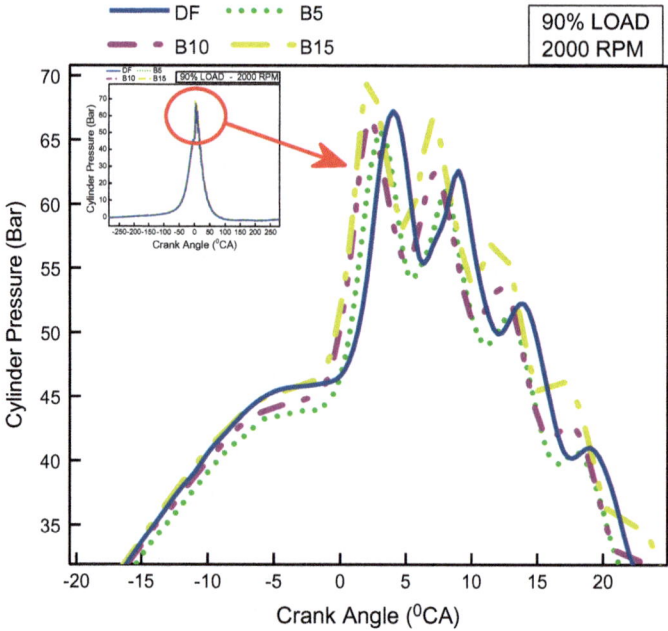

Fig. 15 Cylinder pressure of Jatropha BDF combustion under 90% load through 2000 rpm engine speed

addition, B10 promotes about reasonable CO gas emissions while the rest exhaust emissions were slightly higher than that to DF but at the emissions level below that produced by B5.
(3) B15 promotes a significant reduction of overall engine performance compared to that of DF when operated in a diesel engine. However, there are significant reductions of CO gas emissions, equivalent HC emissions and about analogous CO_2 and NOx emissions promoted by B15 compared to that of petroleum diesel (DF).
(4) The combustion of most BDF experienced higher cylinder peak pressure than DF due to high oxygen content in the BDF molecules leading to an increased rate of combustion, peak temperature and pressure.
(5) The rises of cylinder peak pressure of all fuels tested were proportional to the increases of engine speed as well as the increase of loads.

Acknowledgements This research was supported by Centre of Technology in Marine Engineering (CTME), Ungku Omar Polytechnic through research funded by Department of Polytechnic Education, Ministry of Higher Education Malaysia. The authors also want to oblige Department of Marine Engineering, Ungku Omar Polytechnic and Centre for Energy and Industrial Environment Studies, Universiti Tun Hussein Onn Malaysia for providing facilities, resources and expertise to accomplish the research works.

References

1. Rao, P.V.: Int. Res. J. Eng. Technol. **02**, 643–647 (2015)
2. Mistri, G.K., Aggarwal, S.K., Longman, D., Agarwal, A.K.: J. Energy Resour. Technol. **138**, 1–13 (2016)
3. Nursal, R.S., Zali, Z., Jalil, S., Khalid, A., Hadi, S.A.: Experimental study of the bio-additives effects in biodiesel fuel on performance, emissions and combustions characteristics of diesel engine. ARPN J. Eng. Appl. Sci. **12**(6), 1997–2005 (2017)
4. Kumar, N., Pali, H.S.: Arab. J. Sci. Eng. **16** (2016) (Springer)
5. Oguma, M., Lee, Y.J., Goto, S.: Int. J. Automotive Technol. **13**, 33–41 (2011)
6. Wahab, A.G.: Global Agri. Inf. Net. Reports, (2014)
7. Fattah, I.M.R., Kalam, M.A., Masjuki, H.H., Wakil, M.A.: Biodiesel production, characterization, engine performance, and emission characteristics of Malaysian Alexandrian laurel oil. RSC Adv. **4**, 17787–17796 (2014)
8. Suryanarayanan, S., Janakiraman, V.M., Rao, G.L.N.: SAE Technical Paper, **724** (2008)
9. Lim, S., Teong, L.K.: Renew. Sustain. Energy Rev. **14**, 938–954 (2010) (Elsevier)
10. Rahman, M.M., Masjuki, H.H., Kalam, M.A., Atabani, A.E.: Performance and emission analysis of Jatropha curcas and Moringa oleifera methyl ester fuel blends in a multi-cylinder diesel engine. J. Clean. Prod. **65**, 304–310 (2014)
11. Abedin, M.J., Masjuki, H.H., Kalam, M.A., Sanjid, A., Rahman, S.M.S., Fattah, I.M.R.: Ind. Crops Prod. **59**, 96–104 (2014) (Elsevier)
12. Khalid, A., Nursal, R.S., Tajuddin, A.S.A., Hadi, S.A.: J. Eng. Appl. Sci. **11**, 7424–7430 (2016) (ARPN)
13. Kannan, D., Nabi, N., Hustad, J.E.: SAE Technical Paper, **01**, 1808 (2009)
14. Agarwal, A.K., Dhar, A.: SAE Technical Paper, **01**, 947 (2009)
15. Manieniyan, V., Sivaprakasam, S.: SAE Technical Paper, **01**, 1577 (2008)
16. Jaroonjitsathian, S., Akarapanjavit, N., Sanorh, S.S., Ochanon, R.I., Wuttimongkolchai, A., … Shirakawa, H.: SAE Technical Paper, **01**, 1894 (2009)
17. Tan, P., Hu, Z., Lou, D., Li, B.: SAE Technical Paper, **01**, 2726 (2009)
18. Kumar, M.S., Ramesh, A., Nagalingam, B.: SAE Technical Paper, **01**, 153 (2001)
19. Sani, W., Hasnan, K., Yusof, M.Z.M., Baba, I.: Multi stage transesterifications of high FFA feedstock towards a high conversion of biodiesel in a batch mode production plant. Int. J. Min. Metall. Mech. Eng. **1**(5), 280–285 (2013)
20. Nursal, R.S., Zali, Z., Amat, H.H.C., Ariffin, S.A.S., Khalid, A.: Comparative study of the performance and exhaust gas emissions of biodiesels derived from three different feedstocks with diesel on marine auxiliary diesel engine. ARPN J. Eng. Appl. Sci. **12**(6), 2017–2028 (2017)
21. Sani, W., Hasnan, K.: UTHM biodiesel pilot plant development. In: Proceedings of International Conference on Mechanical & Manufacturing Engineering (2008)
22. PETRONAS Malaysia: Diesel fuel properties. In: Safety data sheet ETRO 4, Petronas Refinery (Melaka), Melaka (2014)
23. Department of Standards Malaysia: Malaysian Standard: Automotive fuels - palm methyl esters (PME) for diesel engines. Requirements and test methods (2008)
24. Department of Standards Malaysia: Malaysian Standard: Automotive fuels - palm methyl esters (PME) for diesel engines. Requirements and test methods (First revision) (2014)
25. Nursal, R.S., Hashim, A.H., Nordin, N.I., Hamid, M.A.A., Danuri, M.R.: CFD analysis on the effects of exhaust backpressure generated by four-stroke marine diesel generator after modification of silencer and exhaust flow design. ARPN J. Eng. Appl. Sci. **12**(4), 1271–1280 (2017)
26. Knothe, G., Sharp, C.A., Ryan, T.W.: Exhaust emissions of biodiesel, petrodiesel, neat methyl esters, and alkanes in a new technology engine. Energy Fuels **20**, 403–408 (2006)

Integrated Full Electric Propulsion System for Tanker Ships with Combined Diesel and Hydro Generator Drive

Wilfredo Yutuc, Hamzah Jamal and Bahktiar Afandi

Abstract In search of new green ship technologies, use of renewable energy, and design of a more energy efficient propulsion system for ships, this study investigated the possibility of combining both diesel and hydro generators in an integrated full electric propulsion (IFEP) system for a very large crude carrier (VLCC) tanker ship. The generators, connected in series, supply to both ship's service power and electric propulsion plant requirement. Unlike other combined electric propulsion system designs for large ships, the configuration makes use of two hydro generators, which operate using a renewable energy source. The hydro generators function under the principle of capturing and converting the kinetic energy and flow-pressure energy of the moving water around the ship's hull into electrical energy. The power output is directed to the ship's main power distribution system to support power generation onboard. This study attempted to compare theoretically the ship's total fuel consumption and overall plant efficiency, when running at full speed on a direct drive diesel-mechanical propulsion system, and on an IFEP system with combined diesel and hydro generator drive.

Keywords Energy efficiency · Green ship · Hydro generator · Integrated full electric propulsion · Tanker ship

---

W. Yutuc (✉) · H. Jamal · B. Afandi
Universiti Kuala Lumpur, Malaysian Institute of Marine Engineering Technology, Jalan Pantai Remis, 32200 Lumut, Perak, Malaysia
e-mail: wilfredo@unikl.edu.my

H. Jamal
e-mail: hamzah@unikl.edu.my

B. Afandi
e-mail: bahktiar@unikl.edu.my

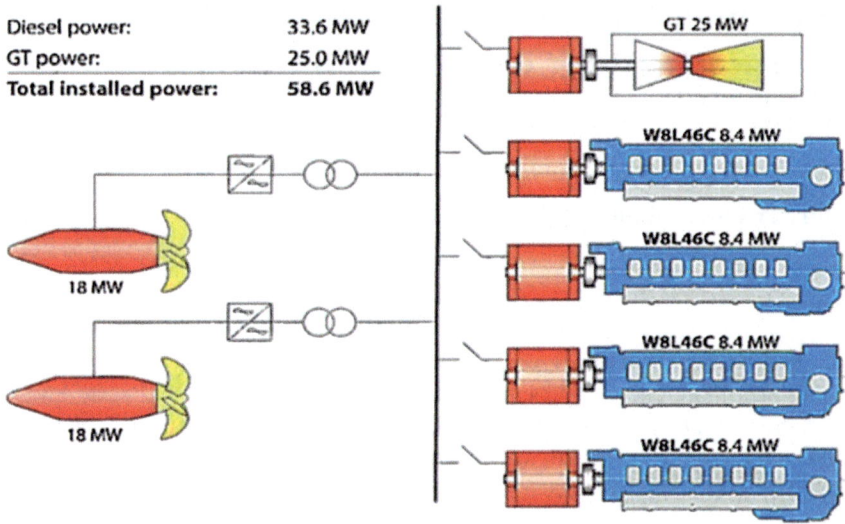

Fig. 1 Queen Mary 2 propulsion plant [3]

1 Introduction

The economic downturn, international regulations, and climate change are among the major driving factors for the design and operation of green ships with energy efficient propulsion systems. In an attempt to address these issues, a number of options have been designed by engine makers and ship builders for the propulsion and electric power plant of ships. Out of the basic mechanical and electrical power plant concepts emerged several configurations combining different types of engines. One, in particular, is the integrated full electric propulsion (IFEP) system or all-electric ship (AES) design. In this arrangement, the electric generators supply to both the electric propulsion and electric power system for ship service [1, 2]. The generator drives, commonly diesel, gas or steam turbines, or a combination of any of these, are connected in series. This is different from the early models of electric drives, wherein the electric propulsion system is separate from the electric power system for ship service, and each system would have its separate electric generator sets and switchboard with no link between the switchboards.

One example of an IFEP system is Cunard Line's passenger ship Queen Mary 2, with a combined diesel and gas turbine (CODAG) drive. In this ship, three diesel generators developing 8.4 MW each are connected in series with one gas turbine generator developing 25 MW output (see Fig. 1). These generators provide electric power to drive two podded propellers and supply to the demands of other electric consumers onboard [3].

2 Background

The use of diesel-electric propulsion nowadays is not only limited to passenger ships, offshore vessels, navy ships and other small specialized vessels. It has already reached the area of large cargo vessels for quite some time already [2, 4]. The Alaska Tanker Company, for instance, has in its fleet four 19,5000 deadweight diesel-electric tanker ships namely: the Alaskan Frontier, the Alaskan Explorer, the Alaskan Navigator and the Alaskan Legend, all of which were in operation since 2006 [5]. A number of dual-fuel diesel-electric propulsion liquefied natural gas (LNG) carriers have also been long built and have been navigating in the open seas for years [4, 6].

Among the benefits of electric propulsion plants worthy of consideration includes:

- flexibility of ship arrangements since direct mechanical connection between the prime mover and the propulsor is not required
- low levels of noise and vibration
- high levels of reliability and redundancy
- high levels of torque possible at low propeller speed
- power generated can be matched to the power required to allow prime movers to operate near their peak efficiency for good fuel economy [7–9].

3 Approach

In this study, the characteristics of an existing VLCC tanker ship, with a direct diesel-mechanical propulsion system driving a fixed pitch propeller, was used. The ship's particulars are shown in Table 1.

Table 1 Ship's particulars

Length, overall	m	329.99
Length, between perpendiculars	m	316.0
Breadth, moulded	m	60.0
Depth, moulded	m	29.7
Draft loaded, moulded	m	19.2
Deadweight	t	259,994
Gross tonnage	t	157,209
Speed	kn	15.5
Main shaft revolution	min^{-1}	78.6
Main engine output (MCR)	kW	25,090.0
Main engine specific fuel consumption at MCR (ISO)	g/kWh	172.0
Diesel engine output (×3 units)	kW	1060
Diesel generator output (×3 units)	kW	960
Diesel engine specific fuel consumption	g/kWh	195.0
Lower calorific value of fuel	$MJ\ h^{-1}$	42.7

Fig. 2 Hydro generator 3D model

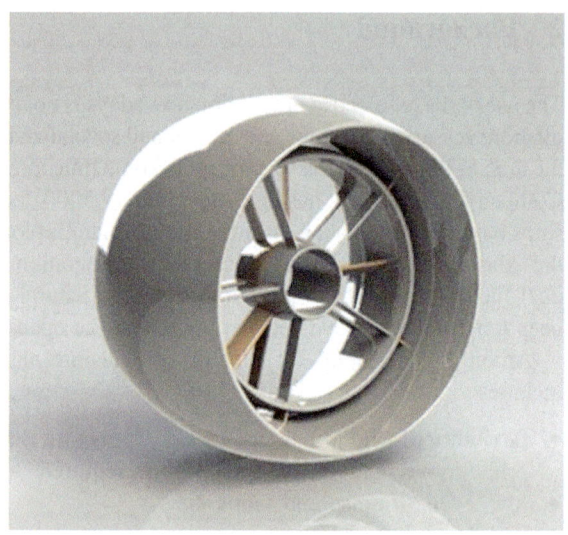

A VLCC tanker ship was considered since vessels of this type usually spent most of their time at sea due to long voyage between ports. This plays a big role in terms of the performance and benefits from the hydro generators (see Fig. 2), which are totally dependent on the ship's movement and speed to generate power.

The hydro generators, which are also termed "tidal generators", operate under the principle of capturing and converting the kinetic energy and flow-pressure energy of the moving water at the bottom of the ship's hull into electrical energy. This energy, converted into electrical power, is directed to the ship's main power distribution system, or main bus bar, to support power generation onboard. Since it is driven by moving water to generate electricity, it does not require the burning of fuel to produce power.

As an alternative power source, two 3-m turbine diameter hydro generators were virtually fitted on the ship's hull, as shown in Fig. 3. The ship's total fuel consumption and overall plant efficiency, with an IFEP system with combined diesel and hydro generator drive, were calculated at full speed of 15.5 knots in calm sea water condition. Results were compared to that of the ship's actual direct diesel-mechanical propulsion system and further evaluated.

3.1 Energy and Power Flow

The energy conversion from fuel to the rotating output shaft of an engine is illustrated in Fig. 4.

The heat input to the engine Q_F, can be obtained by using the formula

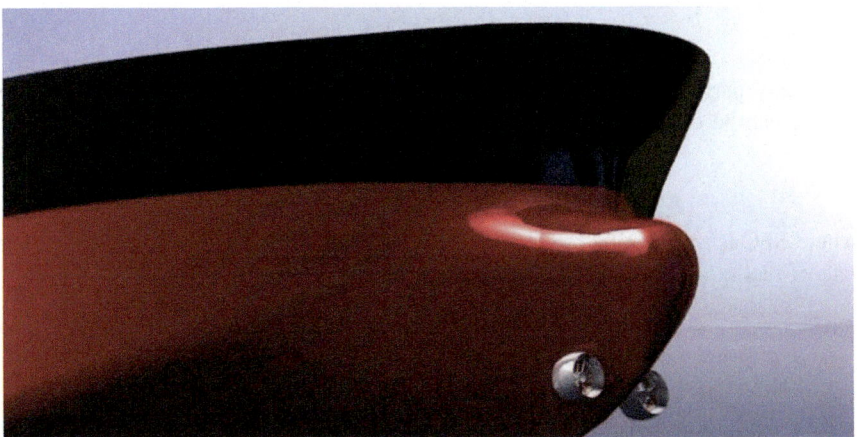

Fig. 3 Graphic conceptualization of the tanker ship with the hydro generators fitted

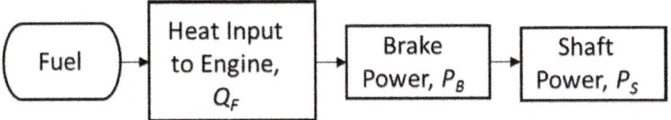

Fig. 4 Energy conversion flow

$$Q_F = m_F \cdot \text{CV} \tag{1}$$

where m_F is the mass flow rate of the fuel and CV is the calorific value of the fuel. The engine brake power P_B, is the power developed by the engine and is defined by

$$P_B = 2\Pi \cdot N_E \cdot T_E \tag{2}$$

where N_E is the engine speed and T_E is the engine torque. The engine effective efficiency, which is defined as the ratio of the engine output and heat input, can be determined by

$$\eta_E = P_B/Q_F \tag{3}$$

The shaft power P_S is the power delivered to the shaft that is connected to the propeller and is defined by

$$P_S = 2\Pi \cdot N_P \cdot T_S \tag{4}$$

where N_P is the propeller speed and T_S is the shaft torque. If there is no gearbox, the brake power equals the shaft power. If however, the transmission includes a gearbox, the gearbox efficiency is obtained by

$$\eta_{GB} = P_S/P_B \qquad (5)$$

At any particular break power output, the energy input to the diesel generator engine can be calculated using the equation

$$E_{IN} = \text{SFC} \cdot P_B \cdot \text{CV} \qquad (6)$$

where SFC is the specific fuel consumption at the corresponding break power. The rate of fuel consumption, m_F, may then be obtained by using

$$m_F = \text{SFC} \cdot P_B \qquad (7)$$

3.1.1 Diesel-Mechanical Propulsion

The propulsion system of the tanker ship considered in this study consists of a slow-speed reversible two-stroke diesel engine directly coupled to a fixed pitch propeller by means of a line shaft, as illustrated in Fig. 5. This is also sometimes referred to as a direct drive diesel-mechanical propulsion. The electric power plant is separate, providing power to the auxiliary systems and other electric consumers.

3.1.2 Diesel-Electric Propulsion

In any electric power generation plant, the generated power must be able to support the maximum power requirement taking also into account the power losses and some amount of power for safety margins. For an electric propulsion system with typical component efficiencies, the power flow can be illustrated as in Fig. 6 [8].

Accounting for the mechanical and electrical power losses in Fig. 6, the electrical efficiency of the system may be found as:

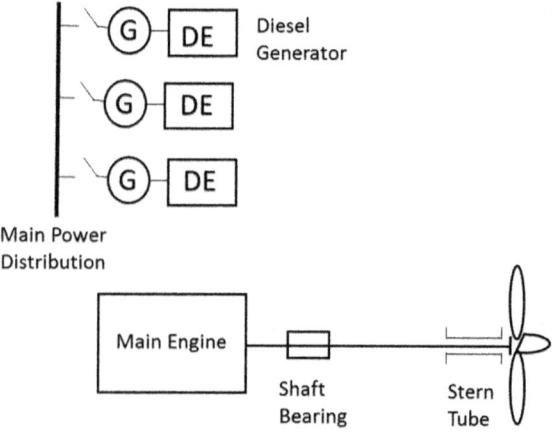

Fig. 5 A typical tanker ship power plant with direct drive diesel-mechanical propulsion

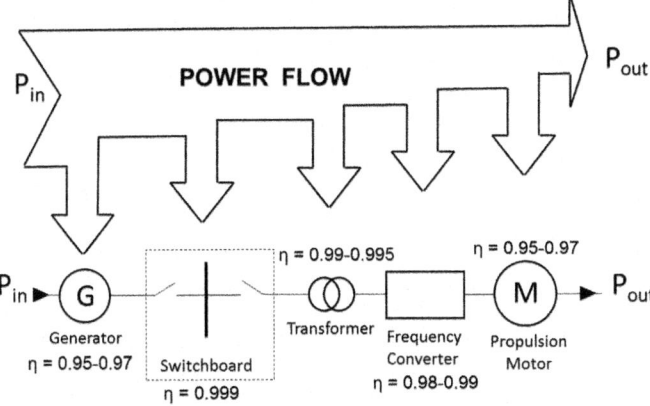

Fig. 6 Power flow in a simplified electric power system

$$\eta_{ELEC} = P_{OUT}/P_{IN} = P_{OUT}/(P_{OUT} + P_{LOSSES}) \quad (8)$$

Hence, with the efficiency range shown in Fig. 6, the efficiency of an electric propulsion system from the prime mover input shaft to the electric propulsion motor output shaft at full load may be seen as between 0.87 and 0.92 [8, 9].

Considering the supplied fuel, the overall efficiency of any complex propulsion plant may be defined as the ratio of the total power output and total power input:

$$\eta_O = \Sigma P_{OUT}/\Sigma P_{IN} \quad (9)$$

3.2 IFEP System Plant Model

In an integrated full electric power plant, a number of electric generators, not necessarily of the same size and power output, may be used [1]. Since the generators supply to a main power distribution system common to the propulsion system and ship service power, the electric generators can be used more efficiently under all operating conditions. This is because only the minimum number of prime movers necessary to meet the required load are operating, all running near their optimum efficiency [1, 2].

The IFEP system plant model for the VLCC tanker ship investigated in this study is composed of four diesel generators connected in series with two hydro generators, as shown in Fig. 7.

Two propulsion motors and a reduction gear box (twin-in-single-out) were considered in the plant layout. This is to fulfil the electric plant's high redundancy

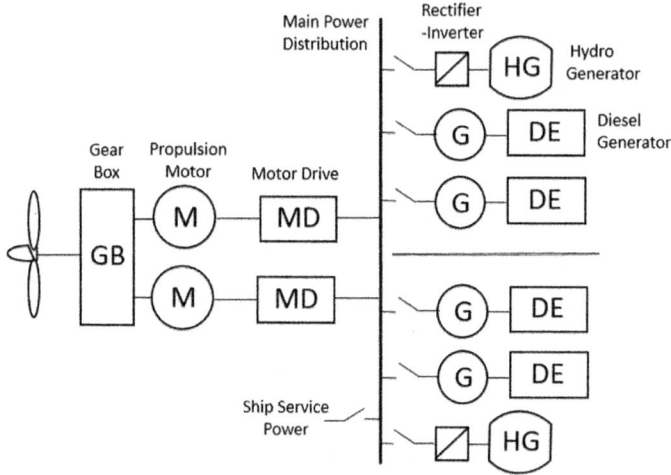

Fig. 7 A simplified IFEP plant layout for the combined diesel and hydro generators

Table 2 IFEP plant diesel generator specification

Engine	Rated power	kW	11,700
	Revolution	min^{-1}	514
	Specific fuel consumption	g/kW·h	189
	Lower calorific value of fuel	MJ·kg^{-1}	42.7
Generator	Rated power	kW	11,290
	Voltage	kV	6.6
	Frequency	Hz	60
	Phase	φ	3

requirement due to high propulsion power output to the propeller shaft and use of higher efficiency synchronous E-motors [9]. To support the 25,090 kW propulsion power requirement of the tanker ship in study when running at maximum continuous rating (MCR), four Wartsila 12V50DF diesel generators with specifications shown in Table 2, were considered [10].

3.2.1 Hydro Generators

The power, P_T, in W harnessed by the hydro generator's turbine, is decided by its sweep area, A, and fluid velocity, V, according to

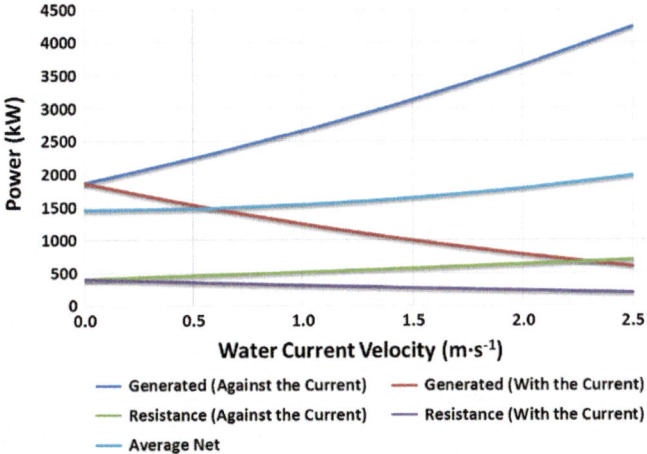

Fig. 8 Generated, resistance and average net power of a 3-m diameter hydro generator at water current velocity of up to 2.5 m·s^{-1} (ship at 15.5 knots)

$$P_T = 1/2 C_p \cdot A \cdot \rho_{fluid} \cdot V^3 \qquad (10)$$

where C_p is the power coefficient and ρ_{fluid} is the fluid density.

The generated electrical power output, P_{OUT}, may then be found as

$$P_{OUT} = P_T \cdot \eta_{blade} \cdot \eta_{gearbox} \cdot \eta_{genr}. \qquad (11)$$

Typical hydro generator efficiency is considered within the range of 0.80–0.85 [11, 12]. But for power output estimation, blade efficiency of 0.90, gearbox efficiency of 0.95 and generator efficiency of 0.95–0.97 may be assumed [8, 13].

An initial computer-generated simulation study conducted using the characteristics of the same VLCC tanker ship virtually fitted with one 3-m turbine diameter hydro generator gives an output data shown in Fig. 8 [14].

In calm water with zero sea water current velocity, the power output of the hydro generator is shown in Table 3 [14].

3.3 Mode of Operation

In actuality, when running at full speed in the open sea, the electric power demand of VLCC tanker ships with diesel-mechanical propulsion is lower and can be supplied by one diesel generator. During maneuvering and in port, however, two diesel generators are operated in parallel to supply the increased power demand due to large machinery used when maneuvering and when cargo operations take place.

Table 3 3-m diameter hydro generator output

Generated power	kW	1864.2
Drag	kN	50.6
Resistance power	kW	403.4
Net power	kW	1460.8

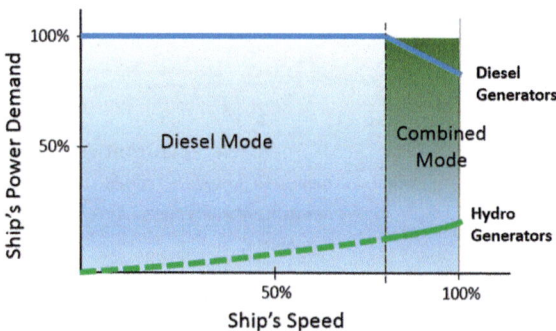

Fig. 9 Combined diesel and hydro generators running mode

This is different from the mode of operation of an IFEP configuration concept, with combined diesel and hydro generators. When the tanker ship is in port, diesel generators are to operate and supply the ship service power requirement, which includes the power for general usage and for the machinery used in cargo operations. This is because the hydro generators are only put in service when the ship is running at a higher speed, when their turbines are turning fast enough, sufficient to generate continuous stable power. Since the hydro generators' operation is affected by the direction and velocity of the water current, a rectifier-inverter module is used to convert the variable frequency and voltage output into fixed frequency and voltage to match with that of the main power distribution system of the ship.

In the open sea, when the tanker ship is at full speed, both the diesel and hydro generators supplement power generation in a combined mode and supplies the entire electric power requirement of the ship. The ship's running mode is illustrated in Fig. 9.

3.4 Power Output at Full Speed

In this concept evaluation, the rating and the loading of the engines for both types of propulsion plants were taken at the highest expected load. For the tanker ship with diesel-mechanical propulsion running at full speed of 15.5 knots, it was assumed that the main engine for propulsion and one diesel generator engine for ship service power are running at MCR. In Table 1, however, the main engine output shown is the engine's brake power and does not take into consideration the losses as power is transmitted to the propeller shaft. Typical shaft efficiency can

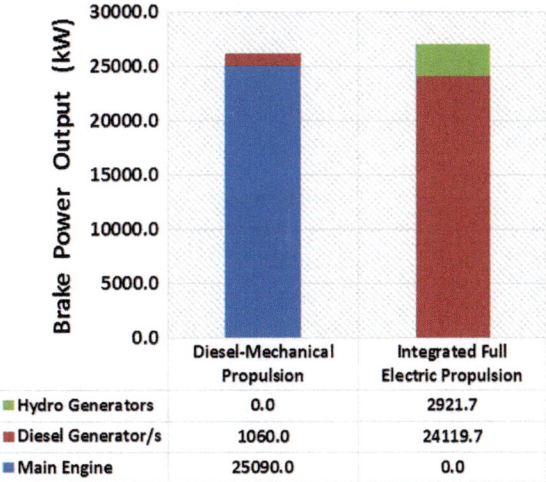

Fig. 10 Power output comparison between diesel-mechanical propulsion and IFEP for the tanker ship in study

vary from 0.96 to 0.995 [15]. With an average transmission efficiency of 0.97, a shaft power of 24,337.3 kW was considered for main engine propulsion.

In the IFEP system configuration, the same equivalent total shaft power of both main engine and one diesel generator engine has to be supplied by the combined diesel and hydro generators. An electric propulsion efficiency of 0.90 is considered [8, 9] to account for the mechanical and electrical power losses.

For the two hydro generators, the net power output was taken from Table 3. Since they operate using a renewable energy source and the power they generate is directly fed to the main power distribution system, the load of the diesel generators will be reduced in proportion to the power produced by the hydro generators when operating in combined mode. The free power produced by the hydro generators is henceforth deducted to the total equivalent load of the ship for propulsion and ship service. To illustrate further, the power output comparison between the two types of propulsion system is shown in Fig. 10.

4 Results

Applying the power, fuel consumption, and efficiency formulas presented earlier, a comparison of the ship's power and overall plant efficiency for diesel-mechanical and integrated full electric propulsion when running at full speed is shown in Table 4.

With the brake power output in Fig. 10, which is at maximum load for propulsion and ship's service power, the fuel consumption was calculated using Eq. (7) and the diesel engine's respective SFC. The result is shown in Fig. 11.

Table 4 Power and overall plant efficiency comparison

No.	Item		Unit	Diesel-mechanical propulsion	Integrated fixed electric propulsion
1	Propulsion	Shaft power output, P_S	kW	24337.3	24337.3
2		Transmission gear efficiency	η_G	0.97	0.90
3		Engine brake power, P_{B1}	kW	25090.0	27041.4
4		Engine SFC	g/kW·h	172.0	189.0
5		Fuel calorific value	MJ/kg	42.7	42.7
6		Power from renewable energy source	kW	0.0	2921.7
7		Engine brake power output considering power from renewable energy source (3–6)	kW	25090.0	24119.7
8		Power input from fuel (4) × (5) × (7)	kW	51186.4	54070.4
9	Ship service power	Electric power for ship (E-load)	kW	960.0	960.0
10		Engine brake power, P_{B2}	kW	1060.0	[a]1000.0
11		Engine SFC	g/kW·h	195.0	189.0
12		Fuel calorific value	MJ/kg	42.7	42.7
13		Power input from fuel (10) × (11) × (12)	kW	2451.7	2241.8
14	Total power output (1) + (9)		kW	25297.3	25297.3
15	Total power input (8) + (13)		kW	53638.1	56312.2
16	Overall plant efficiency (14) ÷ (15)		η_O	0.47	0.45

[a]considering 0.96 generator efficiency

At full speed, the propulsion engine brake power requirement in an IFEP configuration is higher than with the diesel-mechanical propulsion, as shown in Table 4. Even with the free power input from two hydro generators, the IFEP overall plant efficiency is 2% lower. This is mainly attributed to the lower transmission efficiency, i.e. decreased efficiency of energy conversion from the prime mover to the propulsor, due to the number of components in its electrical drive, as seen in Fig. 6. This, in turn, affected the overall plant efficiency which consequently resulted in a higher fuel consumption, as shown in Table 4 and Fig. 11, respectively.

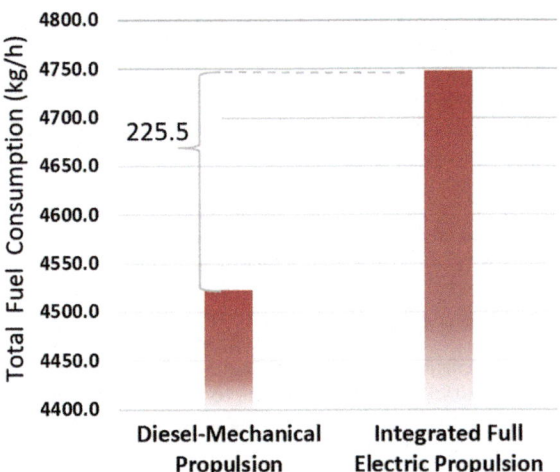

Fig. 11 Calculated fuel consumption of the ship at 15.5 knots with calm sea weather condition

5 Conclusion

From the data and results obtained in this study, it can be seen that the propulsion transmission efficiency played a major role in the overall plant efficiency and fuel consumption of the specific type of ship selected when running at full speed. The ship's principal characteristics, running mode condition, and type of generator drives employed are key contributory factors which determine the benefits of an IFEP plant concept in the design of propulsion and electric power generation for ships.

On another note, the use of hydro generators as an alternative electric power source for ships is worthy of consideration for further research and study. It gives the advantage of using renewable energy that will help address environmental issues related to global warming and sustainability, including compliance with the new regulations for the prevention of air pollution from ships. Its application on another type of ships with increased number of units installed can be further explored.

References

1. Woud, H.K., Stapersma, D.: Design of Propulsion and Electric Power Generation Systems, pp. 103–130. IMarEST Publication, London (2012)
2. Wartsila, Wartsila Encyclopedia of Marine Technology. http://www.wartsila.com/encyclopedia/term/integrated-full-electric-propulsion-(ifep). Accessed 16 Jan 2017
3. Kuiken, K.: Diesel Engines I for ship propulsion and power plants from 0 to 100,000 kW, 2nd Revised Edition. pp. 64–65. Epospress, Zwolle, The Netherlands (2012)
4. Numaguchi, H., et al.: Japan's first dual-fuel diesel-electric propulsion liquefied natural gas (LNG) Carrier. Mitsubishi Heavy Ind. Tech. Rev. **46**(1), (2009)

5. Alaska Tanker Company, Alaska Class Vessels. http://www.aktanker.com/alaska-class-vessels/. Accessed 15 Jan 2017
6. Woodyard, D.: Pounder's Marine Diesel Engines and Gas Turbines, 8th edn. Elsevier, Buttterworth Heinemann (2004)
7. Aprianen, M., et al.: Naval Architecture of Electric Ships—Past, Present and Future. SNAME Trans **101**, 583–607 (1993)
8. Adnanes, A.K.: Maritime Electrical Installations and Diesel Electric Propulsion, ABB Marine (2003). PDF File: http://img1.eworldship.com/2012/0913/20120913041849123.pdf. Accessed 16 Jan 2017
9. MAN, Diesel-electric Propulsion Plants—A Brief Guideline How to Engineer a Diesel-electric Propulsion System. https://marine.man.eu/docs/librariesprovider6/marine-broschures/diesel-electric-drives-guideline.pdf?sfvrsn=0. Accessed 16 Jan 2017
10. Wartsila, Wartsila 50DF Technology. http://cdn.wartsila.com/docs/default-source/product-files/engines/df-engine/wartsila-o-e-w-50df-tr.pdf?sfvrsn=6. Accessed 18 Jan 2017
11. Zimesnick, M.: How efficient is tidal power? Renewable Energy Index (2010). http://renewableenergyindex.com/renewable-energy-questions/how-efficient-is-tidal-power. Accessed 13 Jan 2017
12. New World Encyclopedia. Tidal Power. http://www.newworldencyclopedia.org/entry/Tidal_power. Accessed 13 Jan 2017
13. El-Sharkawi, M.A.: Electric Energy: An Introduction, 2nd edn, p. 107. CRC Press, Boca Raton, Florida (2009)
14. Yutuc, W.E.: Use of Hydro Generator on a Tanker Ship: A Computer-Generated Simulation Study. Lecture Notes in Computer Science (LNCS) 7972, Part II, pp. 207–219. Springer, Berlin (2013)
15. MAN Diesel & Turbo, Basic Principles of Ship Propulsion, p 17. https://marine.man.eu/docs/librariesprovider6/propeller-aftship/basic-principles-of-propulsion.pdf?sfvrsn=0. Accessed 20 Jan 2017

A Stress Analysis and Design Improvement of a Car Door Hinge for Compact Cars

Tajul Adli Abdul Razak, Muhammad Najib Abdul Hamid,
Ahmadiamri Mohd Ghazali, Shahril Nizam Mohamed Soid
and Khairul Shahril Shafee

Abstract The door hinges for automobiles are used to hold the door in a predefined axis during closing and opening operation. Apart from that, the door hinges are also important in automobiles as it holds the door in position during car accidents. The hinge itself is the combination of a pin, bushing, male and female hinge that will form an operating system for automotive doors. The purpose of this paper is to improve the built in stopper design of a stamped hinge assembly to comply with the OEM opening strength test requirement. The design of the door hinge will be developed using CATIA V5. The Hypermesh software will then be used to create a mesh and boundary conditions on the designed parts. Structural analysis will be carried out using ANSYS. In order to validate the results of FEA, an actual test will be carried out on stamped hinges using a test rig specified in the testing requirement. Based on the results further optimization of the built in hinge stopper will be carried out to increase the door hinge assembly performance.

Keywords Stress · Compact car · Design · FEA · ANSYS

T. A. Abdul Razak (✉) · M. N. Abdul Hamid · S. N. Mohamed Soid · K. S. Shafee
Section of Mechanical, Universiti Kuala Lumpur, Malaysian Spanish Institute,
Kulim Hi-Tech Park, 09000 Kulim, Kedah, Malaysia
e-mail: tajul.adli@yahoo.com

M. N. Abdul Hamid
e-mail: mnajib@unikl.edu.my

S. N. Mohamed Soid
e-mail: shahrilnizam@unikl.edu.my

K. S. Shafee
e-mail: khairuls@unikl.edu.my

A. Mohd Ghazali
Vehicle Integration, 42000, Shah Alam, Selangor Darul Ehsan, Malaysia
e-mail: kmadi1st@yahoo.com

1 Introduction

A typical automotive door will use hinges to attach doors to the structural body. Although there is also a type of door that uses track to perform open and close door operations, this paper will discuss mainly the door hinge for automotive applications. The door hinges for automobiles are used to hold the door in a predefined axis during closing and opening operation [1]. Apart from that, the door hinges are also important in automobiles as they hold the door in position while driving in the event of car accidents [2].

The hinge assembly itself is the combination of a pin, bushing, male and female hinge that will form an assembly for automotive doors. In a typical automotive door system, the door hinge system will consist of a pair of hinge assembly that connects the door assembly to the structural body. The Society of Automotive Engineers International (SAE International) defined the hinge system as a system used to position the door relative to the body structure and control the path of the door swing for passenger ingress and engress [3]. Despite the size of the hinge relative to the automotive body itself, it is clear that the hinge plays a major role in providing safety and functional operation for passengers and driver.

During the research and development stage of an automotive door hinge, there will be a number of testing and homologation activity taken place to ensure the reliability and quality of the product. There could be various types of tests depending on the Original Equipment Manufacturer (OEM) requirement. One of the tests under the OEM's requirements is called the Over Bending Test.

In an Over Bending Test, a specific force will be applied onto the newly fabricated door hinge during maximum opening. If the door hinges are able to withstand the exerted force onto it, the door hinges design are considered to be ready for mass production. If the result is contrary, then optimization needs to be done to improve the failed hinge.

1.1 Problem Statement

The typical hinge design that is used in this paper can be traced back to the year 1956 by Hollanworth [4]. Then, other three designs taken from United State Patents were studied to have an idea of effective automotive door hinges [5–7]. The development of new automotive door hinges will be mostly influenced by the available area to mount the hinges onto the structural body. Typically, a new car model will also result in new door hinges design due to different exterior shapes that influence the sufficient area for mounting the hinges. With limited space for newly designed hinge, it require minimal stopper height to hold the door at maximum opening angle. The stopper height will require technical verification based on the OEM's requirement.

Finite element analysis (FEA) using software is part of the activities to reduce error in the development stage. Unfortunately, a common error of conducting the actual experimental work is always associated with the positional accuracy gauge [8] resulting in a more robust hinge design for test compliance. The evolution of material properties has also contributed to the accuracy of classical stress-strain curves which are based on the Ramberg-Osgood expression [9, 10]. Due to these reasons, FEA alone is insufficient for product verification. The over bending test is one from many verification methods related to door hinges required by the OEM. Based on the standard set by specific OEM, the newly designed and fabricated hinge should withstand 400 N of normal force exerted on it during fully open condition. If it is found that the hinges could not withstand the stated force, then the hinges will be considered as not acceptable for mass production.

1.2 Objective

As mentioned previously, an automotive door hinge is highly influenced by the available area for hinge mounting in every car model. In order to suit the hinge with different car models, a new design of door hinge will be developed using CATIA V5. The hinge should consider the manufacturing ability for stamping process which requires a generous radius at corner area but at the same time, provide accurate opening angle as required by the OEM.

Using Abaqus simulation, strength analysis will be performed onto the newly designed hinge to simulate the over bending test and identify the weakest point. Design improvement of the hinge will be conducted to improve any weak point that has the potential to fail during the actual test. A modification of the hinge during the design stage is highly recommended since it will be faster and less hassle. Once the design has been confirmed, then it will proceed to hard tooling for actual parts.

Performing the over bending test by means of simulation is not the ultimate confirmation for verification. The actual parts need to undergo the physical test stated inside the test specification for actual result. This result will influence the decision whether the hinges are fit for mass production or require further improvement.

In a nutshell, the objectives will be as explained previously. To ensure the quality of the door hinge, it is required to design the door hinge, to perform structural analysis using linear static FEA for the over bending test, to verify the analysis and design with actual over bending test and finally, to improve the hinge design based on the information from simulation and actual tests.

2 Experimental Setup

The new door hinge will be designed using CATIA V5 according to the allowable space, suitable material and manufacturing process. A strength analysis simulation will be performed onto the 3D data. A further improvement of the door hinge will be based on the simulation result. The new design will overcome any potential spot that could lead to part failure. Upon completion, the actual test will be conducted using a test rig to verify the simulation result. If the hinge fails to withstand the 400 N force, improvements will take place so that the hinge can pass the test and is available for mass production.

2.1 Material Selection

There are numerous materials in the market that can be used to manufacture door hinges such as ductile iron, aluminum and mild steel. Since stamping method is determined as the manufacturing process for this project, we have selected hot rolled steel grade as SAPH440. A detailed explanation on the material mostly can be extracted form JIS G 3113. It is a Japanese material standard for hot rolled steel plates, sheets, strips for automobile structural usage. SAPH440 is widely used as a material for door hinge due to its high reliability of tensile strength. This characteristic is suitable for automotive applications where more strength is needed and the material performs well under exceeding loading. Table 1 shows the chemical composition of SAPH440. In Table 2, the mechanical properties of the material are shown.

Table 1 Chemical composition of SAPH440

Composition (mass%)				
C	Si	Mn	P	S
0.168	0.020	0.810	0.012	0.008

Table 2 Mechanical properties of SAPH440

Young's modulus	201 GPa
Poisson's ration	0.306
Tensile strength	440 N/mm^2
Vicker's hardness	214 Hv
Elongation to break	44%
Yield strength	302 N/mm^2

2.2 Door Hinge Design

Although there are several concepts of door design that influence the design of the door hinge such as Butterfly, Canopy, Scissors and more, the door hinge used in this project is using a standard door design which is a swing type door. In typical automotive door system, there are 2 units of door hinges that are used to attach the door to the structural body. Both hinges may not have to be the same design as a result to area constraints. Figure 1 shows the design of the door hinge together with the respective exploded view for better understanding.

2.3 Finite Element Analysis

The upper hinge is used to analyse using the finite element method to investigate the value of the maximum stress occurred when subjected to the maximum load. The male and female of the upper hinge is modeled as C3D10 solid deformable elements using ABAQUS CAE 6.14 as shown in Fig. 2. Meanwhile the pin

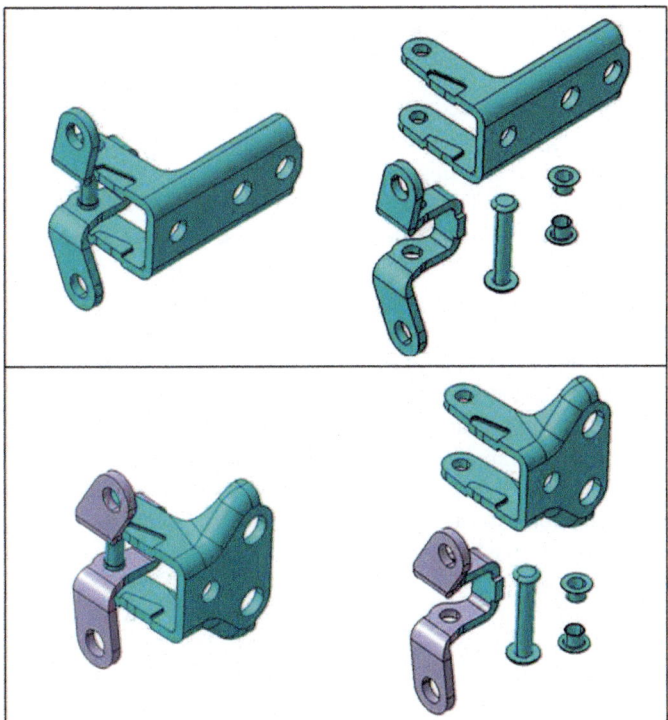

Fig. 1 Upper and lower hinge with respective exploded view on the right

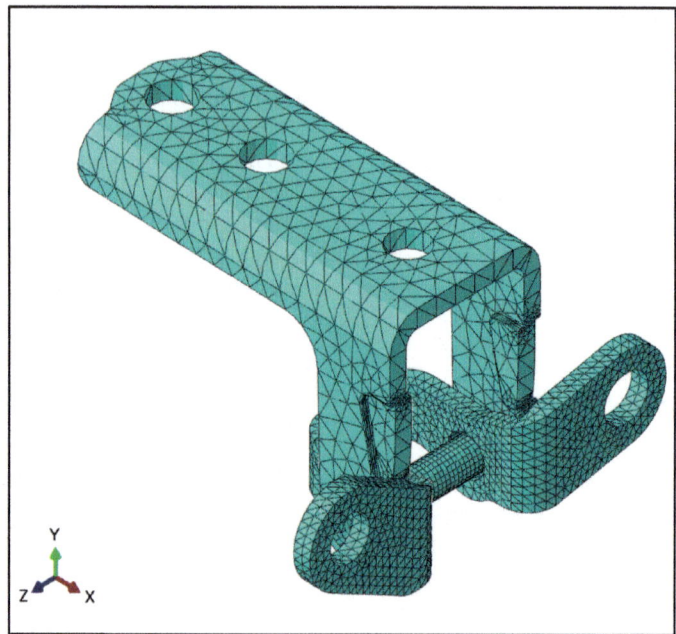

Fig. 2 FE model for the analysis of the door hinge

connected male and female door hinge is modeled as R3D4 rigid elements as this component is not the target to know the displacement or stress in this analysis.

The material properties of this FE model are defined as a linear-elastic material with a yield strength as in Table 2. This material is not modeled completely with the damage criteria model which means that this FE analysis will not produce the material failure such as break or fracture. This is sufficient in order to predict the failure of the component using the maximum stress occurred and the value of the yield strength.

The load and boundary condition for this analysis is the encastre or fixed all degrees of freedom for the base of the female part and the torque is applied to the male part to rotate and push against the stopper. The value of the applied torque is calculated based on the applied load and the geometry of the real door test. The maximum load is set to 400 N. This analysis only used 1 set of the upper hinge so that the maximum load must be divided by two. A static step analysis is used because the effect of dynamics and inertia is neglected due to the real test is assumed as quasi-static.

2.4 Boundary Condition

The static force shall be applied equidistant in the center of the female hinge that acts as a stopper during fully open condition. The contact surfaces of the stopper between the male and female hinge is 0.5 mm at both bottom and upper area making a total contact surface of 1 mm. The requirement from the OEM is, a door hinge assembly, when tested shall be able to withstand an ultimate normal load of 400 N without break. This is considering the fully open condition of the door hinge.

By static analysis of the hinge under extreme conditions;

Max. Deformation: 3 mm
Max. Stress produced: 248.5 MPa

As stresses are well within the limit of the yield stress (302 MPa) and the deformation is quite small, the design is evaluated as safe.

2.5 Test Rig Setup

The position of the force acting on the actual test rig is shown in Fig. 3. The length dimension originated from the axis of both hinges is $l = 830$ mm and the height, h is 415 mm, respectively. The total force used in this experiment is 400 N during the fully open hinge condition.

Actual experimental tests are carried out on the existing door hinge model to validate the FEA results. The static force onto the door hinge during the fully open condition is tested onto the actual car and displacement verses load curves are plotted. The procedure to perform the actual test is:

1. Attach a test fixture to the mounting provision of the hinge.
2. The test fixture is placed in the machine between the grips. The machine itself records the displacement between its cross heads on which the specimen is held.
3. Adjust the load cell to read zero on the computer up to peak load. Once the machine is started it begins to apply an increasing load on the specimen.

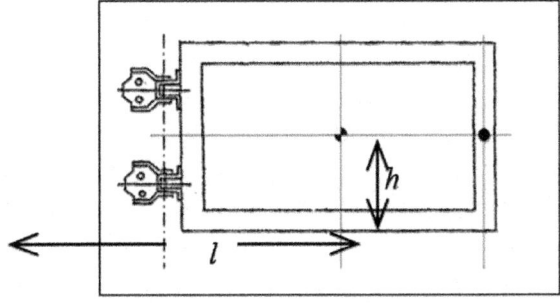

Fig. 3 The position of force in over bending test

4. Throughout the tests the control system and its associated software record the load and displacement of the specimen.
5. According to the results of the experiment, a graph of the displacement variation with load is plotted automatically.

3 Results and Discussion

3.1 Simulation and Actual Test Analysis

The initial design of the door hinge was analysed using FEA to obtain the maximum stress which would exceed the yield strength of the material when subjected to the maximum normal load to the door panel. A stress contour of the male and female door hinge when subjected to the applied load is shown in Fig. 4. For the initial door hinge with a 1.5 mm thickness of the stopper, the maximum stress is 770 MPa which is over the yield strength of this material (302 MPa). This door hinge can be considered as failed when subjected to the maximum load of 400 N but this is not shown in the FE analysis (see Fig. 4) due to limitations of the material model which does not include any damage criterion. From this figure, it can be seen that a high stress concentration occurs at the female of the door hinge.

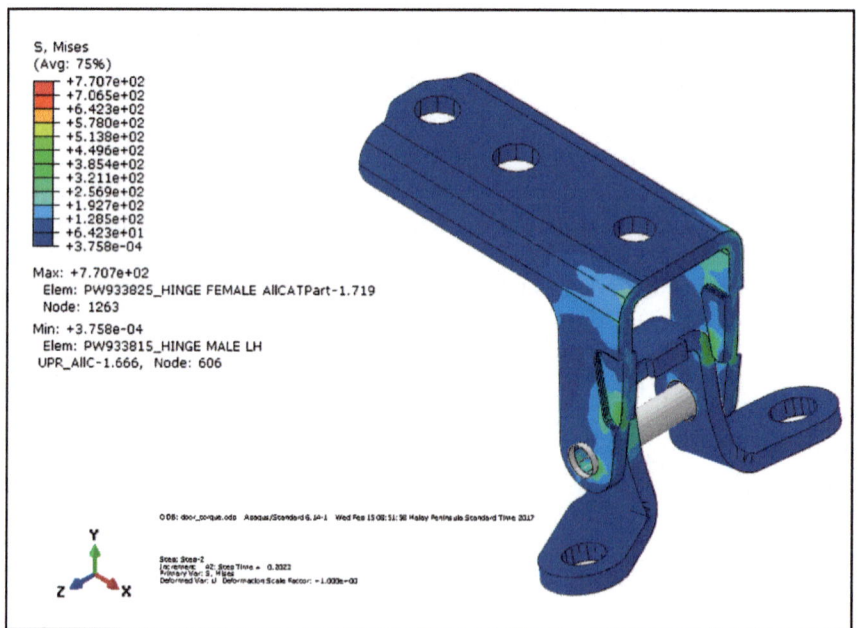

Fig. 4 Stress distribution of one of the door hinge with initial stopper thickness (1.5 mm)

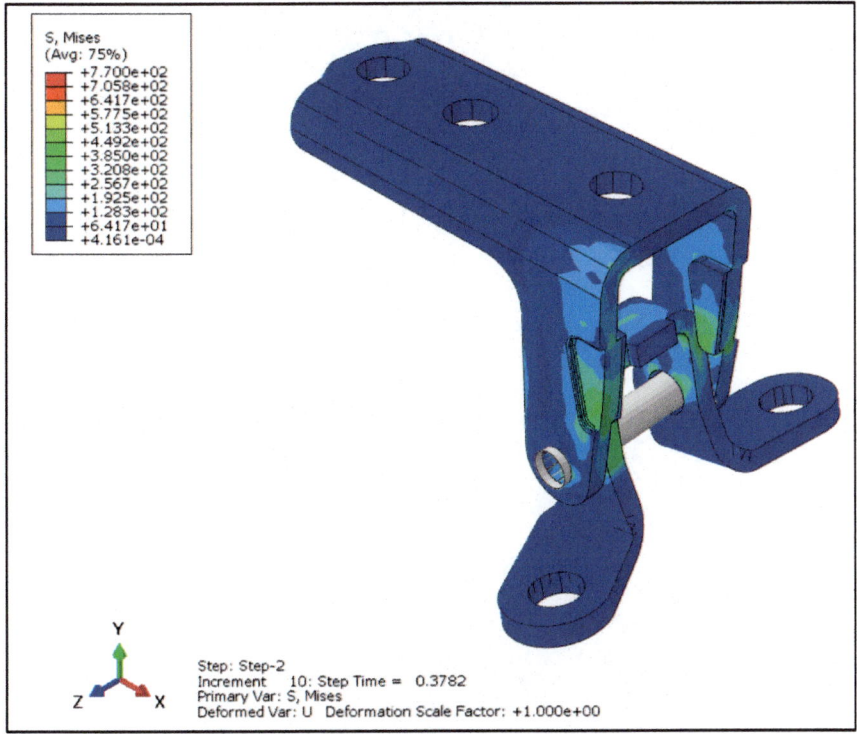

Fig. 5 Stress distribution of one of the door hinge with improved stopper thickness (2.5 mm)

The new design to overcome the failure of the initial design is to increase the thickness of the stopper. The thickness is proposed to increase up to 2.5 mm. The same analysis was done and the result is shown in Fig. 5. When the thickness is increased to 2.5 mm, the maximum stress is reduced to 300.5 MPa which is near and not exceeds the yield strength. This indicates that the door hinge could withstand the maximum applied load of 400 N. A further optimisation process can be performed in order to get the best design and shape of the door hinge but it needs a new process to develop a new design rather than increasing the stopper thickness of the existing design of the door hinge.

After the improvement of the design has taken place with an improved stopper thickness of 2.5 mm, the experiment has been taken to the next level for better evaluation. The manufactured door hinge is ready to undergo the over bending test on the actual test rig. The setup of physical door hinge and test equipment is similar to Fig. 3 in Test Rig Setup chapter. Based on the result taken from the actual test rig, the door hinge reached its yield point during 470 N. Further force increase on the door hinge showed that the hinge deformed and broke exactly at the stopper area as shown in Fig. 6. This condition is predictable since the simulation with Abaqus has shown a similar symptom. The result of the actual test is shown in Fig. 7.

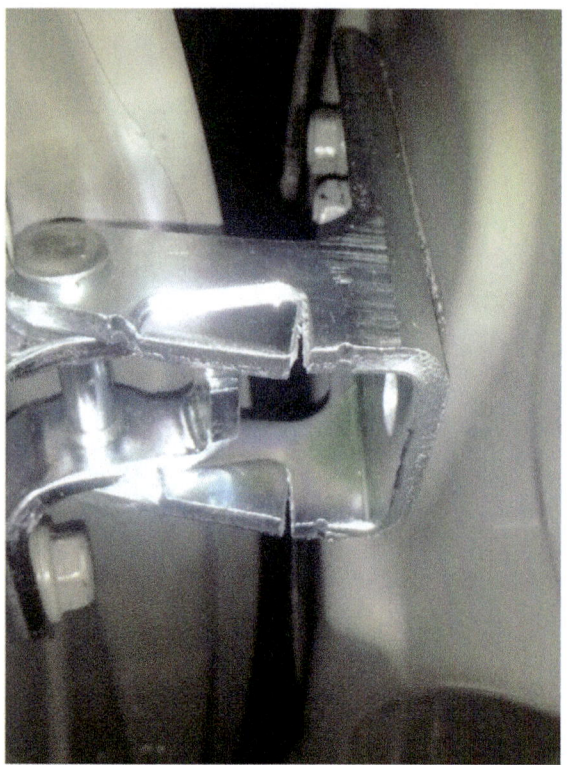

Fig. 6 Broken door hinge on actual test

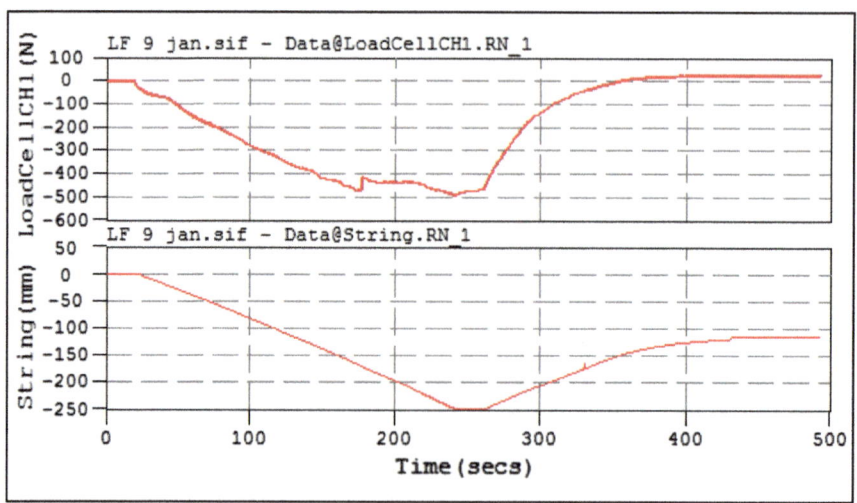

Fig. 7 The result of door hinge from actual test on test rig

4 Conclusion

This paper has been focused on the capability of the current hinge design to withstand a static force of 400 N without break. A 3D model of the component is produced and analysed. A test rig has also been developed according to the test specification form the OEM's test guideline. The actual test is to experimentally validate the results of the numerical analysis. In a three-dimensional analysis of component forces, the result could be the reference in any future door hinge improvement. It could be concluded that the FE analysis done has successfully reflect the actual condition in real life. The improvement in the door hinge design based on FE data has successfully tackled the actual issue of potential hinge failure. The improved design of the door hinge is now considered as suitable for mass production.

Acknowledgements The authors thanked the University Kuala Lumpur Malaysian Spanish Institute (UniKL MSI) and Sciences and Vehicle Integration, Shah Alam for research facilities and financial assistance provided for this project.

References

1. Engineers of SAE-China Congress 2014: Selected Papers. (2014)
2. NHTSA Technical Report and quote: Lives Saved by the Federal Motor Vehicle Safety Standards and Other Vehicle Safety Technologies—Google Search, (1998)
3. SAE J 934: Vehicle Passenger Door Hinge Systems, p. 2. SAE International, Pennsylvania (1969)
4. Hollansworth, M.: United State Patent: Motor Vehicle Door Hinges, 2799042 A (1957)
5. Leung, K.C., Hills, R., Lee, A.G., Hts, S., Kolosick, M.D.: United State Patent: Door Hinge For Vehicle, 6568741 B1 (2003)
6. Richard, E., Mead, B., Zlotnikov, E., Park, H., Us, N.J., Haders, D., Nj, S.: United States Patent: Door Hinge, 6851159 B1 (2005)
7. Morinaga, H.: United State Patent: Vehicle Door Hinge, 9103146 B2 (2015)
8. Lisle, T.J., Shaw, B.A., Frazer, R.C.: External spur gear root bending stress: a comparison of ISO 6336:2006, AGMA 2101-D04, ANSYS finite element analysis and strain gauge techniques. Mech. Mach. Theory **111**, 1–9 (2017)
9. Rasmussen, K.J.R.: Full-range stress-strain curves for stainless steel alloys. J. Constr. Steel Res. **59**(1), 47–61 (2003)
10. Arrayago, I., Real, E., Gardner, L.: Description of stress-strain curves for stainless steel alloys. Mater. Des. **87**, 540–552 (2015)

Development of a Global Warming Impact Prediction Analysis Tool for Mobile Vehicles

Chun Lin Saw, Chee Guan Choong, Yoke Yen Chiang and Azmi Naroh

Abstract Air conditioning has become part and parcel of our daily life, it is used to comfort us by providing the cooled air specifically for mobile vehicles. Nevertheless, the use of refrigerants in air conditioning is harmful to our atmosphere due to direct and indirect emissions to the environment. The direct emission of the refrigerant occurs during the process of car servicing and releases purposely by the untrained technician, while indirect emission occurs from pipe leakage, pipe cracks, storage tank leakage and unintentionally release by the technician. Currently, there is no available analysis tools on global warming impact of refrigerant emissions for the Malaysian context. Data on the amount of gas inject and gas recovery from the recovery machine in grams for 5 months was collected at the Air Conditioning and Refrigeration workshop Politeknik Ungku Omar, Ipoh, Perak. Three types of cars that have been investigated, whereupon the Perodua, Viva car showed the highest CO_2 emission per km compared to the Perodua, Myvi and Alza. A Total Equivalent Warming Impact (TEWI) calculator also was developed to assist to predict the environment impact on global warming due to vehicle air-conditioning refrigerant leakage.

Keywords Air conditioning · Refrigeration · Environmental warming
Vehicle · Perodua

C. L. Saw (✉) · C. G. Choong · A. Naroh
Centre of Air Conditioning and Refrigeration, Politeknik Ungku Omar, Jalan Raja Musa Mahadi, 31400 Ipoh, Perak, Malaysia
e-mail: clsaw78@gmail.com

C. G. Choong
e-mail: cgchoong@gmail.com

A. Naroh
e-mail: narazmi@puo.edu.my

Y. Y. Chiang
Department of Mathematics, Science and Communication, Politeknik Ungku Omar, Jalan Raja Musa Mahadi, 31400 Ipoh, Perak, Malaysia

© Springer International Publishing AG 2018
A. Öchsner (ed.), *Engineering Applications for New Materials and Technologies*,
Advanced Structured Materials 85, https://doi.org/10.1007/978-3-319-72697-7_12

1 Introduction

The mobile vehicle air conditioning is one of the important systems in the vehicle to maintain thermal comfort and improve the ventilation inside the vehicle compartment. However, the mobile vehicle air conditioning refrigerant and the operation of the vehicle air conditioning system possess risks to the environment by producing greenhouse gasses (GHG) [1]. GHG from the mobile vehicle air conditioning can be a refrigerant gas such as R-12 & R-22 is chlorofluorocarbon halomethane (CFC) based gas that has the ozone depletion potential and R134a & R404A is a hydrofluorocarbons (HFC) based gas that has a lower global warming potential, while to operate the vehicle air conditioning system needs the engine to run by consuming gasoline gas that produces carbon dioxide, CO_2.

To reduce the ozone depletion, the CFC-12 refrigerant has been phased out with the HFC 134a refrigerant. In the year of 2015, a total of 180,000 tonnes of HFC 134a refrigerants have been released to the environment, while 270 Mtonnes of CO2-e have been produced globally [2]. The refrigerant used in the air conditioning system can cause the greenhouse effect due to leakage of the refrigerant gas. Total Equivalent Warming Impact (TEWI) is a tool used to gauge the emission of carbon dioxide (CO_2), it can be in terms of direct (refrigerant gas) and indirect emission (energy). Direct emission is due refrigerant losses such as gradual refrigerant loss due to piping leakage, losses during servicing, losses due to accidents and losses due to end of components life. Hence, indirect emission is caused by exhaust emission of the vehicle during transportation. TEWI is a widely used tool to measure the emissions of greenhouse gasses from a certain equipment during its working life [3].

In China, TEWI analysed from government vehicles, public buses, and taxis shows a total of 1.37×10^7 tonnes CO_2-e produced in 2015 that has shown a lot of CO_2 reduction since 2010 [4, 5]. In the UK, hybrid electrical vehicles (HEV) have shown a significant lower CO_2 emission from the vehicle but in terms of generating electricity to charge the HEV increasing GHG emission from the power plant [6]. In the USA, a study on light-duty transportation has been conducted. The study summarized that hydrocarbon based oil consumption vehicles need to be reduced to reduce greenhouse gasses. Introducing advanced technology such ethanol flex-fuel vehicles (FFV), fuel-cell electric-powered vehicles (FCEV) and plug-in electric vehicles (PHEV) can achieved the aim but the feedstocks availability, maturity of the technology and electricity generation from renewable sources is still a concern [7].

Beside emissions of GHG, recovery and reclaim refrigerant gas is another way to reduce the indirect emission of refrigerant gas to the environment. Recovery and reclaim of refrigerant gas has been reported to reduce the indirect emissions, which cause global warming [8, 9]. Table 1 shows that CFC-12 is the most harmful refrigerant that has high ozone depletion potential (ODP) and has a life of 8500 years (GWP) to degrade.

In this paper, a study was conducted on the refrigerant gas leakage to the environment by measuring the refrigerant gas during vehicle air conditioning

Table 1 ODP and GWP of some refrigerants

No.	Refrigerant gas	Ozone depletion potential (ODP)	Global warming potential (GWP) 100 years
1	CFC—12	1	8500
2	HCFC—22	0.055	1700
3	HFC—134a	0	1300
4	HC—600	0	3

system servicing (indirect emission) and CO_2 emission estimation (direct emission) from a Perodua car. The analysis has been done using the TEWI method to predict the carbon dioxide equivalent (CO_2-e) that are produced by the Malaysian manufactured car. The Perodua car model was manufactured in year 2015 and below are selected conditions under this study.

2 Experimental Setup

The data collection of the refrigerants amount for Perodua Myvi, Viva and Alza took place during for 5 months of duration from March until July 2016. The data has been collected at Politeknik Ungku Omar's Air Conditioning and Refrigeration Workshop. A total of 20 cars of Myvi model, 15 units of Viva's car and 11 units of Alza's car have been investigated. For the averaging, 10 units of car from each model have been selected for the TEWI calculation.

The fuel consumption of the selected car has been estimated from the review of the model car users, where averaged fuel consumptions of Myvi, Viva and Alza models are 17 l/km, 11 l/km and 12 l/km, respectively [10–12]. The car air conditioning life time is taken as the average of 11 years as some published articles mentioned the range from 11 to 13 years [13]. Each of the car data such as gas injected, gas reclaimed and CO_2 emissions was inserted in the TEWI calculator developed to get the TEWI CO_2 results as shown in Fig. 1.

A recovery and recharge (R&R) machine is the machine that is used to recycle the refrigerant gas from the car air conditioning system as shown in Fig. 2. The R&R machine will perform the recovery, reclaim and recharge process each time the car air conditioning system is serviced. Recovery means to remove the refrigerant in any condition from a system to the R&R machine. Then, the refrigerant gas available will be reclaimed and stored in an external tank by using a filter.

The recovered gas will be flown back to the air conditioning system through the recharge process. As to improve the awareness of the global warming, an awareness training video on the R&R machine handling and some basic fundamentals of the air conditioning refrigerant gas have been created to introduce to the public on the virtual laboratory to reduce the usage of raw material that can coup the refrigerant gas released to the environment.

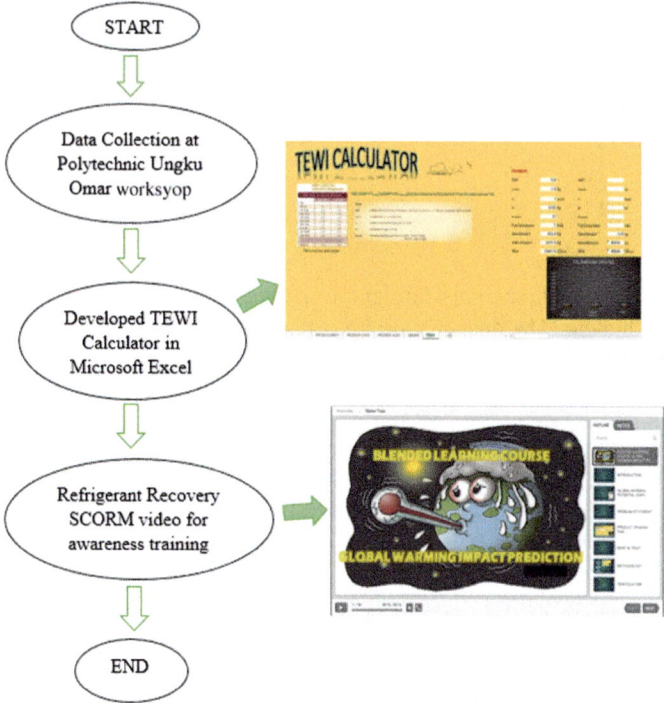

Fig. 1 Process flow on developing Global Warming Impact Prediction Analysis tool

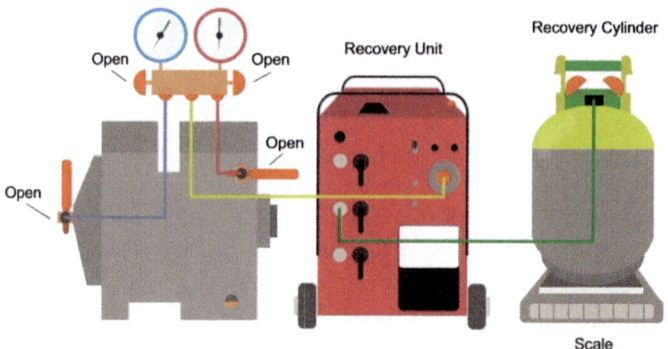

Fig. 2 Recovery and recharge unit

When using a refrigerant, the greenhouse impact is not only caused by refrigerant leakage into the atmosphere, but also by the energy consumption of the refrigeration system. The Total Equivalent Warming Impact (TEWI) which includes direct emissions caused by the leakage of greenhouse gases and indirect emissions caused by the energy consumption of the refrigeration system was then

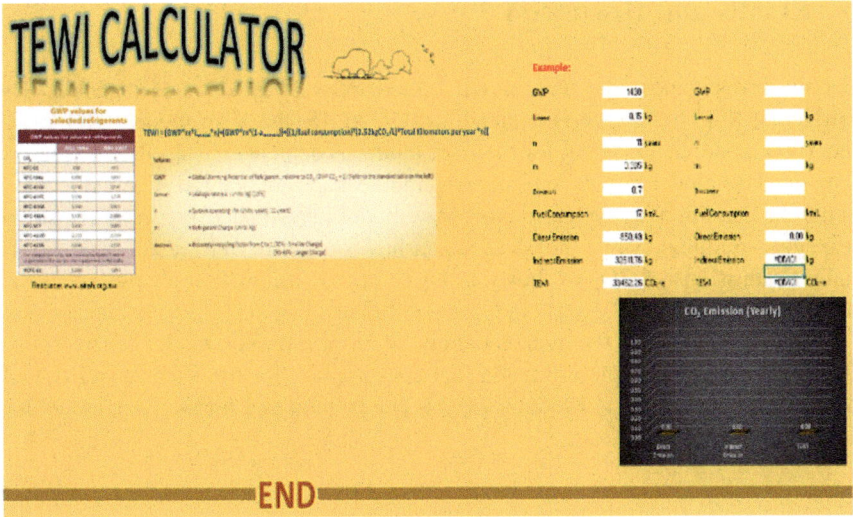

Fig. 3 User interface of TEWI

developed by referring to the best practices guideline [14, 15]. The direct emissions depend on the GWP of the refrigerant, the amount of refrigerant leakage, the equipment running time. While the indirect emissions depend on the energy consumption during operating and the CO_2 emissions per unit energy generation. The developed TEWI prediction tool user interface is shown in Fig. 3.

$$\text{TEWI} = \text{Direct emission} + \text{Indirect emission} \tag{1}$$

Direct emission,

$$(\text{GWP} * m * L_{annual} * n) + (\text{GWP} * m * 1 - a_{recovery}) \tag{2}$$

Indirect emission,

$$[(\text{fuel consumption}) * (2.52 \text{ kg } CO_2/L) * \text{Total Kilometres per year} * n] \tag{3}$$

where:

GWP is the the Global Warming Potential of the refrigerant, relative to CO_2, (GWP CO_2 = 1)
L_{annual} is the Leakage rate p.a. (unit: kg)
n is the system operating life (unit: years) = 11 years
m is the refrigerant charge (unit: kg)
$a_{recovery}$ is the recovery/recycling factor from 0 to 1
Total kilometres per year = 20000 per year
Fuel consumption = litres/kilometres

3 Results and Discussion

For the investigation, the data was collected from 20 cars for each model. The refrigerant R 134a leakage is as much as 0.97 kg for the Viva model, 0.83 kg for the Myvi model and 0.44 kg for the Alza model as depicted in Fig. 4. Two reasons why the Alza model has the lowest refrigerant leakage are that the Alza model has a better air conditioning system or the Alza model investigated is a new car so the leakage is minimal. Since the Viva model has the highest refrigerant leakage, the indirect emissions of CO_2 as shown in Fig. 5 is proportionally higher as calculated using Eq. (2).

Figure 6 represents the fuel economy of a car. Based on the forum of the Perodua Car analyser, the estimated fuel consumption for the Viva model is 12 L per 100 km, for the Myvi Model it is 13 L per 100 km and for the Alza model it is 17 L per 100 km.

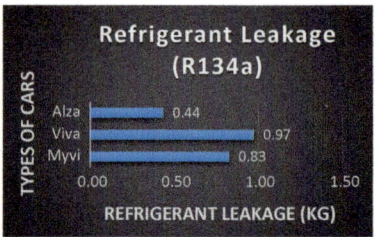

Fig. 4 Refrigerant leakage

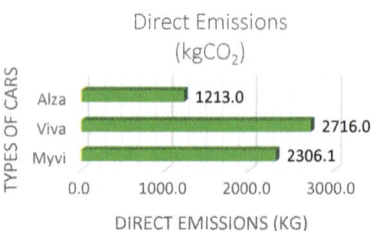

Fig. 5 Direct emissions (kg)

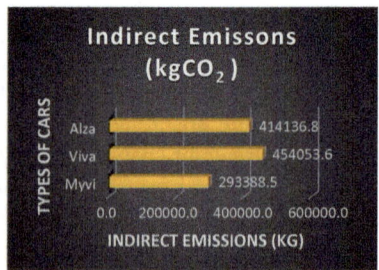

Fig. 6 Indirect emissions (kg)

Fig. 7 Total Equivalent Warming Impact (TEWI) of Perodua car

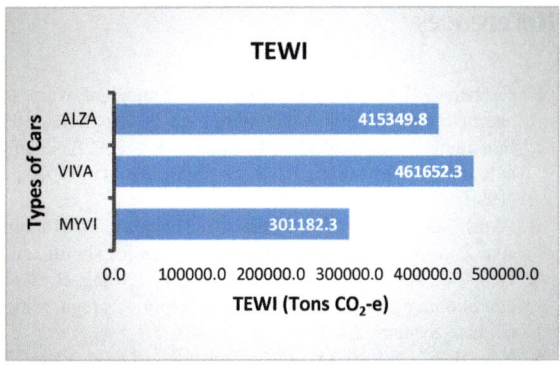

Fig. 8 CO_2 emissions of Perodua car per km

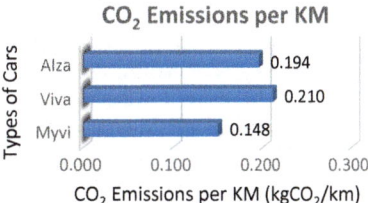

The Viva model is found to be less economical in terms of fuel consumption as the CO_2 emissions is the highest that is 454053.6 kg CO_2 compared to the Myvi model (293388.5 kg CO_2) and the Alza model (414136.8 kg CO_2).

Figure 7 shows the total equivalent warming impact based on kg CO_2 emissions that contributed to the global warming. The Viva model emitted 461.7 tons CO_2, the Myvi model emitted 301.2 tons CO_2 and the Alza model emitted 415.3 tons CO_2. Indirect emissions emitted the most CO_2 compared to direct emissions. Carbon dioxide emission per kilometre can be seen in Fig. 8.

4 Conclusion

Prediction shows that the Alza model has the lowest refrigerant leakage to the environment compared to other models, this is due to a more efficient air conditioning system and leakage mitigation. The Myvi model too showed the least CO_2-e emission. Hence, th eViva model has lower engine capacity but the CO_2-e emissions is the highest due to under power. CO_2-e contribution from indirect emission resulting from the burning of the fuel is the major contributor for global warming.

Acknowledgements The authors would like to be obliged to Centre of Air Conditioning and Refrigeration, Politeknik Ungku Omar for providing laboratory facilities, financial assistance and supports.

References

1. Fischer, S.K., Sand, J.R.: Total environmental warming impact (TEWI) calculations for alternative automotive air conditioning system, SAE (1997)
2. Sumantran, V., Khalighi, B., Saka, K.: An Assessment of alternative refrigerants for automotive applications based on environmental impact, SAE Alternative Refrigerant Forum (1999)
3. Akashi, O., Hijioka, Y., Masui, T., Hanaoka, T., Kainuma, M: GHG emission scenarios in Asia and the world: The key technologies for significant reduction, Energy Econ. **34** (2012)
4. Denis, C., James, B., Chen, J., Hirata, T., Hwang, R., Kohler, J., Suwono, C.P.A.: Mobile air conditioning (IPCC/TEAP Special Report: Safeguarding the Ozone Layer and the Global Climate System, 2003)
5. Yan, H.H., Guo, H., Ou, J.M.: Emissions of halocarbons from mobile vehicle air conditioning system in Hong Kong, J. Hazard. Mater. **278** (2014)
6. Ma, H., Balthasar, F., Tait, N., Riera-Palou, X., Harrison, A.: A new comparison between the life cycle greenhouse gas emissions of battery electric vehicles and internal combustion vehicles, Energy Policy **44** (2012)
7. Andress, D., Das, S., Joseck, F., Nguyen, T.D.: Status of advanced light-duty transportation technologies in the US, Energy Policy (2012)
8. Azmi, N., Zairi, O., Choong, C.G., Saw, C.L.: Reduction of environmental impacts with the application of 3R'S using compact refrigerant recovery machine, iCompEx'15 proceeding (2015)
9. Devotta, S., Astjama, S., Joshi, R.: Challenges in recovery and recycling of refrigerants from Indian refrigeration and air-conditioning service sector, AE International—Asia (2003)
10. Tan, P.: Comparison Perodua Myvi and Perodua viva. Available from: http://paultan.org/2007/05/11/perodua-myvi-vs-perodua-viva/. Accessed 20 Nov 2016
11. Tan, P.: Comparison perodua Myvi and Perodua viva. Available from: https://en.wikipedia.org/wiki/Perodua_Viva
12. Tan, P.: Comparison perodua Myvi and Perodua viva. Available from: http://www.peroduanewcar.com/comparison/perodua-alza-vs-perodua-myvi/
13. Khalighi, B., Fischer, S.: Environmental impact of automotive refrigerants. ASHRAE J. (September 2000)
14. Wu, X., Hu, S., Mo, S.: Carbon footprint model for evaluating the global warming impact of food transport refrigeration system. J. Cleaner Prod. **54** (2013)
15. AIRAH: Methods of calculating Total Equivalent Warming Impact (TEWI): Best Practice Guideline (2012)

Surface Recognition and Volume Generation for Symmetrical Parts Using a Mirror Approach

Ahmad Faiz Zubair and Mohd Salman Abu Mansor

Abstract Automatic feature recognition (AFR) plays an important role in process planning. One of the problems in recognizing cylindrical surfaces via volume decomposition is the problem in generating a uniform thickness for volume decomposition. With the combination of offsetting original surfaces and a mirror approach, the symmetrical cylinder volume decomposition can be generated uniformly. Sub delta volumes for finishing (SDVF) bodies were generated in half section before they being mirrored for full-section SDVF. With full-section SDVF generated, sub delta volume for finishing filled region (SDVF-FR) and sub delta volume for roughing (SDVR) were then generated. Volumes of these bodies were then compared with manual overall delta volume (ODV_{manual}) calculation. Results show that significant volumes were generated and the developed algorithm is adequate for recognizing symmetrical cylinder part models.

Keywords Automatic feature recognition · Symmetrical cylinder
SDVF · ODV

1 Introduction

Computer aided process planning (CAPP) is the bridge between CAD and CAM systems. In CAPP, automated feature recognition (AFR) techniques still have more room to be explored. The volume decomposition method is among many AFR methods an adequate procedure to be used. In order to generate volume decomposition bodies, topological data of part models has to be recognized. To ensure that

---

A. F. Zubair · M. S. Abu Mansor (✉)
School of Mechanical Engineering, Engineering Campus,
Universiti Sains Malaysia, 14300 Nibong Tebal, Pulau Pinang, Malaysia
e-mail: mesalman@usm.my

A. F. Zubair
e-mail: afz15_mec008@student.usm.my

© Springer International Publishing AG 2018
A. Öchsner (ed.), *Engineering Applications for New Materials and Technologies*,
Advanced Structured Materials 85, https://doi.org/10.1007/978-3-319-72697-7_13

the generated bodies from volume decomposition are accurate, the approach on generating must be reliable.

A cylindrical part model has a unique parameter as it has a conical shape in it. A previous method from other researchers using copy face translation in normal directions from the mid-point of the face will work well on a flat surface but shows non-uniform thickness for cylindrical faces. This can be improved by applying an offsetting technique on the cylindrical face rather than normal translation. In order to eliminate topological errors from the developed algorithm, the part model is divided into two sections. Only half of the section is used to generate the sub delta volume for finishing (SDVF). Then to generate other delta volume bodies of sub delta volume for finishing filled region (SDVF-FR) and sub delta volume for roughing (SDVR), the SDVF was mirrored. This approach works well for a symmetrical cylinder part but is limited for an unsymmetrical part as the shape and features will contrast from the other half section.

Research and works on CAPP had been started for more that fives decade ago. More than ten methods of CAPP including feature based, knowledge based, artificial neural networks, genetic algorithms (GA), fuzzy set theory and fuzzy logic, Petri nets (PN), agent, internet, standard for the exchange of product data (STEP)-compliant method, and functional blocks (FB) method/technologies had been reported [1, 2]. Among these methods, the feature based method had been adopted by many researchers due to its ability to facilitate the representation of various types of part data in a significant form to drive automated CAPP [2]. Various applications had used the feature based technology for CAPP including in mill-turn machining [3, 4], sheet metal [5], computer aided inspection planning [6], prismatic micro parts [7]. Beside volume decompositions, other researchers had used the AFR method from other techniques such as from STEP AP203 [8–10], commercial CAD system database [6, 11, 12] to extract features information.

Apart from all the mentioned methods and techniques, past researches show that volume decomposition techniques were also been used. Recently Katavaki and Abu Mansor [13] had used the volume decomposition method in their research, although cylindrical shapes for bosses were covered, whereas overall cylinder shape part models were not in their intentions. Bok and Abu Mansor [14] had used volume decomposition for regular and freeform part models and stated some limitations for cylindrical faces. This paper will focus on symmetrical cylinder part models that are reliable for turning machining feature recognitions. A cylindrical stock model will be considered and conical as well as planar surfaces that are related to turning machining such as tapering, grooving and facing will be recognized as well.

2 Algorithm Framework

Several software packages were needed to develop the algorithm. This included a CAD software to model the part model and export it as .SAT file and an open source C++ language CAD modeller software to develop the algorithm in order to

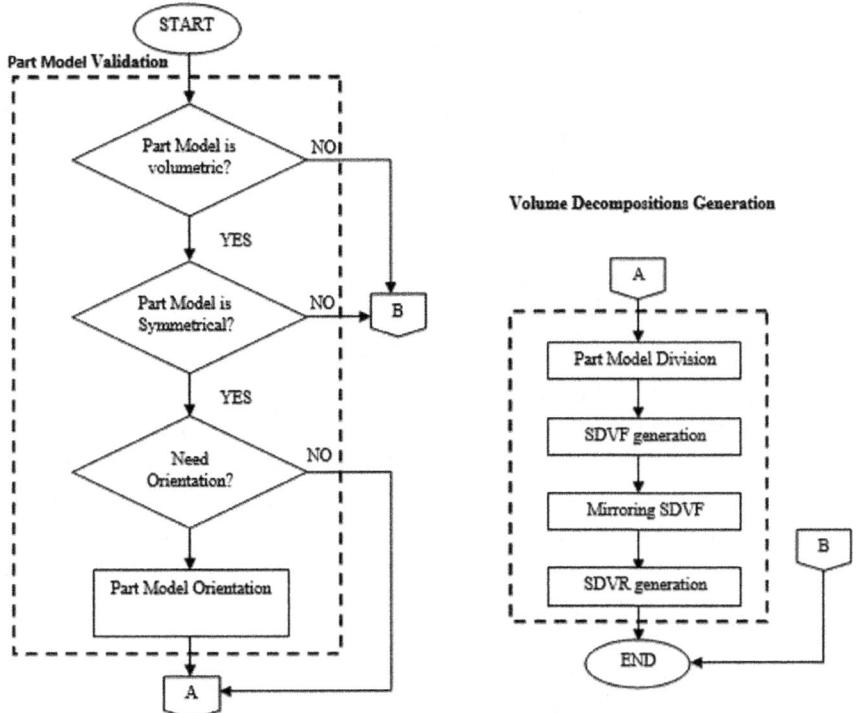

Fig. 1 Algorithm process flowchart

facilitate the recognition of features. The surface recognition will decompose a sub-delta volume that suits the finishing and roughing machining process. SDVF will be generated based on the thickness of the finishing while for SDVR the thickness will be according to the stock model dimensions. Figure 1 shows the process flowchart of the developed algorithm. The algorithm is based on two processes which are; (1) part model validations and (2) the volume decomposition generation.

2.1 Part Model Validation

The part model will be validated in three ways. These include the part model volume, part model symmetry and positioning of the part model. This process is to ensure that the part model is in the correct state and position before the volume decomposition process take place.

2.1.1 Part Model Volume

The part model will be validated in terms of their volume and type of file. If the part model does not represent a body, it means that it does not have any volume. Without volume, the algorithm will end the process. Apart from that if the file is not a .SAT file the algorithm will return an error. The validation process will be using the CAD modeller API function.

2.1.2 Part Model Symmetry

In order to validate if the part model is symmetrical or not, the topological data of the part model needs to be extracted. Topological data such as part model's faces, loops and edges number will be analyzed to find out the symmetry conditions of the part model. A part model is considered to be symmetrical if below conditions are met:

(1) Cylindrical part model, without any features within it, i.e.; holes or pockets.
(2) If the top or bottom plane has more than two cylindrical loops and the midpoint of the loops is at the centre.

2.1.3 Part Model Orientation

The part model orientation will be based on the planar top and bottom faces of the part model. This is explained in the author's previous work [15]. The cylindrical axis of the part model will be identified and it will be reoriented into the desired position of he z-axis if it is not. By the correct positioning of the part model, errors involving part model positioning can be eliminated.

2.2 Volume Decompositions Generation

Features of the part model will be recognized as volume decomposition in terms of SDVF and SDVR. By generating these sub-delta volumes, the exact volume of the material to be removed in the machining process can be calculated and known. For the ease of edge, co-edge and face recognition, the part has to be divided into two as the CAD modeller will work well on a half cylindrical shape rather than a full cylindrical part model. Then after the division, the part model features will be recognized and SDVF will be generated. Next, overlapping and gaps from bodies generated will be eliminated by generating sub delta volume of finishing filling region (SDVF-FR). After the generation of the half section, SDVF will be mirrored to get a full SDVF body. With the generation of full SDVF bodies, the part stock model and SDVR can be generated. All these bodies which are the overall delta volume ($ODV_{algorithm}$) are then compared with the manual overall delta volume (ODV_{manual}) to make sure that the volumes generated are reliable and correct. Equations (1–3) show how the calculation of the comparison of ΔODV being done. These equations are based on the preview work of Bok and Abu Mansor [14].

$$\text{Stock model volume, } Vs = ODV_{manual} + V_{CAD} \tag{1}$$

$$ODV_{algorithm} = \sum_{i=0}^{n} SDVF + \sum_{i=0}^{n} SDVF - FR + \sum_{i=0}^{n} SDVR \tag{2}$$

$$\Delta ODV = \frac{(ODV_{algorithm} - ODV_{manual})}{ODV_{manual}} \times 100 \tag{3}$$

2.2.1 Part Model Division

In order to eliminate the incompatible co-edge error produced by the software, one way to solve it is by dividing the part model into half. This is to eliminate the full round circle when the concave or convex edge of a circular plane meets with the planar or conical plane which is causing the co-edge to be incompatible. Figure 2 shows the process of the part model division. By choosing a middle plane of the xz-plane, the part model is then being cut into two. As the part model is symmetrical in the first place, only the right side of the part model will be chosen to recognize all the features. To make sure that the xz-plane is at the correct position of the middle plane, a calculation of finding the correct position needs to be done. Point n is the coordinate that needs to be chosen to generate the xz-plane.

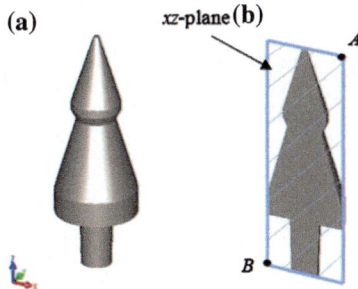

Fig. 2 Part model division. **a** Original part model **b** half section of part model

$$Point\ n = (x, y, z)$$
$$Point\ A = (maximum\ point\ x,\ 0,\ maximum\ point\ z)$$
$$Point\ B = (minimum\ point\ x,\ 0,\ minimum\ point\ z)$$

2.2.2 SDVF Generation

SDVF generation will be according to the three main features recognition. It will be based on planar, cylindrical and conical features attached to the part model. With the recognition of all these features a thin layer of volume decomposition will be generated according to the user input finishing thickness, normally 0.75–2 mm. The method will be using offsetting copy original faces to finishing thickness and generate the volume by lofting the original face and copied face. Figure 3 shows the recognition of cylindrical, conical and planar features of the half section part model. In order to generate the further sub delta volume, for instance SDVR full section of SDVF is needed. As SDVF bodies were generated by offsetting the original face via normal direction, gaps and overlapping between bodies will happen. A similar concept was being done by generating SDVF-FR [15].

2.2.3 Mirroring SDVF

The half-section of SDVF is being called back and analyzed. Topological data of the SDVF half-section body is gathered and being used to do the process. Figure 4 shows the mirror process. Because it is the half-section there is a planar face in the y-axis direction that needs to be selected to be the mirror plane. This plane is selected by determining its face type which is the planar type and normal direction of the plane face. Then the mirror function is being called to mirror all the body.

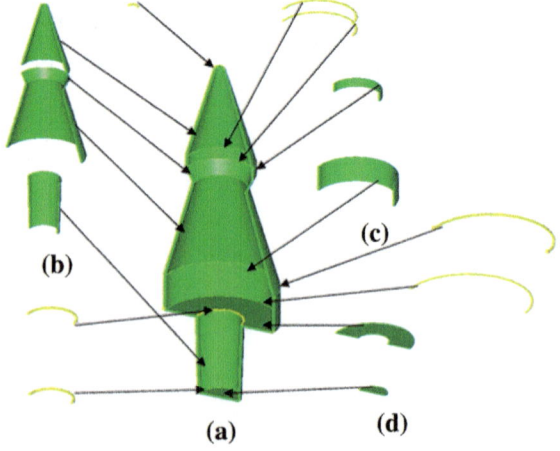

Fig. 3 SDVF generations. **a** Combinations of half-section SDVF **b** conical SDVF **c** cylinder SDVF **d** planar SDVF

Fig. 4 Mirror process. **a** Half-section SDVF before mirror **b** full section SDVF after mirror

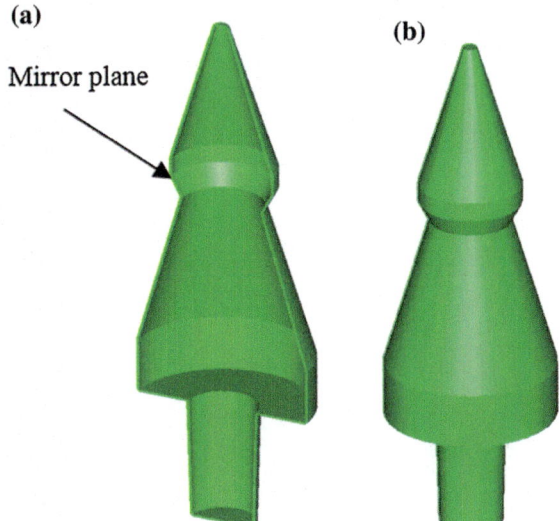

2.2.4 SDVR Generation

As the roughing process will be the first process in machining, the SDVR of the part model will be in the outer layer. Figure 5 shows the SDVR of the part model. SDVR will be the volume that subtracted part stock model, SDVF and the part model itself. The thickness will be based on the user input. The volume of SDVR bodies is then calculated.

In order to generate the part model, the stock model needs to be generated first. Because it is a cylindrical part model, the cylindrical stock model will be used rather than the square box stock model. The boundary thickness, bx for the x-axis direction, by for the y-axis direction and bz for the z-axis direction will be asked from the user. The SDVR will be divided into Left SDVR and Right SDVR for the volume recognition.

3 Results and Discussion

A symmetrical cylinder part model was implemented using the proposed approach in the developed algorithm. Figure 6 shows the exploded view of the ODV bodies generated. ODV bodies were differentiated in different colour codes. Grey colour for the original part model (Fig. 6d), green colour for SDVF bodies, and yellow colour for SDVF-FR and blue colour for SDVR bodies. The left-hand SDVR is shown in Fig. 6a and right-hand SDVR is shown in Fig. 6b. Table 1 shows the volumes of all the ODV bodies. Percentage errors of the different ODV are shown below. The finishing thickness, t is 1 mm and the boundary thickness for each direction, bx, by and bz is 4 mm (Fig. 7).

Fig. 5 SDVR and SDVF of part model in facet view

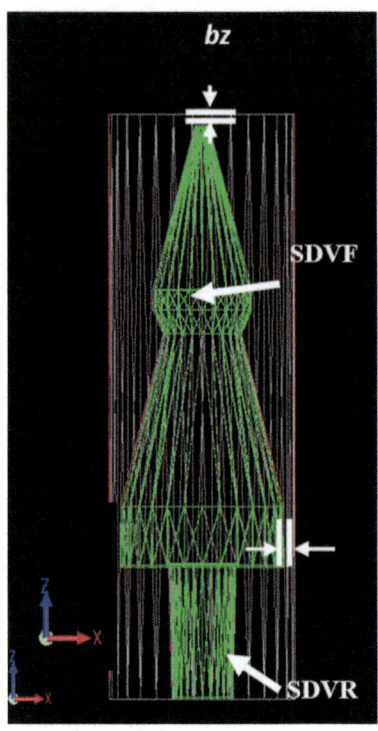

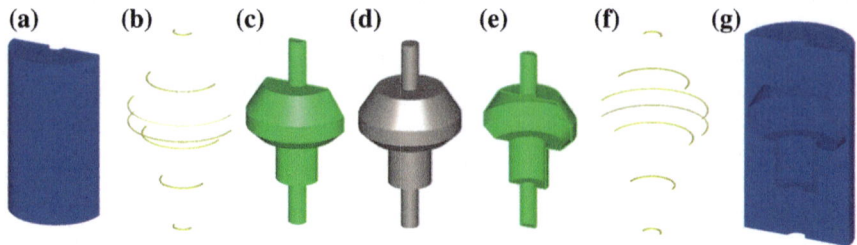

Fig. 6 Exploded view of the generated ODV bodies. **a** Left SDVR, **b** left SDVF-FR, **c** left SDVF, **d** original Part model, **e** right SDVF, **f** right SDVF-FR **g** right SDVR

Calculation of ODV_{manual} for example part model is shown below:

$$Final\ stock\ model\ volume,\ V_s = 3345570\ mm^3$$
$$ODV_{algorithm} = 2553628\ mm^3$$
$$V_{CAD} = 792223\ mm^3$$

Table 1 ODV bodies volume representation

No	Body	Volume (mm^3)
1	Right SDVF	28,042.4
2	Left SDVF	28,042.4
3	Right SDVF-FR	499.8
4	Left SDVF-FR	499.8
5	Right SDVR	1,248,272
6	Left SDVR	1,248,272
	Total $ODV_{algorithm}$	2,553,628

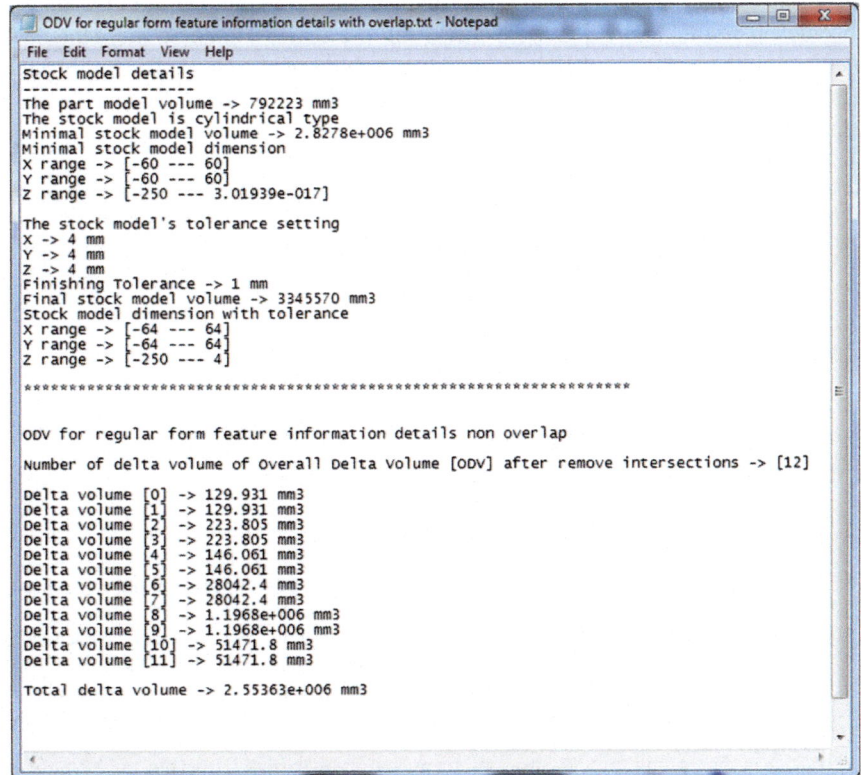

Fig. 7 Output of the volume decomposition generated

From Eqs. (1–3), calculation of ODV_{manual} will be:

$$3345570 = ODV_{manual} + 792223$$
$$ODV_{manual} = 2553347 \text{ mm}^3$$
$$\Delta ODV = 0.011\%$$

4 Conclusion

This paper introduced the mirror approach on generating volume decomposition of symmetrical cylinder part models. The offsetting and division techniques have a propensity to produce a consistent body thickness of SDVF bodies and to produce an efficient result. SDVF-FR were then been created to eliminate the gaps. The blend of all the techniques produced nearly a zero error of ΔODV. The verification part model example shows a significant sum of errors in recognizing surfaces and generating volume decompositions of wanted surfaces.

Acknowledgements This research is supported by Universiti Sains Malaysia under the Research University Grant (No: 814247). The first author also would like to thank the support of the Universiti Teknologi MARA for staff sponsorship.

References

1. Xu, X., Wang, L.H., Newman, S.T.: Computer-aided process planning: a critical review of recent developments and future trends. Int. J. Comput. Integr. Manuf. **24**, 1–31 (2011)
2. Yusof, Y., Latif, K.: Survey on computer-aided process planning. Int. J. Adv. Manuf. Technol. **75**, 77–89 (2014)
3. Campbell, M.I.: Automatic reasoning for defining lathe operations for mill-turn parts: a tolerance based approach. J. Mech. Des. **136**, 1–10 (2016)
4. Dwijayanti, K., Aoyama, H.: Basic study on process planning for turning-milling center based on machining feature recognition. J. Adv. Mech. Des. Sys. Manuf. **8**, 1–14 (2014)
5. Gupta, R.K., Gurumoorthy, B.: Classification, representation, and automatic extraction of deformation features in sheet metal parts. CAD Comput. Aided Des. **45**, 1469–1484 (2013)
6. Kamrani, A., Abouel Nasr, E., Al-Ahmari, A., Abdulhameed, O., Mian, S.H.: Feature-based design approach for integrated CAD and computer-aided inspection planning. Int. J. Adv. Manuf. Technol. **76**, 2159–2183 (2014)
7. Kumar, S.P.L., Jerald, J., Kumanan, S.: CAPP for prismatic micro part.pdf. Int. J. Comput. Integr. Manuf. **28**, 1046–1062 (2014)
8. Yusof, Y.: ISO 14649 (STEP-NC): new standards for CNC machining. Int. J. Integr. Eng. **14649**, 45–52 (2010)
9. Kannan, T.R., Shunmugam, M.S.: Processing of 3D sheet metal components in STEP AP-203 format. Part II : feature reasoning system. Int. J. Prod. Res. **7543**, 1287–1308 (2016)
10. Sivakumar, S., Dhanalakshmi, V.: An approach towards the integration of CAD/CAM/CAI through STEP file using feature extraction for cylindrical parts. Int. J. Comput. Integr. Manuf. **3052**, 561–570 (2013)

11. Oussama, J., Abdelilah, E., Ahmed, R.: Manufacturing computer aided process planning for rotational parts. Part 1 : automatic feature recognition from STEP AP203. Int. J. Eng. Res. Appl. **4**, 14–25 (2014)
12. Deb, S., Parra-Castillo, J.R.: An integrated and intelligent computer-aided process planning methodology for machined rotationally symmetrical parts. Int. J. Adv. Manuf. Syst. **13**, 1479–1506 (2011)
13. Kataraki, P.S., Abu Mansor, M.S.: Auto-recognition and generation of material removal volume for regular form surface and its volumetric features using volume decomposition method. Int. J. Adv. Manuf. Technol. (2016). doi:https://doi.org/10.1007/s00170-016-9394-6
14. Bok, A.Y., Abu Mansor, M.S.: Generative regular-freeform surface recognition for generating material removal volume from stock model. Comput. Ind. Eng. **64**, 162–178 (2013)
15. Zubair, A.F., Abu Mansor, M.S.: Cylindrical axis detection and part model orientation for generating sub delta volume using feature based method. ARPN J. Eng. Appl. Sci. **11**, 13415–13419 (2016)

Influence of Passenger Car Air Conditioner System Thermostat Level Setting to Fuel Consumption and Thermal Comfort

Rawaida Muhammad, Muhammad Khiril Kamaruddin and Yeoh Poh See

Abstract Users would adjust air conditioner knob position to regulate the cool air supply to the cabin. Essentially, adjusting the knob position would affect fuel consumption of the vehicle. Thus, this study is done to measure the fuel consumption of different thermostat level settings of an air conditioning system, in the case of stationary car, without passengers and driver in it. The data in this study were taken in four different modes and two levels of selected fan speed. At the same time, the cabin temperature was monitored by a specific device to record the temperature, relative humidity, and CO_2 emission. Fuel consumption is monitored by filling 5 L of petrol into an external tank. After running for one hour, the total fuel consumption is calculated by measuring the remaining of the fuel amount. At the end of the experiment, the most practical and suitable thermostat level setting is proposed. Besides saving fuel, the outcome of this study showed that the most comfortable condition in the cabin is 25.83 °C, 51.72% relative humidity and CO_2 is 553 ppm. The selected mode is actually at mode 4, low blower fan speed with the fuel consumption is amounted at 1220 mL.

Keywords Air conditioner · Thermostat · Fuel consumption · Thermal

---

R. Muhammad (✉) · M. K. Kamaruddin · Y. P. See
Department of Mechanical Engineering, Politeknik Ungku Omar,
Jalan Raja Musa Mahadi, 31400 Ipoh, Perak, Malaysia
e-mail: rawaida_muhammad@yahoo.com

M. K. Kamaruddin
e-mail: mkhirilk@gmail.com

Y. P. See
e-mail: yeohpohsee@gmail.com

1 Introduction

Malaysian car users desperately need comfort while driving because of the hot and humid weather throughout the year. It is most important during driving in long journeys. The human thermal comfort is defined by the American Society of Heating, Refrigeration and Air Conditioning Engineers (ASHRAE), as the state of mind that expresses satisfaction with the surrounding environment [1].

Thermal comfort of humans corresponds to a simple but permanent motivation that compels them to seek certain climatic situations, to maintain some of them, and to judge them in terms of pleasure or inconvenience. It becomes one of the significant criteria of selection during the acquisition of a car [2].

When the car users need comfort, the thermostat adjuster is set to a low temperature level. This will force the air condition system to use more power and increase the fuel consumption. In the automotive industry, fuel reduction can be done by controlling the use of air conditioning systems, mainly the compressor. Automotive industry today has applied a few modifications to control air conditioning compressors [3]. This system can affect the fuel consumption because the compressor engages with the engine. This will result in an increased load on the engine and fuel consumption also increases. The effect is larger with higher fuel economy vehicles [4]. As the fuel price can get very high depending on demand, the need to minimize fuel usage is a must and at the same time, comfort in the vehicle cannot be ignored.

Normally car users adjust the thermostat without taking into consideration of the fuel consumption. When the thermostat level is at a lower temperature set point, it can cause the compressor to continue to operate and it leads to an increase in the energy consumption. At the same time the temperature in the car becomes too cold that can cause discomfort. The results from this study will provide a temperature adjustment that can reduce fuel consumption and at the same time maintain the car cabin thermal comfort at an acceptable level.

Prediction the thermal comfort in a vehicle is very complex. It is because of the behaviour during a fast cool-down after a hot soak, a non-uniform thermal environment related with a very limited to a small area air velocity and temperature distribution, and solar flux and heat flux of radiation around the inner surface. Analyzing the temperature and velocity distribution in the passenger compartment, coupled with the forecast of the thermal comfort can guide the direction of the design in the early stages of the vehicle development process [5].

Thus, this study is done to measure the fuel consumption regarding to the thermostat level in a car air-conditioning system and also to suggest the most reasonable set point that achieves a good balance between thermal comfort and fuel consumption. It focuses on a Sedan car with approximately 3 m^3 volume of size as a model. Experiments are conducted in the morning around 9.00 a.m to 11.00 a.m. The tested vehicle is static in parking mode and uses 5 L of fuel, supplied to an external tank. The fuel consumption is practically measured after 1 h of experiment together with the temperature and humidity reading.

2 Experimental Setup

2.1 Experimental Parameter

The car cabin temperature environment is complex as it cannot be assessed with temperatures only because it will change from time to time. Air temperature, relative humidity, air velocity, environment radiation are important parameters that affect the thermal comfort [6]. In order to get the appropriate comfort, the temperature and humidity should be at about 72 F (22 °C) to 80 F (26.67 °C) and 30–60% in the summer according to the *ANSI/ASHRAE Standard 55*. Whereas for the CO_2 reading, 1000 ppm would be regarded as the upper limit of healthy condition, and the equipment limitation should be acceptable in the determining of the quality of air in the compartment [7]. Those parameters are actually controlled by using a thermostat.

Thus, this experiment is intended to study the thermostat level setting and also the blower fan speed that affect fuel consumption. Besides, additional parameters that influence the transient thermal comfort during the experiment period are also recorded. The experiments included conditions of 2 blower fan speeds, and 4 thermostat level settings. Each test is repeated in the same manner with different settings.

The procedure to study the temperature-time dependence is done by transferring data from the acquisition to the computer system for every 60 s. The air conditioning system has been used in both situations and the parameters were the same (temperature, flow, type). Values for four cases in the experiment are presented in tables considering between the hours 9:00–10:00 for Case 1 and between 12:00–13:00 for Case 2 [8].

2.2 Experimental Procedures

This research started with the fabrication of the external fuel tank, see Fig. 1. It is made of iron plates and upon completion it has the capacity to store up to 10 L of fuel. The need for an external tank is to enable a more accurate observation and measure of the consumed rate of fuel.

A Proton Persona car as in Fig. 2 is used to collect all the data needed for this study. The Proton Persona is a 2009 model with 1.6 L high line, as shown in Fig. 3. For the purpose of this study, the car's usage is assumed at good conditions and using tinted film on the front side windows at 70% of light transparency, side windows and rear windows at 50% light transparency.

A measuring instrument is used to measure and record the data of temperature and CO_2 emission inside the car. Then, it is installed with a data logger that is capable of logging data every hour, or by minutes or seconds. Other than that, in order to measure the air velocity and volume flow, an anemometer is used.

Fig. 1 External fuel tank

Fig. 2 Experiment car, proton persona, model 2009

SPECIFICATIONS			1.6L BASE LINE		1.6L MED. LINE		1.6L HIGH LINE
			MT	AT	MT	AT	AT
Seating Capacity	(person)		5				
Dimension and weight	Overall Length	(mm)	4477				
	Overall Width	(mm)	1725				
	Overall Height	(mm)	1438				
	Wheel Base	(mm)	2600				
	Kerb Weight	(kg)	1170	1195	1195	1220	1240
Engine	Model		Campro				
	Valve Mechanism		16-V DOHC				
	Displacement	(cc)	1597				
	Maximum Output		82 kW / 6000 rpm				
	Maximum Torque		148 Nm / 4000 rpm				
	Fuel Type		Petrol				
	Fuel Tank Capacity		50 Litre				
Performance	Acceleration 0-100km/h	(sec)	12.0	14.3	12.0	14.3	14.3

Fig. 3 Proton persona specification

Fig. 4 External fuel tank in front of the car

Fig. 5 Instrument positioning; **1**-temperature and CO_2 measuring instrument and RH/temperature data logging kit 2077 7001 and **2**-anemometer

Once the external fuel tank is ready, a new fuel hose is used to supply fuel from the external tank to the engine. By using the new hose, it minimizes the possible contamination of other materials to the system. This ensemble of fuel tank is placed at the exterior front end of the car as shown in Fig. 4. The measuring instrument is placed in the car between the driver and passenger seat; at the left side of the driver; kindly refer to Fig. 5. By using the anemometer to measure the air velocity and volume flow, the rate of air speed at each phase of the speed blower fan is determined.

Once the setup is ready, the car is placed at a suitable area in the workshop where the temperature is at 28 °C ± 1. The experiment is run between 9.00 am and 10.00 am in the workshop. The temperature measured at that particular workshop area around this time ranged between 27 and 29 °C.

Data for this experiment is recorded once the temperature is measured at 28 °C and this is to ensure stable environment temperature readings as referred to ASHRAE readings. Therefore, the initial temperature inside the car's cabin is the same for each phase of experiment. Data collection carried out for after 1 h engine run and ended as soon as it reached 1 h.

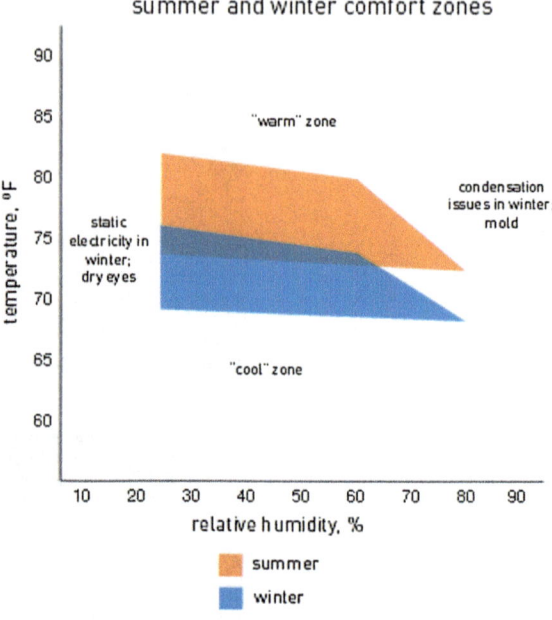

Fig. 6 Comfort zones for summer and winter

All collected and recorded data are presented in the data collection form. Every speed run is recorded in a single form. The data logger also recorded the temperature, humidity and CO_2. Simultaneously, the fuel consumption is obtained using a measuring cylinder. After the experiment ended, the measuring cylinder is used to measure the remaining fuel. This amount is deducted from the initial 5 L in order to obtain the consumed fuel.

Results are then plotted in a graph. From the plotted graph, select the mode, which is most fuel-efficient, and at the same time car's cabin reading as most comfortable. Refer to Fig. 6 for the determination criteria that this research referred to.

3 Results and Discussion

In regards to the results and discussions of the experiment, it is mainly about the effect of the thermostat level setting on the fuel consumption. As the thermostat level knob in this vehicle is not marked with the temperatures, it is assumed that this vehicle is not equipped with an automatic climate control system.

Thus, the thermostat level is believed to indicate the temperature of air supplied to the cabin from the air-conditioning system (which remains fixed without taking the cabin temperature into consideration), and not the temperature of the cabin itself. Such, the decision of the vehicle maker to omit the temperature marking is understandable as any marking will make the passenger think that the air-conditioning system is trying to regulate the cabin temperature as per the displayed temperature marking, which is not true.

3.1 Experiment of Low Fan Speed Performance

The highest temperature as shown in Fig. 7 is 26.16 °C whereas the lowest temperature is at 23.88 °C when the experiment is running at Mode 4. This experiment runs for a span of 5 min or 300 s. The data was not stable in Mode 3 and Mode 4. To overcome this problem, assumptions were made to set these two modes higher than Mode 1 and Mode 2. However, after running the experiment for 1 h, these two modes were still not stable.

To discuss the instability at Mode 3 and Mode 4, it is concluded that a higher heat transfer was needed to decrease the cabin's temperature, thus, longer time recorded. This has resulted in the need that the compressor continuously compresses within the period of experiment. When this happens, extra load is applied on the engine by the compressor and as a result of this, more fuel is consumed.

The graph in Fig. 8 shows how the relative humidity is plotted against time. The highest humidity indicated in this graph is 52.875% at Mode 4 and the lowest temperature is 38.784% at Mode 2. The thermal comfort sensation is optimal when the relative humidity value is about 50% [9]. The ASHRAE guidelines recommend a relative humidity (RH) of 30–60% from *ANSI/ASHRAE Standard 55*, 2010. Through the experiment, data has shown that Mode 1 is the most comfortable because the value of the relative humidity falls in the range of 50–55%. Figure 9 shows that the total CO_2 at Mode 1 and Mode 2 is higher than at Mode 3 and Mode 4.

However, the CO_2 amount is still considered in stable condition. The best starting condition of CO_2 is at Mode 3 because it recorded the lowest reading for the fan speed level. But then again, it is also important to bear in mind, that this data was recorded in a situation whereby there is not a driver or passengers in the car's cabin. It is definitely needed to acknowledge that the effect of CO_2 at this very low

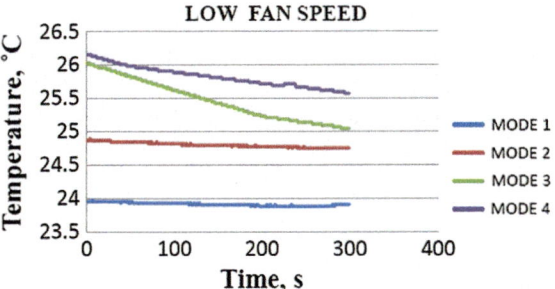

Fig. 7 Graph showing temperature versus time at low fan speed condition

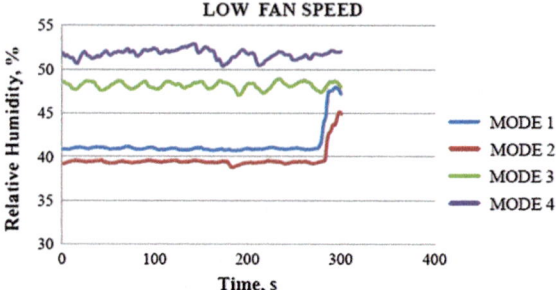

Fig. 8 Graph relative humidity versus time at low fan speed condition

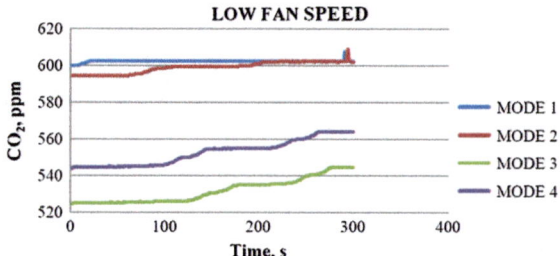

Fig. 9 Graph CO_2 versus time at low fan speed condition

Table 1 The average result for low blower fan in last 5 min data

	Temperature (°C)	Relative humidity (%)	CO_2 (ppm)	Fuel consumption (mL)
Mode 1	23.91	41.30	602	1825
Mode 2	24.75	39.61	599	1770
Mode 3	25.43	48.16	532	1655
Mode 4	25.67	51.72	553	1220

level is also due to non-contributing human factor. The CO_2 is influenced by the air recirculation. The air re-circulation must contain fresh air to reduce the CO_2.

Table 1 presents the average data result collected and recorded for 5 min for low blower fan speed. The parameters shown recorded are temperature, relative humidity, CO_2 emission and fuel consumption. Filling the external tank with 5 L of fuel is used to test the fuel consumption. After running the experiment for 1 h, the amount of remaining fuel is measured. The total of fuel consumption is calculated by deducting the amount of fuel left to the initial 5 L of fuel.

Table 2 The average result for medium blower fan in last 5 min data

	Temperature (°C)	Relative humidity (%)	CO_2 (ppm)	Fuel consumption (mL)
Mode 1	18.5	47.31	650	1660
Mode 2	20.53	50.19	637	1605
Mode 3	22.67	50.14	600	1550
Mode 4	23.4	51.16	602	1440

3.2 The Experiment of Medium Fan Speed Performance

Table 2 presents the average result data in the last 5 min for medium blower fan speed. The parameter shown here are the temperature, relative humidity, CO_2 and fuel consumption.

3.3 Fuel Consumption for Two Different Types of Blower Fan Speed

Figure 10 shows the fuel consumption when two different blower fan speeds are implemented. At Mode 4 for the low blower fan speed, it recorded the lowest fuel consumption at 1220 L. The highest fuel consumption is at Mode 1, for the low blower fan speed with a total of 1825 mL of fuel.

3.4 Temperature History and Fuel Usage During Total Experiment Time (1 h) of Low Blower Fan Speed

At the beginning of the graph in Fig. 11, all modes started at the initial temperature of 28 °C. For this research, the temperature reached the comfort zone at 26 °C when applying Mode 1 in the shortest time i.e. at 1000 s. At that time, the recorded fuel consumption was 550 mL. While Mode 2 was only able to reach the same temperature after 2000 s the air-conditioning turned on. The fuel used was 1000 mL. Mode 3 and Mode 4 were able to reach the comfort zone at 26 °C over 3500 s after the air conditioning was turned on.

However, the lowest fuel consumption is at Mode 4. The temperature at the time was 25.186 °C. This situation is still within the comfort zone. Although the fuel consumption was at its lowest but to reach the comfort zone system, it has actually taken a longer time compared to other modes.

The fastest mode to achieve a comfort level was when applying Mode 1; at 1000 s. Therefore, it is suggested that, this mode is used as an option for the user to achieve faster comfort in a car. But then again, the fuel consumption is still high at this stage recording at 1825 mL of fuel consumed.

Fig. 10 Graph fuel versus mode for two different type of blower fan speed

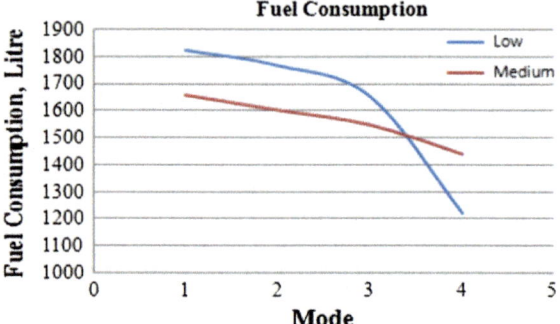

Fig. 11 Temperature and fuel total versus time for low blower fan speed

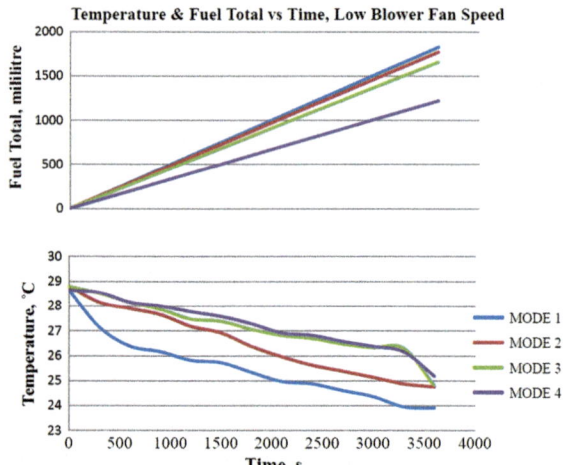

Through the analysis, the user can also reduce the fuel consumption with multiple choice modes. The first option is applying Mode 1. This is because the Mode 1 is able to achieve the level of comfort the fastest. The next selection mode is Mode 2; when the temperature reaches 25 °C at 2000 s. At this point, the fuel consumption was amounted to 1000 mL.

When combined, these modes gave a more efficient result. It is suggested that, first starts the thermostat at Mode 2 for the first 2000 s until 3300 s and then change to mode 4. Total fuel consumption while in Mode 2 amounted to 600 mL at the changed situation, the temperature at Mode 2 recorded at 24.7 °C. At the same time the temperature at the start of Mode 4 was 26 °C, it remained until 3600 s, and the total fuel consumption at Mode 4 amounted to 120 mL. So, the total amount of fuel with diversifies selection mode was recorded and calculated at 1720 mL.

It is more economical for fuel consumption for Mode 1 and Mode 2 but higher for Mode 3 and Mode 4. Therefore, the choice in this range can accelerate the level of comfort achieved and at the same time reduce the fuel consumption.

3.5 Temperature History and Fuel Usage During Total Experiment Time (1 h) of Low Blower Fan Speed

In Fig. 12, the lowest fuel consumption is at Mode 4 with an amount of 1440 mL and a temperature at 23 °C. But at this point the comfort temperature is within the range of 22–26 °C and it is achieved in 1250 s. The fastest mode to reach the comfort temperature of 26 °C is Mode 1. In this mode the fuel consumption is 100 mL. Mode 4 at medium blower fan speed achieved the level of comfort at a faster rate compared to Mode 1 at low blower fan speed.

However, the temperature at Mode 1 as in Fig. 12 shows that it is not within the comfort zone. The temperature ranged from 21 to 18.438 °C at Mode 1 started from 1200 to 3600 s. The temperature in this range is considered to be too cold. Figure 12 shows that Mode 3 achieved the comfort zone because the temperature recorded at that point read 22.011 °C.

From the graph itself, Mode 3 and Mode 4 are selected because they are in the comfort zone. If selections are made with diversify selection modes, then Mode 1 is selected first. Mode 1 reached the level of comfort at 26 °C with a fuel consumption at 100 mL running for 700 s. When Mode 2 reached 26 °C, the thermostat is changed from Mode 1 to Mode 2. The fuel consumption at Mode 2 is 100 mL. Change the thermostat to Mode 3 when it reached a temperature reading of 26 °C. Use Mode 3 for 300 s then switch to Mode 4. For 300 s in Mode 3 only 200 mL fuel was consume. The use of Mode 4 is the longest set point caused of fuel savings going on here. Mode 4 begins at 1300 s and ended at 3600 s, calculated 2300 s used in this mode. Temperature for Mode 4 at 3600 s is 23.016 °C. The fuel that has been used at this mode is 940 mL.

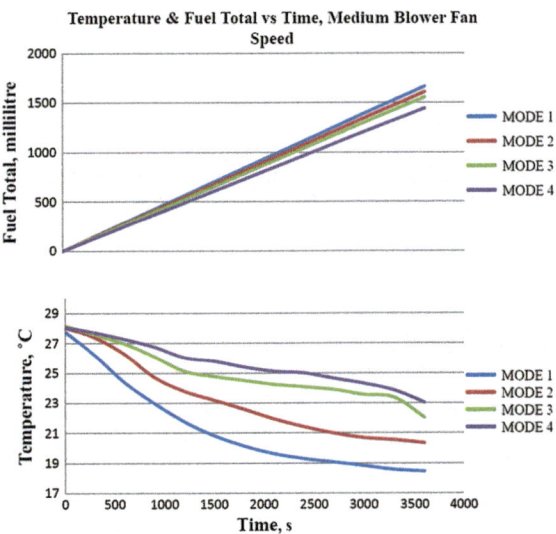

Fig. 12 Temperature and fuel total versus time for low blower fan speed

So, the total amount of fuel with diversifies selection modes was 1540 mL. Definitely the fuel consumption is more economical for Mode 1, Mode 2 and Mode 3 but higher than Mode 4. Therefore the choice in this range can accelerate the level of comfort achieved and at the same time reduce the fuel consumption.

4 Conclusion

This study shows that changing the thermostat level can reduce the fuel consumption. Through data collection and recording, it has been analyzed and discussed that for reducing the fuel consumption, Mode 4 working on the low blower fan speed proved to be the most fuel saving run. The fuel consumed amounted to a mere 1220 mL. The savings earned compared to the highest oil consumption stood at 605 mL. Besides saving fuel, it is also in this mode that the condition of temperature and humidity in the car's cabin were found to be most comfortable. Recorded temperature was 25.83 °C, relative humidity 51.72% and CO_2 is 553 ppm. The selected mode is at mode 4 with low blower fan speed.

In addition to the objective achieved, a choice can also be made to diversify the selection mode. Selection of this mode is to speed up the comfort ability achieved. This situation has been selected by not taking into account the minimum fuel consumption. From experiments made as analyzed and discussed before, at the early stages of Mode 1 on moderate selected fan speed blower until 1200 s; temperature recorded was 22 °C. Right after this, select Mode 2 at medium fan speed blower until 2000 s. At this point the temperature was between 24 °C down to 22 °C. Then, the next selected mode is Mode 3 and Mode 4 at medium blower fan speed or Mode 1 and Mode 2 at low blower fan speed.

It is suggested that this selection/option be used as a last resort; because until 3600 s the situation of temperature is still within the comfortable circumstances. In addition to this, the user can also decide to select Mode 4 at low blower fan speed starting from 3000 s because the use of this mode proved to save fuel but remain and maintain the comfort level.

Acknowledgements The authors thanked the Politeknik Ungku Omar and Department of Mechanical Engineering. Ipoh for research facilities and financial assistance provided for this project.

References

1. Oleesen, B., Brager, G.: A better way to predict comfort: the new ASHRAE standard 55-2004. ASHRAE J. **46**, 20–26 (2004)
2. Mezrhab, A., Bouzidi, M.: Computation of thermal comfort inside a passenger car compartment. Appl. Therm. Eng. **26**(14–15), 1697–1704 (2006)

3. Tenberge, P., Baumgart, R.: Reducing the fuel consumption by optimizing the air conditioning system. In: International Refrigeration and Air Conditioning Conference, vol. 1058. Retrieved from http://docs.lib.purdue.edu/iracc (2010)
4. Johnson, V.H.: Fuel used for vehicle air conditioning: a state-by-state thermal comfort-based approach. SAE Technical Paper (2002)
5. Han, T., Huang, L.: A sensitivity study of occupant thermal comfort in a cabin using virtual thermal comfort engineering. SAE Technical Paper. Reprinted from Climate Control, 724, 2005. (Paper Series)
6. Farzaneh, Y., Tootoonchi, A.A.: Controlling automobile thermal comfort using optimized fuzzy controller. Appl. Therm. Eng. **28**(14–15), 1906–1917 (2008)
7. Sahril, M., Fouzi, M., Othoman, A.M., Sulaiman, S.A.: Effect of recirculation on air quality in a car compartment. Aust. J. Basic Appl. Sci. **8**(4), 466–470 (2014) (Special 2014)
8. Neacsu, C., Ivanescu, M., Tabacu, I.: The influence of the solar radiation on the interior temperature of the car. ESFA 2009, Bucuresti **1**, 173–182 (2009)
9. Musat, R., Helerea, E.: Parameters and models of the vehicle thermal comfort. Acta Univ. Sapientiae, Electr. Mech. Eng. **1**, 215–226 (2009)
10. Boduch, M., Fincher, W.: Standards of Human Comfort. The University of Texas at Austin, School of Architecture, Texas (2009)

Investigation of the Piston Bowl Shape Effect on the Diesel Spray Development

Azwan Sapit, Mohd Azahari Razali, Akmal Nizam Mohammad,
Mohd Faisal Hushim, Azmahani Binti Sadikin
and Md Norrizam Bin Mohmad Ja'at

Abstract The combustion in a diesel engine is a very complex process. Even more when compared to a gasoline engine where the combustion is triggered using spark plugs, since the diesel engine depends on the auto ignition of the fuel-air mixture. This auto ignition phenomenon depends on many variables, especially on the fuel-air mixing condition and this makes the piston bowl geometry an important role on the combustion characteristics of diesel engines. Each piston bowl design was tailored to work well under certain specific conditions. The usage of biomass diesel mixtures in diesel engines increases the fuel viscosity and requires that the piston bowl geometry is carefully studied. The objective of this study is to compare the effect of certain piston bowl shapes, namely the Flat Bottom to the diesel spray development. Simulations were done using the ANSYS FLUENT 16.1 computational fluid dynamics (CFD) software. The simulation was performed on different injection pressures of 40 and 100 MPa, with the ambient temperature in the combustion chamber that is holding the piston at 500 and 900 K. Results showed that if

A. Sapit (✉) · M. A. Razali · A. N. Mohammad · M. F. Hushim
A. B. Sadikin · M. N. B. M. Ja'at
FAST Research Group, Universiti Tun Hussein Onn,
86400 Parit Raja, Batu Pahat, Johor, Malaysia
e-mail: azwans@uthm.edu.my

M. A. Razali
e-mail: azahari@uthm.edu.my

A. N. Mohammad
e-mail: akmaln@uthm.edu.my

M. F. Hushim
e-mail: mdfaisal@uthm.edu.my

A. B. Sadikin
e-mail: azmah@uthm.edu.my

M. N. B. M. Ja'at
e-mail: norrizam@uthm.edu.my

A. N. Mohammad
Faculty of Mechanical & Manufacturing Engineering,
Universiti Tun Hussein Onn, 86400 Parit Raja, Batu Pahat, Johor, Malaysia

the pressure and ambient temperature increase, the spray cone angle increases. In addition, the geometry shape of the piston bowl makes the spray to head downward after the wall impingement to the bottom section and prevents spillage of fuel over the piston bowl.

Keywords Diesel engine · Computational fluid dynamics · Piston bowl

1 Introduction

The fuel injection strategy is important for the diesel engine because it influences the performance and emission. During combustion, the fuel is sprayed in liquid form into a gaseous environment through a nozzle such that the liquid, through its interaction with the surrounding gas and by its own instability, breaks-up into droplets [1]. The spray characteristic is known to significantly affect the combustion and emission processes in diesel engines.

By optimizing spray characteristics, the unwanted emissions from the diesel engine, which are mainly NOx and PM, can be minimized [2–4]. Thus, the study on diesel sprays characteristics is surely very beneficial in the effort to reduce dangerous and unwanted emissions, and in the work of maximizing the efficiency by efficient usage of fuel and combustion [5]. Furthermore, the usage of biodiesel and biomass fuel (e.g. SVO) with increased viscosity makes the study on the diesel engine fuel injection strategy and its spray characteristics very important [4]. There are many studies that deal with diesel engine spray characteristics, due to the nature of the phenomenon; atomization of diesel and fuel mixing are very complex topics. It is understood that the injection pressure has a significant effect on the spray liquid penetration [6].

The spray tip penetration gets longer as the injection pressure increases, until a certain threshold. This result is related to both higher quantity and higher velocity of the droplets at higher injection pressures [7]. Proportional to the injection pressure, the spray penetrates faster at higher injection pressures [8].

The break-up length characterizes the point of discontinuity, where the spray changes from a zone of liquid (bulk liquid, or interconnected ligaments and droplets) to a finely atomized regime of droplets [9]. After the disintegration of the liquid column emerging from the nozzle, the generated droplets may further break-up into smaller ones as they move into the surrounding gas. The development of this deformation leads to the break-up into smaller droplets. The forces associated with the dynamic pressure, surface tension and viscosity control the break-up of a droplet.

However, CFD is a tool being used to predict some aspects of the study that are too difficult to be done through experiment [10]. In this research, the characteristics of the diesel spray are investigated focusing on changing ambient temperature and injection pressure with different bowl shape geometry diameters by using computational fluid dynamics. The obtained results were also compared to actual experimental data from previous research.

2 Methodology

This study used the computational fluid dynamics software ANSYS Workbench 16.1. In the CFD software, the process for the numerical analysis is divided into three stages, which are pre-processing, solution of the system of equations and post-processing. Table 1 shows the various parameters used in this simulation, and Fig. 1 shows the piston bowl shapes. For this study, the geometry of the piston bowl is drawn by using SolidWorks. It is then imported into the Design Modeler (DM) of ANSYS workbench as shown in Fig. 2a. The geometry is then required to be converted into a computational mesh as sown in Fig. 2b. The solver will then calculate the parameter values iteratively until convergence is achieved.

The last step in the ANSYS Fluent software is the post-processing where the animation, production of graphics and reports can be made through the post-processing and the result of the fluid dynamics data becomes legible. In addition, the post-processing can also shade the surface of the transparent model and generate path lines, vector plots, contour plots, and custom defined field variables and scene constructions.

Table 1 Parameter for spray pressure and temperature in combustion chamber

Spray injection pressure (MPa)	Ambient temperature (K)
40	500
40	900
100	500
100	900

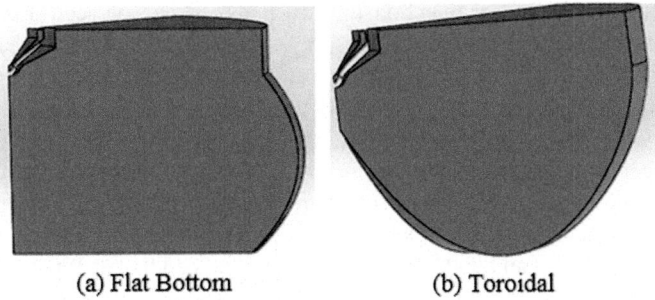

(a) Flat Bottom   (b) Toroidal

Fig. 1 Two types of piston bowl shapes

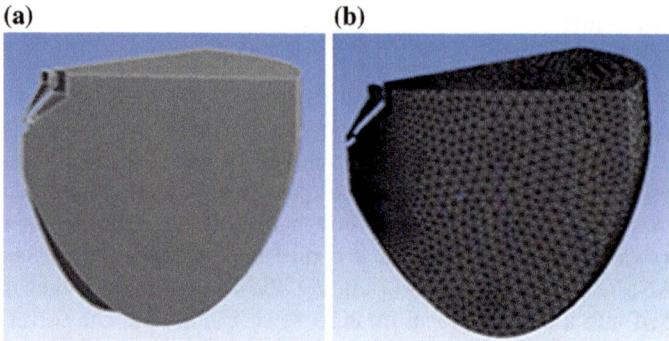

Fig. 2 Model of the piston bowl geometry (**a**) and mesh of the spray and piston bowl (**b**)

3 Results and Discussion

In this result and discussion section, the simulations were conducted at different injection pressures of 40 and 100 MPa, and at temperatures of 500 and 900 K. The fuel properties refer to diesel. The aim is to identify differences in the form of the spray development, spray areas, and spray droplet velocity for each different piston bowl shape.

Figure 3 shows the velocity contour of the diesel spray for (a) a flat bottom and (b) a toroidal shaped piston bowl. Images show the spray development at 0.1–0.3 ms after the start of the diesel fuel injection (SOI).

Results obtained in Fig. 3 show an interesting spray development. The injection pressure, ambient temperature, and the shape of the piston bowl clearly affect the spray characteristics and development. The spray velocity scale is shown in Fig. 4. Red indicates the maximum velocity of the diesel spray at 500 m/s while blue indicates that the velocity was at 0 m/s.

In general, at the initial injection time (0.1 ms), it can be seen that the spray penetrates the chamber with mostly the spray center (liquid core region) having the maximum velocity and the outer region (droplet region) which rapidly interacts with the surrounding air has a lower velocity due to aerodynamic force (air resistance) slowing down the spray particle. As time progresses, the spray core penetrated longer and impinged the piston wall. This is indicated by the longer maximum velocity contour when compared to 0.1 ms. Also, the spray body expands wider across the center axis. This is more significant with the increase in injection pressure and ambient temperature. It can be seen that at 100 MPa injection pressure and 900 K ambient temperature, the spray body has the widest width when compared to other conditions, which indicates that high injection pressure and ambient temperature increase the spray cone angle, as seen in the increase of the spray width.

A detailed comparison of the spray development between the flat bottom and toroidal piston shape indicates that after the spray impinged to the piston wall, the

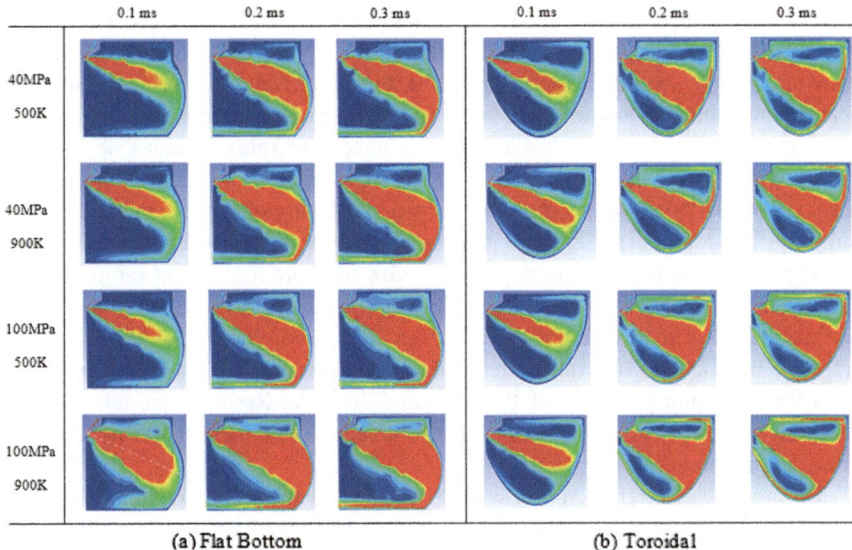

Fig. 3 Spray development on piston bowl shape

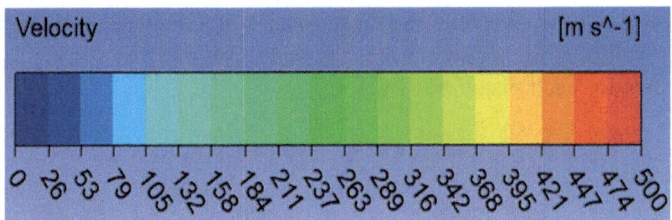

Fig. 4 Scale velocity of diesel fuel spray

spray body will be forced to follow the wall curvature. In the case of a flat bottom piston, the spray impinged at the piston wall and the spray core continues heading downward along the piston wall. In comparison, in the case of a toroidal, after impingement, the spray core seem to be divided into 2, with each separated spray heading upward and downward moving along the piston wall. This is clearly the effect of different piston shape geometries.

The spray development characteristics displayed for each piston bowl shape will definitely influence the effect of the fuel combustion in the chamber. The flat bottom bowl shape design has a longer spray travel distance before wall impingement. Additionally, after impingement, the spray moves downward to the piston bowl bottom section. These characteristics are believed to reduce the SOF generation due to reduction of fuel spillage over and outside of the piston bowl. This is especially helpful if the used fuel has a long ignition delay.

In comparison, the spray in toroidal has a shorter travel distance before wall impingement. In addition, after impingement, the spray is divided into 2, with each moving upward and downward. In general, the spray forced movement due to the piston bowl shape promotes turbulence and mixing. This is beneficial for fast and rapid atomization, but the bowl shape also makes the spray to move up and spill over the piston bowl, which will increase SOF generation. Thus, the use of a toroidal piston bowl shape must be considered thoroughly, and might be more suitable for small diesel engines that operate at high rpm (fast engine), which require very rapid fuel atomization paired with fuel that has a short ignition delay.

Figure 5 shows images of the spray development using rapeseed oil (represents a biomass fuel) for the toroidal and flat bottom shaped bowl. The fuel was allowed to impinge on each piston wall under ambient temperature of T_i = 298 and 700 K with fuel injection pressures of P_{inj} = 40 and 70 MPa. Results were taken from a previous study [4] for comparison. The image was captured at 0.5 ms after the spray impingement on the wall.

The images of the fuel spray, when compared to the results obtained by simulation are very similar. The rapeseed oil spray shows that for toroidal, the fuel spills outward from the piston bowl at higher injection pressure. The spillage becomes much more obvious with higher ambient temperature. As for the flat bottom, although there is some fuel moving upward, most of the fuel after impingement moved downward. When compared with toroidal, the flat bottom piston bowl reduces upward fuel movement and decreases the fuel spillage outside the piston bowl. This validates the simulation results and confirms the spray development characteristics for each piston bowl shape.

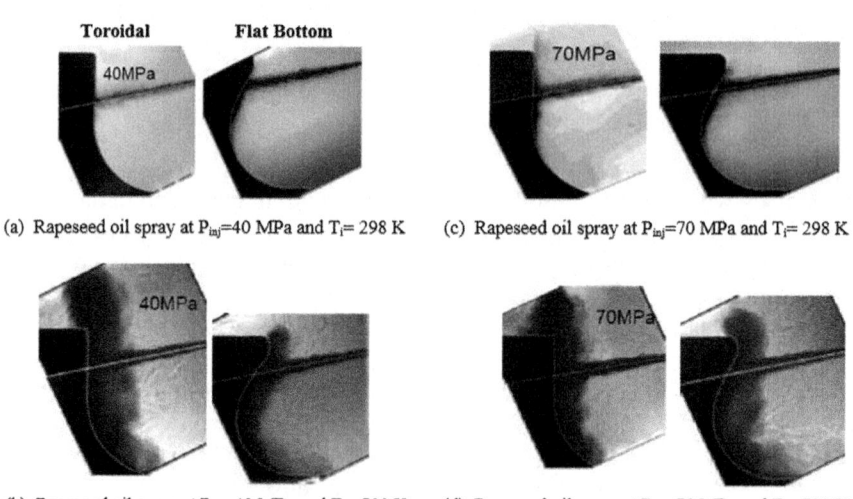

(a) Rapeseed oil spray at P_{inj}=40 MPa and T_i= 298 K   (c) Rapeseed oil spray at P_{inj}=70 MPa and T_i= 298 K

(b) Rapeseed oil spray at P_{inj}=40 MPa and T_i= 700 K   (d) Rapeseed oil spray at P_{inj}=70 MPa and T_i= 700 K

Fig. 5 Images of spray development (rapeseed oil) on piston bowl shape

4 Conclusion

Spray characteristics and development for two types of piston bowl shapes were studied. Each piston bowl shape influences the diesel spray development differently and with considerable effect. The use of a toroidal piston bowl shape must be considered thoroughly, especially in the case when using biomass fuel as after impingement, some of the fuel spills outward from the piston bowl and this could be the cause of SOF. Toroidal might be more suitable for a small diesel engine that operates at high rpm. Flat bottom seems the better choice for biomass fuel as it allows longer spray tip penetration (larger spray area) along the wall, and reduces the fuel spillage over the piston bowl. The usage of each piston bowl design should be paired with a suitable fuel delivery system and strategy to promote the desired combustion characteristic optimum for the required engine operation.

Acknowledgements The authors thanked the Faculty of Mechanical & Manufacturing Engineering, Universiti Tun Hussein Onn (UTHM) and FAST Research Group, UTHM, Johor for research facilities and financial assistance provided for this project.

References

1. Adam, A., Inukai, N., Kidoguchi, Y., Miwa, K. et al.: A study on droplets evaporation at diesel spray boundary during ignition delay period, SAE Technical Paper 2007-01-1893 (2007)
2. Sapit, A., Nagayasu, S., Tsuboi, Y., Nada, Y., Kidoguchi, Y.: A study on improvement of diesel spray characteristics fueled by rape-seed oil. SAE Int. J. Fuels and Lubricants **5** (2011-32-0561), 529–539 (2011)
3. Khalid, A., Tamaldin, N., Jaat, M., Ali, M.F.M., Manshoor, B., Zaman, I.: Impacts of biodiesel storage duration on fuel properties and emissions. Procedia Eng. **68**, 225–230, Elsevier Ltd (2013), Doi:10.1016/j.proeng.2013.12.172
4. Sapit, A., Yano, T., Kidoguchi, Y., Nada, Y.: Effect of wall configuration on atomization of rapeseed oil diesel spray impinging on the wall. Appl. Mech. Mater. **315**, 320–324, Trans Tech Publications (2013)
5. Khalid, A., Hayashi, K., Kidoguchi, Y., Yatsufusa, T.: Effect of air entrainment and oxygen concentration on endothermic and heat recovery process of diesel ignition. SAE Technical Papers, Society of Automotive Engineers of Japan, Inc. and SAE International (2011). Doi: https://doi.org/10.4271/2011-01-1834
6. Khalid, A.: Effect of ambient temperature and oxygen concentration on ignition and combustion process of diesel spray. Asian J. Sci. Res. **6**(3), 434–444, Asian Network for Scientific Information (2013) Doi:https://doi.org/10.3923/ajsr.2013.434-444
7. Hespel, C., Blaisot, J.B., Margot, X., Patouna, S., Cessou, A., Lecordier, B.: Influence of nozzle geometry on spray shape, particle size, spray velocity and air entrainment of high pressure diesel spray, THIESEL (2010)
8. Roisman, I.V., Araneo, L., Tropea, C.: Effect of ambient pressure on penetration of a diesel spray. Int. J. Multiph. Flow **33**(8), 904–920 (2007)
9. Martinez-Martinez, S., Sanchez-Cruz, F.A., Bermudez, V.R., Riesco-Avila, J.M.: Liquids spray characteristic in diesel engine. In: Siano, D. (ed.) Fuel Injection, pp. 20–48. Intech, Sciyo (2010)
10. Mohammed, A.N., Ismail, F.: Study of an entropy-consistent Navier-Stokes flux. Int. J. Comput. Fluid Dyn. **27**(1), 1–14 (2013)

Improvement of the Switching of Behaviours Using a Fuzzy Inference System for Powered Wheelchair Controllers

Jaya Bhanu Rao and Ahmad Zakaria

Abstract Power assisted wheelchairs have become very popular these days among elderly and disabled people. Manoeuvring the power assisted wheelchairs can become very challenging and stressful to the users especially at moderately crowded environments such as shopping complexes and public places. Users have to face various types of obstacles such as walls, static and dynamic obstacles. The wheelchair intelligent controller must be able to change between one behaviour to another behaviour without significant delay and jerk. This paper focuses mainly on the improvement of the switching of behaviours between follow-the-leader and emergency behaviour using the fuzzy inference system (FIS). The algorithm is tested and validated using the Player/Stage open source software under the Ubuntu 12.04 LTS platform. A wireless joystick Saitek Evo 6 axis is used to manoeuvre the powered wheelchair. A Hokuyo laser range finder is also used in this simulation software to scan static and dynamic objects at 180° of viewing angle. The results show significant improvement in the smoothness level in the switching of these behaviours.

Keywords Automation · Wheelchair · Fuzzy inference system

Nomenclatures

F_e Vector of Emergency Stop behaviour
F_l Vector of Follow-the-Leader behaviour
n Number of behaviours
R Single control behaviour for both behaviours
T_e Single control behaviour for Emergency Stop behaviour
T_l Single control behaviour for Follow-the-Leader behaviour

J. B. Rao (✉) · A. Zakaria
Universiti Kuala Lumpur Institute of Product Design and Manufacturing (IPROM), 56100 Kuala Lumpur, Malaysia
e-mail: jayabhanurao@unikl.edu.my

A. Zakaria
e-mail: dzakaria@unikl.edu.my

© Springer International Publishing AG 2018
A. Öchsner (ed.), *Engineering Applications for New Materials and Technologies*,
Advanced Structured Materials 85, https://doi.org/10.1007/978-3-319-72697-7_16

Greek Symbols

β_e Behaviour coefficient for Emergency Stop behaviour
β_i Behaviour coefficients corresponding to individual behaviours
β_l Behaviour coefficient for Follow-the-Leader behaviour

Abbreviations

FIS   Fuzzy Inference System

1 Introduction

Power assisted wheelchairs have become very popular these days among elderly and disabled people. Manoeuvring the power assisted wheelchairs can become a very challenging and stressful experience to the users especially at moderately crowded environments such as bus stations, railway stations, airports, shopping complexes and other public areas either indoors or outdoors. Moreover, users also have to face multiple types of obstacles such as straight walls, static and dynamic objects. Examples of static objects can be a pillar in the building, furniture, flower pots and rubbish bins while dynamic objects can be of people who are walking or running at any locations, children buggies or even pets.

Generally, all first-time power assisted wheelchair users usually struggle in manoeuvring this machine in reaching their destination smoothly. Prolonged use of this wheelchair will eventually increase the psychomotor skills of the user. However, exposure to a new environment will always cause unforeseen and inconsistent arrangements of objects along the path. Users will repeatedly go through the entire learning process with regards to variance in the object locations. As a result of this scenario, the tendency to collide with these objects is very high, in which the safety of both the wheelchair user and the public is always jeopardised.

In this research, the behaviour in a control system is defined based on human actions in overcoming any events such as avoiding obstacles or following a person in any environment. The term "behaviour" is mimicked from human behaviour in responding to the various types of obstacles, restrictions and limitations in reaching their destinations as described by Arkin [1] and Brooks [2]. These authors have discussed the way humans overcome the obstacles along their path by means of cognitive and psychomotor skills in any scenarios. The way humans react and response to the changes of obstacles, has been adopted by researchers and called as "Behaviour Based Control". Hence the following terms have been derived such as "Obstacle Avoidance Behaviour", "Follow-the-Leader Behaviour", "Wall Following Behaviour", "Emergency Behaviour" etc. These behaviours were originally used by researchers specifically for mobile robot navigations [3, 4].

In this paper, only two behaviours will be utilised in the semi-autonomous navigation of a powered wheelchair which are Follow-the-Leader behaviour and Emergency behaviour. In the Follow-the-Leader behaviour, the user wants to go to a specified location passing through the moderately crowded or partly cluttered environment such as shopping complexes or train stations. Unfortunately, due to too many oncoming people from various directions and velocities, the user will be stranded among the crowd, expecting the public to give way. In this situation, the Follow-the-Leader behaviour algorithm will be triggered to identify the potential person who has the same velocity and direction with the user's target. During this period, the user must manoeuvre the wheelchair manually by means of the joystick in order to activate this behaviour or function.

On the other hand, the Emergency behaviour refers to a situation in which the user of the powered wheelchair will be approached by an unexpected oncoming object in the opposite direction at a very close distance that can cause collision. At this juncture, the wheelchair user will not be able to stop the wheelchair on time due to the collision risk zone. This scenario usually happens at the entrance or exit of a door that has been surrounded by two opaque walls with narrow opening. The wheelchair user will not be able to stop or avoid collision immediately due to a situation called "local minima" which was described by researchers in mobile robot navigation. Local minima refer to a situation in which the power assisted wheelchair is surrounded by static and dynamic obstacles. The assistive intelligent controller of this wheelchair will not be able to divert its path to any direction; hence it comes to halt. In this context, the Emergency Behaviour will be activated to avoid or to minimise the wheelchair exposure to local minima.

The manual navigation of a powered wheelchair by the user will lead to many behaviours along its path. During this journey, the assistive intelligent wheelchair controller must be able to switch from one behaviour to another behaviour smoothly and without any significant delay. Many researchers have successfully designed and implemented the switching methods between behaviours in mobile robots. However, more research has to be conducted to adopt this algorithm in semi-autonomous power assisted wheelchairs. Firstly, one of the most significant differences between mobile robots and powered wheelchair is the velocity. Most mobile robots have a maximum range of speed from 0.5 m/s to 0.9 m/s while in powered assisted wheelchairs the maximum speed range varies from 1.5 m/s to 1.7 m/s. This is in compliance with the Road Transport Act in the United Kingdom. Secondly, almost all the mobile robots are fully autonomous with complete self-navigation systems while powered wheelchairs are semi-autonomous in which the navigation will be handled by the user. Upon reaching any potential obstacles the behaviour based assistive controller will overwrite the user's manual navigation.

In this paper, switching between Follow-the-Leader behaviour and Emergency behaviour is discussed in detail to improve the switching performance. Improvement on the switching time between these two behaviours has been simulated using Player/Stage open source software with the aid of a Saitek Evo 6-axis wireless joystick [5].

2 Methodology

Many researchers have ventured in designing a robust power wheelchair controller to assist the elderly or disabled people in manoeuvring their wheelchair to the desired destination [6–15]. However, sudden change in the switching of behaviours can lead to discomfort ride and fatigue to wheelchair users. Therefore, in order to minimise these effects an improved behaviour switching algorithm has been designed in this study. Figure 1 shows the overall block diagram of the implementation process for this project that is employed in this paper. The modified architecture for the powered wheelchair control is shown in Fig. 2. It is used to accommodate an improvement to the switching behaviour between Follow-the-Leader and Emergency stop. In order to increase the level of smoothness in the switching of behaviours, a Fuzzy Inference System (FIS) has been incorporated in the wheelchair control architecture.

In this paper, two switching behaviours are used for the improvement in the switching namely, the Follow-the-Leader behaviour and the Emergency stop behaviour as highlighted in Fig. 2. In this structure, the vector output of direction, velocity and detection angle of the laser sensor are fed into the fuzzy inference module to obtain the weightage values from both behaviours. Eventually, these values are blended together to obtain the final output vector which will be executed in the microcontroller unit. These signals are fed into the motor controller for the task accomplishment in switching of behaviours. The FIS module accepts input signals from both behaviours prior to computation of weighting values, as indicated in Fig. 3, with respect to the following Fuzzy Inference System rules.

Fig. 1 Powered wheelchair controller implementation process

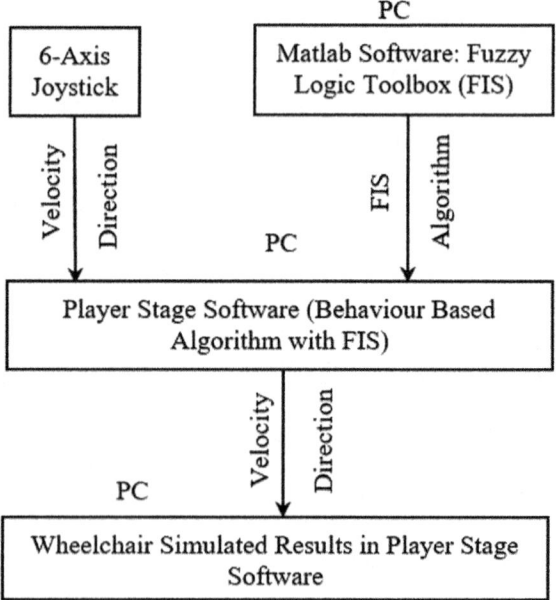

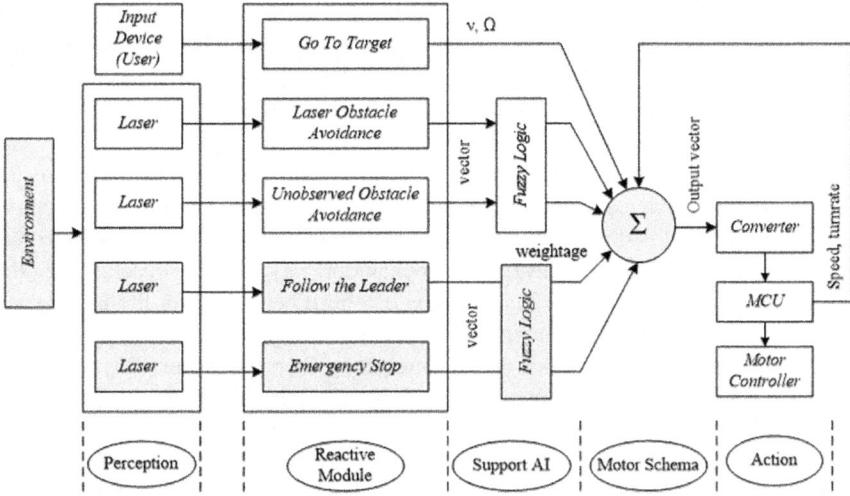

Fig. 2 Architecture of powered wheelchair controller

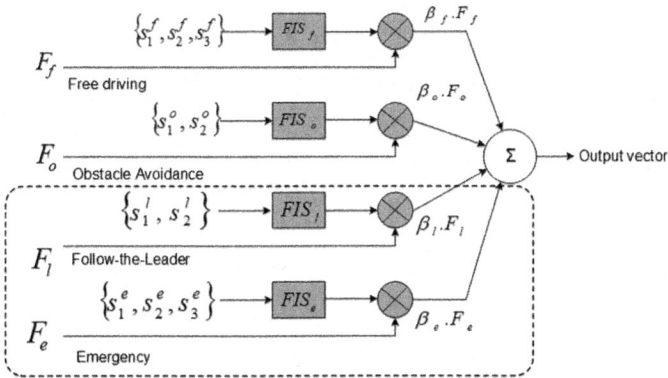

Fig. 3 Switching of behaviour module with FIS

If Distance is Close AND Angle is Small then Behaviour is Average.
Else if Distance is Close AND Angle is Average then Behaviour is High.
Else is Distance is Close AND Angle is Large then Behaviour Very High.
Else if Distance is Halfway AND Angle is Small then Behaviour is Low.
Else if Distance is Halfway AND Angle is Average then Behaviour is Average.
Else if Distance is Halfway AND Angle is Large then Behaviour is High.
Else if Distance is Far AND Angle is Small then Behaviour is Very low.
Else if Distance is Far AND Angle is Average then Behaviour is Low.
Else if Distance is Far AND Angle is Large then Behaviour is Average.

Figure 4 shows the FIS membership functions for the output variables of switching between these two behaviours. There are two input variables, namely the "Distance" and the "Angle", while the output variable is the "Behaviour" value. The input and output membership values are ranging from 0 to 1 in the shape of triangular form.

The following equations have been employed to compute the final output vector for the assisted powered wheelchair control system. Where T_l is the single control behaviour for Follow-the-Leader behaviour, β_l is the behaviour coefficient corresponding to the Follow-the-Leader behaviour, F_l is the vector of the Follow-the-Leader behaviour, T_e is the single control behaviour for the Emergency Stop behaviour, β_e is the behaviour coefficient corresponding to the Emergency Stop behaviour, F_e is the vector of the Emergency Stop behaviour, R is the single control behaviour for both behaviours as above, β_i is the behaviour coefficient corresponding to individual behaviours, F_l is the vector of individual behaviours and n is the number of behaviours. In this case, n can be represented as 2 for two behaviours namely, the Follow-the-Leader behaviour and the Emergency Stop behaviour.

$$T_l = \beta_l \cdot F_l \quad (1)$$

$$T_e = \beta_e \cdot F_e \quad (2)$$

$$R = \sum_{i=1}^{n} \beta_i \cdot F_i \quad (3)$$

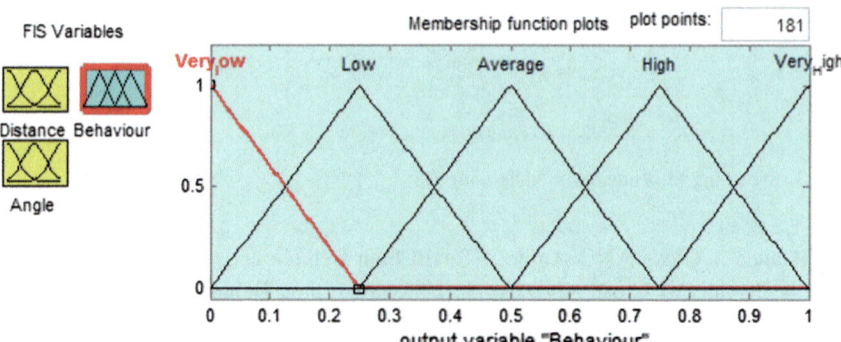

Fig. 4 FIS variables for switching of behaviours

3 Results and Discussion

The Fuzzy Inference System algorithm has been tested and validated using the open source Player/Stage simulation software under the Ubuntu 12.04 LTS platform. Figure 5 shows the graph of wheelchair speed versus time which has been captured in the software when the user was manoeuvring the wheelchair manually via wireless joystick. The results from the graph shows the jittering and uneven movement of the wheelchair in the partly cluttered environment.

3.1 Object/Obstacle Detection

Further experimental tests have been conducted by deliberately introducing two square obstacles in the scenario. This is to investigate the pattern of detection view of 180° by the Hokuyo Laser Range Finder, as illustrated in Fig. 6. The laser range finder is mounted in front of the powered wheelchair and it is capable to scan the environment in 2D single layer. In this study, the scanning distance has been set to 5 m even though it is capable of scanning objects up to 30 m.

The distance and angle of scanned view by this sensor for two square objects have been recorded and plotted in a graph, as shown in Fig. 7.

3.2 Follow-the-Leader Behaviour

The Follow-the-Leader behaviour has been simulated by deliberately introducing another dynamic wheelchair user as a leader in a scenario as illustrated in Fig. 8.

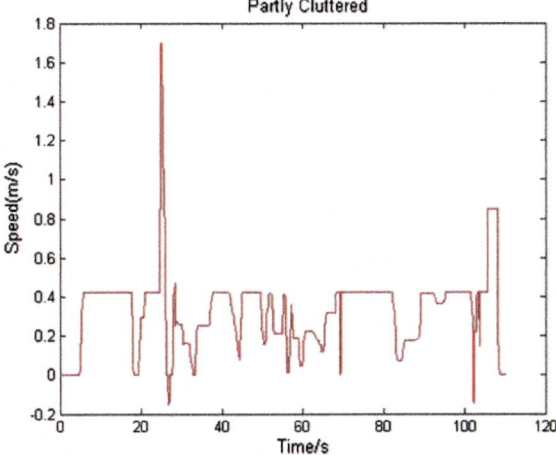

Fig. 5 Wheelchair travelling speed in partly cluttered area

Fig. 6 Object detection with Hokuyo laser rangefinder

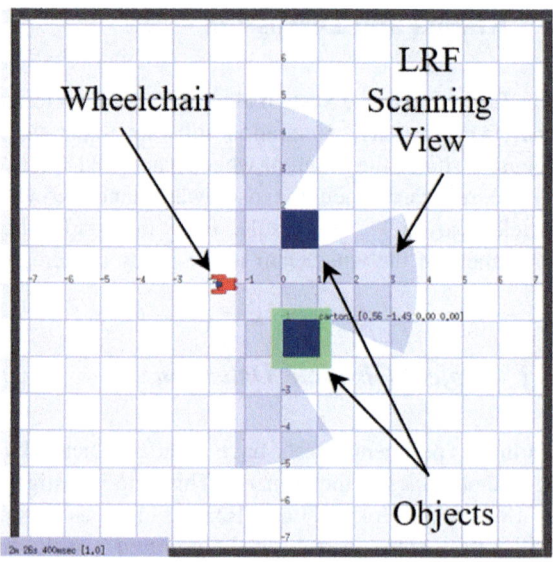

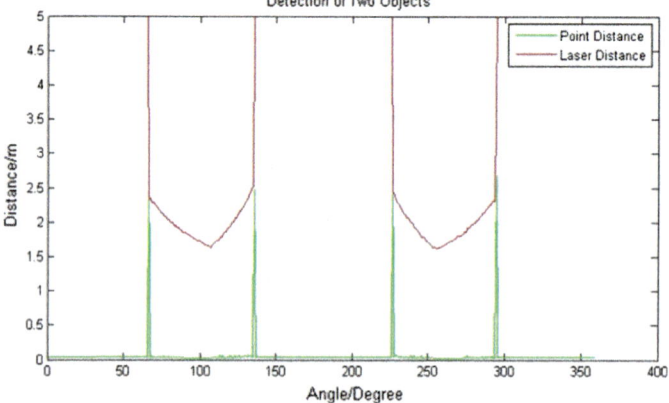

Fig. 7 Graph of object detection with laser rangefinder

The velocity and the travelling time of the wheelchair movement have been recorded. It is clearly indicating that this behaviour assists the user by following the front wheelchair to a certain distance and immediately overtakes upon clearance of partly crowded areas. In case if the user wishes to stop at any location, the user just has to release the joystick to the mid position. At this juncture, all the behaviour controllers will be halted. In order to resume the operation of the powered wheelchair, the user has to shift the joystick pedal to any desired direction and

Fig. 8 Travel path of Follow-the-Leader behaviour

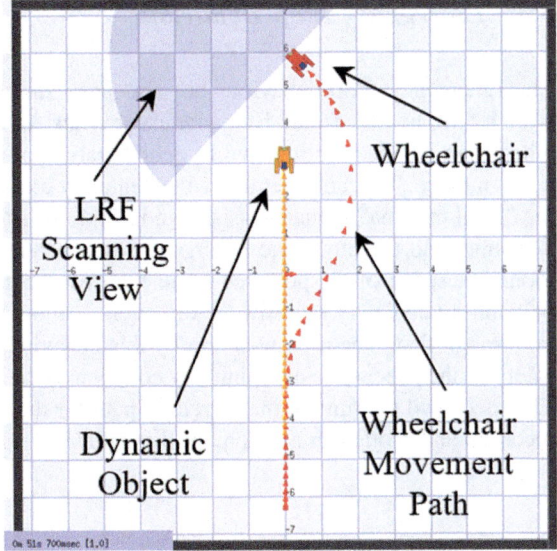

determine the required velocity to the destination. Similarly, if the user prefers to change the destination, it can be easily accomplished by shifting the joystick pedal to new direction and velocity. The Follow-the-Leader behaviour will be terminated and manual operation supersedes. The speed and time taken for the entire journey have been recorded and plotted in a graph as shown in Fig. 9.

Fig. 9 Travelling speed of Follow-the-Leader behaviour

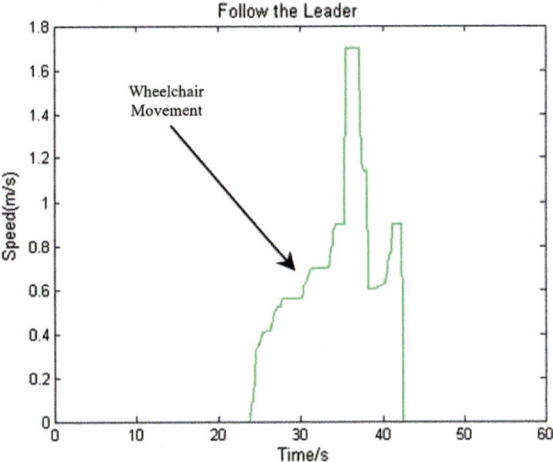

3.3 Emergency Stop Behaviour

The Emergency Stop behaviour has also been simulated in the Player/Stage software by means of a Saitek Evo wireless joystick. In this scenario, another powered wheelchair has been introduced to deliberately approach the first wheelchair at the door entrance. This arrangement will eventually create a potential collision and also the "local minima" scenario, as shown in Fig. 10. At this point, the collision risk is very high and therefore, the Emergency Stop behaviour is activated. In this behaviour, the speed of the powered wheelchair decreases gradually before it comes to halt and immediately reverses the direction to allow the approaching wheelchair to go through first. The prompt action by this behaviour ensures the comfort stop and safety of the user while avoiding any collision to the approaching dynamic objects. The speed and the time of the travelling path for the powered wheelchair have been recorded and plotted in a graph, as shown in Fig. 11.

3.4 Switching Between Behaviours

The implementation of the Fuzzy Inference System algorithm to perform the switching of behaviours between the Follow-the-Leader behaviour and the Emergency Stop behaviour has been simulated in the Player/Stage software. Figure 12 shows the generated image of the FIS surface viewer for switching between behaviours which has been simulated in Matlab software with the aid of the Fuzzy Logic toolbox.

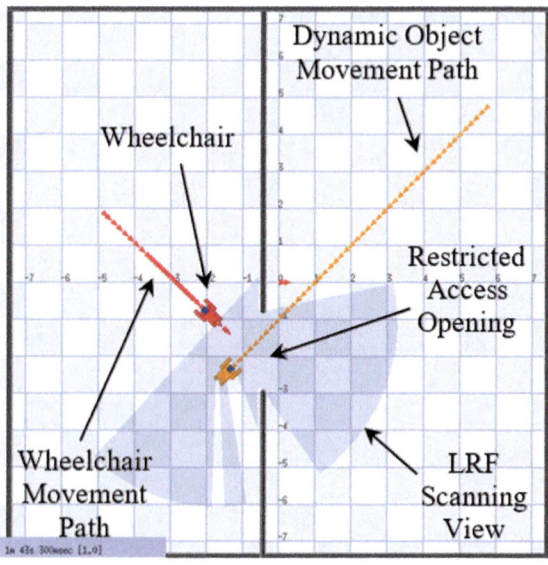

Fig. 10 Simulation of Emergency Stop behaviour

Fig. 11 Speed pattern for Emergency Stop behaviour

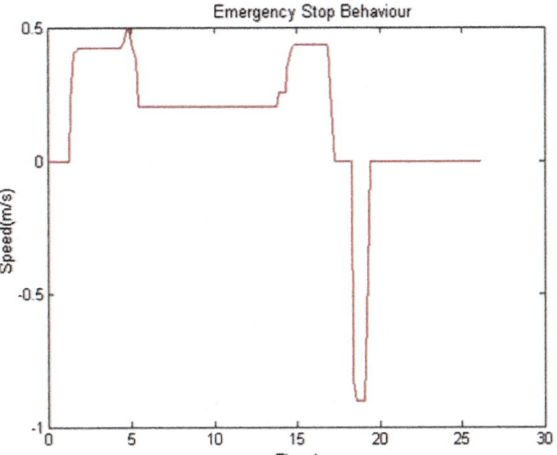

Fig. 12 FIS surface viewer for switching of behaviours

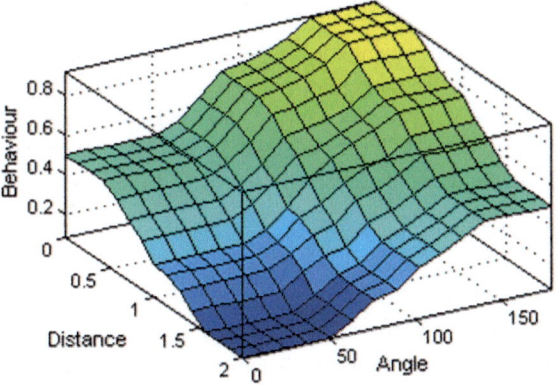

Figure 13 shows the graph of speed versus time for the switching between behaviours. It shows a significant improvement in the switching of behaviours between the Follow-the-Leader and the Emergency Stop behaviour.

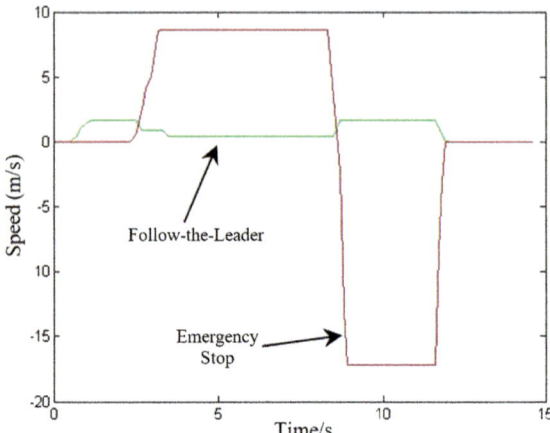

Fig. 13 Switching characterises between behaviours

4 Conclusions

Based on the results, it can be clearly seen that the FIS algorithm has significantly improved the switching behaviour, i.e., velocity and response time, between the Follow-the Leader behaviour and the Emergency Stop behaviour. However, further research has to be conducted to incorporate more than two behaviours such as obstacle avoidance behaviour and follow-the-wall behaviour.

References

1. Arkin, R.C.: Robotics and automation. Proceedings, IEEE International Conference (March 1987)
2. Brooks, R.A.: A robust layered control system for a mobile robot. J. Robot. Autom. **RA-2**(1), 14–23 (1989)
3. Nattharith, P.: Mobile robot navigation using a behavioural control strategy. Ph.D Thesis, Newcastle University, UK (2010)
4. Alsaab, A.: Behavioural strategy for indoor mobile robot navigation in dynamic environments. Ph.D Thesis, Newcastle University, UK (2014)
5. Gerkey, B.P., Vaughan, R.T., Howard, A.: Player User Manual Version 1.5. from http://playerstage.sourceforge.net/ (2004)
6. Berkvens, R., Rymenants, W., Weyn, M., Sleutel, S., Loockx, W.: Autonomous wheelchair: concept and exploration. In: The Second International Conference on Ambient Computing, Applications, Services and Technologies, pp. 38–44 (2012)
7. Ding, D., Cooper, R.A.: Electric powered wheelchairs. IEEE Control Syst. **25**(2), 22–34 (2005)
8. Cooper, R.A., Corfman, T.A., Fitzgerald, S.G., Boninger, M.L., Spaeth, D.M., Ammer, W., Arva, J.: Performance assessment of a pushrim-activated power-assisted wheelchair control system. IEEE Trans. Control Syst. Technol. **10**(1), 121–126 (2002)
9. Hall, K., Partnoy, J., Tenenbaum, S., Dawson, D.R.: Power mobility driving training for seniors: A pilot study. Assistive Technol. **17**(1), 47–56 (2005)

10. Matsumoto, O., Komoriya, K., Hatase, T., Yuki, T., Goto, S.: Intelligent Wheelchair Robot "TAO Aicle". In Tech Open, Croatia (2008)
11. Pires, G., Nunes, U. Almeida, A.T.d.: Robchair—a semiautonomous wheelchair for disabled. In: 3rd IFAC Symposium on Intelligent Autonomous Vehicles (IAV'98) Proceedings, pp. 648–652, Madrid (1998b)
12. Yanco, H.A.: Development and testing of a robotic wheelchair system for outdoor navigation. In: Proceedings of the 2001 Conference of the Rehabilitation Engineering and Assistive Technology Society of North America, RESNA Press (2001)
13. Lin, C.T., Euler, C., Wang, P.J., Mekhtarian, A.: Indoor and Outdoor Mobility for an Intelligent Autonomous Wheelchair, pp. 172–179. Springer, Berlin, Heidelberg (2012)
14. Levine, S.P., Borenstein, J. and Koren, Y.: The NavChair control system for automatic assistive wheelchair navigation, RESNA 13th Annual Conference, vol. 10, pp. 193–194. Washington D.C. (1990)
15. Špacapan, I., Kocijan, J., Bajd, T.: Simulation of fuzzy-logic-based intelligent wheelchair control system. J. Intell. Rob. Syst. Theory Appl. **39**(2), 227–241 (2004)

Mesh Filtering Algorithm for Virtualisation of Rapid Prototype Models Based on Digitised Data

Nur Ilham Aminullah Abdulqawi and Mohd Salman Abu Mansor

Abstract This research focuses on the process planning stage of rapid prototyping (RP) models in order to reduce material wastage in the actual prototype building process. Meshes which are redundant will translate to more material usage in the actual physical prototype. Eliminating the unnecessary meshes prior to the actual process will reduce material usage since a cleaned model generally has less unwanted features in the prototype compared to the one from the raw model. This paper will introduce an algorithm focusing on mesh filtering by percentage sizing in order to generate cleaned models as output and hence eliminate waste in the RP process by volumetric measurement, and the percentage of reduction in materials is calculated based on ABS plastics material which typically is used in rapid prototyping.

Keywords Rapid prototyping · Computer-aided process planning
Mesh data · Materials usage analysis

Nomenclatures

A Area
g Gram
M Mesh
mm   Millimeter
P_n Patches
V Vertices

N. I. A. Abdulqawi · M. S. Abu Mansor (✉)
School of Mechanical Engineering, Engineering Campus,
Universiti Sains Malaysia, 14300 Nibong Tebal, Pulau Pinang, Malaysia
e-mail: mesalman@usm.my

N. I. A. Abdulqawi
e-mail: nia14_mec011@student.usm.my

© Springer International Publishing AG 2018
A. Öchsner (ed.), *Engineering Applications for New Materials and Technologies*,
Advanced Structured Materials 85, https://doi.org/10.1007/978-3-319-72697-7_17

Abbreviations

3D	3-dimensions
ABS	Acrylonitrile butadiene styrene
CAD	Computer aided design
NaN	Not a number
RE	Reverse engineering
RP	Rapid prototyping
STL	StereoLithography
STEP	Standard for the Exchange of Product model data

1 Introduction

Rapid prototyping is a necessary phase mainly used for evaluation purpose in product design especially with reverse engineering (RE) methods. RE on the other hand produced CAD models in various formats such as STL, STEP and so on, where these formats have a meshed based formation. The process planning in rapid prototyping varies from time, quality and material cost, while RP of the RE method usually consumes more time and material costs in order to produce the finished product. Meshes even though are necessary to form the basic features of an object, they are sometimes appeared to be unnecessary when the nearest neighbour search algorithm is not robust enough to differentiate the distance of each vertices. Hence, the mesh from point cloud must be applied to an additional function in order to filter out the unnecessary meshes. This research hence will attempt to remove the unnecessary meshes for eliminating waste of material usage prior to the actual prototyping process.

2 Related Works

Rapid prototyping (RP) was originally developed for product realisation purpose during the product design and development stage [1]. Recent CAD/CAM/CAE software packages have integrated RP functions in order to stay relevant and competitive especially in manufacturing industries [2]. Reverse engineering (RE) on the other hand produces CAD models prior to RP but usually in a non-standard form, also known as free-form [3]. When associate with digitisation, RE will not produce standard geometries, but instead a point cloud in 3D form [4, 5]. A point cloud is an intermediate type of output type via the RE method which resulted from the digitisation process and it will have rather a scatter pattern than an organised pattern associate with noise [6] and outliers [7]. Hence, reproducing the RP via RE method can be challenging. The mesh is another form of digitised data related to a point cloud, typically a point cloud will be acting as vertices in order to form the mesh, the most common mesh formation is the triangulation of a point cloud [8]. Both point cloud and mesh had been studied from time to time in areas of computers and graphics. The final mesh models, when comes to RP, are supposed to be cleaned and preserving the original features of the original objects.

3 Methodology

Reverse engineering (RE) after the digitisation phase will produce point clouds or mesh surfaces. Point clouds and mesh surfaces are interchangeable in term of vertices and triangulation. Each 3 nearest vertices by default will form a mesh based on triangulation so that the surface area of the meshes when added up will correctly represent the surface geometry of the intended shape. The relation of mesh and vertices is explained in Fig. 1.

In order to create a mesh from a point cloud, patches are created at first based on Delaunay triangulation as shown in Eq. (1), where P_n represents patches.

$$P_n = DT(x, y, z) \qquad (1)$$

In certain circumstances, RE of objects with open-ended features will cause the point cloud to self-triangulate more meshes than necessary. Figure 2 shows the

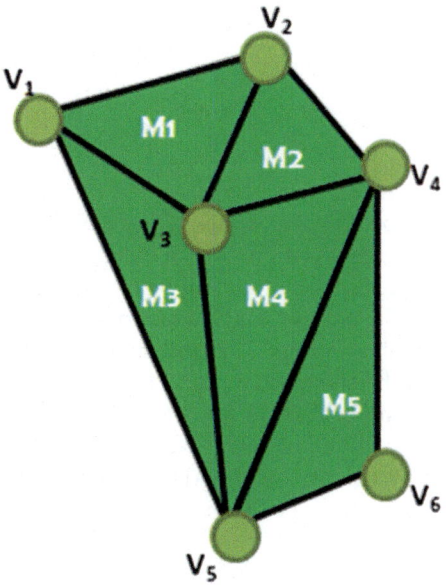

Fig. 1 The relation of vertices of V_1–V_6 and triangulated meshes of M_1–M_5

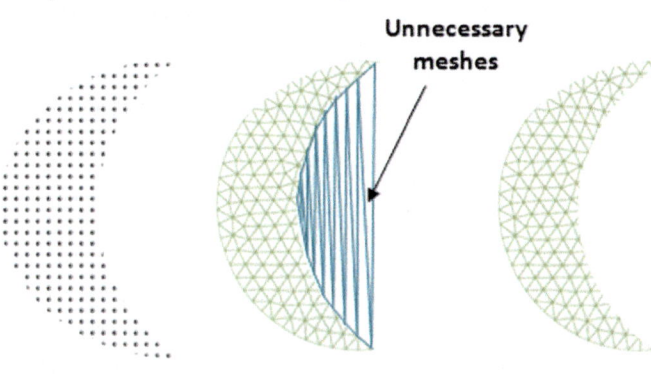

Fig. 2 The case of open-ended features that causes unnecessary meshes in the raw mesh from digitised data

open-ended feature case of certain geometries that can lead to unnecessary mesh formation.

To filter out the unnecessary meshes, some criteria are required to be input into the raw mesh model. The raw mesh model contains full triangulated meshes and the unnecessary mesh on open-ended features. A true and false conditioning program based on absolute number of the 3rd vertices less the 1st vertices, 3rd vertices less the minus 2nd vertices and 2nd vertices less the 1st vertices has to be larger than the predetermined constant factor of 9% of the overall mesh model's surface area size, A.

$$P_n = [\text{NaN }] \begin{cases} |V_3| - V_1 > 0.09A \\ |V_3| - V_2 > 0.09A \\ |V_2| - V_1 > 0.09A \end{cases} \qquad (2)$$

Patches after being assigned empty values will be treated as false as they return as not a number, (NaN). The NaN value stored in the vertices will be compressed into only vertices without patching. After running the algorithm coding, the unnecessary meshes will be cleaned off.

The process flow in general starts with a raw point cloud converted to meshes, and then the algorithm will process the separated meshes, then the separated meshes will be combined together into a final clean mesh as shown in Fig. 3.

The final clean mesh model can be saved in triangulation file format such as .STL, .STEP or. IGS, so that it can be utilised for rapid prototyping process through the interaction with CAD software.

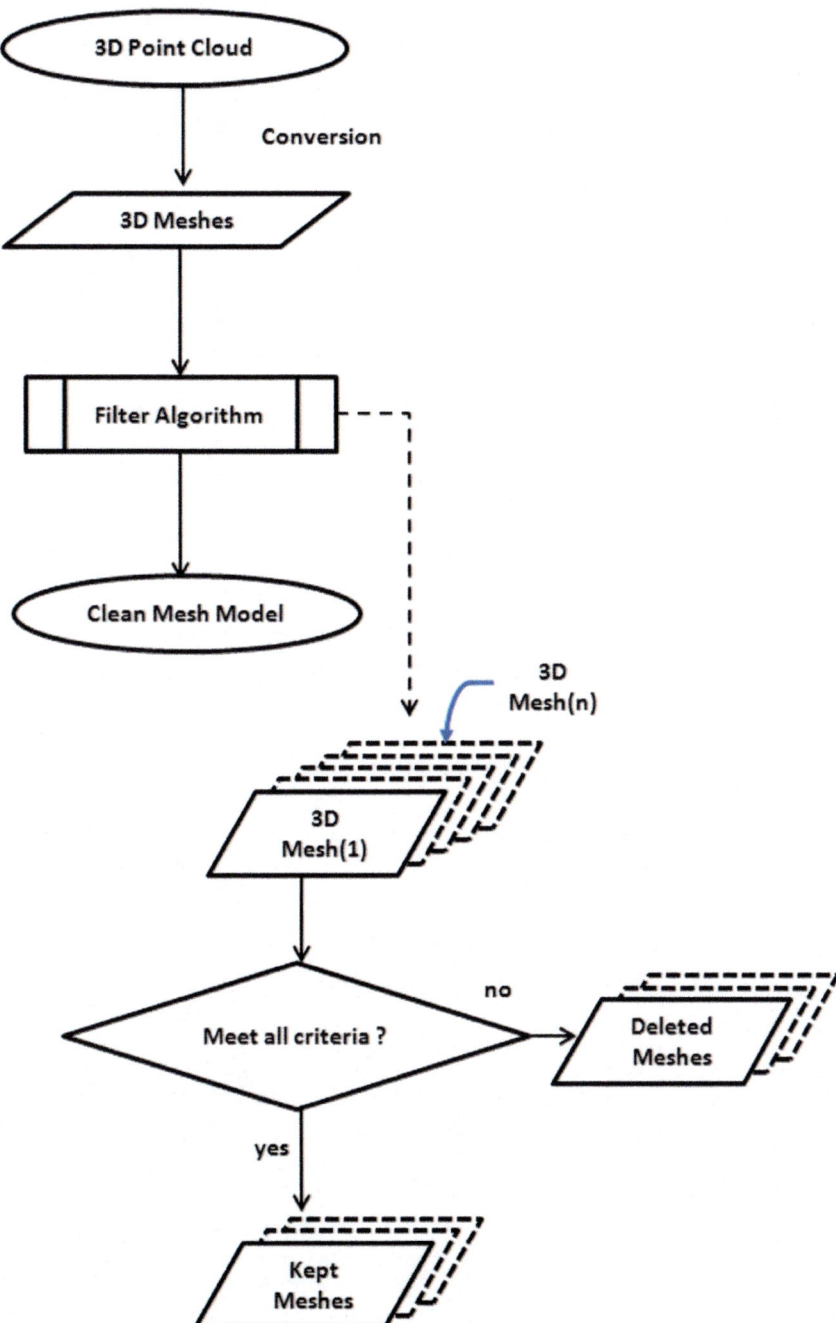

Fig. 3 The process flow of the method of filtering mesh from point cloud

4 Implementation

The mesh filtering algorithm is tested on several actual RE objects namely the bush cutter cover, mini spoon and iron's back part and handheld razor. The chosen objects each has more than one open-ended feature and can be identified by visual observation as shown in Fig. 4.

The 3D digitising method used is surface digitisation and the algorithm is written with the MATLAB 2014a complier. After implementing the codes towards the raw digitised data, clean meshes are produced. Figure 5 shows the before and after results of the bush cutter cover by applying the mesh filtering algorithm.

The following Fig. 6 shows the before and after results of the mini spoon by applying the mesh filtering algorithm.

Following is Fig. 7, which shows the before and after results of the iron's back part by applying the mesh filtering algorithm and the following Fig. 8 shows the before and after results of the handheld razor by applying the mesh filtering algorithm.

Besides of implementation, the volumetric changes are recorded via the STL format and opened in the SolidWorks CAD software, in order to obtain the measurement reports of each files.

Figure 9 shows the mass properties from SolidWorks that contained the volumetric measurement reports calculated based on the input mesh model.

Fig. 4 The chosen objects with open-ended features for testing of the algorithm

Fig. 5 Effect of the filtering algorithm on the raw mesh model for bush cutter cover (**a**) before and (**b**) after

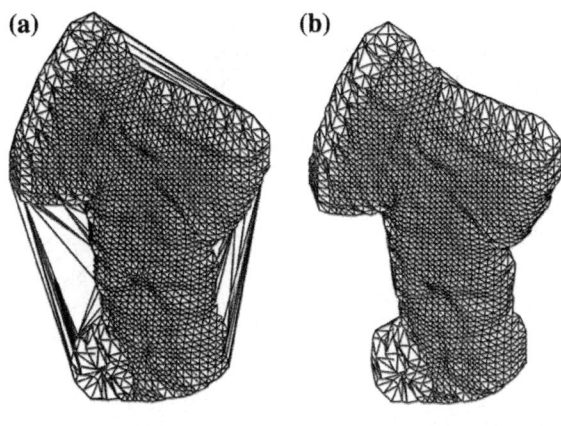

Fig. 6 Effect of the filtering algorithm on the raw mesh model for mini spoon (**a**) before and (**b**) after

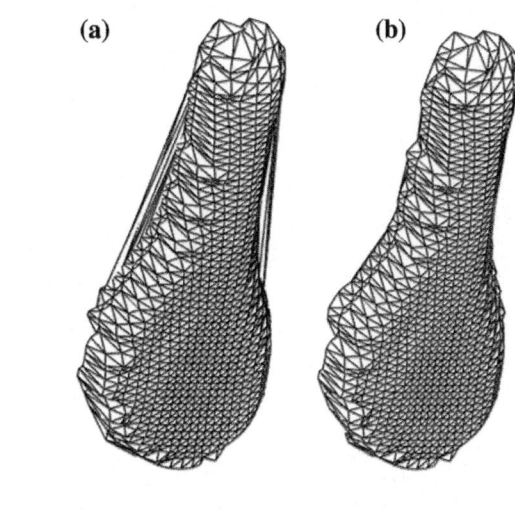

Fig. 7 Effect of the filtering algorithm on the raw mesh model for iron's back part (**a**) before and (**b**) after

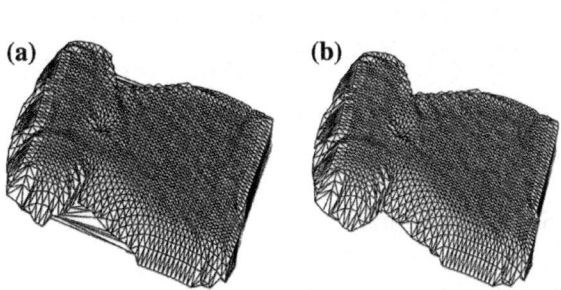

All the measurable variables are recorded and tabulated as Table 1. The results are all based on ABS PC material with a density of 0.00107 g/mm^3 with 1 mm material thickness, auto-generated by the CAD software.

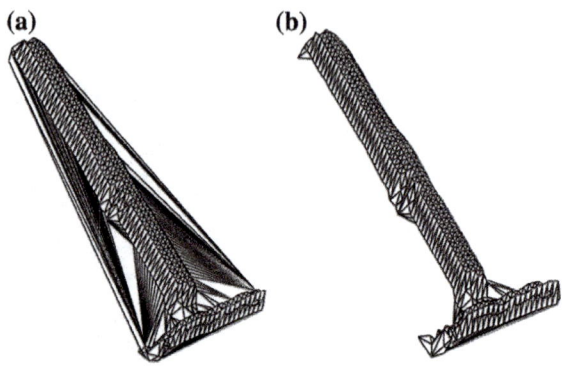

Fig. 8 Effect of the filtering algorithm on the raw mesh model for handheld razor (**a**) before and (**b**) after

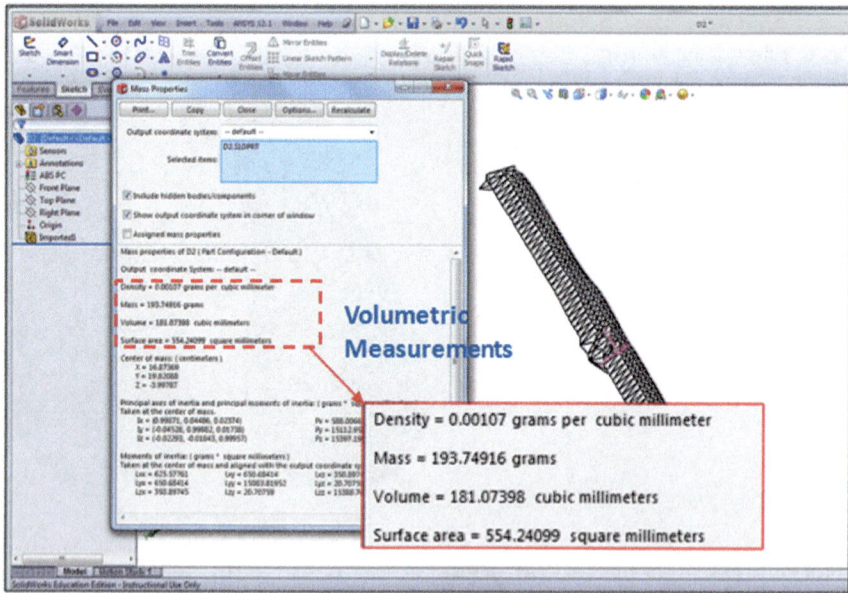

Fig. 9 The calculated volumetric measurements based on the generated mesh model via SolidWorks CAD software

5 Discussions

Based on the result shown according to the implemented objects, all filtered mesh models are found to have a lower mass, volume and surface area compared to the original raw mesh models. This shows that a reduction of mass in the mesh model is possible via the mesh filtering algorithm, hence this implementation will potentially be used for eliminating unnecessary ABS plastic material during the actual RP process.

Table 1 The volumetric measurements of mass, volume and surface area recorded

Objects		Mass (g)	Volume (mm^3)	Surface area (mm^2)
Bush cutter cover	Before	981.49588	917.28587	2413.09432
	After	902.65171	843.59973	2069.05370
Mini spoon	Before	381.77848	356.80232	910.20154
	After	373.14088	348.72979	821.30750
Iron's back part	Before	815.51824	789.82096	1817.39646
	After	762.16658	738.15043	1749.90051
Hand-held razor	Before	301.67772	281.94180	862.75062
	After	193.74916	181.07398	554.24099

Each of the variables are also further analysed for trend and percentage details. As for mass reduction and volume reduction, both variables are grouped together as in analyses because the values are very close to each other. Due to this fact, both mass and volume are plotted together for visual analyses. Figure 10 shows a combined bar chart comprising mass and volume for the tested objects.

As referred to the combined chart, percentages of mass reduction of the tested objects are found interrelated with volume reduction. For example, the bush cutter cover has 981.50 g before and 902.65 g after the reduction. The percentage of reduction by mass is 8.03%, the same as the percentage of reduction by volume. The other objects are also found the same as the bush cutter cover each at a calculated reduction by 2.26, 6.54 and 35.78% by mass and also volume. This is due to the density of the material used are assumed constant at all time, with ABS plastics material having a density of 0.00107 g/mm^3.

Furthermore, for the gap of each scattered points representing the mass reduction, the before and after distances are similar to the distances of each volume bars in the chart, indicating the volumes of each object. This trend also shows that the reduction amount of the objects by mass and volume is the same.

Meanwhile, Fig. 11 shows another bar chart comprising surface areas of each object.

Although there is no correlation found in the volume and mass reduction versus surface area reduction in term of percentage, the main determining factor of ABS materials usage in the actual prototype is based on the volumetric measurement rather than the surface area measurement. This algorithm in fact had shown that open-ended features in RE objects lead to more reduction of the material used. The highest reduction of 35.78% by volume had been recorded in the case of the handheld razor, it is because of the object's initial mesh has more unnecessary meshes compared to the other two, hence it is justified that this algorithm is useful whenever open-ended features are found in RE objects.

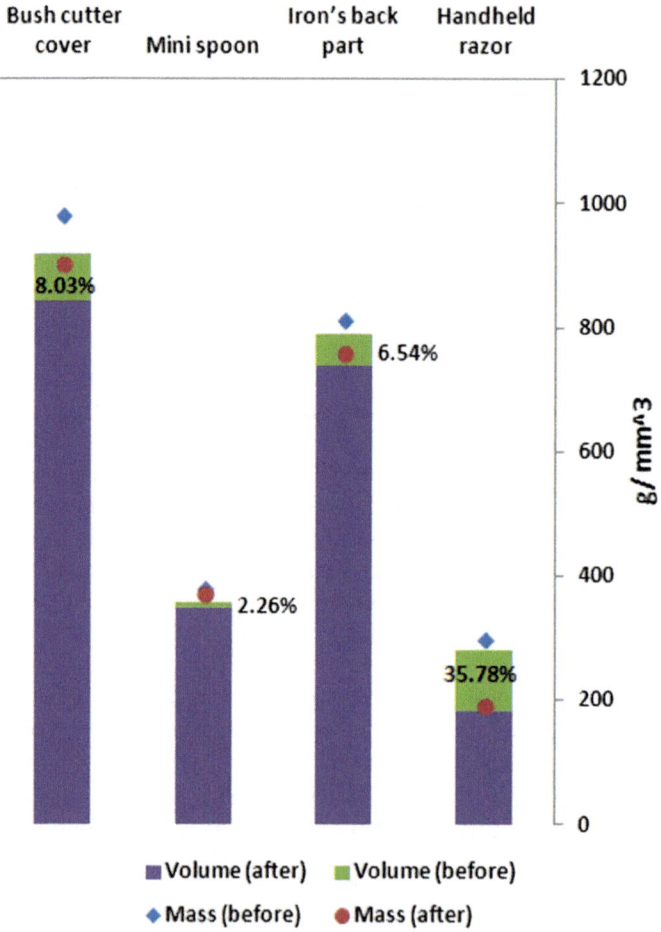

Fig. 10 Reduction percentage of mass and volume of each object at 8.03, 2.26, 6.54 and 35.78%

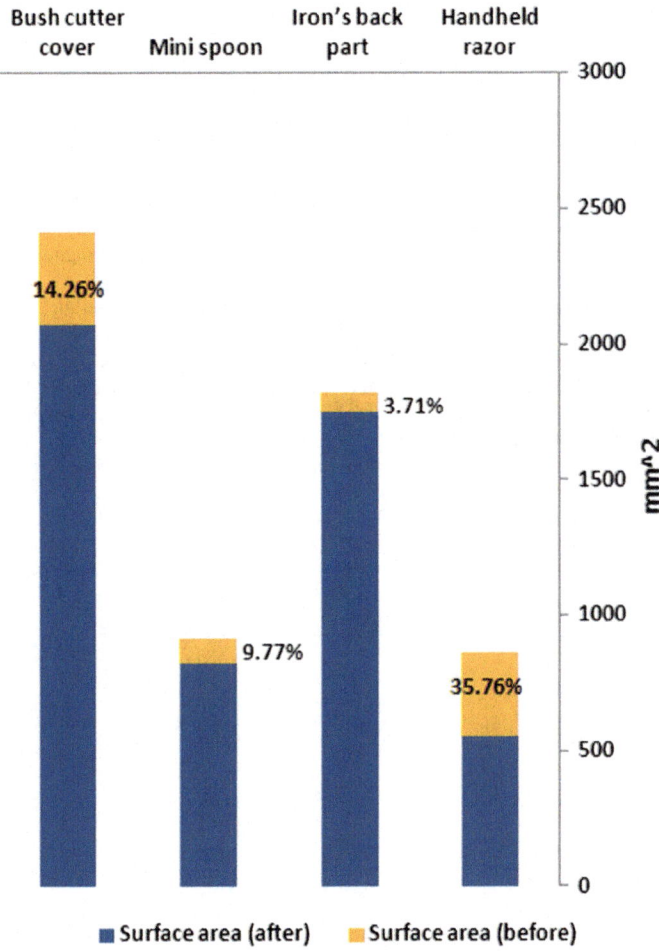

Fig. 11 Reduction percentage of surface area of each objects at 14.26, 9.77, 3.71 and 35.76%

6 Conclusions

The proposed algorithm is able to filter out unnecessary meshes in order to perfect the prototype model, and hence reducing material wastage. This algorithm can be used to improve process planning of rapid prototyping prior to the actual build. Material reduction up to 35.78% volumetrically is possible based on test objects.

Acknowledgements This research is supported by Universiti Sains Malaysia under the Research University Grant (No: 814247). The first author also would like to thank the tuition fee sponsorship of My Master by the Ministry of Higher Education Malaysia.

References

1. Wang, G., Li, H., Guan, Y., Zhao, G.: A rapid design and manufacturing system for product development applications. Rapid Prototyping J. **10**, 200–206 (2004)
2. Matta, A.K., Raju, D.R., Suman, K.N.S.: The integration of CAD/CAM and rapid prototyping in product development: a review. Mater. Today Proc. **2**, 3438–3445 (2015)
3. Salman, M., Mansor, A.: Free-form surface models generation using reverse engineering technique—an investigation. Project Report, Universiti Sains Malaysia/Proceedings of Malaysian Research Group International Conference 2006 United Kingdom, pp. 379–385 (2006)
4. Bagci, E.: Reverse engineering applications for recovery of broken or worn parts and re-manufacturing: three case studies. Adv. Eng. Softw. **40**, 407–418 (2009)
5. Abdulqawi, Nur Ilham Aminullah, Mansor, Mohd Salman Abu: A computer method for generating 3D point cloud from 2D digital image. J. Image Graph. **4**(2), 89–92 (2016)
6. Mehra, R., Tripathi, P., Sheffer, A., Mitra, N.J.: Visibility of noisy point cloud data. Comput. Graph. **34**, 219–230 (2010)
7. Wang, Y., Feng, H.-Y.: Outlier detection for scanned point clouds using majority voting. Comput. Aided Des. **62**, 31–43 (2015)
8. Yang, Z., Seo, Y.-H., Kim, T.: Adaptive triangular-mesh reconstruction by mean-curvature-based refinement from point clouds using a moving parabolic approximation. Comput. Aided Des. **42**, 2–17 (2010)

The Development of a Mobile Campus Information Sharing Android Application

Shareen Adlina Shamsuddin, Mohammad Arif Kader,
Norazlina Abdul Nasir, Nurul Akmal Radzi
and Fatimah Abdul Hamid

Abstract The development of a campus information sharing Android application is basically an improvement project of the traditional way of spreading information with the technology approach using smart devices. The relevance to the idea is the encountered problems faced with the vulnerability of papers, and also the related costs. There are a few projects around the world, which have a similar concept as this project. Some of the similarities, including the usage of the server as the storage element. Other than that, there is an Android application which uses the date and time to store information in a specific category. The development of this project is entirely using the MIT App Inventor, a free open source Android programming base. As a result of the development, the Android application functions in three separate pages; the home page, the editor mode, and the user mode. This application helps the users to spread and retrieve information anywhere, anytime equipped with security features to avoid misleading or false information. The scope of the application usage is targeted to university students to help them to increase the efficiency of the information sharing process.

Keywords MIT App Inventor · Android · Campus information

S. A. Shamsuddin (✉) · M. A. Kader · N. A. Nasir · N. A. Radzi · F. A. Hamid
Malaysian Institute of Marine Engineering Technology, Universiti Kuala Lumpur, Jalan Pantai Remis, 32200 Lumut, Perak, Malaysia
e-mail: shareen@unikl.edu.my

M. A. Kader
e-mail: ariff@unikl.edu.my

N. A. Nasir
e-mail: norazlina@unikl.edu.my

N. A. Radzi
e-mail: nurulakmal@unikl.edu.my

F. A. Hamid
e-mail: fatimah@unikl.edu.my

© Springer International Publishing AG 2018
A. Öchsner (ed.), *Engineering Applications for New Materials and Technologies*,
Advanced Structured Materials 85, https://doi.org/10.1007/978-3-319-72697-7_18

Nomenclatures

.apk Android application package
f-test Population variances comparison
P N/a
t-test Population variances comparison for smaller samples

Greek Symbols

☐ N/a

Abbreviations

API Application programming interface
ID Identification
MIMET Malaysian Institute of Marine Engineering Technology
SMS Short message service

1 Introduction

Annually, countless numbers of papers are used in order to disseminate information to students, yet most of them are not concerned with the information placed on the informational notice boards. This shows that the main function of the notice board, that is to share and spread information, is no longer important and has started to gradually weaken. This statement was taken from a previous work in the primary stage of the project [1].

Potentially, one of the downfalls is the hasty intensification of technologies that has infused a pessimistic impact towards its effectiveness. However, the purpose of this android application development is to implement a digital system by using an android device in information sharing with students. This device also has the exact and similar roles as the current system in spreading information by using informational notice boards within the campus. The design of the mobile campus information sharing android application involves several features in ensuring the security and the reliability of the content of the application.

The android application consists of two different sections: the user mode, and the editor mode. The user mode is only able to retrieve information without the ability to change the information saved in the application. But, it is different from the user mode; the editor mode has the ability to create new information, and also to alter the information within the application. In order to access to the editor mode, the user needs to enter a valid ID and password. This kind of feature is to ensure that the information posts in the application is from reliable sources.

An observation was made and it could be concluded that almost every organization and institution use informational notice boards to disseminate information to its intended audiences. However, there are several problems that could be seen. The first problem is that papers are extensively used in spreading information by using the notice boards, yet students are still unaware of the information placed. Rather, they still prefer their latest technological devices. Aside from that, an expensive and electrically powered electronic board has also now replaced the traditional notice boards as it is able to attract the attention of any passers-by. But there are also several setbacks to the use of these electronic boards. A study on the effectiveness of the informational board in the Marine Institute of Marine Engineering Technology (MIMET) was conducted and it shows that 90% of the students were not affected at all by the existence of the notice boards.

This project was created in order to develop an android application to disseminate information within the campus. It was also created as a device to be implemented in a real-life setting, and also to be further evaluated for its effectiveness by the application users.

This project used an android-based software that was developed by using the MIT App Inventor; an open-source web application. This kind of web application uses a block code programming language in helping fledgling developers with less knowledge of the Java textual language in creating codes.

2 Methodology

The concept of the mobile campus information sharing android application is to save, store, and retrieve various information within one application. This idea was established from one of the android application development projects created in 2013 [2]. As compared to the application, this project used a web-based server. Figure 1 illustrates the implementation of the server in the application's system operation.

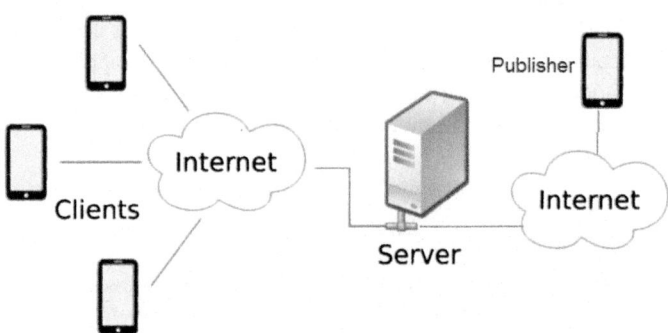

Fig. 1 Application's system operation

The system operates in a manner where the publisher is able to create or alter information placed in the web-based server. All the information stored in the server could be deleted by using the mobile application or the normal web browser. On the other hand, the user will also be able to retrieve information from the server through the user mode within the application with an internet connection. As for the graphical user interface, the homepage of this application shows two different buttons that lead to two different sections; the user and the editor section. In the user section, the user will only be able to read the saved information, and it does not need to be verified; meanwhile the editor mode is only accessible to the holders with verified IDs and passwords. The security features included are to ensure that the information posted within this application is from reliable sources. The notification feature in this application will ensure that it outruns the orthodox system of disseminating information using the physical information board. Other users will also be notified through SMS once the editor publishes or edits the information in the editor section [2].

Another function featured in this application is the List Picker that is used to view the list of departments in the campus to categorize the information. It is directly aimed at the Tiny Web DB component, which represents the web server used to store and retrieve information [3]. Figure 2 indicates one of the List Picker and Tiny Web DB blocks components with the mobile campus information sharing android application.

The blocks component in Fig. 2 shows an example of the relation between the List Picker and the Tiny Web DB components. The editor chose the blocks on the left side to store one of the four departments that are viewed through the interface within the application. Meanwhile, the blocks on the right side are to store the information based on the department in the web server as chosen by the editor that are not viewed within the interface of the application.

The two-major hypothesis addressed in this study were the awareness of students towards new information published on the informational board, and whether the students are able to obtain new information or not.

Two surveys were conducted to obtain information for this project. The purposes of this study are to identify the level of efficiency of the students towards the existing system of the information sharing system by using the information board and conducted before the implementation of the mobile campus information sharing Android application, and also to analyse whether the mobile application is able to influence the students' level of awareness towards information within the campus after the implementation of the mobile Android application system.

Fig. 2 Blocks component of list Picker and Tiny Web DB

Three tests were conducted for the pre-implementation survey: an f-test and two t-tests. An f-test is used to compare population variances, meanwhile, a t-test is used for smaller samples. A statistical test was used for the f-test with a test statistic that has an f-distribution under a null hypothesis. A 10-question survey was organized for 250 students, equivalent to 10% of MIMET's student's population. A t-test was conducted as the second approach to determine the significant difference in the sets of data. The value of alpha is associated with the confidence level of the tests. For results with a 90% level of confidence, the value alpha is $1 - 0.90 = 0.10$ is used. The value of alpha is 0.1, indicating a 10% risk to conclude the existence of difference with no actual difference.

A similar approach was used for the post-implementation survey to obtain the P value. A group of 20 mobile campus information sharing Android application users from among MIMET students was chosen for this study. It was conducted to study the efficacy of the mobile Android application on the students' alertness on the new, updated information. A ten-question questionnaire was distributed. The significant or the alpha value, however, remains unchanged to indicate the risk of concluding the existence difference with no actual difference.

3 Results and Discussions

To run the application within a real-work environment, the Android Application Package (.apk) file format is required. However, to translate the programmes into the .apk format, the whole programme has to be error-free. This is to ensure the smoothness in running the application without any kind of interference. Once the .apk file is produced and saved in the computer, the file is then transferred to an Android working device. In this case, a Jelly Bean 4.3 version is running in the application programming interface (API) level 18. The level of API is to determine the variety of features in the interface of an application [4]. Once the application is successfully installed, it is available to run as normal. At the beginning of the operation, the homepage will appear with a brief description of the application. Two out of three buttons will lead the users directly to two different sections as mentioned earlier. Figure 3 illustrates the homepage interface of this application. In the user mode section, there are two buttons on the interface which need to be specified in order to retrieve a particular information. The first button is to determine the departments, while the second button is to select the title of a specific information as similar from the selected departments. Once these buttons are specified, the content of the selected title in the selected departments will be shown in the content viewing section. Figure 4 shows the screenshot results from the user mode.

In the editor mode, the accessible users are given options to create new information, edit information, or delete information. These options are in the forms of buttons that could only be selected after the user has entered a verified ID and password. This is also similar to the user mode, whereby the editor mode needs the user to primarily select the department and the title first before they are able to

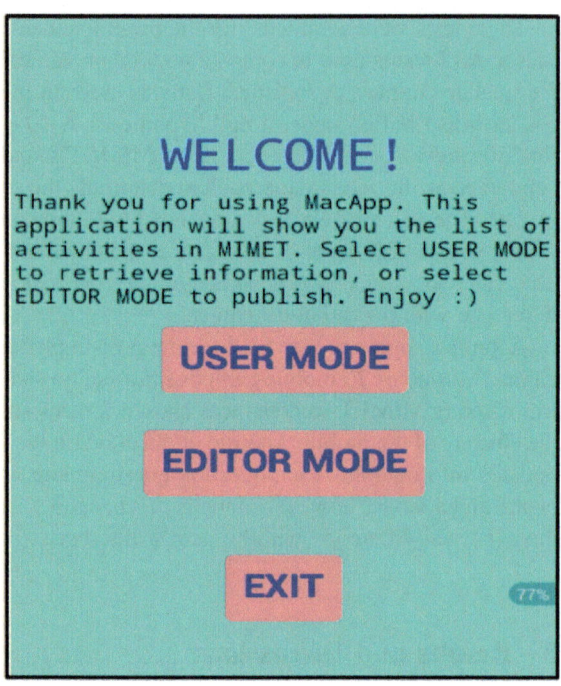

Fig. 3 The homepage GUI of the application

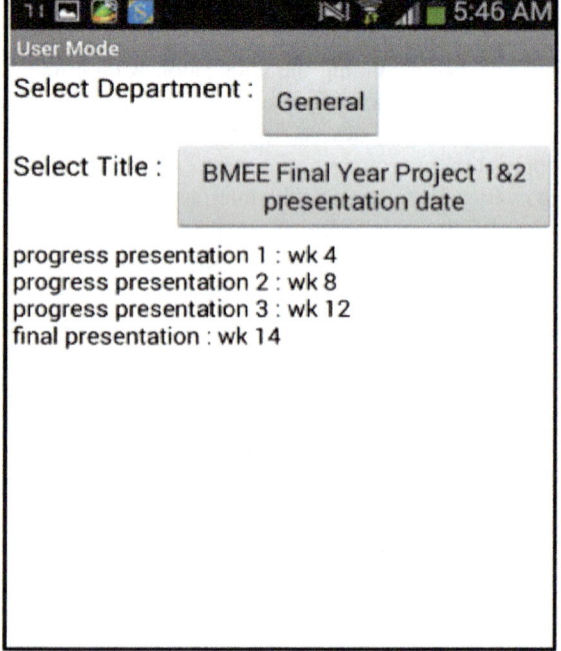

Fig. 4 The user mode interface

create or edit a specific post. Another function included in the editor mode is to erase information stored in the server. This feature is to ensure that the application will not be packed with an old information that is no longer necessary. Another reason is to keep the application light to ensure smooth operation for the users to retrieve information.

Data was collected based on the survey with a tabulated data of mean, variance, and critical values for all ten questions that were created. A pilot test was conducted to test the null hypothesis. However, it rejects the null hypothesis, which means students are not aware when a new information is posted on the informational notice board. The P value obtained for both f-test and t-test is 0.8. For the pre-implementation survey, the null hypothesis is accepted, which means that students do not receive any new information from the informational notice boards. The P value obtained is 0.42. The same approach as in the pre-implementation survey was also used. In the post-implementation survey, it also accepts the null hypothesis saying that students are aware of the new information posted on the mobile campus information sharing android application. The P value obtained is 0.32.

4 Conclusion

As a conclusion, the users are offered with mobility and flexibility to retrieve information regardless of place and time with the development of the android mobile application by transforming a current system of spreading information into a more technology-based system. This MIT App Inventor application utilizes various functions of operation blocks which offer similar functions as the textual operation. Besides the potential to reduce the number of paper usage, this application hypothetically is able to increase the competency of the information to be received by the students with the notifications included within the application. This eye-opener implementation should be able to point out that the campus does follow the flow of technologies and change the way in delivering effective information.

References

1. Schervish, M.J.: P values: what they are and what they are not. Am. Stat. **50**(3), 203–206 (1996)
2. Hong Liu, J., Chen, J., Yi-Li, Wu, Li Wang, P.: Android Application Server for Mobile Platforms. Department of Electrical Engineering, National Cheng Kung University, Taiwan, Institute of Computer and Communication Engineering (2013)
3. Abeywardena, I.S.: Educational app development toolkit for teachers and learners. (2015)
4. McDonnell, T., Ray, B., Kim, M.: An empirical study of api stability and adoption in the android ecosystem. In 29th IEEE International Conference on Software Maintenance (ICSM), pp. 70–79. Sept 2013

Service Restoration Based on Simultaneous Network Reconfiguration and Distributed Generation Sizing for Loss Minimization Using a Modified Genetic Algorithm

W. M. Dahalan, H. Mokhlis, M. K. P. Zarina, A. G. Othman and N. M. Salih

Abstract The main purpose of service restoration is to restore as many loads as possible by transferring loads in the out of service areas to other distribution feeders via changing the status of normally close and open switches which is known as distribution network reconfiguration (DNR). In the event of any fault occurrence within the system, immediate restoration is indeed required in the particular area. Therefore, the distribution system must be equipped and planned in such a way that it will continuously supply the power without any interruption during the out-of-service condition. The primary idea in this work is to have the reconfiguration process embedded with distributed generation (DG) and being operated simultaneously to reduce power losses by using a modified genetic algorithm (MGA). A detail performance analysis is carried out in 33-bus systems demonstrate the effectiveness of the MGA. The proposed method is adopted and its impacts on the network real power losses and voltage profiles are investigated as well as improving the voltage profile while fulfilling distribution constraints. The results obtained show that the improve MGA method will reduce the power losses up to 50%.

W. M. Dahalan (✉) · M. K. P. Zarina · A. G. Othman · N. M. Salih
Malaysian Institute of Marine Engineering Technology, Jalan Pantai Remis,
32200 Lumut, Perak, Malaysia
e-mail: wardiah@unikl.edu.my

M. K. P. Zarina
e-mail: puterizarina@unikl.edu.my

A. G. Othman
e-mail: sweetanis70@yahoo.com

N.M. Salih
e-mail: noorazlinams@unikl.edu.my

H. Mokhlis
Department of Electrical Engineering, Faculty of Engineering,
University of Malaya, 50603 Kuala Lumpur, Malaysia
e-mail: hazli@um.edu.my

Keywords Distributed generation · Optimization techniques · Power loss reduction · Reconfiguration · Genetic algorithm

Nomenclatures

I_t	Current of line t
k	Value of power generation
kW	Kilowatts
l_t	Topology status of line
MW	Megawatts
N_{dg}	Total number of DG
N_{switch}	Number of switches
P_{dg}	Power generation by DG
P_{ld}	Total load power
P_{loss}	Total power loss
$p.u$	Power up
$P_{substation}$	Power from substation
R_t	Resistance of line t
t	Switch's number
S	Line number
s	Seconds
V	Voltage

Abbreviations

ASC	Ant search colony
CPU	Central processing unit
DG	Distributed generation
DNR	Distribution network reconfiguration
GA	Genetic algorithm
MGA	Modified genetic algorithm
PSA	Particle swarm optimization
TS	Tabu search

1 Introduction

Fault events are almost unavoidable, especially in any complex electrical power distribution system. It could happen due to short circuit, aging factor of equipment or human error. After the occurrence of the fault, the operator will then find the location of faults, isolate the fault and then restore the service to the healthy components at the out-of-service area. Finding a new network configuration is not the only concern in service restoration. It is also important to find the optimal sequence of switching operations. In order to not interrupt loads while changing configurations, the switching operations must be performed sequentially. Hence, the task of reaching a particular configuration can be regarded as scheduling of switching operations that are closing and opening switches. These are sectionalizing switches (normally closed switches) and tie switches (normally opened switches).

One of the most significant control schemes in electrical systems is distribution network reconfiguration (DNR) which can bring optimal minimization of real loss, improvement in voltage profile, and relieving of overloads in the network. However, the DNR technique has its own limitation. If the improvement of power loss is too great, it will make the solution not reliable. Therefore, the implementation of distributed generation (DG) in the distribution network is still needed to improve the distribution system. The DNR process is affected by the interconnection of DGs. The growth of DG insertion is obviously increasing recently due to its significant impacts to the power system performance. Consequently, changing the environment of power systems design and operation have necessitated the need to consider active distribution network by incorporating DG units sources [1–7]. The implementation of DG and other devices in the distribution system required a huge installation costs. Therefore, appropriate size and location of DG is highly important in maintaining the network stability and reduce the power losses in the system. Meanwhile, the DNR technique only required some additional lines and controlling methods to improve the reliability of the system. Furthermore, the installation of DNR is much simpler and cost efficient compared to other techniques. The reconfiguration result obtained was able to reduce the power losses and improve the reliability of the network compared to other available algorithms in [8–13].

Due to the increasing utilization of DGs in the distribution network, the researchers have been employed to solve the reconfiguration problem with DG units of distribution network such as by using the Tabu Search (TS), Genetic Algorithm (GA), Particle Swarm Optimization (PSO) and Ant Search Colony (ASC). However, only a few of them [14–17] associate DG in their research and foresee DG's impact on the reconfiguration distribution network. On top of that, their reconfiguration system and DG sizing operates in a different situation and sequentially.

Therefore, in this work the researcher is trying to solve the problem by service restoration via the optimal reconfiguration and DG sizing simultaneously after fault is isolated from the system which is explained in detail in Sect. 4. In order to achieve that, a proposed method based on a modified genetic algorithm (MGA) has been proposed in reconfiguration problems due to its excellent capability for searching optimal solutions to a complex problem. Three different scenarios have been considered during the analysis, which consists of a distribution network with and without DGs. In this work, we only determine the optimal size of DG while the location of DG is fixed. The location of DG is assumed has been determined based on the suitability of the geographical location or any optimal location methods such as in [18]. The details of Mathematical problem are discussed in Sect. 2. Meanwhile, Sect. 3 show the performance of this algorithm using standard test function. The simulation and results analysis in term of power loss and voltage profiles are discussed in Sects. 4 and 5. Last but not least, the conclusion for the whole analysis of the study is conclude in Sect. 6.

2 Mathematical Model for DNR Problem

The connection of DG and reconfiguration techniques in the distribution network will change the direction of power that flow throughout the network. Both techniques are reciprocally complementing each other to reduce the power losses and improve the voltage profile of the system. In this study, the main objective for doing reconfiguration, either with or without DG units, is to obtain the minimum power losses after the service restoration process. Therefore, the objective function of this study is to obtain the minimum power losses in the system based on current formulation as shown in Eq (1):

$$\{P_{loss} = \sum_{t=1}^{Nline} |I_t|^2 l_t R_t\} Nline = \text{Total Line} \qquad (1)$$

where

- t line number.
- It current of line t.
- Rt resistance of line t.
- Lt the topology status of line t (1 = close, 0 = open)

The power system constraints are also being considered in the analysis. The constraints are:

(a) *Generator operation constraint*:

The value of active power generation ($P_{dg,k}$) at DG k (k = 1, 2, …, total DG) should be within the stipulated range. This condition is shown in Eq. (2):

$$p_k^{min} \leq P_{dg,k} \leq p_k^{max} \qquad (2)$$

(b) *Power injection constraint*:

In the effort of avoiding power injection by the DG to the main grid (substation), the output of DG cannot be more than the total load (P_{ld}) and the total power loss (P_{loss}) in the network. This condition is shown in Eq. (3):

$$\sum_{k=1}^{Ndg} P_{dg,k} < (P_{ld} + P_{loss}), N_{dg} = \text{Total number of DG} \qquad (3)$$

(c) *Power balance constraint*:

The total power generated by the DG unit (P_{dg}) and the power from substation ($P_{substation}$) must be equalled to the amount of total load (P_{ld}) and power loss (P_{loss}) as in Eq. (4):

$$\sum_{k=1}^{Ndg} P_{dg,k} + P_{substation} = P_{ld} + P_{loss} \qquad (4)$$

(d) *Voltage bus constraint*:

The voltage magnitude for all buses in the system should be operated within the acceptable limit as in Eq. (5). In this work, the value is assumed between 1.05 and 0.95 (±5).

$$V_{min} \leq V_{bus} \leq V_{max} \qquad (5)$$

(e) *Radial configuration constraint*:

The network must always be in the radial form after the reconfiguration process. A set of rules has been adopted for selection of switches. Those switches that do not belong to any loop, connected to the sources and contributed to a meshed network have to be closed.

3 The Implementation of MGA in Service Restoration via DNR and DG Size

From the literature, the length of the chromosome of GA is the main factor in the efficiency or robustness in the operating system. Too long string or chromosome will increase the time consuming in the searching space for the optimum especially when the system operates in the larger and more complex system. New generation or offspring are generated with the help of three types of GA operators which are selection, Crossover and Mutation.

Mutation is the process to improve the position slightly in chromosome or where some individuals are randomly modified. It means that the elements of genes will experience a little bit change. These changes are mainly caused by errors in copying genes from parents. On the other hand, mutation is supposed to help the exploration of the whole search space and also help escape from local minima's trap and maintains diversity in the population. The existing methods mostly used binary coded as a chromosome. However, in some works, it will convert the chromosome from real coded in binary coded first before converting the final outcomes back to real coded. A mutation process in binary coded is slightly different from real coded in which in binary coded the genes will change the position whether '0' or '1' and

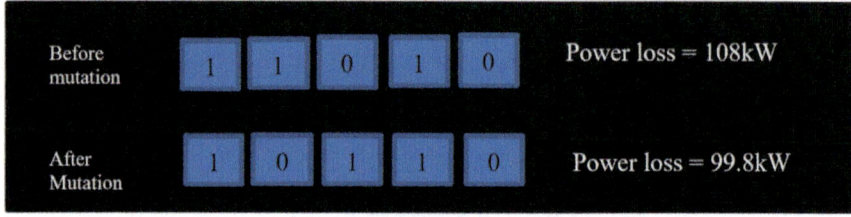

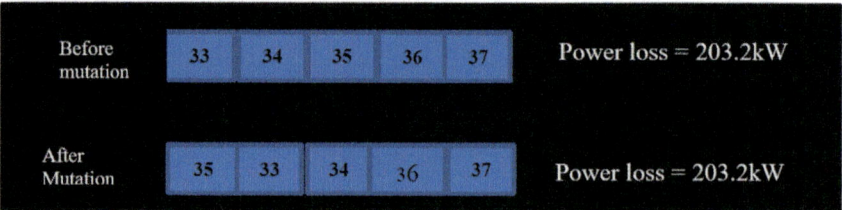

Fig. 1 Example chromosomes of binary coded and real coded

the fitness value will be changed (Fig. 1a). However, in real coded, the genes of chromosome consist of real numbers. If we only change the position of the chromosome, the fitness value will not give any changes (increase or decrease) as shown in Fig. 1b.

Since conventional GA takes longer time during the simulation process, a new idea using a simultaneous approach has then been proposed in this work in order to increase the efficiency and reduce the computational time. The proposed algorithm is known as the modified genetic algorithm (MGA). Basically, the steps involved in MGA are mostly similar to GA steps except a slight difference in the mutation process. The chromosomes which consist of tie-switches and DG size are represented in numbers. The best way to make the value of fitness change is by reducing or adding the value of the genes in chromosomes directly through subtracting one (−1) or addition one (+1) randomly.

We have been testing the real coded chromosomes in the research through the simulation process by using the parameters starting with −4 to +4. The results obtained show that the best solution falls within the range $-2 \leq x \leq 2$. However, if the parameter is further increased, the fitness value becomes higher and has the possibility to exceed the initial fitness value as depicted in Fig. 2. Therefore, this new algorithm will be tested in finding the optimal reconfiguration and DG size in the distribution system. The example of the mutation process with the parameter used −1 is as shown in Fig. 3.

The advantage of MGA is that the user can accelerate the searching speed of the results because it is not necessary to convert the encoding and decoding process as required in binary coded. Furthermore, it is a simple design tool to treat complex constraints because the method is close to problem spaces. MGA will ensure only

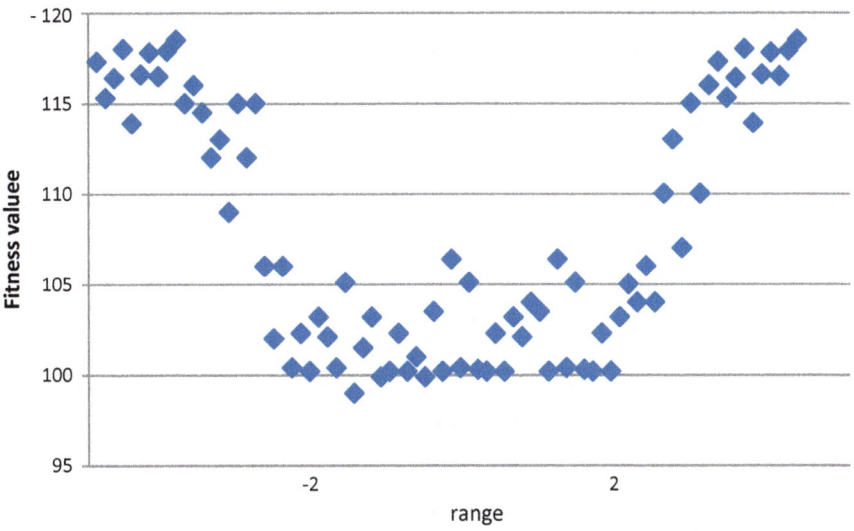

Fig. 2 The best solution within the range (−2 and 2)

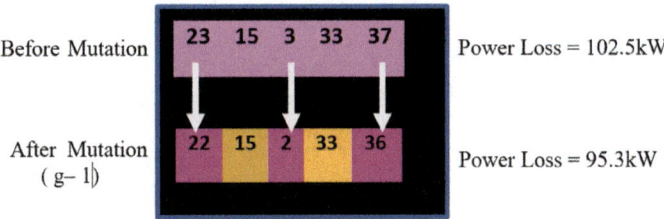

Fig. 3 Mutation process with the parameter −1

the highest potential chromosome are selected for the next iteration. Thus, the results obtained will be nearest to the optimum solution. A population of chromosomes representing the open switches and DG size is generated in this step. The chromosome is determined by selecting randomly the opened switches and size of DG as shown below:

$$x = [S_1, S_2, \ldots, S_{Nswitch}, P_{dg1}, P_{dg2}, \ldots, P_{dg,Ndg}] \quad (6)$$

where S representing the switch's number that is open (Examples; 1, 2, …) with a total number of switches that will be opened is given by *Nswitch*. Only the chromosomes that satisfy all the constraints in (2), (3), (4) and (5) will be considered as the initial population. Finally, Graph Theory is applied to check whether the network is in radial structure. Distribution systems must meet each mesh. Radial structure as it is considered very efficient. If all switches are closed in a distribution system, a meshed network will be formed. For a network with meshed form, the switches must be opened to retain a radial network structure. Therefore, the configuration is done through the changing of the network status of switches in such a way that radiality is always re-established to simplify the over current protection [19–22]. The summarized function of the MGA is as shown in Fig. 4.

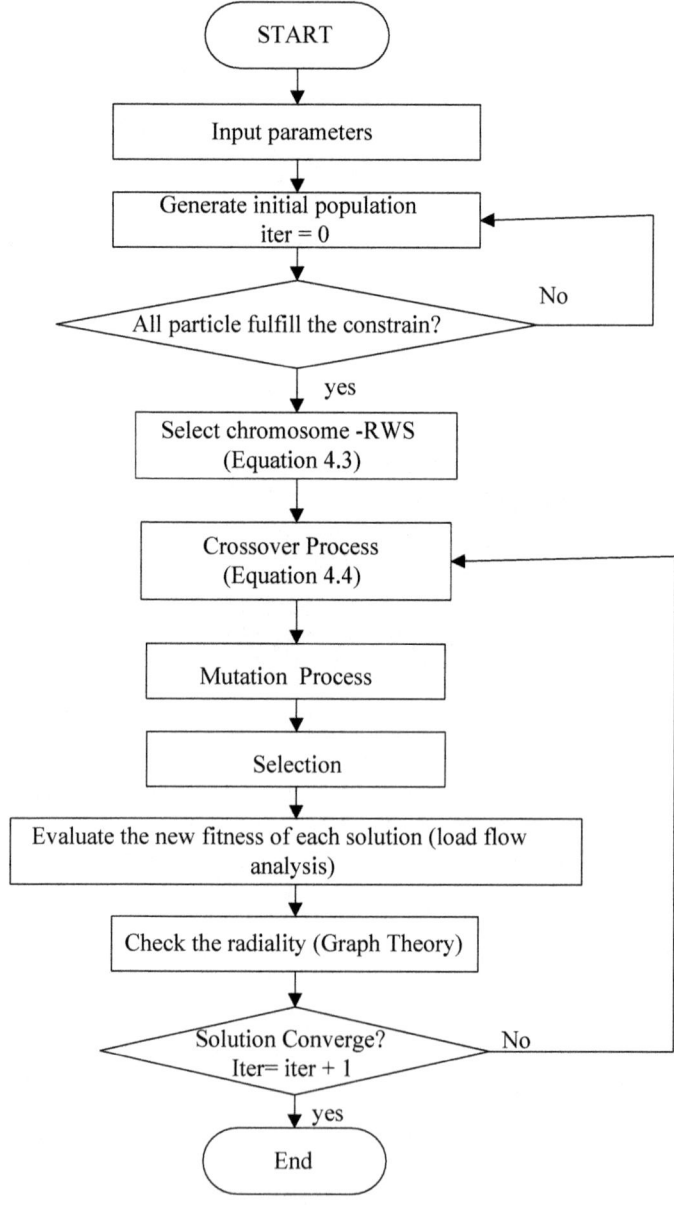

Fig. 4 Flow chart of modified genetic algorithm

Table 1 Descriptions on three different scenario studies

Scenario 1	The system is operated with service restoration via DNR
Scenario 2	The system is operated with service restoration via DNR and DG sequentially
Scenario 3	The system is operated with service restoration via DNR and DG simultaneously

In this paper, the problem is solved and analysed into three scenarios as shown in Table 1 which is service and restoration with and without DG. The standard 33-bus radial distribution systems are considered to test the effectiveness of the algorithms. Initial diagrams of 33-bus are applied in the reconfiguration distribution systems are illustrated in Fig. 5. The network consists of 33 buses, 38 lines, 5 switches and 3 branches (excluding the main branch). The total load of the system is 37.15 MW while the initial system power loss 202.6 kW. The size of the population used is 50. The location of DG used in this analysis are 6, 16 and 25.

4 Analysis of the Results on Power Loss

Fault in fact can happen at any branch in the system. For instance, Scenario 1, Scenario 2 and Scenario 3 are referred to fault occurs at branch 7.

4.1 Scenario 1—The System Is Operated by Service Restoration via DNR

In this scenario, the system operates via DNR without DG. The explanation about this scenario is referred to the fault occurs between bus no. 7 and 8, the feeder from 8 to 18 is considered as the out-of-service area. The out-of-service area can be restored through the network closure either by switch 33, 35 or 36. A path which undergoes the shortest distance and the fastest to restore the outage area will be selected. Besides being the most efficient method, it is also needed to solve the problem quickly. After service restoration, the opened switches have been changed to 7, 10, 34, 36 and 37 which is considered to be the best path (combination switches). It means, switch 33 and 35 is closed due to the short distance and time. Feeder 8 and 9 get the power supply through switch 33, while feeder 11 until 18 get the power supply through switch 35 as shown in Fig. 6. The minimum power loss is reduced from 202.3 to 138.6 kW which is 31.5%. Meanwhile, the number of switching operations is reduced to 4.

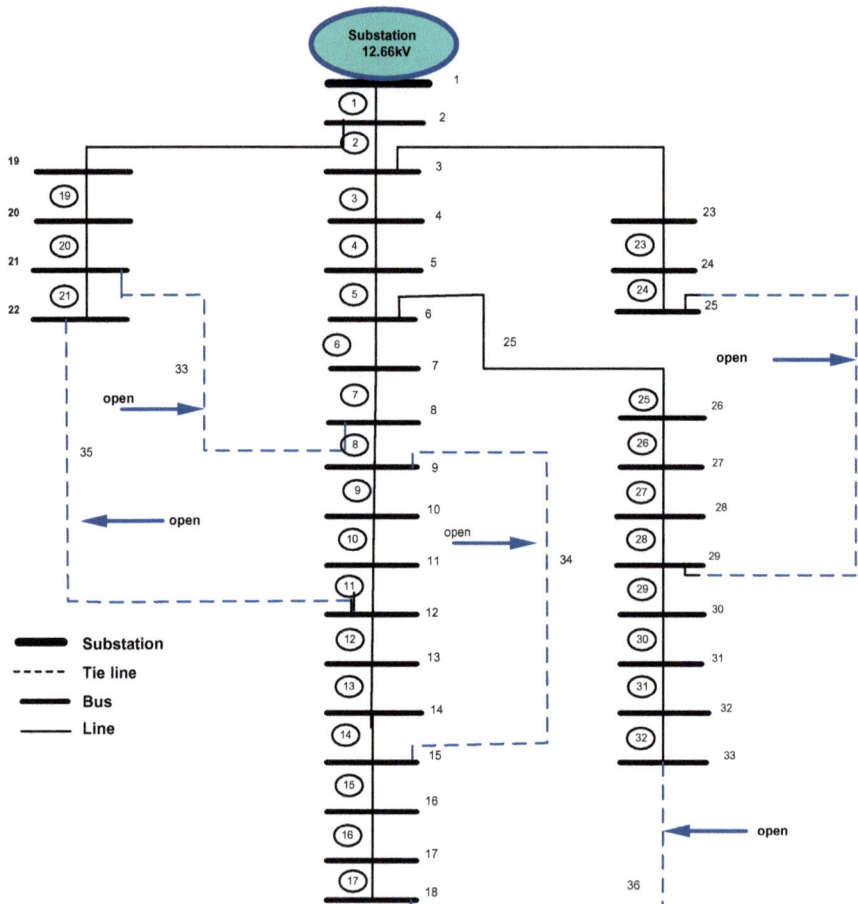

Fig. 5 33-bus distribution network with 5 tie switches

4.2 Scenario 2—The System Is Operated by Service Restoration via DNR and DG Sequentially

In this scenario, the system is operated through DNR and DG sequentially. The optimal location of the DG is identified and installed at bus no. 6, 16 and 25. The process of service restoration is started after the optimal DG size is obtained through MGA which are 0.7050, 0.7230 and 1.165 MW with the minimum voltage profile is 0.9810 p.u. The minimum power loss after service restoration with DG sequentially is 104.6 kW and un-restored load is zero. While, the new opened switches have been changed to 7, 9, 34, 36 and 37 which is to be the best path (combination switches). The adjustment on the individual DG size and configuration are not only affecting the power loss value, but it also changing the total DG

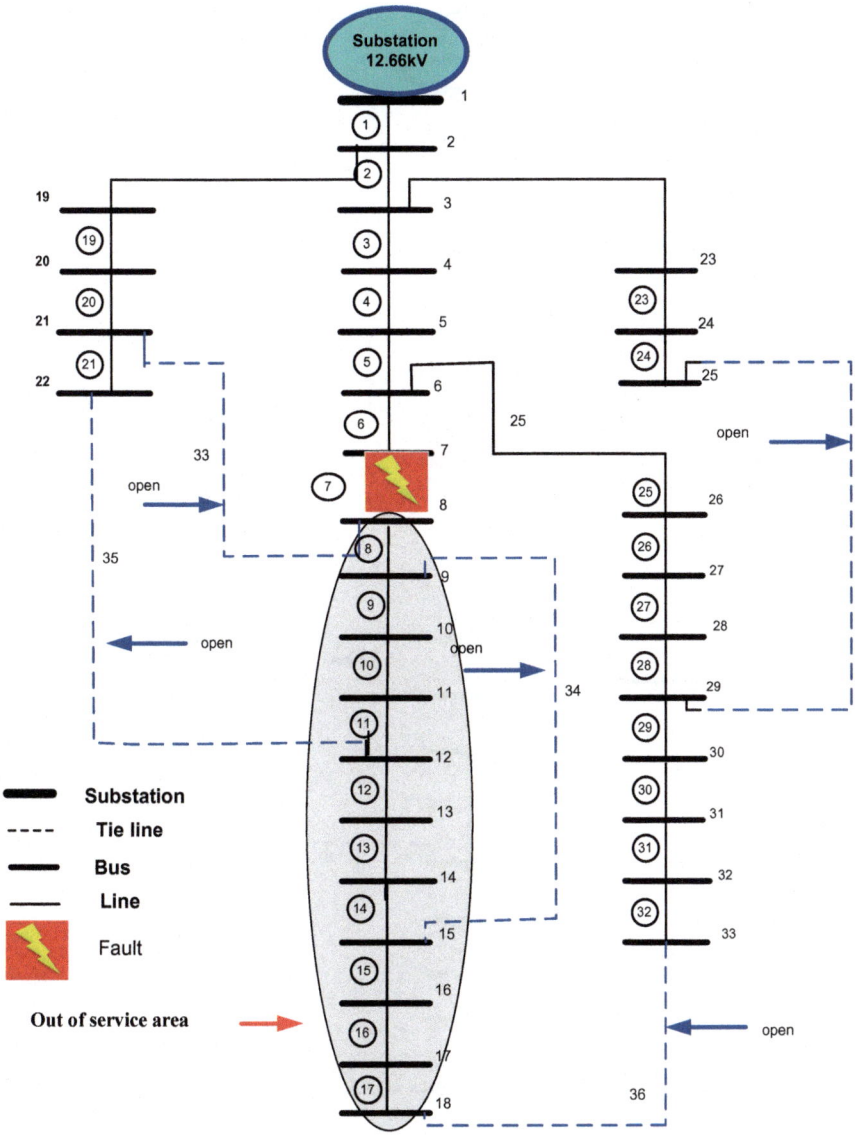

(a) Network before service restoration.

Fig. 6 Network before and after service restoration

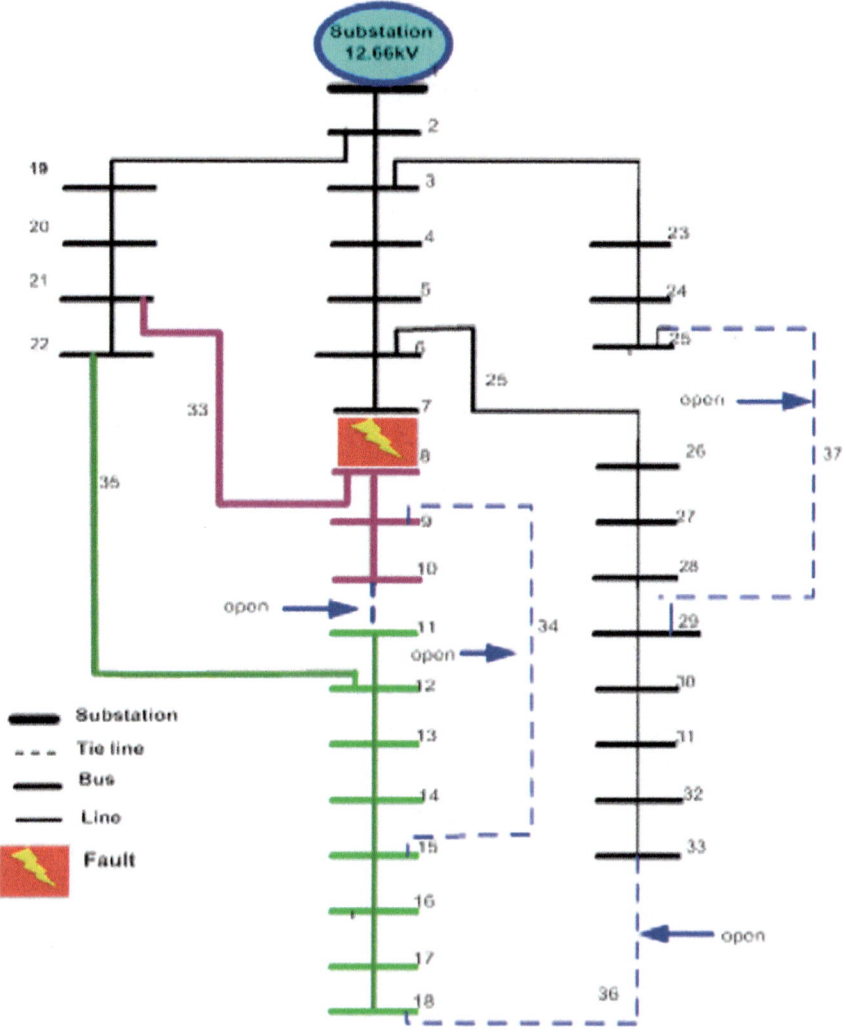

(b) Network after service restoration.

Fig. 6 (continued)

size in the system as well as the number branches in the network. All feeders have been restored quickly after a service restoration process which only takes about 13.8 s. Thus, the present of DG obviously assist to reduce the most power loss in the distribution system.

4.3 Scenario 3—The System Is Operated with Service Restoration via DNR and DG Simultaneously

In Scenario 3, the system is operating with service restoration through DNR and DG simultaneously. The explanation about this scenario is also referred to Fig. 7 as base scenario except that the DG has been installed at bus no. 6, 16 and 25. In this

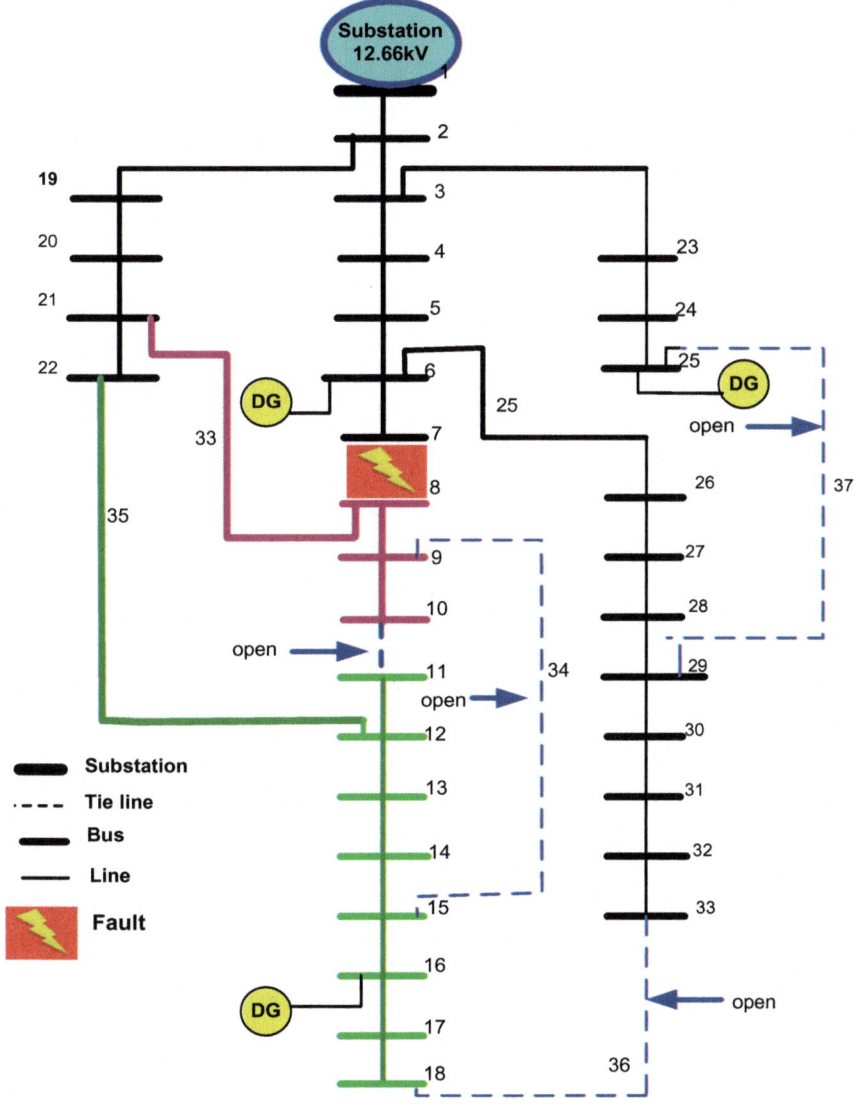

Fig. 7 Network after service restoration with DNR and DG simultaneously

Table 2 Opened switches for branch 7 of 33-bus system

Fault occurred	Possibilities opened switches (manually)	Power loss (kW)	Service restoration via DNR and DG using MGA	Power loss (kW)	CPU time (s)
Branch 7	7, 9, 12, 36, 37	96.9			
	7, 8, 13, 36, 37	95.8			
	7, 9, 14, 36, 37	96.5			
	7, 10, 14, 33, 37	95.2	**7, 10, 34, 36, 37**	**92.7**	**8.9**
	7, 11, 33, 34, 37	96.2	(Simultaneously)		
	7, 8, 12, 36, 37	97.2			
	7, 11, 13, 36, 37	97.9			
	7, 9, 14, 17, 37	93.0	7, 9, 34, 36, 37	104.6	13.8
	7, 8, 16, 34, 37	96.4	(Sequentially)		
	7, 10, 17, 34, 37	95.7			
	7, 14, 21, 32, 37	96.4			
	7, 13, 21, 22, 33	158.3			
	7, 13, 25, 35, 36	117.6			
	7, 9, 13, 36, 37	93.5			
	7, 10, 14, 31, 37	99.5			
	7, 9, 32, 34, 37 ↓ 435 897 (combination switches)	100.9			

scenario, there are 435 897 possibilities of opened switches when fault occurred at branch 7. After service restoration made through DNR and DG using MGA method, the result showed a speedier outcome with a minimal total power loss as shown on Table 2. The opened switches have been changed to 7, 10, 34, 36 and 37 which is to be the best path (combination switches) and provide the lowest power loss is as shown in Fig. 7. The minimum power loss is reduced from 202.6 to 92.7 kW, which is 54.2% and size of DG are 1.1690, 0.7757 and 0.5732 MW.

Meanwhile, the number of switching operations is reduced to 4. In this case, the switches change from 33 and 35 to switch 7 and 10 (4 switches involved in operation) whereas switches 34, 36, and 37 remained unchanged. The adjustment on the individual DG size and configuration are not only affecting the power loss value, but it also changes the total DG size in the system as well as the number branches in the network. All feeders have been restored quickly after a service restoration process which only takes about 8.9 s, which is faster than Scenario 2 (13.8 s). Therefore, if a fault occurred at any branch, there are many ways to choose the opened switches through try and error. However, with the proposed method (MGA), the best results obtained which is lowest power loss can get faster. Meanwhile, Table 3 shows the analysis of the lowest power loss of fault occurred after service restoration with Scenario 3 at each branch in the 33-bus system.

Table 3 Analysis of fault occurred on 33 bus system

Fault occurred at branch	Switch opened	Power loss (kW)	No. of switching operation	DG size (MW)		
				6	16	25
2	2, 7, 14, 35, 37	106.1	6	1.0080	0.9070	0.6543
3	3, 12, 34, 35, 36	101.6	4	1.1027	0.8180	0.7290
4	4, 11, 14, 17, 37	94.9	8	1.1519	0.9478	0.6680
5	5, 15, 34, 35, 37	105.3	4	1.1319	0.9335	0.6578
6	6, 10, 34, 36, 37	86.4	4	1.1690	0.9757	0.6732
7	7, 10, 34, 36, 37	92.7	6	1.0410	0.9050	0.7001
8	7, 8, 13, 36, 37	92.8	6	1.0439	0.9061	0.7012
9	7, 9, 14, 36, 37	91.5	6	1.0499	0.9098	0.7099
10	7, 10, 14, 33, 37	91.2	6	1.3004	0.5300	0.7054
11	7, 11, 33, 34, 37	91.9	4	1.0480	0.9070	0.7001
12	7, 8, 12, 36, 37	97.2	6	1.1127	0.9180	0.7291
13	7, 11, 13, 36, 37	93.9	6	1.0489	0.9118	0.7312
14	7, 10, 14, 32, 37	89.5	8	1.1590	0.9658	0.6732
15	7, 9, 15, 17, 37	93.0	8	1.1490	0.9427	0.6332
16	7, 8, 16, 34, 37	96.4	6	1.1523	0.9545	0.5312
17	7, 10, 17, 34, 37	94.7	6	1.1519	0.9378	0.5680
18	11, 13, 15, 18, 37	138.2	8	1.1330	0.8510	0.6220
19	8, 19, 31, 33, 37	107.7	6	1.1519	0.9335	0.6678
20	20, 34, 35, 36 37	133.8	2	1.1590	0.9747	0.6632
21	7, 14, 21, 32, 37	96.4	8	1.1733	0.9651	0.6363
22	7, 13, 21, 22, 33	158.3	8	1.1019	0.7575	0.7780
23	7, 13, 23, 35, 36	117.6	6	1.1604	0.5310	0.7740

(continued)

Table 3 (continued)

Fault occurred at branch	Switch opened	Power loss (kW)	No. of switching operation	DG size (MW)		
				6	16	25
24	9, 24, 33, 35, 36	101.1	4	1.1490	0.9427	0.6332
25	7, 10, 13, 25, 33	102.4	8	1.1604	0.6310	0.7740
26	7, 8, 14, 26, 33	96.8	8	1.0480	0.9070	0.7001
27	9, 27, 32, 33, 34	93.7	6	1.1127	0.9180	0.7290
28	7, 14, 21, 28, 36	97.5	8	1.0489	0.9118	0.7312
29	10, 12, 29, 33, 37	137.9	6	1.1604	0.531	0.674
30	7, 9, 13, 36, 37	93.5	6	1.1490	0.9427	0.6332
31	7, 10, 14, 31, 37	89.5	8	1.1523	0.9545	0.6312
32	7, 9, 32, 34, 37	100.9	6	1.1419	0.9378	0.5680

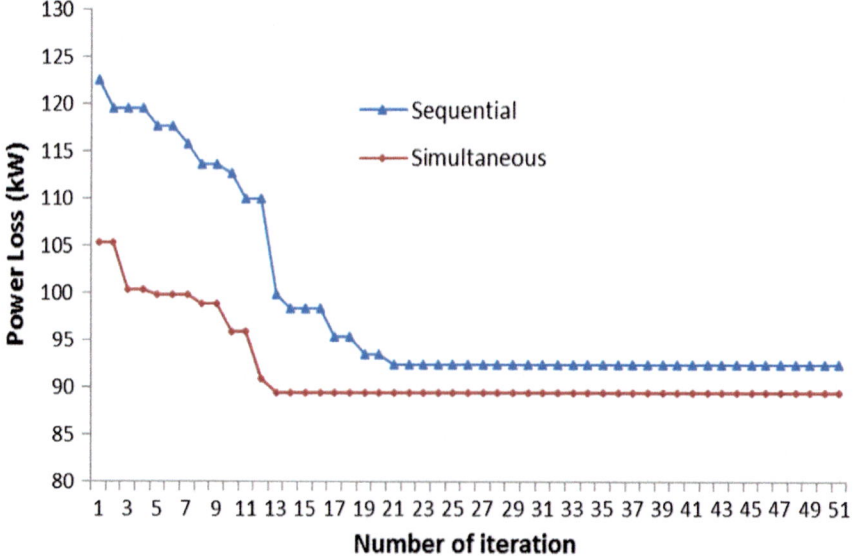

Fig. 8 Convergence characteristics of the proposed method of 33-bus system

As a conclusion, the system operates thru service restoration with DNR and DG simultaneously really gives great impacts on the power loss. It certainly reduces more of the total power loss as compared to the system operated sequentially.

5 Analysis of the Results on Voltage Profile

By observing the results, we can conclude that the voltage profile has improved and the average bus voltages that have reached exceed 0.950 p.u from 0.91 p.u. In Scenario 1, the system operated with the presence of reconfiguration, the voltage profile of the system has been improved considerably with a minimum node voltage of 0.9618 p.u. The voltage profile is improved clearly for bus 5 until 16. However, the voltage profile looks almost similar for bus 17 until 19. The rest of the bus only shows a slight improvement. Therefore, the implementation of the service restoration via reconfiguration technique has given a better voltage profile compared to service restoration without reconfiguration. The voltage profile can be improved further through applying service restoration and DG sizing sequentially as mentioned in Scenario 2. However, the voltage profile has been improved more effectively whenever the service restoration and DG in the system is operated simultaneously as shown in Scenario 3 where all bus voltages satisfy the 0.95 p.u. voltage constraints and near to 1 p.u. The voltage profiles in Scenario 3 for 33 bus systems show the significance of the voltage profile changes between bus 15 and 18. The voltage profile improves to 0.9996 p.u. The voltage profile shows the greatest improvement after service restoration with DG simultaneously technique applied.

Meanwhile, the convergence curve summarizes the capability and efficiency of each scenario and the speed of the algorithm in reaching the optimal point. Figure 8 shows the convergence characteristics of the proposed method of the 33-bus system. With the updated technique, the value of power loss is improved until the best solution is reached. From the observation, the Scenario 3 is the fastest (13 iterations) to reach the optimal solution followed by Scenario 2.

Finally, the switch opened in the system after service restoration for Scenario 2 and 3 when a fault occurrence between bus 7 and 8 can be roughly modelled as an undirected graph and checked the radiality of the distribution system by using Graph Theory as shown in Fig. 9. The distribution network system after reconfiguration must be in radial structure in order to simplify the over-current protection and simple design. The convergence is guaranteed for any radial distribution network due to the efficiency of distribution the load flow and no feeder section can be left out of service.

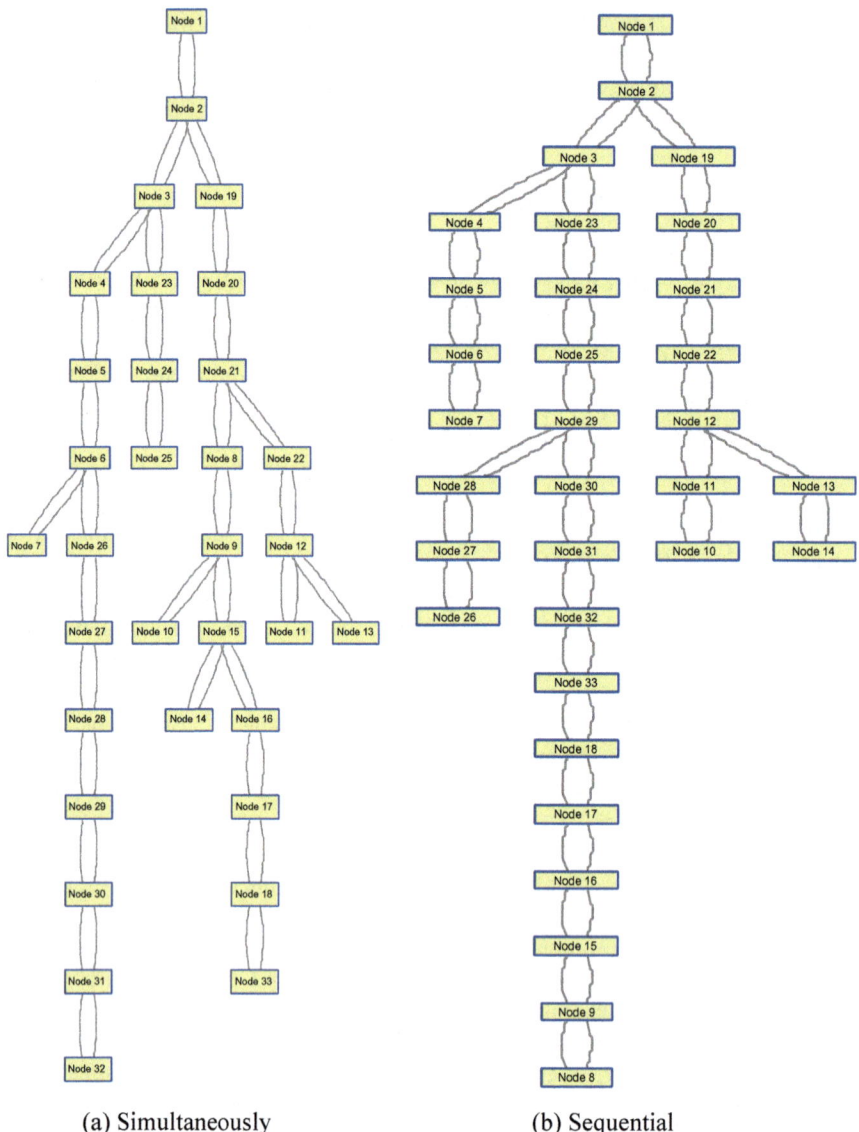

(a) Simultaneously   (b) Sequential

Fig. 9 Checking radiality after service restoration at branch 7

6 Conclusion

The objective of service restoration is to restore as many loads as possible by transferring loads from the out of service areas to other distribution feeders via DNR. In this work, DNR and DG are run simultaneously in a 33-bus system to assess the effectiveness of the MGA algorithms. Even the solutions have many possibilities, but by using the MGA algorithm after service restoration, the fault occurs will be overcome quickly in forming a new configuration with the shortest time and lowest power loss. Furthermore, restoration plan is not only reducing the amount of power loss, but at the same time to minimize the number of switching operations involved in each configuration. Thus, implementing DNR with DG simultaneously has indeed contributed great power loss improvement in the network distribution system.

References

1. Abu-Mouti, F.S., El-Hawary, D.M.E.: Optimal distributed generation allocation and sizing in distribution systems via artificial bee colony algorithm. IEEE Trans. Power Deliv. **26**(4), 2090–2101 (2011)
2. Nerves, A.C., Roncesvalles, J.C.K.: Application of evolutionary programming to optimal siting and sizing and optimal scheduling of distributed generation, TENCON 2009. In: IEEE Region 10 Conference, pp. 1–6, 23–26 Jan 2009
3. Yasin, Z.M., Rahman, T.K.A., Musirin, I., Rahim, S.R.A.: Optimal sizing of distributed generation by using quantum-inspired evolutionary programming. In: Power Engineering and Optimization Conference (PEOCO), 2010 4th International, pp. 468–473, 23–24 June 2010
4. El-Zonkoly, M.: Optimal placement of multi-distributed generation units including different load models using particle swarm optimization. IET Gener. Transm. Distrib. **5**(7), 760–771 (2011)
5. Moradi, M.H., Abedini, M.: A combination of genetic algorithm and particle swarm optimization for optimal DG location and sizing in distribution systems. Int. J. Electr. Power Energy Syst. **34**(1), 66–74 (2012)
6. Soroudi, A., Ehsan, M., Caire, R., Hadjsaid, N.: Hybrid immune-genetic algorithm method for benefit maximisation of distribution network operators and distributed generation owners in a deregulated environment. In: IET Generation Transmission and Distribution, vol. 5, no. 9, SMGAt. 2011, pp. 961–972 (2011)
7. Jamian, J.J., Musa, H., Mustafa, M.W., Mokhlis, H., Adamu, S.S.: Combined voltage stability index for charging station effect on distribution network. Int. Rev. Electr. Eng. IREE **6**(7), 3175–3184 (2011)
8. Subrahmanyam, J.B.V., Radhakrishna, C.: A simple method for feeder reconfiguration of balanced and unbalanced distribution systems for loss minimization. Electr. Power Compon. Syst. **38**(1), pp. 72–84
9. Gupta, N., Swarnar, A., Nizai, K.R.: Reconfiguration of distribution systems for real power loss minimization using adaptive particle swarm optimization. Electr. Power Compon. Syst. **39**(4), 317–330
10. Sivanagaraju, S., ViswanathaRao, J., Raju, P.S.: Discrete particle swarm optimization to DNR for loss reduction and load balancing. Electr. Power Compon. Syst. **36**(5), 513–524
11. Su, C.T., Chang, C.F., Chiou, J.P.: Distribution DNR for loss reduction by ant colony search algorithm. Electr. Power Syst. Res. **75**(2–3), 190–199 (2005)
12. Sathish Kumar, K., Jayabarathi, T.: Power system reconfiguration and loss minimization for an distribution systems using bacterial foraging optimization algorithm. Electr. Power Energy Syst. **36**, 13–17 (2011)
13. Fang Zong Wang, J.Y.: A refined plant growth simulation algorithm for distribution DNR. IEEE Trans. Power Syst. 4244–4738 (2009)

14. RugithaicharoencheMGA, N., Sirisumarannukul, S.: Feeder reconfiguration for loss reduction in distribution system with distributed generators by tabu search. GMSARN Int. J. **3**, 47–54 (2009)
15. Li, Q.W., Ding, W.: A new reconfiguration approach for distribution system with distributed generation. In: International Conference on Energy and Environment Technology ICCET 2009, pp. 24–26, 16–18 Oct 2009
16. Olamie, J., Niknam, T., Gharehpetian, G.: Application of Particle Swarm Optimization for Distribution feeder Reconfiguration Considering Distributed Generators
17. Olamie, J., Niknam, T., Gharehpetian, G.: Application of particle swarm optimization for distribution feeder reconfiguration consideringdistributed generators. Appl. Math. Comput. 575–586
18. Wu, Y.K., Lee, C.Y., Liu, L.C., Tsai, S.H: Study of reconfiguration for the distribution system with distributed generators. IEEE Trans. Power Delivery **25**(3), 1678–1685 (2010)
19. Lalitha, M.P., Reddy, V.V., Usha, V.: Optimal DG placement for minimum real power loss in radial distribution systems using PSO. J. Theor. Appl. Inf. Technol. **13**(2), 107–116 (2010)
20. Arun, M., Aravindhababu, P.: A new reconfiguration scheme for voltage stability enhancement of radial distribution systems. Energy Convers. Manage. **50**, 2148–2151 (2009)
21. Sivanagaraju, S., et al.: Enhancing voltage stability of radial distribution systems by network reconfiguration. Electr. Power Compon. Syst. **33**(5), 539–550 (2005)
22. Jouybari, B.R., Hosseini, M., Byagowi, Z.: Optimal placement of distributed generators and reconfiguration of distribution systems for loss reduction using genetic algorithm. Int. Rev. Electr. Eng. **6**(2) (2011)

Improved Design of the UniKL Amphibious Research Crawler II for Underwater Exploration

Ahmad Makarimi Abdullah, Khairul Arieff Abu Jalil, Hisyam Hamid, Nursyahida Izzati Zakaria, Norhafizah Othman, AshmanYusoff Iqmal Mohd Haidhir and Mohd Iqram Mohd Kamro

Abstract A crawler is an instrument, which is widely used for seabed exploration and even for underwater operation. A crawler system has the potential to expand the research and development on seafloors with irregular stepped terrains and multiple types of seafloors. The characteristic parameters in the water, such as added mass, buoyancy, and hydrodynamic forces, considerably affect the stability and decrease the mobility of crawler-type remotely operated vehicles (ROVs). To study and evaluate the mobile performance of a crawler system, it is important to investigate the design of the crawler system by considering these effects. This paper presents multiple results of an underwater crawler system done by researcher to show the dynamic effects on the vehicle's motion. Experiments were conducted on a crawler-type ROV climbing over a bump in a water tank to examine the traction characteristics of the crawler. The model presented in this paper simulates the effects of added mass to the dynamic motion and the traction characteristics very well, and it reveals the physics of the crawler-type ROV's motion.

A. M. Abdullah (✉) · K. A. A. Jalil · H. Hamid · N. I. Zakaria · N. Othman
A. I. M. Haidhir · M. I. M. Kamro
Universiti Kuala Lumpur Malaysian Institute of Marine Engineering Technology,
Jalan Pantai Remis, 32200 Lumut, Perak, Malaysia
e-mail: makarimi@unikl.edu.my

K. A. A. Jalil
e-mail: khairularieff90@gmail.com

H. Hamid
e-mail: hishamhamid@unikl.edu.my

N. I. Zakaria
e-mail: syahida_izzati92@gmail.com

N. Othman
e-mail: norhafizhothman6@gmail.com

A. I. M. Haidhir
e-mail: ash_man93@hotmail.com

M. I. M. Kamro
e-mail: iqramkamro@gmail.com

© Springer International Publishing AG 2018
A. Öchsner (ed.), *Engineering Applications for New Materials and Technologies*,
Advanced Structured Materials 85, https://doi.org/10.1007/978-3-319-72697-7_20

Keywords Chassis analysis · Performance analysis · Traction analysis

1 Introduction

Underwater vehicles nowadays are widely used such as for underwater research, taking samples, welding operations, even in oil and gas offshore platforms. There are many types of underwater vehicles such as remotely operated vehicles (ROV), autonomous underwater vehicles (AUV), and submarines. This paper presents the design and development of crawler, which is an extension of remotely operated vehicles for seabed explorations. It is very crucial for the ROVs-crawler type to withstand environmental effects, hydrodynamic effects, which will affect the stability and mobility of the crawler. The hydrodynamic effect is one of the major factors that affects the crawler during seabed exploration. This is due to the high pressure and high sea current which will cause the crawler to face difficulties to move or to maintain posture during underwater operation [1]. In order to handle a large amount of stress that is added to the crawler and forces that affect the crawler, suitable materials are important as all the forces that will directly create tension to the main chassis of the crawler [2].

Some sea floors with irregular stepped terrain contain steep terrains having slopes of more than 30° in some areas, and the seafloor can be bumpy with assumed irregularities of 20–50 cm. To conduct the work or survey operations there, ROVs need functions to move and bear the reaction force of operations. In addition, ROVs sometimes need to remain stationary or in static position and keep their posture stable during operations such as taking sample, which is an important issue for conventional ROVs [3]. In regards to the irregular terrain, it is important for the crawler to have suitable dimensions of main chassis as well as proper type of tire or wheel. Based on many researches, it is essential for the crawler type ROV to have good traction between the tire and seafloor, which helps the crawler to grip during underwater operations. This enhances the underwater mobility thoroughly.

Some conventional ROV such as those that hover, generally will bear the reaction force by their weight and thrusters. In this case at which the crawler is implementing some elements of hovering, it will also bear the same reaction force as the conventional ROV that hovers. However, this is sometimes insufficient for bearing large forces, which limits the crawler operation. The aim is to integrate the crawler with the remotely operated vehicle for underwater inspection because conventional ROVs are equipped with thrusters to move in water but they do not have the function of move on the seafloor. By developing an attachment of the crawler, it also helps to improve the mobility of the crawler on a variety of seafloors such as sand, rock reefs with inclinations of more than 30°, and seafloor covered with cobbles or gravel or even to conduct experiments in a water tank and at sea. This confirmed that the flipper-type crawler system has advantages in performance; it has the potential to operate on seafloors with irregular steep terrains and climb over bumps higher than those climbed over by a conventional crawler system.

2 Problem Statement

Subsea vehicles, which are well known as ROVs, are widely used in many scopes such as, inspection, exploration and even search and rescue missions. Some ROVs with crawler type systems are so advanced with robotic systems that they allow underwater operations such as underwater welding and repairing of underwater piping without sending humans to the work field. Different levels and types of seafloor require a stable crawler to operate where it is important for the crawler to maintain posture during the operation [4]. As to expend the research on the various types of seafloors such as mud, sand, gravel and irregular steep terrain, the crawler must be able to handle or withstand the hydrodynamic effect such as added mass, buoyancy and center of buoyant. This refers to the integration of robotic arms, subsea ports or the power management for the crawler and thruster, where this changes will affect the crawler itself. Due to the changing of mass and modification to the main chassis, it is essential to analyze the changes as for the crawler to maintain performance for optimum operation. The aim is to study and evaluate the crawler type ROV integrated with a robotic arm, i.e. the performance such the characteristic and principle of operation. The installation of the tethered system (cable) reduces the mobility and its performance, because the type of cable used will affects the size of the umbilical cord (tethered system). Thus, it is important to know the exact type of cable that is going to be used for the crawler in order to minimize the hydrodynamic effect to the crawler. This paper presents the development of the underwater crawler and also a study of the hydrodynamic effect on the crawler. Experiments were conduct on land and underwater to stimulate the results in order to make a comparison and also to enhance the project

3 Literature Review

In this part, most of the research that has been carried out by researchers will be analyzed in order to find the gap of technology or to implement some of the researcher's ideas. Figure 1 shows the output traction for each wheel [5]. The researchers showed from the result that each wheel could produce the same traction even when climbing obstacles.

In Fig. 2, the researcher showed the fabricated crawler when climbing obstacles. However, a plural type wheel is required in the crawler as to solve the slip problem encountered by the crawler [6]. Figure 3 shows the output traction using track wheel over time which superior than wheel types. However, the researcher encountered a problem of handling the pitch angle of the crawler as shown in Fig. 4 [7]. Figure 5 shows the fabricated crawler climbing an obstacle during testing.

In this section, the result obtained from the previous version will be analyzed for the improvement of the UARC II main chassis. Figure 6 shows the isometric plan view of the previous version occupied with top view, side view, front view and 3D

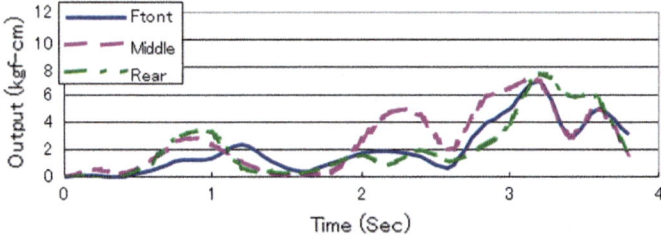

Fig. 1 Output traction versus time. *Source* Chugo [7]

Fig. 2 Fabricated product. *Source* Chugo [7]

view. The dimensions of the crawler are approximately 40 cm in width, 60 cm in length and 15 cm in height, with a total volume of 360 m³. Figure 7 shows the fabricated crawler with a total weight of approximately 15 kg integrated with the controller. The main material used to fabricate the crawler is aluminum in marine grade which suitable for underwater applications.

The underwater crawler will start to operate as soon as the power cable is being plugged to the power supply and the switch is moved to "ON". The system will initialize the cooling fan as to maintain the power supply unit at safe temperature (room temperature) during operation to get ready. Besides that, the red LED will turn on to show that there is 220 VAC power supply being switched ON and two yellow LEDs indicate that the PSU 12 VDC is operational. When the system is initialized, pull the trigger as to power up the motor and position the switch at the

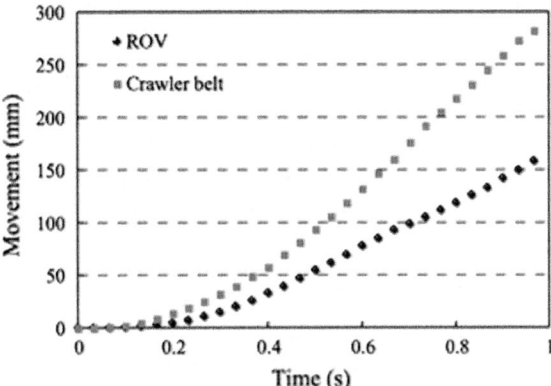

Fig. 3 Output movement versus time. *Source* Inoue [3]

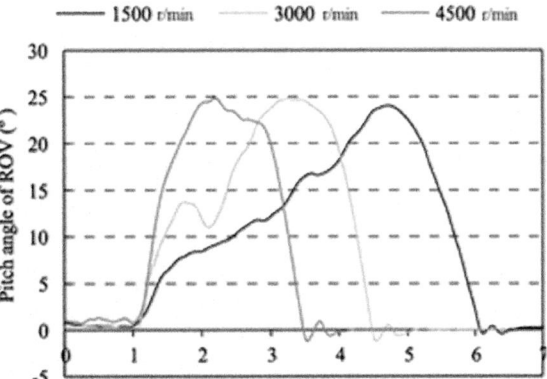

Fig. 4 Pitch angle versus time. *Source* Inoue [3]

Fig. 5 Output movement versus time. *Source* Inoue [3]

top of the trigger to control the motor either forward or reverse. The voltage will be channeled to the connector in the subsea port on the crawler and perform the given task by the activated motor needed by the user. The output of this motor will

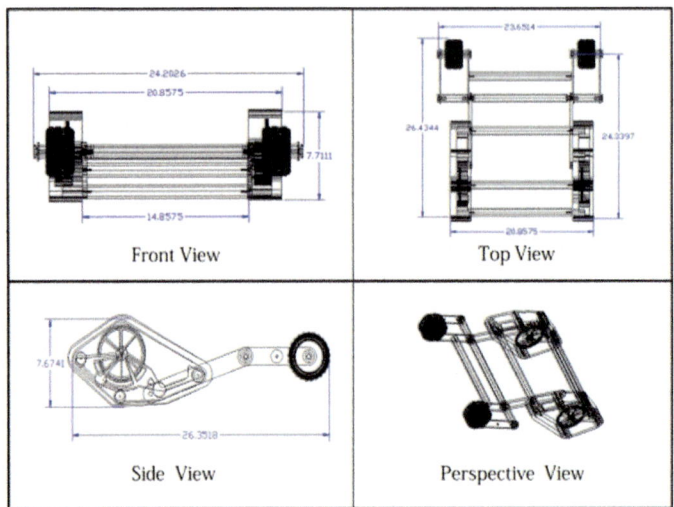

Fig. 6 Isometric plan view of UARC I (version I)

Fig. 7 Fabricated UARC I (previous version)

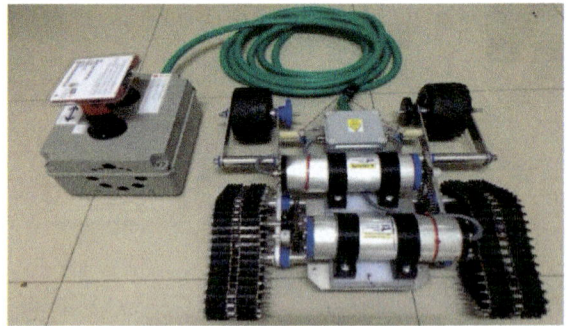

conduct the given task. After the output is successfully activated, the crawler will move as the user wants or needs.

Figure 8 shows the result obtained from experiment that had been conducted. The distance tested is fixed to 3 m and loads are manipulated from 0 kg up to 15 kg. From the result it can be seen that the speed of crawler is reduced significantly with the additional load either on land or underwater. The percentage of uncertainty in speed for the UARC 1 based on the underwater testing at a depth ∼15 m has been calculated. The on land result shows that the speed percentage of uncertainty is ∼6.36% while for the underwater testing at 15 m depth ∼5.54%. Based on this calculation the underwater speed reduction at 15 m depth is ∼0.82%.

Load (KG)	Result (on land)			Result (underwater)		
	Distance (cm)	Time taken (second)	Speed (cm/s)	Distance (cm)	Time taken (second)	Speed (cm/s)
0	3000	6.38	47.02	3000	6.51	46.08
5	3000	6.50	46.15	3000	6.62	45.32
10	3000	6.63	45.24	3000	6.79	44.18
15	3000	6.80	44.12	3000	6.88	43.60

Fig. 8 Result of speed with variations of load of UARC I (previous version)

4 Preliminary Results

Figure 9 shows the improved version of the main chassis of the UARC II. The improved version of the main chassis will be 30% larger than the UARC I (previous version). Thus, the main chassis will have a greater stability compared to the previous version of the main chassis. The position of the thruster will also be analyzing for maximum performance of the thrust when overcoming an obstacle during underwater operation.

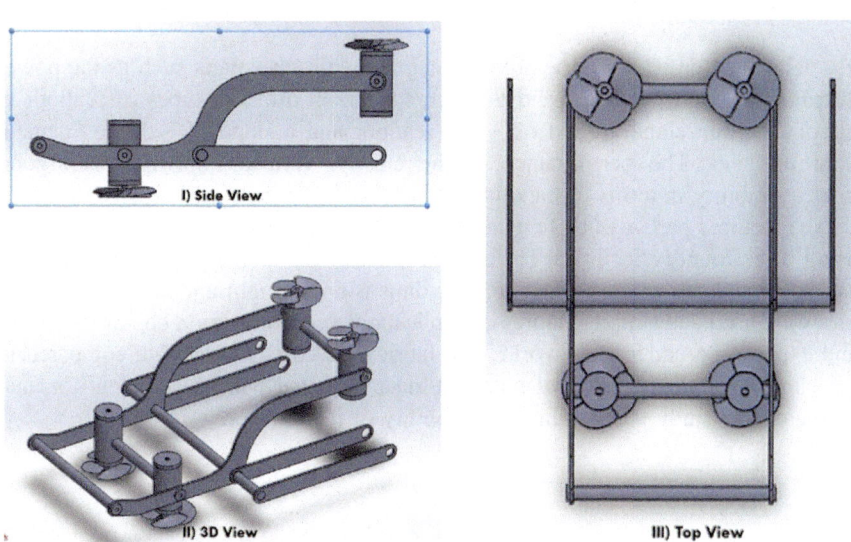

Fig. 9 Isometric plan view

5 Methodology

This section presents the process of fabricating UARC II as to study the stability and mobility of the crawler.

Figure 10 shows that the first step is to design and fabricate the circuit and the main chassis of the underwater crawler to integrate with the remotely operated vehicle (ROV) for underwater inspection. This is required for the crawler to be able to move and maneuver at different seabeds or seafloors. The main circuit as shown in Fig. 1 illustrates the connection of the crawler from the power supply to the motor. For the option of tasks that can be done, a design of the crawler is also been constructed as shown in Fig. 9. The next step is to fabricate the crawler chassis and to place all the component of the crawler. The chassis is fabricated with a combination of aluminum and 3D printing (ABS) machineries' as shown in Fig. 9. After that, troubleshoot the circuit to avoid that any dysfunction or fault occur during the installation process. Lastly, the prototype will be decorated and cleaned up as the final finishing.

6 Expected Outcome

The fabricated crawler will be able to integrate with other parts such as the power distribution, control system, and thruster, for use in different types of seafloor at irregular terrain slope around of $\sim 30°$ seafloor and a slope of $\sim 30°$ on cement floor on land. The performance of the crawler is also improved with good maneuverability in terms of handling of the crawler on land and underwater.

The required parts will be fabricated using the technologies that are locally made using basic foundry facilities, basic electronic devices and 3D technology, which can be considered as cost effective. The dimensions, weight and simple technologies were used and simple features of application are suitable for a beginner to learn and explore the underwater world. The fabrication cost is affordable compared to the existing products. Last but not least, data collected will be sufficient for analyzing and to study the stability and mobility of the crawler.

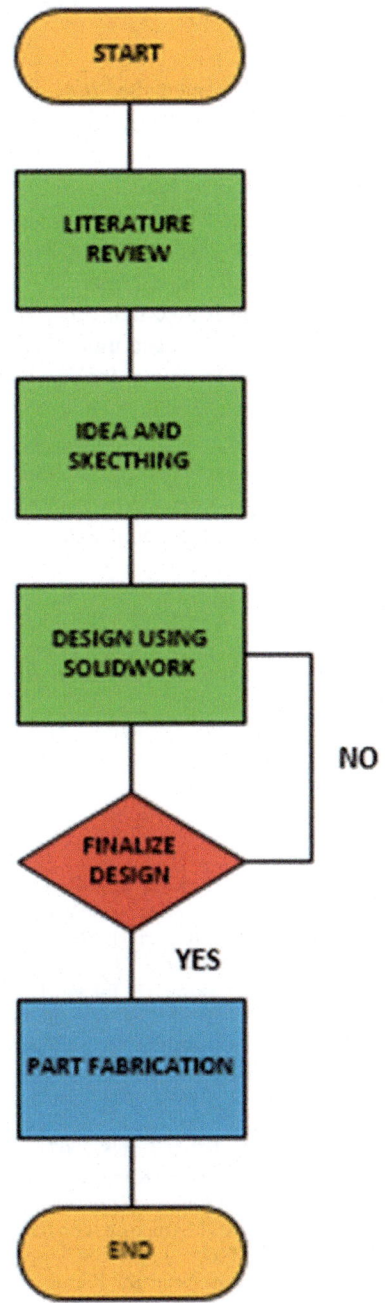

Fig. 10 General process flowchart of UARC II

7 Conclusions

This paper presented the study and process of designing a new main chassis for UARC II and also the dynamic model of the underwater crawler system to show that characteristic parameters in water affect and lower the mobility considerably. Dynamic motion simulations will be conducted for climbing over a bump and for slip measurement when running on the sand seafloor. Also, the experiments will be conducted using a small-size ROV in the water tank and at sea.

The results of the experiments of climbing over the bump are consistent with the simulations until a certain time. However, it was observed in the experiment that the attitude was changed more than the maximum changing attitude obtained by the simulation. Furthermore, the results of the experiments of slip were consistent with the simulations.

This reveals that the dynamic model established and presented in the paper can express the dynamic motions of the ROV with the crawler system, will enable to continue the investigation regarding the moving performance of an underwater crawler system and develop an advanced crawler system for survey and exploration on the seafloor.

Acknowledgements This project has been approved by UniKL Mimet and supported with Short Term Research Grantt (STRG) Deepest thanks to, Mr. Ahmad Makarimi bin Abdullah as our main supervisor for excellent guidance, caring, patience, and providing us with an excellent atmosphere for doing our final year project. Furthermore, the contribution of lecturers and technicians at Unikl Mimet is acknowledged.

References

1. Lee, S., Suh, I.H., Kim, M.S.: Recent Progress in Robotics: Viable Robotic Service to Human: An Edition of the Selected Papers from the 13th International Conference on Advanced Robotics, vol. 370. Springer Science and Business Media (2008)
2. Rodocker, D., Rodocker, J.: U.S. Patent Application No. 11/710,013 (2007)
3. Inoue, T., Shiosawa, T., Takagi, K.: Dynamic analysis of motion of crawler-type remotely operated vehicles. IEEE J. Oceanic Eng. **38**(2), 375–382 (2013)
4. Chatzakos, P., Papadmitriou, V., Psarros, D., Nicholson, I., Gan, T.H.: On the development of an unmanned underwater robotic crawler for operation on subsea flexible risers. In: IEEE Conference on Robotics Automation and Mechatronics (RAM), 2010, pp. 419–424, June 2010
5. Waldmann, C., Richter, L.: Traction properties of the wheels of an underwater crawler on different soils. In: OCEANS 2007, pp. 1–7. IEEE, September 2007
6. Yu, S.C., Hyun, J., Kim, D.Y., Park, M.K., Pyo, J.H., Cho, P.C., Hwang, C.S.: The development of multi-functional autonomous amphibious vehicle. In: Underwater Technology (UT), 2011 IEEE Symposium on and 2011 Workshop on Scientific Use of Submarine Cables and Related Technologies (SSC), pp. 1–6, April 2011
7. Chugo, D., Kawabata, K., Kaetsu, H., Asama, H., Mishima, T., Takase, K.: Slip reduction control using plural wheel information for step-climbing vehicle. In: Industrial Electronics, 2009. IECON'09. 35th Annual Conference of IEEE, pp. 2373–2378, November 2009

Preliminary Design and Analysis Study of Propeller for Autonomous Underwater Vehicle (AUV)

Mohd Amin Hakim Ramli, Mohd Iqram Bin Mohd Kamro,
Muhamad Fadli Ghani, Ahmad Makarimi Abdullah,
Norhafizah Othman, Ashman Yusoff Iqhmal Mohd Haidhir,
Nursyahida Izzati Zakaria, Khairul Arieff Bin Abu Jalil,
Muhammad Haziq Bin Noor Zaharil Ehsan,
Muhammad Farhan Bin Mohamad Noor
and Fatin Zawani Binti Zainal Azain

Abstract A propeller is an important part in marine vehicle as it provides the thrust and transmits the power generated by engine or motor to move the vehicle either forward or backward and in some circumstances upward and downward. The study and design of propeller have been an important point since the vehicles were manufactured to improve the vital part of the propulsion force. This study will review the designation of propeller, the effect of hub diameter, number of propeller blade, the rake and pitch of propeller effect and efficiency of propeller.

Keywords AUV · Lathe · 3D printing · Propeller · Bollard Pull Test

---

M. A. H. Ramli (✉) · M. F. Ghani · A. M. Abdullah · N. Othman · N. I. Zakaria ·
M. H. B. N. Z. Ehsan · M. F. B. M. Noor · F. Z. B. Z. Azain
Universiti Kuala Lumpur, Malaysian Institute of Marine Engineering Technology,
Jalan Pantai Remis, 32200 Lumut, Perak, Malaysia
e-mail: mohdamin@unikl.edu.my

M. F. Ghani
e-mail: muhamadfadli@unikl.edu.my

A. M. Abdullah
e-mail: makarimi@unikl.edu.my

N. Othman
e-mail: norhafizahothman6@gmail.com

N. I. Zakaria
e-mail: syahida_izzati92@gmail.com

M. H. B. N. Z. Ehsan
e-mail: haziqlekir@gmail.com

1 Introduction

Autonomous Underwater Vehicle (AUV) is being made as an inspections device to inspect the condition of the on sea, navigation, mapping and so on. This is done to save the cost to departure human to the unknown part of the ocean and put the life of human at risk. This vehicle can collect data according to the setting made by coding that are coded to the AUV systems. AUV also known as underwater drones that can operates independently with direct human input that have been install into the AUV before launching [1].

The AUV was inspired by the tragedy of MH370 in Malaysia. During that time, Malaysia does not own any device or expertise for search and rescue (SAR) [2]. Without any option, Malaysia government had to import the technology from other country which involved a high cost. The AUV is one of the solutions for our country to do inspection in the seabed for the future.

The developments of research are being made to improve the ability of the device and adding some new installation to make the AUV limitless. So, it is the researchers responsibility to design analysis of propeller for AUV. The AUV are only on the design stage where there are none prototype being made, but all of the simulation for the design have been done. In this paper the propulsion systems are taken into account because it is one of the major parts of the AUV. Other than this research about the propulsion system, motion controlling itself for the AUV are also challenging but will not be discuss in this paper [1].

Improvement that will be implemented to the prototype AUV is an installation of propulsion system with optimum performance. A propulsion system consists of a few mechanisms that produce thrust to push an object forward or backward with the presence of propeller.

The propulsions system gives the AUV the ability to propel in order to move from one point to another point. This propulsion will be able to produce optimum thrust without consuming over power.

M. F. B. M. Noor
e-mail: farhan2489696@gmail.com

F. Z. B. Z. Azain
e-mail: fatinzawani@unikl.edu.my

A. Y. I. M. Haidhir · K. A. B. A. Jalil
Universiti Kuala Lumpur, Malaysia France Institute (MFI), Section 14, Jalan Teras Jernang, Seksyen 14, 43650 Bandar Baru Bangi, Selangor, Malaysia
e-mail: ash_man93@hotmail.com

M. I. B. M. Kamro
Institute of Product Design and Manufacturing, No.9, Jalan Perdana 7/3, Taman Shamelin Perkasa, 56100 Kuala Lumpur, Selangor, Malaysia
e-mail: iqramkamro@gmail.com

Currently, researchers are working on to improve and upgrade the ability and performance of autonomous underwater vehicle for sea exploration and mapping. This paper is a review of the preliminary design and analysis study of propeller for the AUV based on some parameter.

First parameter that needs to be determined is the propellers diameter. Diameter of propeller is the vital part in propeller section. The propeller diameter will determine the efficiency of a propeller and the power installation generate by the motor or engine. The increases in propeller diameter will decrease the revolution per minute; this generally increases the efficiency of propeller. Usually, diameter is limited by draft and underwater appendages but it is best to fit the largest diameter propeller with appropriate clearance.

Next parameter to be studied is the propeller pitch. Conversely, a higher pitch will deliver a top speed but slower acceleration. The thickness of the hub plays a vital part to adjust the propeller pitch. Third parameter is the number of propeller blade. Three bladed propellers have generally proven to be the best comprise between blade area and efficiency. Four propellers however would seldom be as efficient as three bladed because the closer blades create additional turbulence and reduce vibration.

2 Methodology

There are two stages in completing this project which are designing and fabricating.

2.1 Design

There a few designs of propeller that were created by using the Inventor CAD [3, 4]. Some of the propellers that we designed were used widely in marine vehicle such as spoon design. Propeller that we used for this AUV is different compare to the commercials propeller. The parameters for the propeller are; hub diameter 25 mm, height of the hub 30, diameter of propeller 140 mm (Figs. 1, 2, 3 and 4).

2.2 Fabrication

2.2.1 Lathe Machine

The propeller are not available on the market, this is because the propeller are design based on the resistance calculated for the AUV, and suitable with the amount of power generated to propel the AUV. So, the shaft for the propeller was fabricated by using the lathe machine. There are two methods used to fabricate the shaft using

Fig. 1 Design A

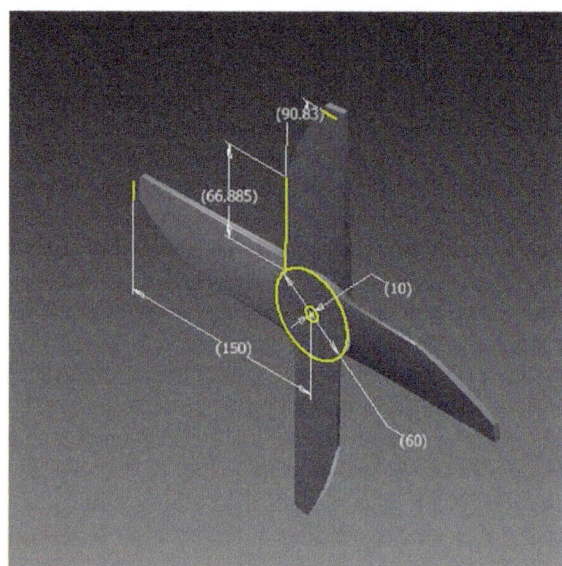

Fig. 2 Design B

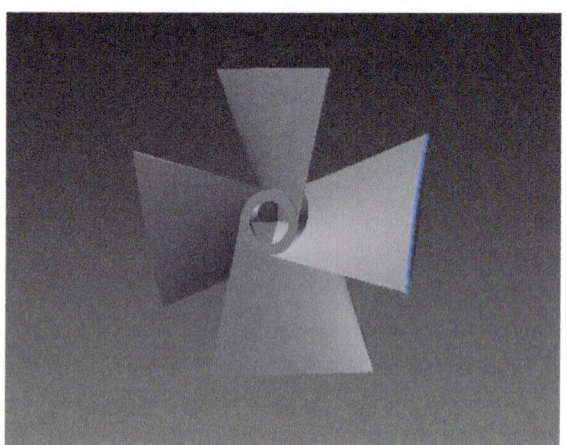

lathe machine which are facing method and turning step method. The motor casing are also are fabricated from solid aluminum, the reasons we are not using water proof motor are because of the budget, when we fabricate the motor casing, a lot of cost can be save. First the solid aluminum is clamp to the chucks jaw in order for the machine to spin it [5]. Then, the aluminum will be carve from 25 mm are cut until the diameter are 12 mm. The surface of the shaft then smooth by using sand paper.

Fig. 3 Design C

Fig. 4 Design D

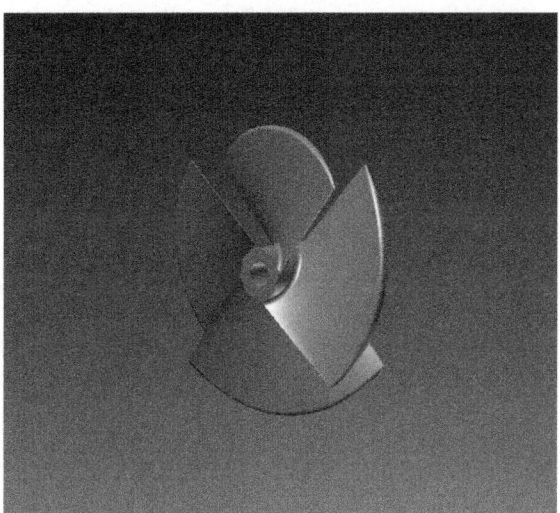

2.3 3D Printing

As a starting study for development of propeller, this review uses 3D printer. The filament used for 3D printing were made of Polylactic acid or Polylactide (PLA) [6, 7] with size 3 mm. This is only the prototype for the real propeller for AUV because this method is cheaper compare to the conventional propeller [8–10].

The end product of 3D printing creates rough surface that is needed to be smoothed by using sand paper, this is done to reduce the resistance to water [11].

2.3.1 Tap and Die Tool

After the shaft have been lathe and smooth using sand paper, the end of the shaft needed to be thread by using die method. This is done so that a nut can be used to secure the propeller to the shaft. In front of the propeller is the placement of a drive dock so that the propeller does not move when the shaft rotates. The drive dock needed to be secured to the shaft, so a hole is drilled and the hole is threaded using tap method.

3 Results

This review only focused on the final design of the propeller (Figs. 5 and 6). This propeller id 3D printed as shown in Fig. 7. The demonstration test was conducted to analyze the operation of the propeller when in underwater. The test started by testing the propeller that can operate. The test included the Digital Spring Balance to get the result using this equipment.

The other testing was done by operating the speed of propeller for three stroke as slow speed, medium speed, and high speed. The testing at different stroke made

Fig. 5 Front view

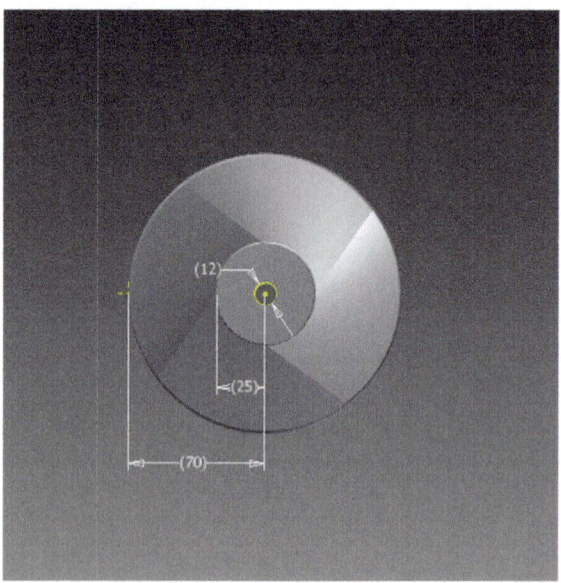

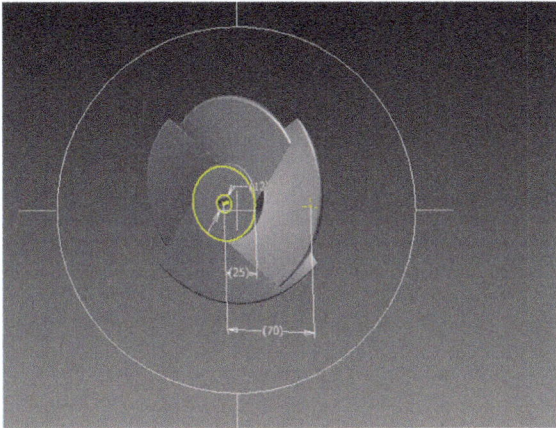

Fig. 6 Front view 2

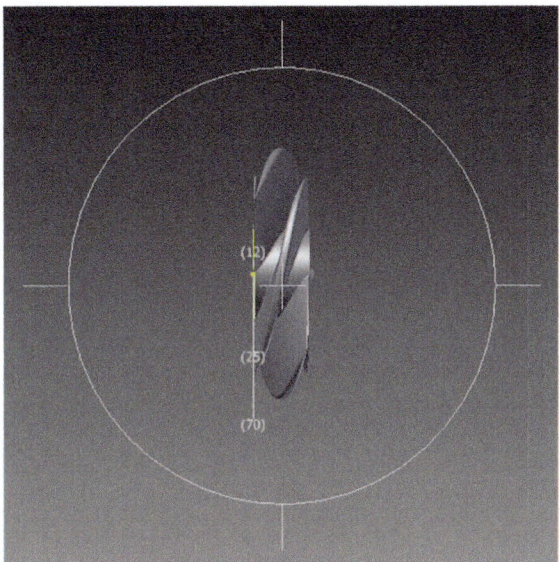

Fig. 7 Side view

Table 1 Bollard Pull Test result

Speed (m/s)	Weight (kg)	RPM (rev/min)	Current (I)	Voltage (V)	Power (W)	Torque (N/m)
Low	0.15	650	1.56	4.81	7.5×10^{-3}	8.128×10^{-5}
Medium	0.60	840	2.00	8.30	1.66×10^{-2}	1.392×10
High	0.80	1120	3.20	13.43	4.3×10^{-2}	27.035

different reading on Digital Spring Balance. The result as shown in Table 1, Bollard Pull Test results.

Table 1 was the data collected during the Bollard Pull Test [12]. The speed of the propeller rotation increased as the load applied was increased. To calculate the torque of propeller, the torque equation was used:

$$\text{Voltage} = \text{I (current)} \times \text{R (resistance)} \quad (1)$$

$$\text{Ampere} = \text{V (voltage)} / \text{R (resistance)} \quad (2)$$

$$\text{Power} = \text{V (voltage)} \times \text{I (current)} \quad (3)$$

$$\text{Torque} = [\text{Hp (horsepower)} \times 5252]/\text{RPM} \quad (4)$$

$$1\,\text{Hp} = 754.7\,\text{Watt}$$

4 Discussion

From this review, the thrust generates by the propeller are calculated by using the Eqs. 1, 2, 3 and 4. To find the voltage of the motor used, Eq. 1 was used. Equation 2 was used to find the power generated and then the value of the power will be substituted into Eq. 4 in order to find the torque produce.

The problem occurred during using lathe machine process was process accuracy (Figs. 8, 9 and 10). From this study it can be said that center hole drilled cannot be achieved without precision to the center diameter of the shaft (Fig. 11). Poor maintenance of the machinery can cause error on the material. Error on the material can cause other problems that can lead to other major problems.

Fig. 8 Process of lathe

Fig. 9 Complete shaft with drive dock

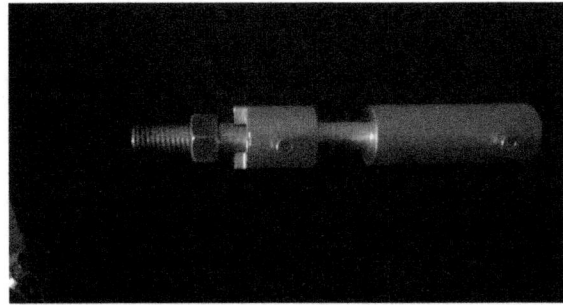

Fig. 10 Complete propeller from 3D printer

Fig. 11 Process of threading the shaft

The troubleshooting when using the tap tools to get inlet threads. Inlet of threads can be troubleshot when it does not pressure the tools uniformly. Threads on screw can deform on the shaft. Using the die tools, threads on screw can deform when rotate the die tools uniformly.

For the first phase of the printing process of 3D printing, the printed model stick to heat base of the platform make it difficult to withdraw the model. After several observation and experimentation, the problem occurred were identified. The model that was applied to the platform due to the chemical solution was also applied on the platform before the printing process. To prevent the occurred problem, the chemical is not applied before the printing process.

Acknowledgements The authors thanked the University Kuala Lumpur Malaysian Institute of Marine Engineering Technology (UniKL MIMET), Lumut for research facilities and financial assistance provided for this project.

References

1. Waghmare, G.V.L.L.M.: Robust maneuvering of autonomous underwater vehicle: An adaptive fuzzy PI sliding mode control. Intell. Serv. Robot. **10**(3), 195–212 (2017)
2. Pleter, O.T., Constantinescu, C.E., Jakab, B.I.: Reconstructing the Malaysian 370 flight trajectory by Optimal Search. 2016, 1–23
3. Porter, L.A., Washer, B.M., Hakim, M.H., Dallinger, R.F.: User-Friendly 3D Printed Colorimeter Models for Student Exploration of Instrument Design and Performance. ACS Publications, Washington, DC (2016)
4. Porter, L.A., Chapman, C.A., Alaniz, J.A.: Simple and Inexpensive 3D Printed Filter Fluorometer Designs: User-Friendly Instrument Models for Laboratory Learning and Outreach Activities. ACS Publications, Washington, DC (2017)
5. Finkelstein, N., Aronson, A., Tsach, T.: Toolmarks made by lathe chuck jaws. Forensic Sci. Int. **275**, 124–127 (2017)
6. Giovanna, M., et al.: Polylactic acid-based porous scaffolds doped with calcium silicate and dicalcium phosphate dihydrate designed for biomedical application. Mater. Sci. Eng. C **82**, 163–181 (2018)
7. Nguyen, T., Vayer, M., Sinturel, C.: PS-b-PMMA/PLA blends for nanoporous templates with hierarchical and tunable pore size. Appl. Surf. Sci. **427**, 464–470 (2018)
8. Guo, J., Guo, S., Li, L.: Design and characteristic evaluation of a novel amphibious spherical robot. Microsyst. Technol. **23**(6), 1999–2012 (2017)
9. Huynh, J.Q.T.: Detailed Design of a Thruster Solution for a Small Mass-Market Remotely Operated Underwater Vehicle, pp. 7. (2016)
10. Maljaars, P.J., Kaminski, M.L.: Hydro-elastic analysis of flexible propellers : An overview. In: Fourth International Symposium on Marine Propulsors, 2015
11. Bollig, L.M., Hilpisch, P.J., Mowry, G.S., Nelson-cheeseman, B.B.: 3D printed magnetic polymer composite transformers. J. Magn. Magn. Mater. **442**, 97–101 (2017)
12. Song, Y.S.: Thruster modeling for a hovering autonomous underwater vehicle considering thruster-thruster and thruster-hull interaction. In: Underwater System Technology: Theory and Applications (USYS), IEEE International Conference on, pp. 100–104. IEEE, 2016

Preliminary Study on the Development of Two Degree of Freedom Robotic Arms for Underwater Applications

Noorazlina Mohamid Salih, Ashman Yusoff Iqmal Mohamad Haidhir, Muhamad Fahezal Ismail, Ahmad Makarimi Abdullah, Khairul Arieff Abu Jalil, Norhafizah Othman, Muhamad Fadli Ghani, Muhamad Faizal Muhamad Isa and Mohd Iqram Mohd Kamro

Abstract A robotic arm is a mechanical arm that is designed to work similarly to a human arm. The first robotic arm was introduced by a scientist named George Devol Jr. in the 1950s. During that time, robotics development was a bit slow because most of the robotics applications were used for space exploration. Then, in 1980s the robotic arms began to be integrated in automotive and manufacturing assembly lines. The aim of robotic arm development is to design an arm that can be stored without the arm dangle outside of the ROV body for underwater research and discovery. In order to meet the objectives of the study, the structure of the study was divided into three main stages, i.e. the modelling, development of hardware and simulation/testing. During the modelling stage, the SolidWorks software was used to design the structure of the underwater robotic arm. After the modelling stage, we continued with the hardware stage. In simulation/testing, the parameters of the arm are the extension and retraction, the maximum load that can be lifted (up to one hundred (100) gram) and the ability to work up to thirty (30) centimetre depth. There are so many advantages using a robotic arm in such places. One of the

N. Mohamid Salih (✉) · A. M. Abdullah · M. F. Ghani · M. F. M. Isa
Universiti Kuala Lumpur, Malaysian Institute of Marine Engineering Technology, Jalan Pantai Remis, 32200 Lumut, Perak, Malaysia
e-mail: noorazlinams@unikl.edu.my

A. M. Abdullah
e-mail: makarimi@unikl.edu.my

M. F. Ghani
e-mail: muhamadfadli@unikl.edu.my

M. F. M. Isa
e-mail: Syahida_izzati92@gmail.com

A. Y. I. M. Haidhir · M. F. Ismail · K. A. A. Jalil
Universiti Kuala Lumpur, Malaysia France Institute (MFI), Section 14, Jalan Teras Jernang, Seksyen 14, 43650 Bandar Baru Bangi, Selangor, Malaysia
e-mail: ash_man93@hotmail.com

© Springer International Publishing AG 2018
A. Öchsner (ed.), *Engineering Applications for New Materials and Technologies*, Advanced Structured Materials 85, https://doi.org/10.1007/978-3-319-72697-7_22

advantages is to lower the costs which means to reduce the labour costs. A robotic arm can work for 24 h daily without break. This leads to a big output of production if it is used in production lines. A robotic arm is also fast and accurate in performing specific tasks programmed by the user.

Keywords Robotics · Robotic arms · Manipulators · Degree of freedom Underwater applications

1 Introduction

These days, people are looking for adventure activities to explore. One of the activities is to explore the underwater world. People are able to go underwater and investigate the underwater world, but they are unable to operate at long hours and only at a limited certain depth [1]. From this limitation and difficulties, we have invented a remotely operated vehicles (ROV) to help to encounter this problem [2]. The robotic arm is one of main features for the ROV. Many kinds of work can be operated by the robotic arm, even beyond of the human hand capabilities [3, 4]. Every undertaking movement of the arm involved in a mode system where it can be defined as one degree of freedom (DOF) [4]. It moves using mechanical parts connected at joins, allowing either rotational motion or linear motion.

In this study, the underwater robotic arm is designed based on the structure of the underwater crawler and its application. The underwater robotic arm must be able to grip and lift up a sample from underwater. This robotic arm consists of two motors which is a stepper motor and an antenna motor. The stepper motor is used for the gripper to open and close. The antenna motor is used for the arm to extend and to retract. The data will be taken based on the minimum and maximum load that can be lifted up from the water, the length of depth that the robotic arm can go, and the time for the arm to extend and retract.

M. F. Ismail
e-mail: fahezal@unikl.edu.my

K. A. A. Jalil
e-mail: khairularieff90@gmail.com

N. Othman
Universiti Kuala Lumpur, British Malaysian Institute (BMI),
Bt. 8, Jalan Sungai Pusu, 53100 Gombak, Selangor Darul Ehsan, Malaysia
e-mail: norhafizahothman6@gmail.com

M. I. M. Kamro
Universiti Kuala Lumpur, Institute of Product Design and Manufacturing, No.9, Jalan Perdana 7/3, Taman Shamelin Perkasa, 56100 Kuala Lumpur, Selangor, Malaysia
e-mail: iqramkamro@gmail.com

2 Methodology

2.1 Initial Requirements Study

In this initial stage, the latest research activities that are related to the underwater robotic arm, methods and techniques are reviewed. The scope will focus on the mechanism of the gripper and a suitable type of motor that will be used underwater. The literature review based on journals, articles and books will be used to collect the information for the hardware, mechanism of gripper, motor and controller. Every factor must be measured from DOF, parameters, materials, method, and more.

The specifications of the motor are choose based on the requirement needed for the motor to support the arm [5, 6]. The load target is set to make sure that the robotic arm can be achieve [2]. The material of the robotic arm also needs to be measured as it is important to avoid any leaking and short circuit to the arms.

2.2 Designing and Performing the Experimental Modeling

For the second stage, the findings from the initial requirement study will be used for designing a prototype model. This model will consider many factors such as the type of mechanism gripper, use of waterproof motor, wiring method, cost and materials of the hardware [5, 7]. After that, the best selection of mechanism, hardware, controller, wiring method will be selected. The SolidWorks 2014 software will be used for the design of the underwater robotic arm.

2.3 Development of Models

For the last stage, a model of the robotic arm will be developed using a 3D printer and it will be controlled by Arduino. The microcontroller can be selected from various sorts of controllers such as Raspberry Pi and Arduino. This models is controlled by Arduino as its microcontroller [8]. The expected outcome for this stage is that the model is able to function underwater.

2.4 Block Diagram

Figure 1 shows the main important part of the robotic arm system. To build the robotic arm, the main parts such as the controller, programming, mechanism, and hardware must function very well to make sure that the underwater robotic arm can

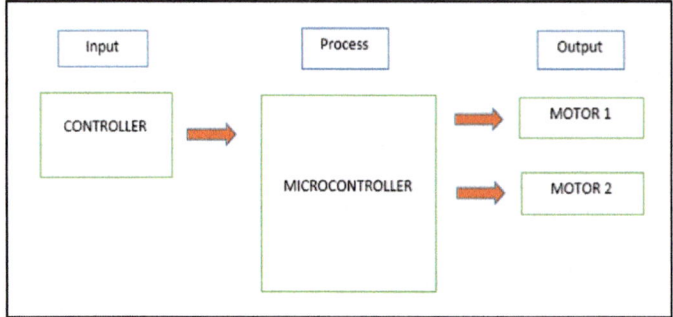

Fig. 1 Block diagram of the system

function as intended [1]. The controller consists of an on-off button that functions to control the movement of the robotic arm.

The button "on" functions to retract the gripper, and the button "off" for the gripper to return back to the initial condition. The programming is designed to give the signal to the antenna motor to decide to retract or to return to the normal condition.

This mechanism is the concept of the gripper to be function. And for the hardware, the gripper is designed to move an object from one place to another place.

2.5 Flowchart of Study

This study undergoes four stages to make sure that this study can undergo smoothly. Stage 1 is the process design, i.e. to sketch some ideas to find a suitable design of the underwater robot arm. Stage 2 is the experimental part, i.e. to select the mechanism and programing used for the underwater robot arm. Stage 3 is to integrate the hardware, to continue with the fabricating of the hardware and to develop the controller for the robotic arm operation. Finally stage 4 is the analysis, i.e. testing and analyzing the robotic arm for completion of the final report of this study (Fig. 2).

2.6 Design of Robotic Arms

Figure 3 shows the full-scale model of the robotic arm. The robotic arm consists of three parts, i.e. arm, gear, and gripper. The length of the arm is about one hundred and twelve millimeters, the length of the gripper is seventy-five millimeters, and the

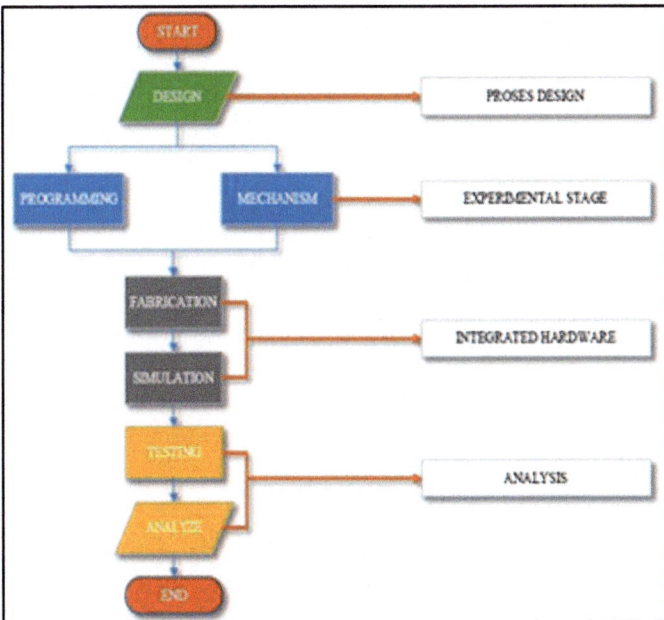

Fig. 2 Flowchart of study

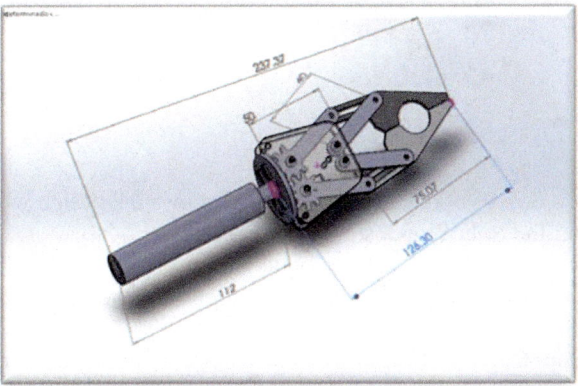

Fig. 3 Full scale of robotic arm

length of the gear is fifty millimeters. The maximum length the gripper can open is hundred millimeters.

The robotic arm is designed based on the requirement and specification of the ROV. The aim is to design an arm that can be stored in the ROV without the arm dangle outside of the ROV body. Figure 4 shows the design when the gripper is

Fig. 4 Gripper closed

Fig. 5 Gripper open

closed using the SolidWorks software. Figure 5 shows the design when the gripper is opened using the SolidWorks software.

Figure 6 shows the design of the robotic arm in the normal condition using the SolidWorks software. The blue color in the figure above shows the arm in the retracted condition.

The maximum length that the arm can be extended is three-hundred millimeters. The arm consist of three parts, at which every part has one-hundred millimeters in length. So, when the arm is back to the initial condition, the length of the arm will be one-hundred millimeters (Fig. 7).

Figure 8 shows the full design of the robotic arm when the function of arm is extending. This robotic arm has two functions. First, the arm can be extending and retracting. Secondly the griper can be opened and closed.

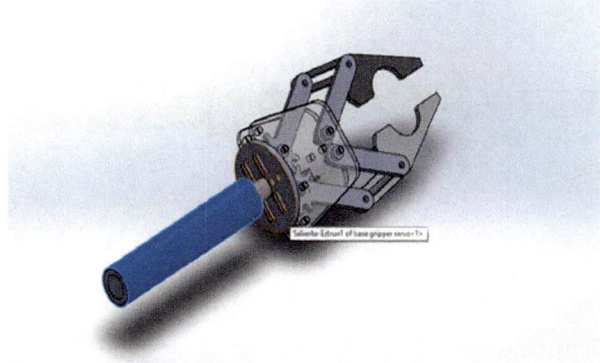

Fig. 6 Arm in retract condition

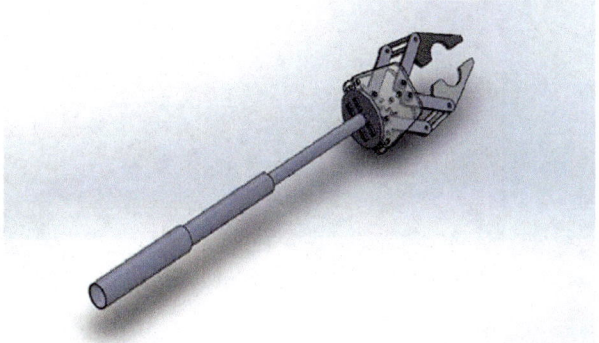

Fig. 7 Arm in extend condition

3 Testing and Results

The model has been tested to lift loads based on six different weights.

3.1 Underwater Lifting Load Test Result

The robotic arm was tested to lift several loads to evaluate the maximum load that the arm can lift. The range for the arm also is considered as how far can the arm reach.

Figure 9 shows the underwater load test. For the first try, the gripper is tested to lift up to 20 g of load. For the result, the robotic arm successfully lifted the considered load.

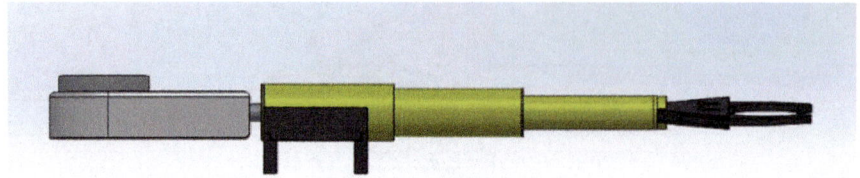

Fig. 8 Full design of robotic arm

Fig. 9 Underwater 20 g load test

Figure 10 shows the on-land load test. For the second test, the gripper is tested to lift up to 60 g of load. For the result, the robotic arm successfully lifted this load.

Figure 11 shows the on-land load test. For the third test, the gripper is test to lift up to 100 g of load. For the result, the robotic arm successfully lifted the 100 g load.

3.2 Result for Underwater Load Test

The model has been test to lift load. Table 1 shows the lifting result in underwater conditions based on six different weights.

Table 1 shows the data collected during the underwater testing. The starting load was 10 g, continued with 40 g, 60 g, 100 g, 120 g, and lastly 150 g. During these tests, the load of 120 and 150 g could not be lifted up because there was the influence of the underwater pressure. The arm and gripper cannot support any weight which is exceeding 100 g.

Fig. 10 Underwater 60 g load test

Fig. 11 Underwater 100 g load test

Table 1 Underwater lifting result

Trial	Weight (g)	Achieved
1	10	Yes
2	40	Yes
3	60	Yes
4	100	Yes
5	120	No
6	150	No

3.3 On-land Extend and Retract Arm Test Result

Figures 12 and 13 show the arm in the normal condition and when the arm is extended. The data and results for time to extend and retract the arm are shown in Table 2. For the second test in experiment two, it shows how many times the robotic arm can extend and retract using a 12 V battery.

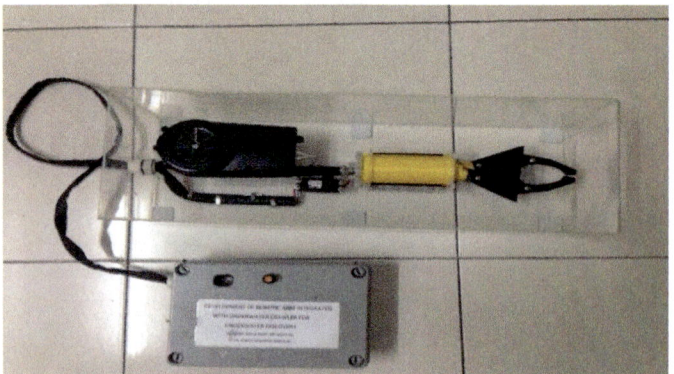

Fig. 12 Arm in normal condition

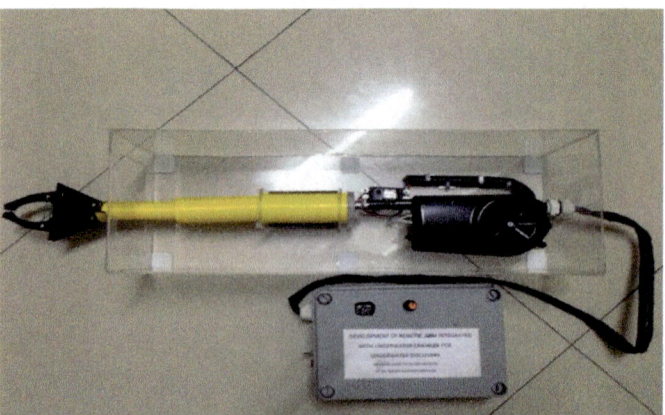

Fig. 13 Arm extend

Table 2 Extend and retract result

No	Data	Result
1	– Time to arm extend using 12 V	3 s
2	– Time to arm retract using 12 V	3 s
3	– Total time robotic arm can extend and retract using 12 V	45 time

4 Recommendation

This robotic arm can be developed in many ways in the future in terms of material, method and more. Various methods can be usd besides using a stepper motor and an antenna motor. A servo motor, or a hydraulic or pneumatic system can also be considered as methods that are suitable for the robotic arm to perform underwater [2, 9, 10].

By increasing the number of DOF, one can also consider that more work can be done [11–13]. The higher the DOF, the more movement of the arm can do. Large power consumption is needed for the motor running continuous in long duration [11, 14].

Besides that, the motor needs to be sealed for more compatible mobility. Finally, to ensure the underwater robotic arm is able to operate in deeper conditions, a higher grade material and a servo motor must be used.

5 Conclusion

This study successfully designed and constructed a small scale of an underwater robotic arm that can be integrated with an underwater crawler. It was able to function in one meter depth and can lift up to one-hundred grams of load to the surface, while on-land this robotic arm can lift up to one hundred and twenty grams. This is due to the pressure difference between land and water. This study can be improved more in the future by using different methods, materials and parameters. Based on study, the robotic arm was successfully designed, fabricated and tested.

References

1. Mohammed, A.A., Sunar, M.: Kinematics Modeling of a 4-DOF Robotic Arm, pp. 87–91 (2015)
2. Gonc, R., Wisniewski, E., Gomes, M., Corazzim, M., Beaulieu, A.: Development and Implementation of a Natural Interface to Control an Industrial Hydraulic Robot Arm (2015)
3. Muslim, M.A.: Design of Geometric-based Inverse Kinematics for a Low Cost Robotic Arm, pp. 82–86 (2014)
4. Al-halimi, R.K., Moussa, M.: Performing Complex Tasks by Users With Upper-extremity disabilities Using a 6-DOF Robotic Arm : A Study, vol. XX, no. XX, pp. 1–8 (2016)
5. Almurib, H.A.F., Al-qrimli, H.F., Kumar, T.N.: On the Development of a 7-DoF SPP Robotic Arm, pp. 930–934 (2011)
6. Guo, Q., Yu, T., Jiang, D.: Adaptive Backstepping Design of Electro-hydraulic Actuator based on State Feedback Control, pp. 888–891 (2015)
7. Kumra, S., Saxena, R. Mehta, S.: Design and Development of 6-DOF Robotic Arm Controlled by Man Machine Interface (2012)

8. Afaq, A., Ahmed, M., Masood, V., Shahzaib, M., Rashid, N., Tiwana, M., Awan, A.: Development of FPGA-based system for control of an Unmanned Ground Vehicle with 5-DOF Robotic Arm, no. Iccas, pp. 724–729 (2015)
9. Iulia, C., Muntean, I.: Internal Model Control for a Hydraulically Driven Robotic Arm, pp. 5–9 (2014)
10. Rivas, D., Manjarrés, F., Mena, J., Carrillo-medina, J.L., Bautista, V., Erazo, M., General, A.: Inverse Engineering Design and Construction of an ABS Plastic, Six DOF Robotic Arm Structure, pp. 352–357 (2015)
11. Xu, X., Ananthanarayanan, H.: Design and Construction of 9-DOF Hyper-Redundant Robotic Arm, vol. 1, pp. 321–326 (2014)
12. Pol, R.S.: Labview Based Four DoF Robotic ARM, pp. 1791–1798 (2016)
13. Kuijing, Z., Pei, C.U.I., Haixia, M.A.O.: Basic Pose Control Algorithm of 5-DOF Hybrid Robotic Arm Suitable for Table Tennis Robot, pp. 3728–3733 (2010)
14. Moradkhani, A., Ahmadi, K., Ghalambor, S., Habibizadeh, E. Yahyaei, M.: Mechanism with Four Degrees of Freedom, Design and Construct with Controller for Increasing the Power Output of Solar Cell, pp. 542–545 (2011)

Design and Analyses of a Ship Floating Dry-Dock

Roslin Ramli and Anwar Faris Sobri

Abstract The aim of this paper is to design and analyze a ship floating dry-dock. A ship floating dry-dock is a structure that can be submerged to permit the entry and docking of a ship and then be raised to lift the ship from the water for repairs or maintenance. Advancements in ship floating dry docks and transfer systems enable the floating to meet new requirements of launching and repairing the vessels and are very important in today's shipping. This paper aims to prepare a design of a ship floating dry dock with a length 250 ms for two types of the hull which are the monohull and the twin hull types and also to compare the ballast water displacement of both hulls. The innovations lie in separating the docking process from the maintenance platform and the ballast water displacement analysis is performed by using the Maxsurf software. This study will include details of the general arrangement and stability requirement.

Keywords Ship floating dry dock · Ballast water displacement
Maxsurf software

1 Introduction

A ship floating dry-dock is a structure that can be submerged to permit the entry and docking of a ship and then be raised to lift the ship from the water for repairs or maintenance [1]. It is a dock that is held in place laterally, as by pilings, but rests on floats and is free to rise and fall with changes in the water level. In the nautical term,

---

R. Ramli (✉)
Universiti Kuala Lumpur, Malaysian Institute of Marine Engineering
Technology, Jalan Pantai Remis, 32200 Lumut, Perak, Malaysia
e-mail: roslin@unikl.edu.my

A. F. Sobri
Petronas Maritime Service Sdn Bhd, Universiti Kuala Lumpur,
24300 Kerteh, Terengganu, Malaysia
e-mail: inianwarfaris@gmail.com

a large boxlike structure that can be submerged to allow a vessel to enter it and then floated to raise the vessel out of the water for maintenance or repair.

The aim is a ship floating dry-dock system with a simple way to raise vessels for quicker turn-around of maintenance and repair [2]. For this purpose, the dock is flooded and submerges enough to allow a ship to sail in. Then, the water is pumped out of the tanks and the floating dock rises, lifting the ship out of the water. In this way work can be carried out underneath the ship's hull.

In most conditions, after docking that the operator could secure the vessel onto a stabilised raft, they either tow the raft to a mooring, or a servicing or survey facility, or else perform the work on the raft offshore. This would clear the dock for the next vessel, reducing queuing, while allowing more flexibility to maintenance workers by freeing the vessel on the raft from the floating dry-dock schedule. This would be a production system that could be used to lift and transfer vessels, or any other items of equipment, in the marine environment [2].

This research is focusing on the design of a ship floating dry dock by using a related software, to analyse the ballast water displacement of a monohull and a twin hull.

2 Methodology

The flow of the design of the ship floating dry-dock will start with generating the hull form by developing the hull by using the Maxsurf Modeler software. Then, the large angle data will be collected by using the Maxsurf Stability module. Afterwards, the properties of both hull shapes for upright floatation are determined and the ballasting analysis is performed. From Maxsurf, the model is transfered into a STL file to create the 3D arrangement. Lastly, the combination is done by using the AutoCAD software. Lastly, the results of the ballasting stability and structural analysis are compared. The flow chart is shown in Fig. 1.

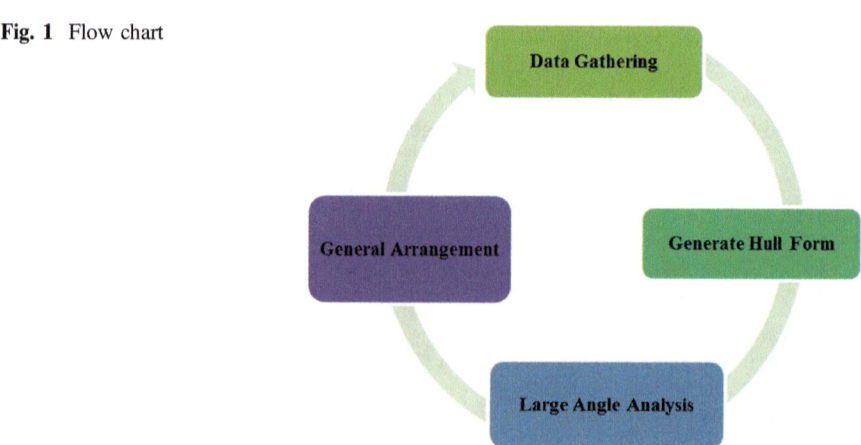

Fig. 1 Flow chart

a. Generate hull form

Based on data gathering, the principal dimensions will set for both hulls which have the same size as shown in Table 1.

i. *Monohull*

The software that was used for generating the hull form was Maxsurf Modeler. The Maxsurf Modeler is a professional level system used by most professional commercial designers. It is a powerful three-dimensional surface modelling system for use in the field of marine design. It provides a clear and familiar environment in which to work, allowing for systematic experimentation and rapid optimisation of any new design.

Several steps are required for generating the hull form of the monohull ship floating dry-dock, which starts with setting the type of vessel to a monohull. Then, add the box and set up the size surface, move the control point to create the hull in U shape or call the displacement hull. Once, the hull is completed, check the frame of reference, size surface and the design grid as shown in Fig. 2.

ii. *Twin hull*

The similar software was used to generate the hull form for the twin hull. It involves several steps, which start with setting the type of vessel to a catamaran. Then, add the box and set up the size surface. Add a control point shaping the hull outer and inner when finished bond the control point which bonds edges with no tangency as shown in Fig. 3. Once completed, create the tunnel bottom to combine

Table 1 Principal dimensions

Dimension	Meter (m)
Length overall	250
Breadth	70
Depth	22

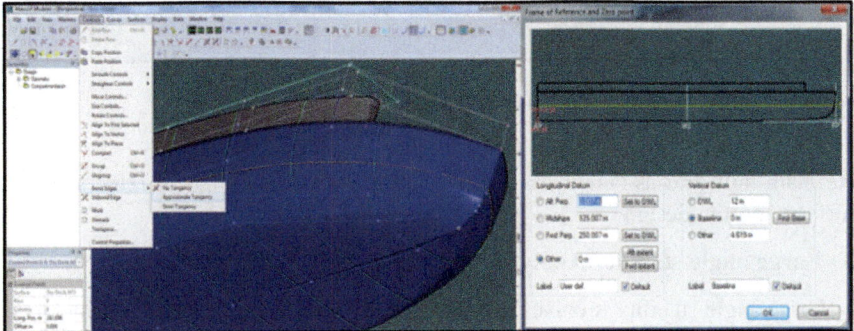

Fig. 2 Generate hull form for monohull

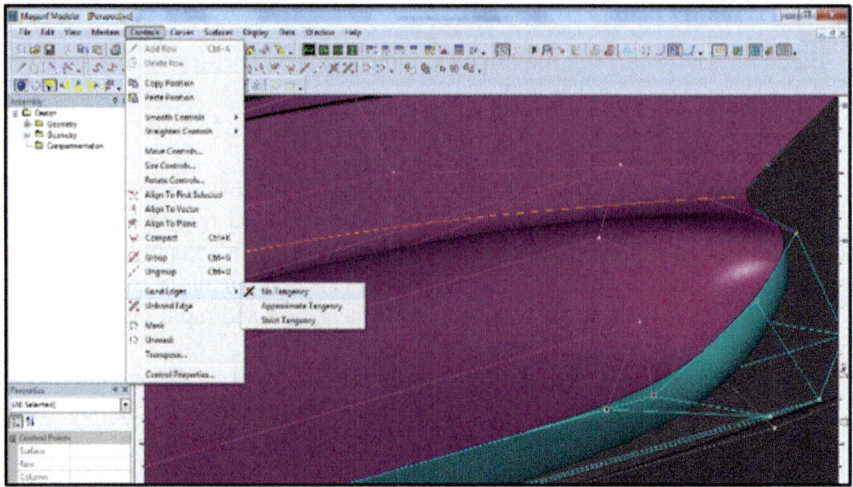

Fig. 3 Bond edges with no tangency for 2 control point

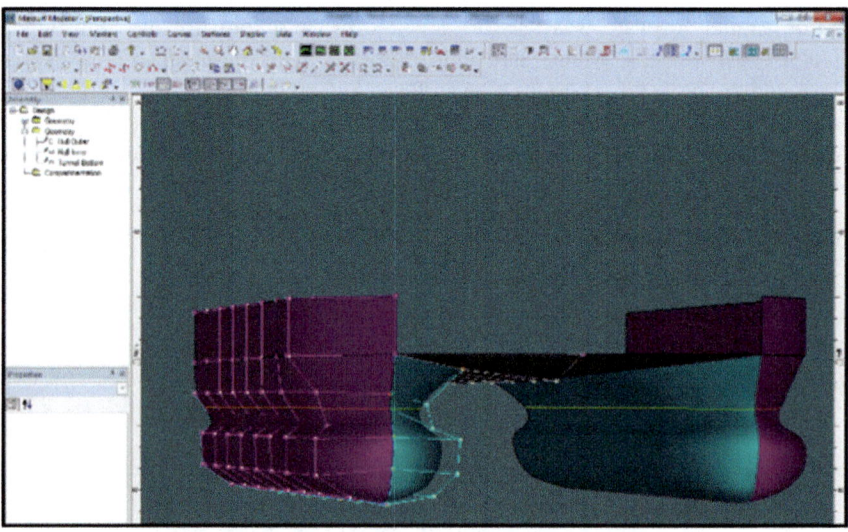

Fig. 4 Main hull combined with tunnel bottom

the main hull grid as shown in Fig. 4. Finally check the frame of reference, size surface and the design.

b. Large angle stability Analysis

Large angle stability involves the determination of the hydrostatic parameters of the hull at a range of heel angles either with or without trim or free-to-trim. The heel angle that was used to analyse the stability ranged from $-30°$ to $30°$ in steps of $10°$.

Fig. 5 Criteria IMO A.479 (18) code on intact stability

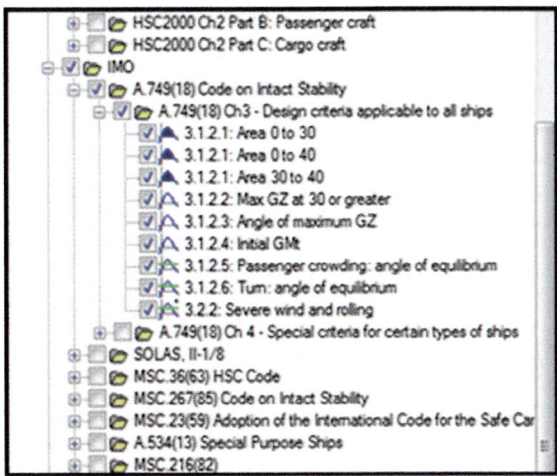

The vessel is heeled and gives the value of every single heeling condition. The analysis of large angle stability requires the displacement (lightship) and the center of gravity to be specified in the load case table. The value of lightship was obtained at the weight estimation stage. The load case condition has been placed with five conditions; lightship condition, 10% load condition, 25% load condition, 50% load condition and 100% condition loading of the ship.

The criteria were used according to the IMO A.749 (18) code on intact stability which is ch3 design criteria application to all ships as shown in Fig. 5.

c. **General arrangement**

The general arrangement (GA) drawing is a drawing that shows the arrangement of all spaces on the ship floating dry-dock. The process of designing the general arrangement is related to the function of the ship that must follow its rules and regulation that has been decided by the classification board. The general arrangement must be economical, fulfil the owner's requirement demand, being logically and solve the existing problem about the arrangement of the machinery.

Nevertheless, the general arrangement for the ship floating dry dock either for a mono-hull or twin hull only has the ballast tank, side wall completed with the substation, switch gear room and buoyancy chamber, and four cranes. The complete design in max surf modeler will be exported to a DXF file, which can be imported to Auto-CAD to generate the general arrangement.

3 Result and Discussion

The result and discussion have focused on the design of the ship floating dry-dock and the analysis of the large angle stability. The design of the 3D monohull and twin hull with complete general arrangement will be discussed. The large angle stability will discuss on the ballast water tank and draft increment in various conditions.

a. **Design of ship floating dry-dock**

The design process of the ship floating dry-dock focused on two types of the hull which are the monohull and the twin hull and to compare the ballasting analysis of both hulls. The dimensions follow the research which is the length of the ship floating Dry-Dock, i.e. 250 m, 22 m of depth and 70 m of the beam.

In the Maxsurf Modeler software, the 3D design of the hull shape was generated. Then, to complete the design for both configurations, the hull shape was transferred into AutoCAD to complete the general arrangement and converted to a STL file for the 3D design.

The monohull of the ship floating dry-dock completed with the ballast tank, side wall and crane is shown Fig. 6. The twin hull of the ship with the ballast tank in the tunnel, side wall and crane is shown in Fig. 7. The complete hull was analyzed in the max surf modeller in regards to the large angle stability.

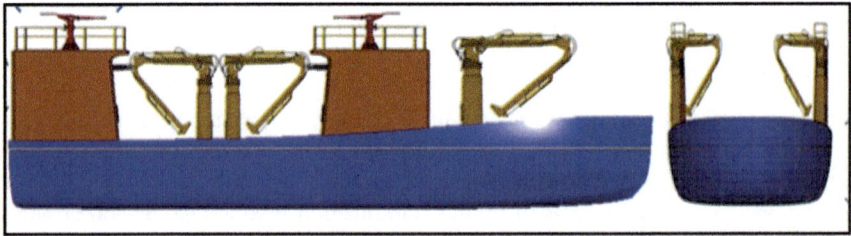

Fig. 6 Monohull of ship floating dry dock

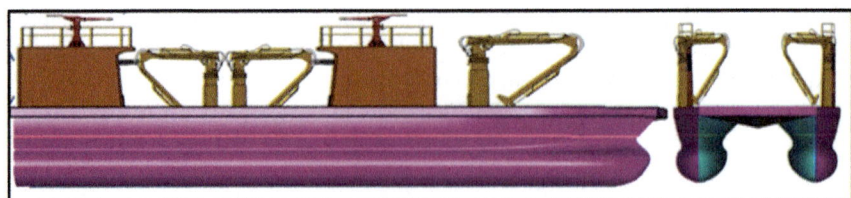

Fig. 7 Twin hull of ship floating dry dock

b. **Analysis of large angle stability**

Based on the maxsurf modeler, both hull designs were analyzed in the maxsurf stability in regards to the large angle stability. This analysis was carried out using the large angle stability with loading conditions, which contain of five different conditions for the twin hull and monohull. These results were automatically checked by the software according to the criteria which are mentioned in MO A.749 (18) Chap. 3.

As a result for the monohull ballast water tank, a maximum volume of 314,359.92 m³ and for the twin hull, a maximum volume of 139,321.23 m³ was obtained for the ballast water tank. It can concluded that the monohull is more suitable for the ship floating dry-dock. The difference of both hulls for the ballast tank water was 44%, which it almost equal to 50%.

Other than that, the draft of the monohull was 33.542 m and the draft of the twin hull was 35.799 for 100% condition. Its is shown that the monohull is suitable for the ship floating dry-dock with 14–22% increment according to the various conditions. For the twin, the hull has 20–40% increment according to the various conditions.

All a result, both hulls have passed the criteria mentioned in IMO A.749 (18) Chap. 3. The overall result of large angle stability analysis is shown in Fig. 8 for the monohull and in Fig. 9 for the twinhull.

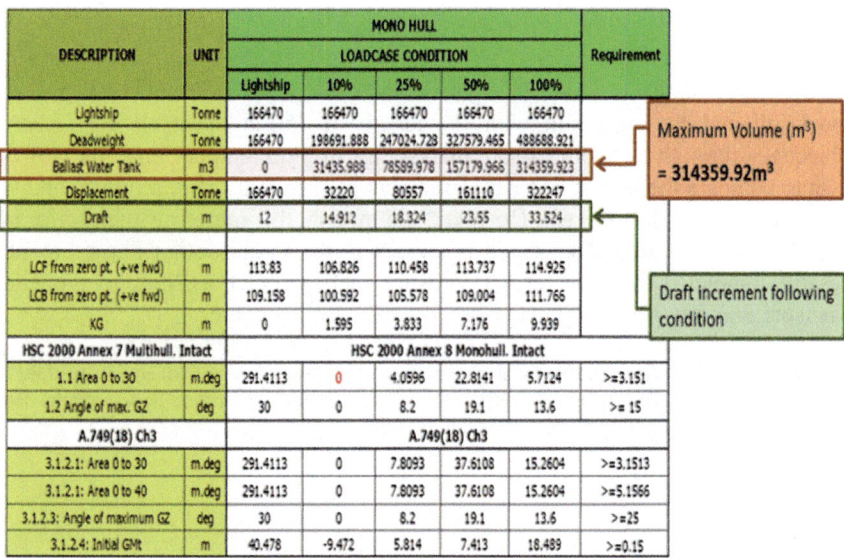

Fig. 8 Monohull of ship floating dry dock result

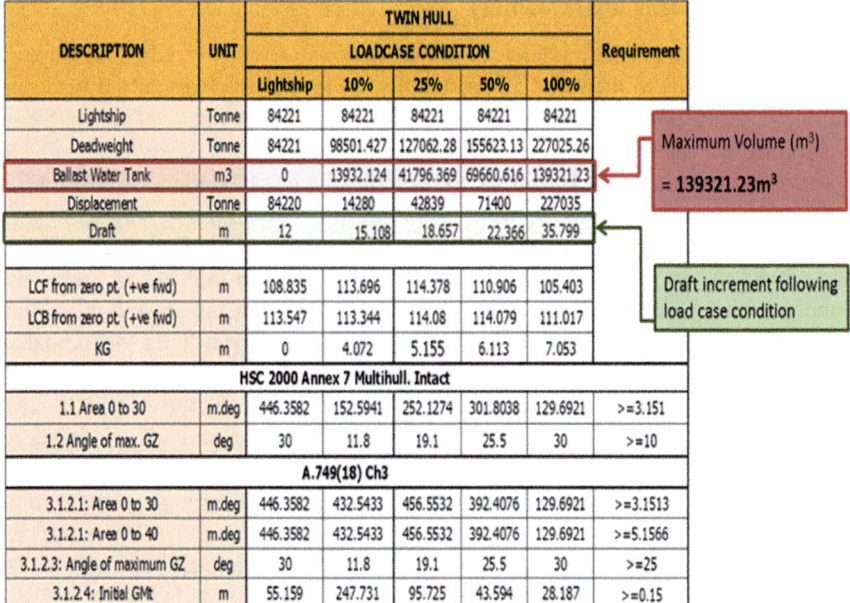

Fig. 9 Twin hull of ship floating dry dock result

4 Conclusion

This research showed that the design models of two types of hulls, which is the monohull and the twin hull with the length of 250 m, meet all the research objectives through the comparison analysis by using a stability analysis. All the data to compare the ballast water displacement of both hulls were achieved and the monohull was finally selected. However, the monohull is more suitable use for the ship floating dry-dock. The whole processes in the initial design phase are successful completed by using related software packages such as Maxsurf modeller, Maxsurf Stability and AutoCAD.

Acknowledgements The authors thanked the University Kuala Lumpur Malaysian Institute of Marine Engineering Technology (UniKL MIMET) and Petronas Maritime Services Sdn. Bhd, Kerteh for research facilities provided for this project.

References

1. Heger, R.: Dock master training manual. Dry Dock Inc, by (June 2005)
2. Hudson, W.R.: The ship lift floating dry-dock system (June 2010)

3D Design of a Ship Floating Dry-Dock by Using Simulation Software

Roslin Ramli, Nur'Aqilah Mohd Sabri and Nurul Asima Zainon

Abstract This paper presents the 3D design of a ship floating dry-dock by using simulation software. The design is carried out by using the Autodesk Inventor 2015 software, which has some functionality needed to design this type of vessel. Two models are designed where each is different in terms of their type of hull, i.e., a monohull and twin hull. This software is chosen due to its strength in the parametric area, dimensional and geometric constraints, and parameter adaptation besides its handling in the assembly process, which is a perfect solution to design the ship floating dry-dock. In this paper, we created a 3D design of the ship floating dry-dock, including each part of both types of hull which were assembled in order to give a deeper understanding of the simulation. Throughout the Autodesk Inventor application, the design is fabricate using a 3D printing process. The basic printing process that was used was fused deposition modeling (FDM) and PolyJet for creating objects directly.

Keywords 3D design · Ship floating dry-dock · Fused deposition modelling PolyJet

1 Introduction

A floating dry dock is a vessel, which is mainly purposed to repair a ship [1]. The dry-dock is a structured area wherein construction, repairs and maintenance of merchant vessels and boats are carried out [2]. The type of vessel that was designed in this project was a mono and twin hull configuration and this gave a different

---

R. Ramli (✉) · N. Mohd Sabri · N. A. Zainon
Malaysian Institute of Marine Engineering Technology, Universiti Kuala Lumpur,
Jalan Pantai Remis, 32200 Lumut, Perak, Malaysia
e-mail: roslin@unikl.edu.my

N. Mohd Sabri
e-mail: aqilah.sabri@ymail.com

N. A. Zainon
e-mail: nurulasima@unikl.edu.my

© Springer International Publishing AG 2018
A. Öchsner (ed.), *Engineering Applications for New Materials and Technologies*,
Advanced Structured Materials 85, https://doi.org/10.1007/978-3-319-72697-7_24

design of the ship floating dry-dock. In line with the title "3D Design of a Ship Floating Dry Dock by using Simulation Software", the design is created by using Autodesk Inventor 2015 where an assembly is carried on the vessel part by part.

This software was chosen due to its strength in the parametric area, dimensional and geometric constraints, and parameter adaptation besides its handling in the assembly process, which is a perfect solution to design the ship floating dry-dock. The vessel is equipped with a ship hull instead of a pontoon type floating dry-dock as it possesses less resistance from the wave energy.

This will theoretically ease the towing of the vessel across the sea. At the end of the project, an understanding is achieved on the floating dry dock and its difference between one with and without ship hull.

The ship floating dry-dock is a new vessel that was created with two kind types of hull, i.e., the mono hull and the twin hull as a SWATH. Usually a floating dry-dock does not have a hull but it still can be moving smoothly with the low speed when brought using tugboats [3].

This project was focusing on some objectives. The first objective was to determine the ship floating dry-dock in terms of its shape, function and size. Second was to design the ship floating dry-dock by using a simulation software which is Autodesk Inventor 2015. By using this software, each part of both types of hulls was assembled in order to give a deeper understanding of the simulation. The final objective was to produce the prototype by using 3D printing methods. The prototype was printed without assemble, once the prototype's parts were completely printed, they were assembled according to the 3D design ship floating dry-dock.

2 Methodology

The flow of methodology is divided into several parts such as data gathering, design in the Inventor software, assemble part by part of the 3D design of ship floating dry-dock, 3D printing and prototyping.

The gathering of data is used to design in Autodesk Inventor 2015. The 3D design is divided into two components which is a main and a sub-main component. The component is divided part by part because of the exsistence smaller parts. Afterwards, everything is assembled.

The complete 3D design produced the prototype by using a 3D printer which used the used deposition modelling (FDM) and polyjet approach. The flowchart is shown Fig. 1.

2.1 Data Gathering

There are several types of dry-docks, such as basin dry-docks, floating dry-docks, marine railways, marine travel lifts, etc. The main focus here is on floating dry-docks which use a u-shape for monohull and a swath hull for twin hull. The

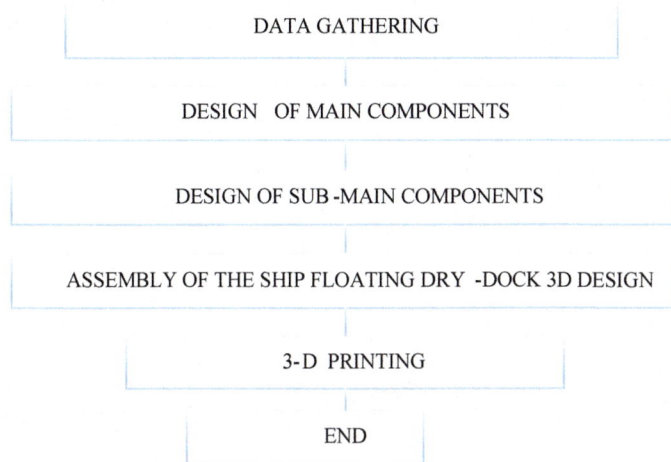

Fig. 1 Flow chart

Table 1 Principal dimension

Dimension	Meter (m)
Length overall	200
Breadth	50
Depth	22

principal dimensions are shown in Table 1. However, these principal dimensions were scaled for 3D printing by the ratio 10:41.67. Once the data was completed, the design process proceeded.

2.2 Design of Main Components

The main component of the ship floating dry-dock in this project consists of the hull and the deck of the vessel. Both are represented in detail where the design started from where the tools are placed, how to put data, and the dimensions. The tools for each drawing are told until the step to build the real component. The typical steps on how to design the monohull ship floating dry-dock are shown in Fig. 2.

The design process of the monohull started with a 2D sketch. Once the 2D sketch was finished, the sketch will extruded to create the 3D view. These steps are shown in Fig. 3. A similar design process was adopted for the twin hull which started with a 2D sketch until the extrusion to create the 3D design as shown in Fig. 4.

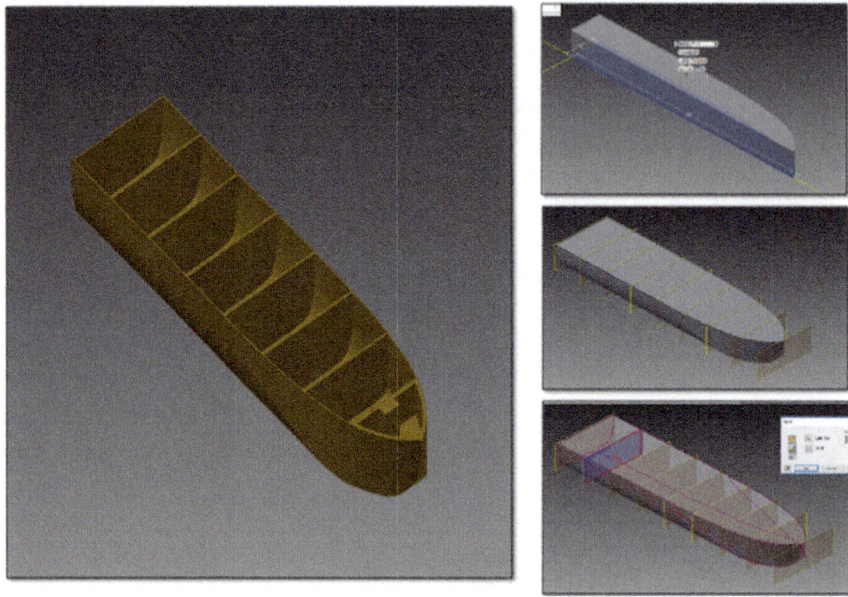

Fig. 2 The typical step on how to design the monohull ship floating dry dock

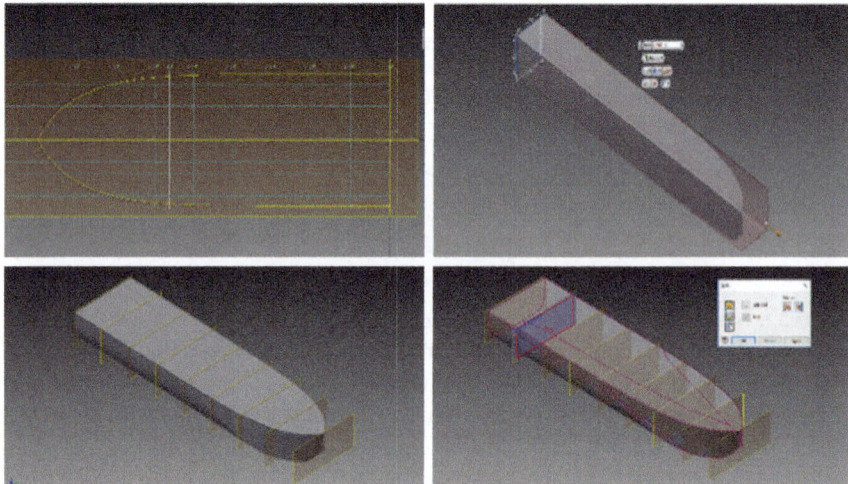

Fig. 3 Step design the monohull ship floating dry dock

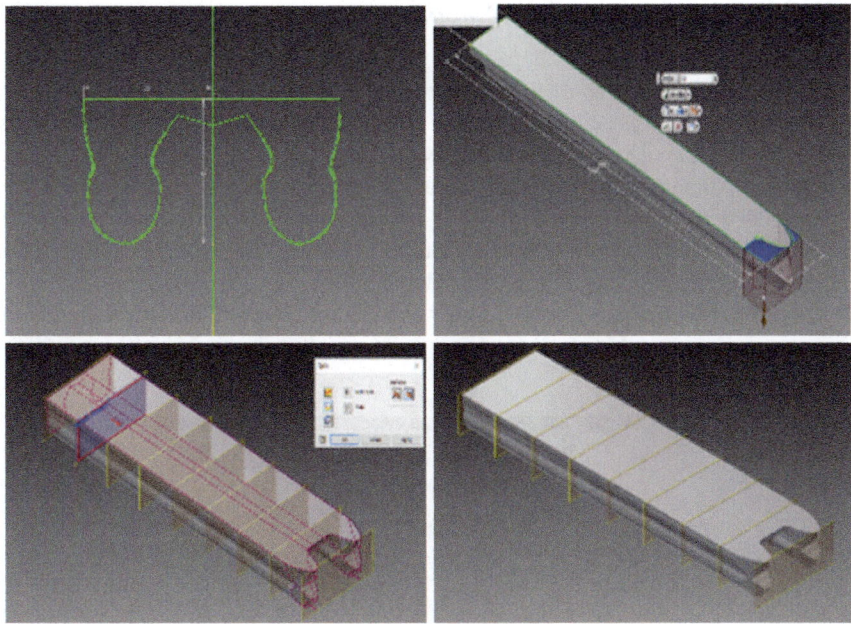

Fig. 4 Step design the twin hull ship floating dry dock

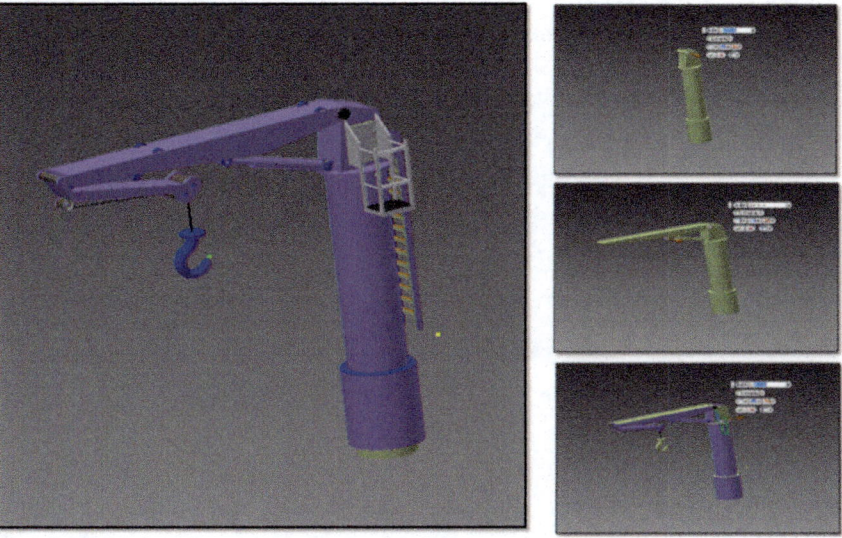

Fig. 5 Typical steps on how the crane part is designed

2.3 Design of Sub-main Component

The sub-main components of the ship floating dry-dock consist of the cranes and the tanks which are assembled together with the hull and the deck. This part is explained such that it is not as detailed as the main component as the way to draw each part is explained only in a general way. The typical steps on how the crane part is designed is shown in Fig. 5.

The design process of the sub main component started with the 2D sketch of the half part. Once the 2D sketch was finished, the sketch was extruded to create the 3D view. In this case, the gantry crane used was mirrored to get complete gantry crane. This step is shown in Fig. 6.

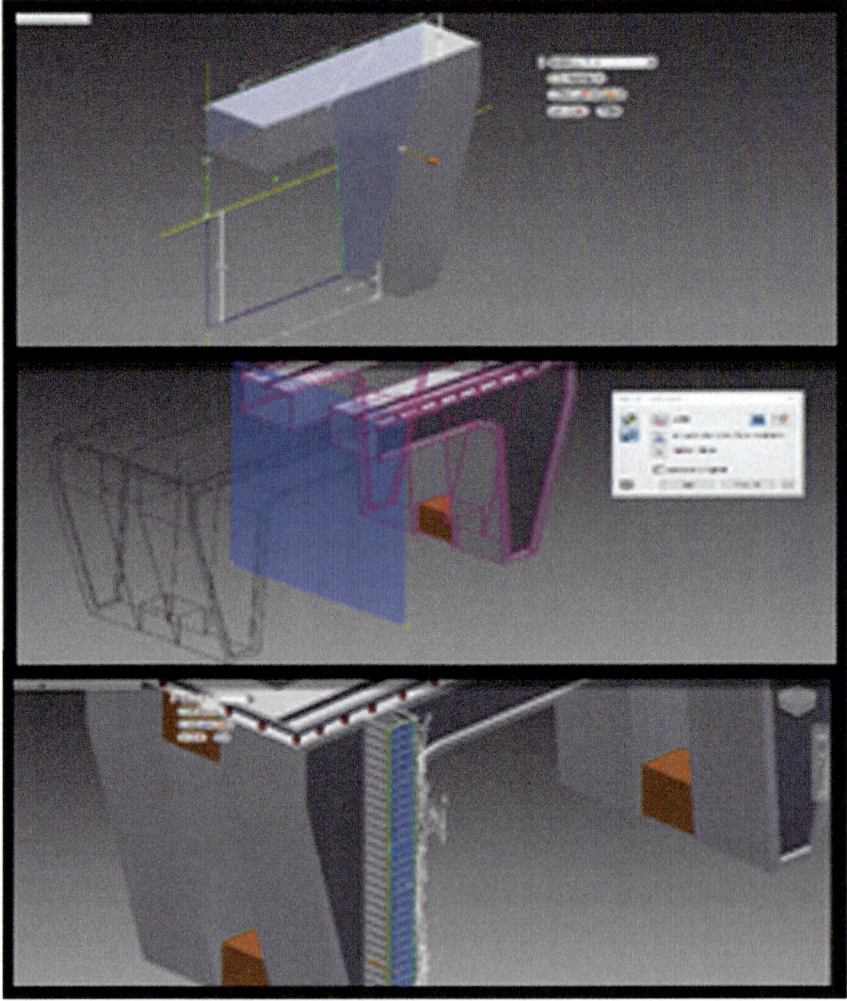

Fig. 6 Gantry crane

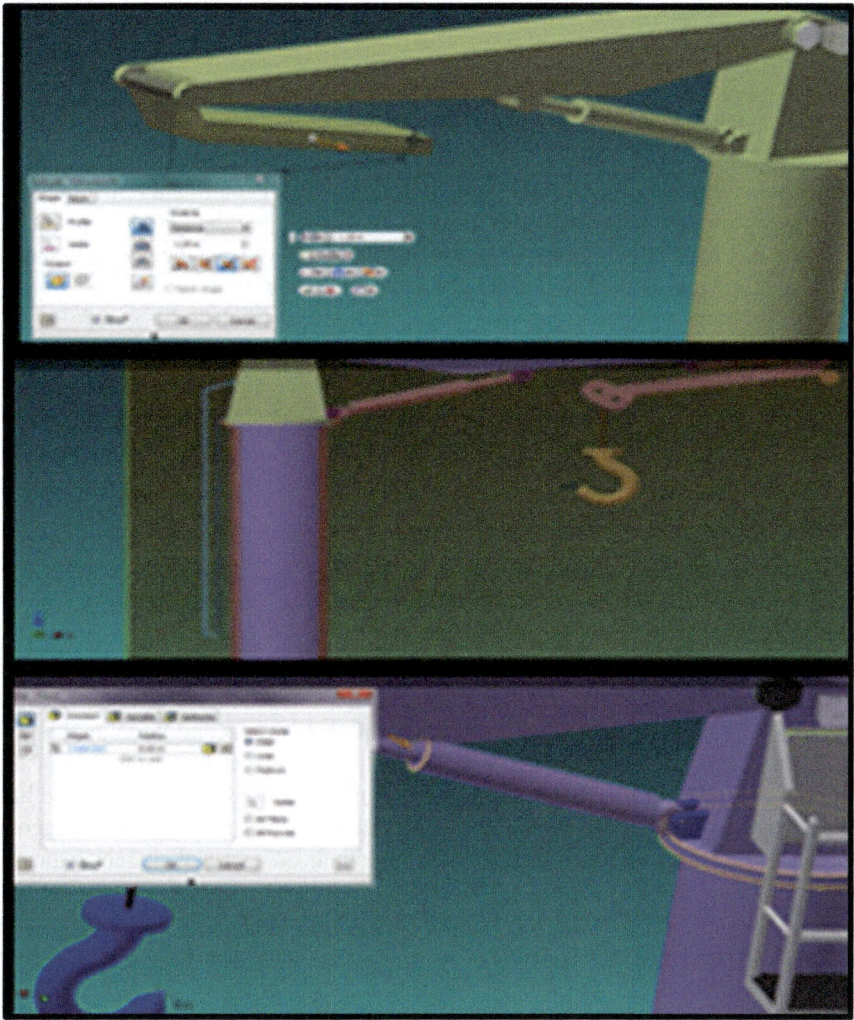

Fig. 7 Knuckle boom crane

The knuckle boom crane started with a 2D sketch of each part. It started with the column part until the crane arm part. After finishing a part, it was extruded to create the 3D design as shown in Fig. 7. This is a similar design process as for the jib crane.

2.4 Assembly of the Ship Floating Dry-Dock 3D Design

An assembly puts multiple parts together to show how each part is used in the design. The main component part has an assembly with sub-main components. The process of assembling parts is important for the specific sizes and can control by dimensioned sketches. It also an adaptive process, which controls the relationships to other assembly components. This is done by associating the geometry included from another parts and by designating the part as adaptive. The new part can change the size when constrained to a fixed geometry in the assembly.

Once, the assembly is completed, it can proceed to print the prototype by 3D printing. An assembly monohull and twin hull is shown in Figs. 8 and 9, respectively.

2.5 3D Printing

The prototype of the vessel was created by 3D printing with FDM and PolyJet as the main technologies. The final design was converted into the STL file format before connecting to the printing software. Then, the 3D printing file was ready to print the prototype.

The material and process that was used in the prototype stage of the ship floating dry-dock was based on the 3D printing hardware that was used to build up the ship structure. There are two types of technologies used to print this prototype of the ship floating dry-dock which are the PolyJet and the fused deposition modelling (FDM) technologies. Both technologies use materials that are very advantageous for customers who use service printing because their own special characteristics of the printing result.

PolyJet is a 3D printing technology that can produce smooth surfaces and accurate parts for prototypes. Besides, it also can produce thin walls and structures with an accuracy in the dimension down to 0.1 mm by using microscopic layer resolution [4]. In this case, the ship floating dry-dock prototype used the PolyJet printing technology for the cranes such as the jib crane, gantry crane and knuckle boom crane. Because of the crane structures are more complicated, the PolyJet printing technology can create smooth surfaces to prototypes and achieve complex shapes, intricate details and delicate features. The printing duration is very short especially for small dimensions although it can create complicated structures with the full detailed design.

Fused deposition modeling (FDM) is a professional 3D printing technology that uses production grade thermoplastics for parts and results in superior mechanical, thermal and chemical strength [4]. 3D printers that run on the FDM technology build the component part by part layer wise where its start form the bottom up by heating and extruding thermoplastic filament. In the ship floating dry-dock prototypes the ship structures used the FDM printing technology because it is easier to

3D Design of a Ship Floating Dry-Dock … 307

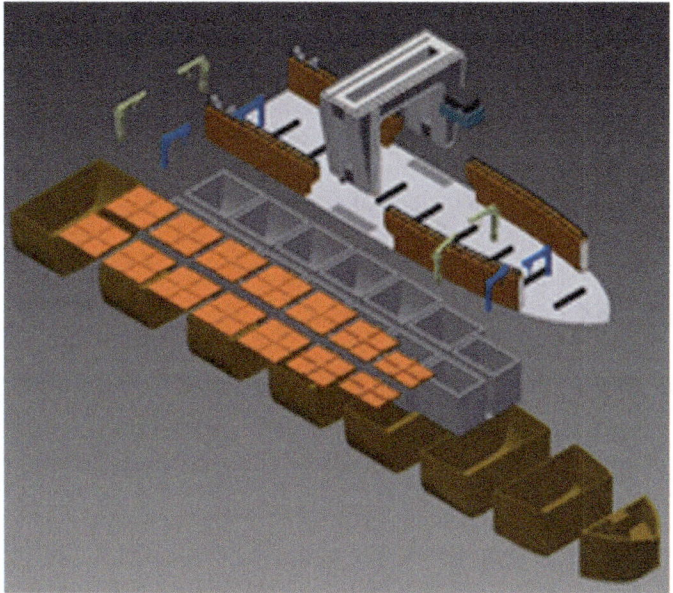

Fig. 8 An assembly process for monohull

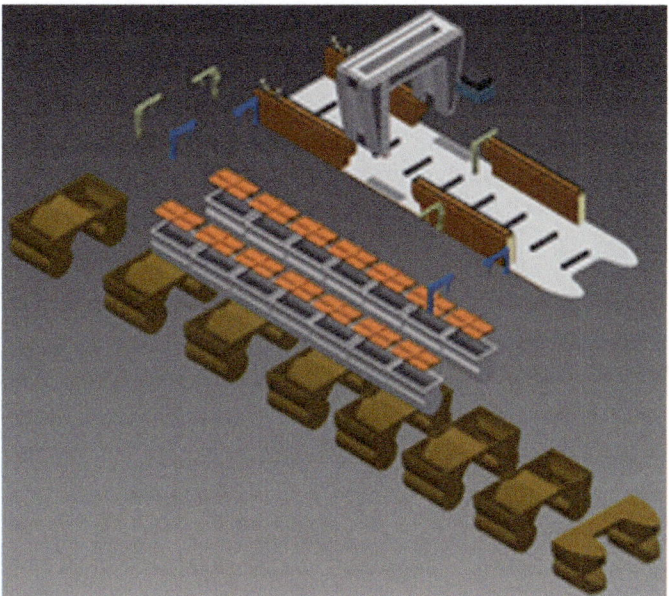

Fig. 9 An assembly process for twin hull

construct a smooth surface of the outside view. This is commonly used due to advantages of the material in the FDM technology, i.e., clean, simple to use and eco-friendly.

Hence, the material for making the prototypes of the ship floating dry-dock is a high quality printing ink due to the material's toughness and reliability. Besides that, this printing technology s commonly used in printing industrial parts because of their good conditions and method of construction.

3 Result and Discussion

3.1 Design of Monohull and Twin Hull

The principal dimension as a standard dimension for this ship floating dry-dock (refer to ship hull) is 200 m overall in length, 50 m beam and 22 m depth. By adding the wing wall and the crane, a rough estimate for total the depth is 86 m. With this dimension, the maximum size of a ship that can fit into the ship floating dry-dock is around 1–180 m in overall length, 1–40 m beam and 1–30 m total in depth. Any type of ship is possible to be repaired in the ship floating dry-dock but there is limitation to their size.

The 3D design of the floating dry-dock for the monohull is completed with 14 ballast tanks, decks with the gantry crane on the midship and a wing wall at the port and starboard side. Each wing wall is fixed with a knuckle boom crane and a jib crane. It also has nine dry dock blocks on the deck. The complete design of the monohull floating dry docks is shown in Fig. 10.

The 3D design of the floating dry dock the for twin hull is complete with 14 ballast tank, decks with a gantry crane on the midship and a wing wall at the port and starboard side. Each wing wall is fixed with a knuckle boom crane and a jib

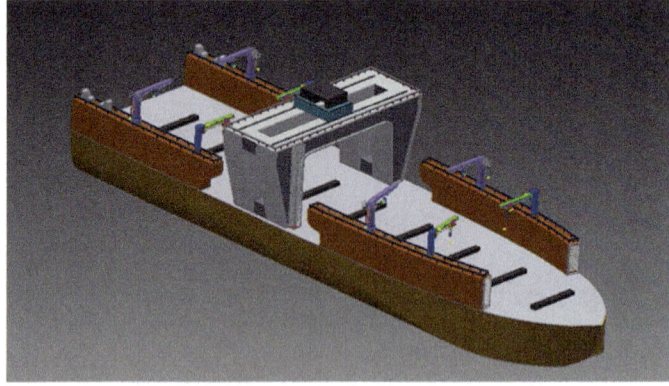

Fig. 10 Complete design of monohull floating dry doc

3D Design of a Ship Floating Dry-Dock ...

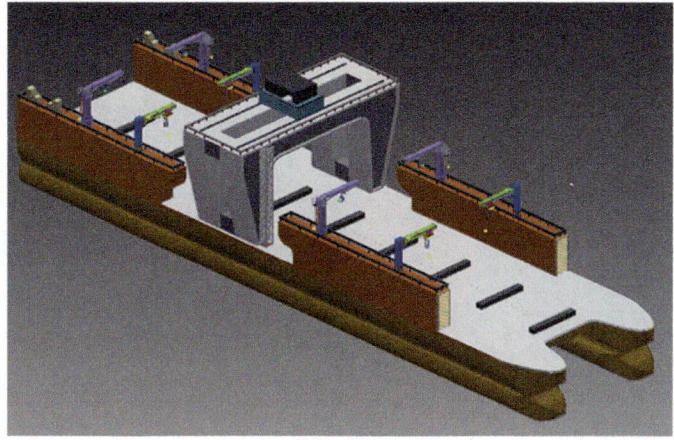

Fig. 11 Complete design of twin hull floating dry dock

crane. It also has nine dry dock blocks on the deck. The complete design of the monohull floating dry docks is shown in Fig. 11.

3.2 Prototype of Ship Floating Dry Dock

Once, the 3D design of the ship floating dry-docks is completed, it is ready to be 3D printed as a prototype. The parts were printed with different printing technologies, i.e., FDM and poly jet 3D printing because there were small and larger parts. The total duration of the complete printing for the monohull was two days. After completing, the parts were ready for assembling as shown in Fig. 12. A ship

Fig. 12 Prototype of monohull ship floating dry-dock

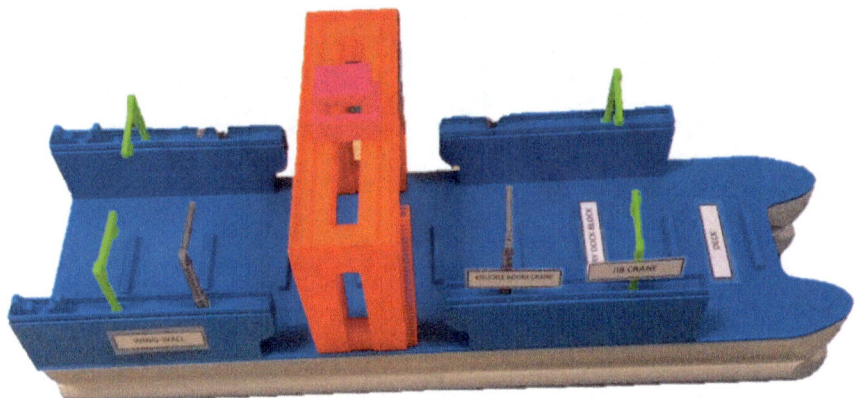

Fig. 13 Prototype of twin hull ship floating dry-dock

floating dry-dock takes three days for a complete printing because of the hull part was difficult to print with smooth faces. When completed, it is ready for assembly. The complete assembly is shown in Fig. 13.

4 Conclusion

As the project was planned, it already successfully achieved all the objectives of this project. The process of the project is completed smoothly according to the project schedule.

Monohull and twin hull of ship floating dry-dock had been successfully presented by using the inventor Professional software according to the defined dimensions. Choosing this software to complete the designs was the right decision because Inventor is strong in the parametric area, dimensional and geometric constraints and parameter adaptation. The inventor software also good in the assembly area because it has a separate area for part models and assembly models.

The fabrication of the prototype has been done by 3D printing based on the FDM and Polyjet technology. 3D printing was able to fabricate each part and layers of the prototype. The parts produced have a smooth surface for both models with different colours.

References

1. Paul, P.W.C., Bertelsen, J.: The Design of a Floating Dry Dock, Massachusetts Institute of Technology, Department of Naval Architecture and Marine Engineering, 1921 (second author),

and Thesis (B.S.)–Massachusetts Institute of Technology, Department of General Engineering, 1921 (first author) (1921)
2. Anish: Dry Dock, Types of Dry Docks and Requirements for Dry Dock. Marineinsigth (2016)
3. James, R., Clynch, Thomas, A., Rago, Curtis, A., Collins: Comparison of SWATH and Monohull Vessel Motion for Regional Class Research Vessels, Naval Postgraduate School (2005)
4. Stratasys: Precision 3D Printing in a Wide Range of Materials. Stratasys, US (2016)

Hybrid Combination Product Between Aluminum Can with Reinforcement Fiberglass for Autonomous Underwater Vehicles

Mohd Amin Hakim Ramli, Syajaratunnur Yaaakup,
Fatin Zawani Zainal Azaim, Khairul Anuar Mat Saad,
Mohd Faizal Abdul Razak, Muhammad Fadli Ghani, Naqib Idris
and Ahmad Shaharibudin Dato Abd. Aziz

Abstract Composite materials have been a subject of interest with various special types of advanced materials during the last decades. However, increasing demand in marine industry for high-performance and lightweight structures have stimulated a strong trend towards the development of refined models for hybrid composite materials known as fibre metal laminates [1]. Fibre metal laminates are hybrid composites built up from aluminium can and fibre reinforced plastics. GLARE is the best example to compare that has been introduced in aircraft industry. Aluminium layers and fibre reinforced laminates can be bonded by mechanical and adhesive techniques. Based on these approaches, the adhesive technique with plain aluminium sheets is used to determine its effect as a fibre metal laminate [2]. The

M. A. H. Ramli (✉) · S. Yaaakup · F. Z. Z. Azaim · K. A. M. Saad
M. F. A. Razak · M. F. Ghani · N. Idris · A. S. Dato Abd. Aziz
Malaysian Institute of Marine Engineering Technology, Universiti Kuala Lumpur,
Jalan Pantai Remis, 32200 Lumut, Perak, Malaysia
e-mail: mohdamin@unikl.edu.my

S. Yaaakup
e-mail: syajaratunnur@unikl.edu.my

F. Z. Z. Azaim
e-mail: fatinzawani@unikl.edu.my

K. A. M. Saad
e-mail: fatinzawani@unikl.edu.my

M. F. A. Razak
e-mail: fatinzawani@unikl.edu.my

M. F. Ghani
e-mail: muhamadfadli@unikl.edu.my

N. Idris
e-mail: naqib.idris91@gmail.com

A. S. Dato Abd. Aziz
e-mail: shahsal_1737@yahoo.com

© Springer International Publishing AG 2018
A. Öchsner (ed.), *Engineering Applications for New Materials and Technologies*,
Advanced Structured Materials 85, https://doi.org/10.1007/978-3-319-72697-7_25

design orientation used had 7 layers where the first layer was an aluminium can and then laminated with 2 layers of 430 chopped strand mat (CSM). Next, it was combined with woven roving followed by a CSM and then woven roving and lastly we applied 2 sheets of CSM. Tests were conducted to determine the mechanical properties of the fibre metal laminate by tensile testing and the results were tabulated [3]. This testing followed the ASTM standard which is D3039s. The weight optimization of fibre metal laminates with the combination is lighter than laminated wood.

Keywords Chopped strand mat (CSM) · Fiber metal laminates (FML) Glass fiber reinforced aluminum (GLARE) · Tensile testing · Woven roving

1 Introduction

The changes to fiberglass in marine industries began the needs of search for fiberglass materials. In addition, due to the growth of the marine industry over the past decades, the needs of fiberglass demand were extremely high in this industry to manufacture and to construct vessels using this method.

The demand for fibre glass increases and people get to explore for fibre glass in deeper research and it has caused the rapid growth of marine industry especially in fibre glass boat construction. It has resulted in a widely use of making the hull form construction mostly for small vessels such as fishing boats, coastal vessels, catamarans, chine hulls etc. In addition, fibre glass has increased its function to be adapted in oil and gas industry.

In this research, the aim is to study the strength of the material when it is combined (hybrid) with other materials for the reinforcement fiberglass [4]. Most important in this industry is to find suitable materials in regards to weight, costs, strength and others properties to improve their products and to increase their market chances. This hybrid method is one of the ways for the development of this industry. It will gain some advantages and disadvantages. More in depth study is needed to improve the fiberglass function to be more functional.

Many research and studies were conducted to find new material compositions with desirable properties. Today's engineers are posed with a big challenge for the right selection of a material and the right selection of a manufacturing process. It is difficult to study individually all the materials. Therefore, a broad type of classification is necessary to simplify the characterization. By the selection of different types of laminate components, and the possibility to vary the volume fraction of the composite and fiber orientation, a wide range of material properties of the resultant product can be produced [5].

Composite materials are usually classified based on the matrix materials. Generally, there are three main groups of composite materials, which are polymer matrix composites (PMC), metal matrix composites (MMC), and ceramic matrix composite (CMC) [6]. Among these three, PMC is the most common composite that is usually used and is known as fiber reinforced polymers (FRP). It uses a polymer-based resin as the matrix while the reinforcement is usually provided by fibers such as glass, carbon or aramid [7]. MMC is a composite material which uses metal as a matrix such as aluminum. This type of material is frequently used in automotive industry and aeronautics industry but never applied in the maritime industry [8]. CMC uses ceramic as the binder for the reinforcement which is usually made of silicon carbide or boron nitride. A ceramic composite is usually used in a very high temperature environment [9].

2 Experimental Setup

There are three types of hybrid composites that were created: 7 layers laminated with 1 layer aluminum can, 7 layers laminated with 2 layers aluminum can, and lastly just for 7 layers laminated fiberglass. For the 8 layers' configuration, the first layer constitutes the aluminum can followed by 2 layers of 430 chopped strand mat, the woven roving, followed by a layer of 430 chopped strand mat, woven roving and lastly 2 layers of 430 chopped strand mat. The difference between the 2 layers of aluminum can is that the 2 layers of aluminum can configuration is fabricated with an aluminum can sheet at the last of the orientation after the 2 layers of 430 chopped strand mat. For the fiberglass layers, it is laminated with 7 layers of standards as used to the hybrid laminated but without the aluminum can. Hand layup technique was chosen as it is ideally suited to manufacture low volumes without tooling costs. The directions of the lamination have followed the standards. The standards that was used in this research is D3039.

2.1 Experiment Preparation

Following the standard used, the specimens were cut with dimensions of 250 mm in length and 25 mm in width with 8 mm in thickness. The tensile properties of the glass fiber reinforced aluminum laminate composites were determined according to the ASTM 3039 test standard specification. The average tensile properties were determined from 5 test specimens for each type of configuration. The tensile tests were performed on a universal testing machine at a crosshead rate of 2 mm/min.

The specimens were mounted in the grips of a universal testing machine and gradually loaded in tension while recording the load. The ultimate strength of the material was determined from the maximum load carried before failure and various failure modes were analyzed. After that, the strain was monitored with displacement transducers and then the stress-strain response of the material was determined, from which the tensile strain, modulus of elasticity was derived.

3 Results and Discussion

Stress versus strain plots for the 3 different composite modifications in this research were generated. For the 1 layer of aluminum can, the stress result obtained is 59.75 MPa for the average of the maximum load applied. For the 2 layers aluminum can testing, the result that was calculated for the average of the maximum load is 35.80 MPa while the result for the fiberglass laminate is 77.36 MPa.

The tensile properties of various combinations of the laminates were evaluated and the results were tabulated as above. Tables 1, 2 and 3 show the average calculation for all the specimens.

From the calculation that was tabulated, below are the graphs which compare all the types of design configurations based on the plotted stress versus strain data.

Table 3 Fiber reinforced plastic layer

No. specimen	Tensile stress at maximum load (MPa)	Tensile strain at maximum load (mm/mm)
1	97.29056	0.03526
2	91.6863	0.01696
3	50.50824	0.01373
4	91.36143	0.02361
5	55.95259	0.05117
Average	77.359824	0.028146

Table 2 2 Layer aluminum can

No. specimen	Tensile stress at maximum load (MPa)	Tensile strain at maximum load (mm/mm)
1	17.77833	0.00244
2	26.01601	0.01181
3	72.41367	0.02058
4	26.76891	0.02004
5	36.0443	0.02594
Average	35.804244	0.012046

Table 1 1 Layer aluminum can

No. specimen	Tensile stress at maximum load (MPa)	Tensile strain at maximum load (mm/mm)
1	91.53606	0.02309
2	71.41995	0.05912
3	30.11445	0.0108
4	27.83829	0.00528
5	77.83631	0.0787
Average	59.749012	0.035398

4 Conclusions

The tensile strength is higher in the plain fiberglass specimens compared to the hybrid combinations. This is because of the failure of the layers of the hybrid materials when the load was applied during the tensile testing. Compared to the reference results, the hybrid material should give the higher strength than the fiber glass materials. The obtained result for the 1 layer configuration in regards to the maximum tensile load was 59.75 MPa, while for 2 layers the result was 35.80 MPa, and lastly for the fiberglass specimen the result was 77.36 MPa. Hence, the finite elements analysis result does not agree with the experimental results.

References

1. Vlot, A., Gunnink, J.W. (eds.): Fibre Metal Laminates—An Introduction, Delft University of Technology, Faculty of Aerospace Engineering Delft, The Netherlands. Vogelesang, L.B., Vlot, A.: Development of fibre metal laminates for advanced. J. Mater. Process Technol. **103**, 1–5 (2000)
2. Botelho, E.C., Silva, R.A., Pardini, L.C., Rezende, M.C.: A review on The development and properties of continuous fibre/epoxy/aluminum hybrid composites for aircraft structures. Mater. Res. **9**(3), 247–256 (2006)
3. Kaufman, J.G.: Fracture toughness of 7075-T6 and—T651 sheet, plate and multilayered adhesive-bonded panels. J. Basic Eng. **89**
4. Vogelesang, L.B., Marissen, R., Schijve, J.: A new fatigue resistant material: aramid reinforced aluminumlaminate (ARALL) In: Proceedings of the 11th ICAF Symposium (1981)
5. Roebroeks, G.: The metal volume fraction approach, TD-ROO-003, Structural Laminates Industries, Delft University of Technology, Delft (2000)
6. Schwartz, M.M.: Composite Materials Handbook. McGraw Hill, New York (1988)
7. Bader, M.G.: "Hybrid Effect" Handbook of Polymer-fibre Composites, pp. 225–230. Longman Scientific Technical, Harlow (1994)
8. Higgins, A.: Adhesive bonding in aircraft structures. Int. J. Adhes. Adhes. **20**, 367–376 (2000)
9. Vermeeren, C., Beumler, T., De Kanter, J., Van Der Jagt, O., Out, B.: Glare design aspects and philosophies. Appl. Compos. Mater. **10**, 257–276 (2003)

The Comparison of Impact Energy and Three Point Bending Properties on Coconut Fiber Composite for Marine Application

Amirrudin Yaacob, Jaswar Koto and Mohd Yazid Bin Yahya

Abstract The development of high performance materials made from natural resources is increasing worldwide. The interest in natural fiber reinforced polymer composite materials is rapidly growing both in terms of their industrial applications and fundamental research. They are renewable, cheap, completely or partially recyclable, and biodegradable. Coconut fiber can be a potential candidate to replace the industrial core and foam and it may be applied worldwide. Tougher materials such as coconut fiber need higher energy or impact to break or fracture. So, this means that it can absorb more energy applied on it. A specimen with lower absorbed energy means it is brittle and has a lower toughness, which can break easily and cannot withstand sudden high loads. The higher the resulting numbers the tougher the material, which is coconut fiber with 2.035 J followed by 3D core PET foam and infusion grooved PVC foam which are 0.977 and 0.95 J, respectively. The flexural strength for coconut fiber shows the highest value which is 65.306 MPa even though the thickness of the specimen is lower compared to others. The 3D core PET foam shows 9.661 MPa and the infusion grooved PVC foam is 7.102 MPa. The Young's modulus reflects the stiffness of a material. Therefore, the coconut fiber exhibited a little lower stiffness than the 3D Core PET foam in flexural testing. It should be mentioned that the lower characteristic of elasticity for the coconut fiber improved the impact strength but reduced the stiffness, which might help to explain the lower Young's modulus of the coconut fiber.

A. Yaacob (✉)
Universiti Kuala Lumpur, Malaysian Institute of Marine Engineering Technology,
Jalan Pantai Remis, 32200 Lumut, Perak, Malaysia
e-mail: amirrudin@unikl.edu.my

A. Yaacob · J. Koto
Department of Aeronautics, Automotive and Ocean Engineering,
Faculty of Mechanical Engineering, Universiti Teknologi Malaysia,
Johor Bahru, Johor, Malaysia
e-mail: jaswar@utm.my

M. Y. B. Yahya
Department of Applied Mechanics and Design, Faculty of Mechanical Engineering,
Universiti Teknologi Malaysia, Johor Bahru, Johor, Malaysia
e-mail: b-yahya@utm.my; yazid@mail.fkm.utm.my

It can be proven that coconut fibers are suitable to be used as one of the laminating materials for fiberglass boat building.

Keywords Point bending properties · Coconut fiber · Laminating material

1 Introduction

Fiber is a class of material that exists as continuous filaments or in discrete elongated pieces, similar to the lengths of thread. They can be spun into filaments, string, or rope. It can be used as a component of composite materials, or matted into sheets to make products such as paper. Fibers are often used in the manufacture of other materials. The majority of improvements in marine composite construction over the last fifty years have been made through better resin chemistry. A fiberglass boat is much more attractive due to its strength such as light weight, high vibration damping capability, high impact resistance, low construction costs, ease of fabrication and ease of maintenance. Fiberglass boat constructions have expanded over the years with the various methods for the sole purpose of improving the techniques and skills for the boatbuilding process including the vacuum infusion process for fiberglass interceptor boat in Malaysia [1]. The objective in the fiberglass boatbuilding is to achieve lightness in weight, vibration damping, corrosion resistance, and impact resistance, low construction cost and ease of construction. Composite is currently a new development in technology. Nowadays, many industries such as automotive, marine industries, materials and building have developed composite structures as a new alternative to replace the usage of hard materials such as aluminum alloys. First of all, a composite is as a combination of two or more materials that results in better properties than those of the individual components used alone. In contrast, each material retains its separate chemical, physical, and mechanical properties. The two constituents are the reinforcement and a matrix. The main advantages of composite materials are their high strength and stiffness, combined with low density, when compared with bulk materials, allowing for a weight reduction in the finished part. Recently, there has been a rapid growth in research and innovation in the natural fibre composite (NFC) area. Interest is warranted due to the advantages of these materials compared to others, such as synthetic fibre composites, including low environmental impact and low cost and support their potential across a wide range of applications [2]. Much effort has gone into increasing their mechanical performance to extend the capabilities and applications of this group of materials. Their availability, renewability, low density, and price as well as satisfactory mechanical properties make them an attractive ecological alternative to glass, carbon and man-made fibers used for the manufacturing of composites. The natural fiber-containing composites are more environmentally friendly, and are used in transportation (automobiles, railway coaches, aerospace), military applications, building and construction industries [3]. This research is conducted to investigate the mechanical properties of coconut fiber in the manufacturing of fiberglass boats.

1.1 Infusion and Vacuum Technique

Resin infusion is a specialized advanced laminating technique that greatly improves the quality and strength of fiberglass parts as opposed to conventional hand layup. Applying laminate engineering and resin infusion technology simultaneously allow for the optimization of the parts in terms of strength and weight. The use of resin infusion will likely become the standard in yacht construction and has been in use since the 1960s. However, a lot of factors need to be taken into consideration such as the vacuum pressure. In tensile testing, increases in vacuum pressure tend to increase the tensile properties of composites structures. This process also has far fewer vacancies and the dominant failure here is fiber extraction which gives the vacuum infusion process samples much more strength and also stiffness [4]. The preparation for the vacuum infusion is to install a disposable vacuum bag, securing it tightly to mold flanges and the main purpose is to eliminate air leaks. A series of resin infusion lines is then added with each line number-coded according to which area of the hull they will feed. This is very important just to ensure that the resin will flow in all areas. That is the main reason why the high-performance resin system, specially formulated for use on large parts, has a water-like viscosity that accommodates controlled resin flow throughout the laminate need to be used. This ensures complete wet-out of reinforcing fabrics without resin-starved areas even in the notched sections of the stepped hull. After infusion, the epoxy cures at room temperature and is then post-cured under a tarp with a heat blanket that maintains a temperature of 1608 F for eight hours [5]. The outcome from the research has revealed that the vacuum infusion technique, in its various forms, remains a dominant process for the manufacture of wind turbine blades and other sandwich cored composite products. Methods have been sought that improve the processes production time and quality by maintaining repeatability and consistency in the laminate structure [6]. This manufacturing method also can be optimized by using pin loaded composites which can improve the properties of the materials [7].

1.2 Natural Fiber

Environmental awareness is growing day by day and worldwide researchers triggered by this reason to implement and utilize materials which are eco-friendly like coconut fiber. Besides that, high mechanical performance, low cost and ease of processing, cheap, pose no health hazards and offer solution to environmental pollution by finding new uses for waste materials [8]. The natural fiber composites can be used as a cost effective material for building and construction industries, electrical devices such as switches, storage devices, military vehicles where high strength and stiffness required and many more [9]. However composite materials made with the use of unmodified plant fibers frequently exhibit unsatisfactory mechanical properties and to overcome this, in many cases, surface treatments need to be used prior to composite fabrication [10]. The properties can be improved both by physical treatments which are cold plasma treatment, corona treatment and chemical treatments such as maleican hydride organosilanes, isocyanates, sodium hydroxide permanganate and peroxide [11].

1.3 Core Materials

Based on the Standard for Certification No. 2.21, April 2010 for Craft by Det Norske Veritas (DNV), the sandwich construction can be fabricated either by lamination on core, application of the core against wet laminate, by bonding the core against a cured skin laminate using core adhesive or by resin transfer moulding of the core together with one or both of the skin laminates [12]. So this research was conducted for determination of the performance of coconut fiber as the core material for fiberglass boatbuilding.

1.4 Coconut Fiber

Coconut fiber comprises of 30% by weight of fiber and 70% by weight of pith material. The fibers are extracted from the husk by several methods such as traditional retting, decortications, using bacteria and fungi and also mechanical and chemical processes [13] as shown in Fig. 1.

The fiber is abundant, non-toxic in nature, biodegradable, low density and very cheap. The fibers, instead of going to waste are explored for new uses, which in turn provide gainful employment to improve the standard living condition of individuals [14, 15]. Coconut fiber has the highest percentage by volume of lignin, which makes the fiber very tough and stiffer compared to other natural fiber. This can be

Fig. 1 Coconut fiber core

Table 1 The properties of coconut fiber [18]

Property	Coir
Density (g/cm^3)	1.2
Modulus (Giga Pascal)	4–6
Tensile strength (Mega pascal)	175
Elongation to failure (%)	30
Water absorption (%)	130–180

attributed to the fact that the lignin helps provide the plant tissue and the individual cells with compressive strength and also stiffens the cell wall of the fiber where it protects the carbohydrate from chemical and physical damage. The lignin content also influences the structure; properties, flexibility and rate of hydrolysis. At high lignin content, it appears to be finer and also more flexible [16]. The main importance is that the mechanical properties of coconut fiber have a strong association with the dynamic characteristics and greatly dependent on the volume percentage of fibers and the best fiber volume is 5% [17]. The properties of coconut fiber are shown in Table 1.

1.5 3D Core Foam

3D core foam is especially suitable for vacuum injection [19]. The matrix facilitates the infusion of the resin into the laminate. The basis of the process always is a well vented resin system with appropriate curing time. Very good results are also achieved using the hand lay-up technique. Many foams are also suitable for other methods like RTM, RTM-light, RIM, etc. In most cases the design of the laminate is calculated with CLT based on standard values. ESC provides a basis of standard values for cores of 10 mm. 3D core foam covered with two layers of certified 110 g/m^2 twill glass fabric and 600 g/m^2 quadraxial glass fabric on top and bottom. This core is filled with a standardized epoxy resin system from the French producer Sicomin as shown in Fig. 2.

1.6 Infusion Grooved PVC Foam

EasyCell 75G infusion grooved foam is drilled to allow resin to transfer from one side of the foam to the other and grooved on one side to facilitate the distribution of resin on the underside of the foam during a vacuum assisted resin infusion. The 75 kg/m^3 density is ideally suited to be used as a core material in range of resin infused panels such as decks, bulkheads and other applications. EasyCell 75G is a simple core material to be used. The sheet is placed with the grooves down against the mold side of the laminate. EasyCell 75G is relatively inexpensive, offers

Fig. 2 3D core foam [20]

Fig. 3 Infusion grooved PVC foam [21]

excellent mechanical performance and is compatible with all standard resin systems including polyester, vinylester and epoxy. The closed cell structure also means that the PVC foam can be used in vacuum manufacturing processes and it very well suited to RTM, resin infusion and vacuum bagging as well as conventional open lamination [21].

The infusion grooved PVC foam [21] is shown in Fig. 3.

2 Experimental Setup

2.1 Experiment Preparation

This experiment has been divided into two phases; (1) to construct the flat testing panels for three types of core and foam, (2) to conduct the mechanical testing for three types of core and foam. The construction of testing panels and the general setup of resin infusion method are shown in Fig. 5 until 8 which are using the same principle of vacuum pressure, used to drive resin into a laminated layer of fiber mat.

2.2 Preparation of Testing Panels

The synthetic fiber used in this research was a chopped strand mat (CSM) 450 which means 450 g for the weight of the fibers per square meter. It was made up of 1″–2″ long fiberglass strands that were randomly oriented and typically held together with a styrene-soluble binder that acts like glue connecting the fibers. This laminating schedule for testing panels as shown in Table 2.

The binder was designed to dissolve upon contact with styrene in polyester resin or vinylester resin. Once dissolved, the fabric softened, allowing it to drape around curved shapes [22]. The coconut fiber turned into core by using the needle punch process [23]. These results are shown in Fig. 4.

From the inside of the mold, after the usual mold release wax was applied, the gel coat and skin coat of thin fiberglass reinforcements were applied in the conventional manner and allowed to cure. From here on everything differs. Next, in the infusion process, the outer skins of the fiber reinforcement fabrics were carefully fitted into the mold over the top of the skin coat. These were put to dry and held in place with a spray contact adhesive which shown in Fig. 5.

Vacuum packing is a method of packaging that removes air from the package prior to sealing. This method involves (manually or automatically) placing items in a plastic film package, removing air from inside, and sealing the package. The vacuum plastic bag will ensure there is no air bubble inside the mold. Once the process was completed, both inlet and outlet hoses were clamped shut before the

Table 2 Laminating schedule for testing panels

Panel	1st layer	2nd layer	3rd layer
Panel 1	2 × CSM450	10 mm infusion grooved PVC foam	2 × CSM450
Panel 2	2 × CSM450	10 mm 3D core foam	2 × CSM450
Panel 3	2 × CSM450	10 mm coconut fiber	2 × CSM450

Fig. 4 Kitting process for CSM and coconut fiber

Fig. 5 Vacuum infusion preparation

Fig. 6 Vacuum infusion process completed

vacuum pump was turned off. The infused panel was left to the cure process for 8 h as it shows in Fig. 6.

2.3 Preparation of Specimens

A series of tests were conducted to determine the performance of the products under testing scope of impact test and flexural test. Five specimens taken from Panel 1, five specimens from Panel 2 and another five specimens from Panel 3 and the total specimens for the impact test are 15 specimens and for the flexural test are 15 specimens, these series of data are shown in Table 3.

Table 3 Total number of specimens

Test	Methods	QTY
Impact test	ASTM D256-10	15
Flexural test	ASTM D790-07	15

2.4 Mechanical Properties Testing

Two types of mechanical testing were chosen to test the material and were done at UniKL MIMET Technical Foundation Lab. All the testing was performed using the ASTM standard as reference.

2.5 Standard Test Methods for Determining the Izod Pendulum Impact Resistance of Plastics—ASTM D256-10

Notched izod impact is a single point test that measures a materials resistance to impact from a swinging pendulum. Izod impact is defined as the kinetic energy needed to initiate fracture and continue the fracture until the specimen is broken. Izod specimens are notched to prevent deformation of the specimen upon impact. This method is shown in Fig. 7. The specimens are placed into the impact tester fixture with the notched side facing the striking edge of the pendulum. The pendulum is released and allowed to strike through the specimen until break. The starting angle before impact is 120°. The standard specimen for ASTM is 64 × 12.7 × 3.2 mm. The most common specimen thickness is 3.2 mm, but the preferred thickness is 6.4 mm because it is not as likely to bend or crush. However, for this study, the thickness of specimens was 10 mm.

Fig. 7 Impact test arrangement

2.6 Flexural Properties of Unidirectional and Reinforced Plastics and Electrical Insulating Materials—ASTM D790-07

The flexural test measures the force required to bend a beam under three point loading conditions. The data is often used to select materials for parts that will support loads without flexing. Flexural modulus is used as an indication of materials stiffness when flexed. Most commonly for ASTM D790 the specimen lies on a support span and the load is applied to the center by the loading nose producing three points bending at a specified rate. The test is stopped when the specimen reaches a certain deflection or breaks. A typical test speed for standard test specimens is 2 mm/min (0.05 in/min). A variety of specimen shapes can be used for this test, but the most commonly used specimen size for ASTM is 3.2 mm × 12.7 mm × 125 mm (0.125″ × 0.5″ × 5.0″). However, for this study, the thickness of specimens was 10 mm as shown in Fig. 8.

3 Results and Discussion

The ASTM impact energy is expressed in J/m or ft-lb/in. Impact strength is calculated by dividing the impact energy in J (or ft-lb) by the thickness of the specimen. The test result is typically the average of 5 specimens. ISO impact strength is slightly different and expressed in kJ/m^2 because the impact strength is calculated by dividing the impact energy in J by the area under the notch. The test result is typically the average of 10 specimens. The higher the resulting numbers the tougher the material. By comparing all the specimens, it can be concluded that the best specimen that gives highest impact energy is coconut fiber. Coconut fiber is tougher

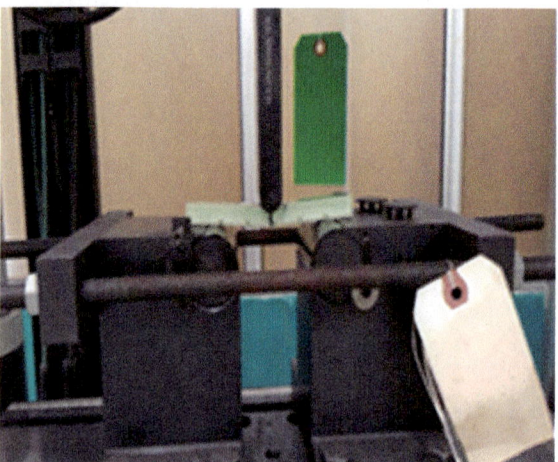

Fig. 8 Flexural test arrangement

Table 4 Impact strength for coconut fiber core

	Resultant energy (X)	Energy absorbed	Thickness of specimens	Impact strength (E/t)
	J	E = 304.44−X	mm	J
1	294.12	10.32	6	1.720
2	290.97	13.47	6	2.245
3	290.97	13.47	6	2.245
4	290.97	13.47	6	2.245
5	294.13	10.31	6	1.718
Mean	292.23	12.208	6	2.035

Table 5 Impact strength for 3D core PET foam

	Resultant energy (X)	Energy absorbed	Thickness of specimens	Impact strength (E/t)
	J	E = 304.44−X	mm	J
1	297.27	7.17	11.2	0.640
2	294.13	10.31	11.2	0.921
3	290.97	13.47	11.2	1.203
4	294.13	10.31	11.2	0.921
5	290.97	13.47	11.2	1.203
Mean	293.49	10.946	11.2	**0.977**

Table 6 Impact strength for infusion grooved PVC foam

	Resultant energy (X)	Energy absorbed	Thickness of specimens	Impact strength (E/t)
	J	E = 304.44−X	mm	J
1	297.27	7.17	10	0.717
2	294.13	10.31	10	1.031
3	294.13	10.31	10	1.031
4	294.13	10.31	10	1.031
5	297.27	7.17	10	0.717
Mean	295.39	9.054	10	**0.905**

than the other core specimens as it is more compact and solid. Tougher materials such as coconut fiber need higher energy or impact to break or fracture. So, this means that it can absorb more energy applied on it. Table 4 shows the impact strength for coconut fiber core. While impact strength for 3D core PET foam are shown in Table 5. Table 6 shows the impact strength for infusion grooved PVC foam and Table 7 shows the comparison of the impact strength (Fig. 9).

Table 7 Comparison of impact strength

Panel	Resultant energy (X) J	Energy absorbed J	Impact strength (E/t) J
Coconut fiber	292.23	12.208	2.035
3D core PET foam	293.49	10.946	0.977
Infusion grooved PVC foam	295.39	9.054	0.95

Fig. 9 Comparison of impact strength

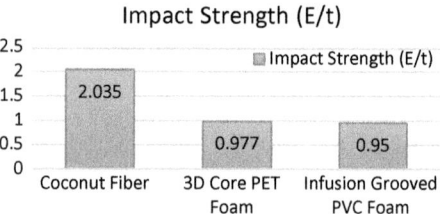

Thus, the hypothesis that can be made from this experiment is the more energy absorbed by the specimen, the more toughness the materials will have. In this experiment, the coconut fiber specimens broke completely into two parts with smoother broken surface while the other two broke into a few pieces. This is due to the stiffness and compactness of the coconut fiber. For the specimen with the lowest absorbed energy, it means it is brittle and has the lowest toughness, i.e. it can break easily and cannot withstand sudden high loads which is in this test obtained for the infusion grooved PVC foam. For the specimen with the highest absorbed energy means, it is ductile and has the highest toughness, which can withstand sudden high loads which in this test obtained for coconut fiber. For flexural strength, the data such as maximum applied load, known as "P", material span length between points in the test setup, known as "L", width of the material specimen, known as "b"; and average depth of the specimen, known as "d" are collected from the load experiments. Replace or substitute the numerical values for this data in the equation $R = 3P*L/2b*d^2$ for manipulating the flexural strength. In the equation, R is the flexural strength, in units of lbs. per square inch and quantity 'd' squared. Table 8 shows the flexural strength for the coconut fiber core. While Table 9 shows

Table 8 Flexural strength for coconut fiber core

	Width (mm)	Thickness (mm)	Support span (mm)	Max. load (N)	Flexural strength (Mpa)
1	12.7	6	25.4	823.19	68.599
2	12.7	6	25.4	732.45	61.038
3	12.7	6	25.4	891.45	74.288
4	12.7	6	25.4	776.69	64.724
5	12.7	6	25.4	694.56	57.880
Mean	12.7	6	25.4	783.67	**65.306**

Table 9 Flexural stress for 3D core PET foam

	Width (mm)	Thickness (mm)	Support span (mm)	Maximum load (N)	Flexural strength (Mpa)
1	12.7	11.2	25.4	343.48	8.215
2	12.7	11.2	25.4	400.37	9.575
3	12.7	11.2	25.4	473.8	11.331
4	12.7	11.2	25.4	413.54	9.890
5	12.7	11.2	25.4	388.67	9.295
Mean	12.7	11.2	25.4	403.97	**9.661**

Table 10 Flexural stress for infusion grooved PVC foam

	Width (mm)	Thickness (mm)	Support span (mm)	Maximum load (N)	Flexural strength (Mpa)
1	12.7	10	25.4	242.19	7.266
2	12.7	10	25.4	232.2	6.966
3	12.7	10	25.4	211.11	6.333
4	12.7	10	25.4	238.76	7.163
5	12.7	10	25.4	259.38	7.781
Mean	12.7	10	25.4	236.73	**7.102**

the flexural stress for the 3D core PET foam and Table 10 shows the flexural stress for the infusion grooved PVC foam.

From Table 11, a tremendous gap could be observed for the value of flexural strength between these 3 types of core. The flexural strength for coconut fiber shows the highest even though the thickness of the specimen is the lowest compared to others. Besides that, from this result, it is proven that coconut fiber has highest value in terms of maximum load in Newton. The mean for Young's modulus of the coconut fiber sample was slightly lower than the 3D Core PET foam sample while the infusion Grooved PVC foam sample was the lowest. The Young's modulus reflects the stiffness of a material. Therefore, the coconut fiber exhibited a lower stiffness than the 3D Core PET Foam in flexural testing mentioned that the lower characteristic of elasticity in the coconut fiber improved the impact strength

Table 11 Comparison of flexural strength

Panel	Maximum load	Modulus (Automatic Young's)	Flexural strength
	N	MPa	MPa
Coconut fiber	783.67	3529.76	65.306
3D core PET foam	403.97	3618.06	9.661
Infusion grooved PVC foam	236.73	2886.02	7.102

but reduced the stiffness, which might help to explain the lower Young's modulus of the coconut fiber. The mean for flexural strength of the coconut fiber sample was significantly higher than the other core sample. During the flexural test, compressive stresses were generated at the middle of the upper surface of the tested strip, whereas tension stresses were generated at the lower surface. The flexural strength is related to the distance between the two supporting bars, test speed, and dimensions (width and thickness) of the tested strip. However, the coconut fiber sample has the highest maximum load applied which makes it to possess a high flexural strength.

4 Conclusion

By comparing all the specimens for impact testing, it can be concluded that the best specimen that gives the highest impact energy is the coconut fiber. These specimens are tougher than the other core specimens as they are more compact and solid. Tough material such as coconut fiber need a higher energy or impact to break or fracture. So, this means that it can absorb more energy applied on it. A specimen with a lower absorbed energy means that it is brittle and has a lower toughness which can break easily and cannot withstand sudden high loads. The higher the resulting numbers the tougher the material. For the coconut fiber it is 2.035 J followed by the 3D Core PET Foam and the infusion grooved PVC foam which are 0.977 J and 0.95 J respectively as shown in Fig. 9.

The flexural strength of coconut fiber is the highest value, which is 65.306 MPa, even though the thickness of the specimen is lower compared to the other types. The 3D Core PET foam shows 9.661 MPa and the infusion grooved PVC foam is 7.102 MPa. The Young's modulus reflects the stiffness of a material. Therefore, the coconut fiber exhibited a lower stiffness than the 3D Core PET foam in flexural testing mentioned that the lower characteristic of elasticity in the coconut fiber improved the impact strength but reduced the stiffness, which might help to explain the lower Young's modulus of the Coconut Fiber. However, the coconut fiber sample has the highest maximum load applied which makes it to possess a high flexural strength. It can be proven that coconut fiber is suitable to be used as one of the laminating materials for fiberglass boatbuilding as shown in Fig. 10.

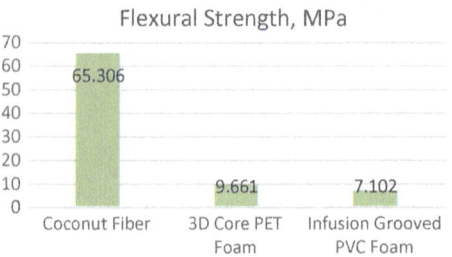

Fig. 10 Comparison of flexural strength

References

1. Yaacob, A., Zakaria, Z. A., Zarina, M.P., Koto, J., Kidd, P.: Production process of fiberglass fast interceptor boat in Malaysia. Sci. Eng. **19** (2015)
2. Pickering, K.L., Efendy, M.A., Le, T.M.: A review of recent developments in natural fibre composites and their mechanical performance. Compos. A Appl. Sci. Manuf. **83**, 98–112 (2016)
3. Chandramohan, D., Marimuthu, K.: A review on natural fibers. Int. J. Res. Rev. Appl. Sci. **8**(2), 194–206 (2011)
4. Kim, D., Hennigan, D.J., Beavers, K.D.: Effect of fabrication processes on mechanical properties of glass fiber reinforced polymer composites for 49 meter (160 foot) recreational yachts. Int. J. Naval Architect. Ocean Eng. **2**(1), 45–56 (2010)
5. Hoge, J., Leach, C.: Epoxy resin infused boat hulls. Reinf. Plast. **60**(4), 221–223 (2016)
6. Broad, A.I.: Development of Vacuum Assisted Composites Manufacturing Technology for Wind Turbine Blade Manufacture, Doctoral dissertation, University of Central Lancashire (2012)
7. Sevkat, E., Brahimi, M., Berri, S.: The bearing strength of pin loaded woven composites manufactured by vacuum assisted resin transfer moulding and hand lay-up techniques. Polym. Polym. Compos. **20**(3), 321 (2012)
8. Bongarde, U.S., Shinde, V.D.: Review on natural fiber reinforcement polymer composites. Int. J. Eng. Sci. Innovative Technol. **3**(2), 431–436 (2014)
9. Sailesh, A., Prakash, S.: Review on recent developments in natural fiber composites. Int. J. Eng. Res. Technol. **2**(9), ESRSA Publications (September 2013)
10. Taj, S., Munawar, M.A., Khan, S.: Natural fiber-reinforced polymer composites. Proc. Pak. Acad. Sci. **44**(2), 129 (2007)
11. Luo, S., Netravali, A.N.: Mechanical and thermal properties of environment-friendly "green" composites made from pineapple leaf fibers and poly (hydroxybutyrate-co-valerate) resin. Polym. Compos. **20**(3), 367–378 (1999)
12. Veritas, D.N.: Standard For Certification No. 2.21 (2010)
13. Kao-Walter, S., Mfoumou, E., Ndikontar, M.: Mechanical properties and life-cycle sustainability aspects of natural fibre. Adv. Mater. Res. **347**, 1887–1893, Trans Tech Publications (2012)
14. Reddy, N., Yang, Y.: Biofibers from agricultural byproducts for industrial applications. Trends Biotechnol. **23**(1), 22–27 (2005)
15. Vavrina, C.S., Armbrester, K., Mireia, A., Pena, M.: Coconut Coir as an Alternative to Peat Media for Vegetable Transplant Production. University of Florida, Southwest Florida Research and Education Centre P.O. Drawer 5127, Immokalee, FL 33934
16. Rajan, A., Senan, R.C., Pavithran, C., Abraham, T.E.: Biosoftening of coir fiber using selected microorganisms. Bioprocess Biosyst. Eng. **28**(3), 165–173 (2005)
17. Bujang, I.Z., Awang, M.K., Ismail, A.E.: Study on the dynamic characteristic of coconut fiber reinforced composites. In: Regional conference on engineering mathematics, mechanics, manufacturing & architecture, pp. 185–202 (January 2007)
18. Priya, N.A.S., Raju, P.V., Naveen, P.N.E.: Experimental testing of polymer reinforced with coconut coir fiber composites. Int. J. Emerg. Technol. Adv. Eng. **4**(12), 453–460 (2014)
19. Information at http://www.3dcore.com/en/downloads/3D_Flyer_Take_less_en.pdf
20. Information at http://www.3d-core.com/en/3dcore3d-core-info.html
21. Information at http://www.easycomposites.co.uk/#!/core-materials/closed-cell-foam-and-3dcore/easycell75g-infusion-grooved-closed-cell-pvc-foam-core.html
22. Information at http://www.westsystem.com/ss/assets/Upload/Ew21chopped.pdf
23. Information at http://textinfo.files.wordpress.com/2012/01/needle-punching1.pdf

Tensile and Hardness Analysis of Dissimilar Friction Stir Welding Between AA6061 with AA5083 and Mild Steel

Wan Mohd Syafiq Wan Sulong, Mohd Afendi Rojan and Mohd Noor Mazlee

Abstract In this paper, the friction stir welding (FSW) technique was used to create dissimilar alloy joints of AA6061-AA5083, and dissimilar material joints of AA6061-S235JR mild steel. Joints were fabricated by a conventional belting milling machine. Plates of 5 mm thickness were welded in a butt joint configuration. The joints were evaluated by extracting tensile testing specimens to study the failure strength at different rotational speeds. Tunnel defects were formed under certain parameters and changing the parameter values had different effects on the tunnel defects depending on the type of dissimilar joint welded. Vickers micro hardness measurements were taken on cross sections of welded joints to study the hardness pattern across the weld zones.

Keywords Friction stir welding · Dissimilar joint welded · Hardness pattern

1 Introduction

Friction stir welding (FSW) is gaining interest in the welding and joining industry because it does not suffer from the flaws and defects that typically affect conventional fusion welding problems such as porosity and solidification cracking [1, 2]. This is due to the fact that FSW is a solid state process where the process temperatures do not reach melting temperatures of the workpiece. This welding

---

W. M. S. Wan Sulong (✉) · M. A. Rojan
School of Mechatronic Engineering, Pauh Putra Campus, Universiti Malaysia Perlis (UniMAP), 02600 Arau, Perlis, Malaysia
e-mail: wmsyafiq10@gmail.com

M. A. Rojan
e-mail: afendirojan@unimap.edu.my

M. N. Mazlee
School of Materials Engineering, Centre of Excellence for Frontier Materials Research, Universiti Malaysia Perlis (UniMAP), 01000 Kangar, Perlis, Malaysia
e-mail: mazlee@unimap.edu.my

technology also gets rid of the need to utilize shielding gases, filler metals and numerous safety precautions that conventional welding techniques, such as gas tungsten arc welding and gas metal arc welding, demand. The FSW technology also made it achievable to weld several aluminum alloys that in the past were not recommended for welding such as the 2xxx series and 7xxx series [3, 4].

The joining of aluminum alloy with steel has been attracting a lot of interest for a long time in the manufacturing industry. This is due to the combination of strong and robust characteristics of steel together with the high formability of lightweight aluminum alloys. As such, the joining has been long sought after in the automotive industry in an attempt to reduce payload and fuel consumption of vehicles. However, the main difficulty in producing these type of joints for conventional or even advanced welding techniques is the strong tendency to form brittle intermetallic compound (IMC) at the joint interface, as well as the difference in thermomechanical and chemical properties of the two base materials [5]. Several techniques have been tested to manufacture an aluminum-steel joint such as friction welding, diffusion welding and ultrasonic welding. However, friction welding can only be applied for tubular parts, whereas ultrasonic welding are only applicable for very thin sheets [6–8]. The main advantage that the friction stir welding process offers over other conventional welding techniques is its solid-state process where the welding temperature does not reach melting points for the materials welded. This relatively low heat input means IMC layers grow less, thus improving welding quality. The versatile FSW process is also able to join plates with different thickness, as well as dissimilar aluminum alloys [9–11].

In this paper, the friction stir welding between dissimilar aluminum alloys and dissimilar materials, that is to say aluminum alloy and mild steel are investigated. Mechanical properties of the joint such as tensile strength and hardness of the two types of dissimilar FSW joints are considered and evaluated.

2 Experimental Setup

In this study, 5 mm thick aluminum alloy 6061 plates were friction stir welded with aluminum alloy 5083 plates or S235JR mild steel plates in butt joint configuration. The rotating tool used to execute FSW was fabricated from H13 tool steel. Heat treatment was applied to the tool so that its hardness was increased to 52 HRC. The tool is shown in Fig. 1. To carry out the FSW, a conventional milling machine was used.

In FSW, tool rotational speed and tool travel speed are two parameters that play a large role in determining the properties of the welded joint. The parameter values used to carry out the welding process are tabulated in Table 1. It is important to note that for FSW of dissimilar material (aluminum and steel), a "tool offset" parameter was included. This is to avoid excessive interaction between the rotating tool and mild steel, which could lead to extreme frictional heat and short lifespan of the rotating tool. No offset was applied during the FSW of dissimilar alloys between AA6061 and AA5083. The setup for friction stir welding on a conventional milling

Fig. 1 Shoulder and pin of the FSW tool

Table 1 Welding parameters and their respective values

Parameters	Value
Rotational speed, ω (rpm)	1110, 1750
Travel speed, v (mm/min)	30
Tool offset, TO (mm)	0.2

machine is demonstrated in Fig. 2. The samples and parameters list are tabulated in Table 2.

After welding has been completed, tensile testing specimens were extracted from the weld piece by machining using a wire cut electrical discharge machine (EDM). Dimensions of the tensile testing specimens were taken from ASTM E8/E8 M-13a for sheet-type standard specimen. A universal testing machine (UTM) was used to carry out tensile testing using a crosshead speed of 1 mm/min. Vickers microhardness profiles were taken at mid thickness on the cross section normal to the welding direction. The indenter was applied for 15 s at a load of 200 N.

Table 2 Samples and parameters list

Samples	Parameters		
	ω (rpm)	v (mm/min)	TO (mm)
AA6061-AA5083			
A1	1110	30	–
A2	1750		–
AA6061-S235JR			
D1	1110	30	0.2
D2	1750		0.2

3 Results and Discussion

3.1 Macrostructure

3.1.1 Dissimilar Alloy (AA6061 and AA5083)

Figure 3 shows the tensile specimens for dissimilar alloy FSW. AA5083 is positioned to the left, whereas AA6061 is positioned to the right of the figure. Figure 4

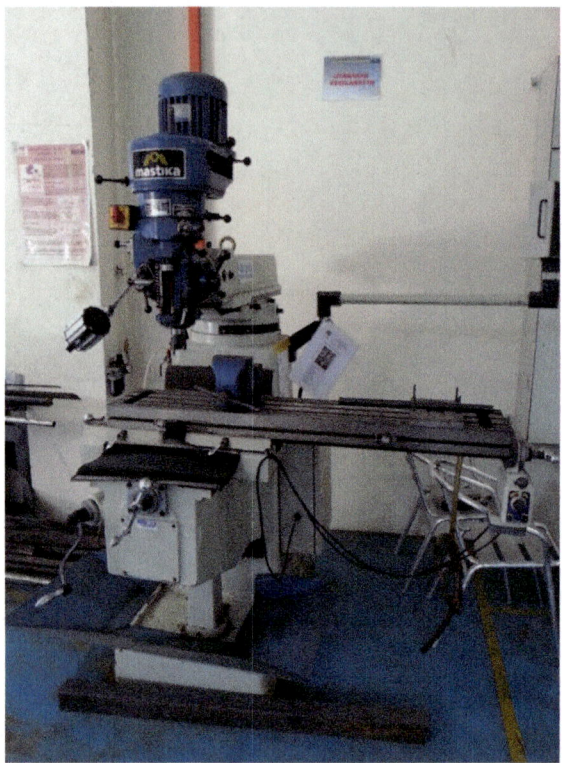

Fig. 2 Milling machine used for FSW

Fig. 3 Tensile specimens A1 and A2 for dissimilar alloy FSW

shows the failure sites and tunnel defect in A1 and A2, respectively. The tunnels were formed in the middle, or nugget zone of the welded joint in both A1 and A2. Although the tunnel seen in A2 was much smaller. According to Zhang's model to describe the tunnel defect formation, increasing the tool rotational speed or reducing the tool travel speed will increase the tool's ability to push plasticized material into the void left by the advancing tool pin [12]. As demonstrated by samples A1 and A2, increasing tool rotational speed indeed helps in getting rid of

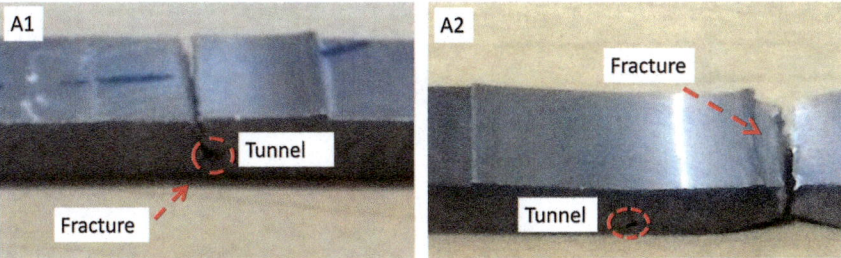

Fig. 4 Failure site and tunnel defect for A1 and A2

tunnel defects. The low rotational speed used in welding A1 led to the tunnel defect formation due to inadequate material flow [13].

For A1, the failure can be seen in Fig. 4 and occurred at the interface, coinciding with the tunnel defect position. The presence of the tunnel defect in the sample created a stress concentration point thus making it the weakest point in the joint. On the other hand, for sample A2 the failure happened at the heat affected zone (HAZ) on the AA6061 side. This can be attributed to the softer material generally found and aluminum alloy's HAZ, therefore it is the first site to experience necking thus failure happened here. The phenomenon was also seen by Rodriguez et al. [11] in FSW of dissimilar AA6061-AA7075, where the specimen failed at the HAZ of AA6061 side due to low microhardness values recorded in this zone. Failure in A2 did not happen at the tunnel because the tunnel size was sufficiently small enough that it did not negatively affect joint strength.

3.1.2 Dissimilar Material (AA6061 and Mild Steel)

Figure 5 shows the tensile specimens D1 and D2, while Fig. 6 shows a closer view of the failure site and tunnel defects for both specimens respectively. It was seen that the tunnel defect size found in D2 is larger than D1, but failure of the specimen

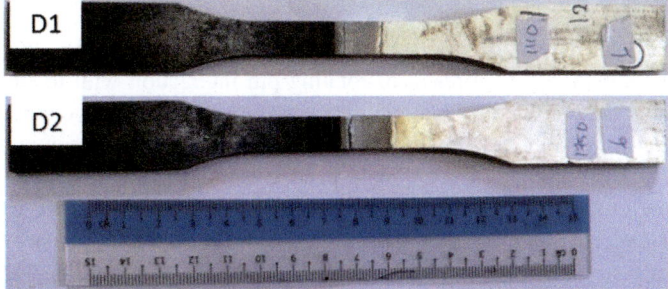

Fig. 5 Tensile specimens D1 and D2 for dissimilar material FSW

Fig. 6 Failure site and tunnel defect for D1 and D2

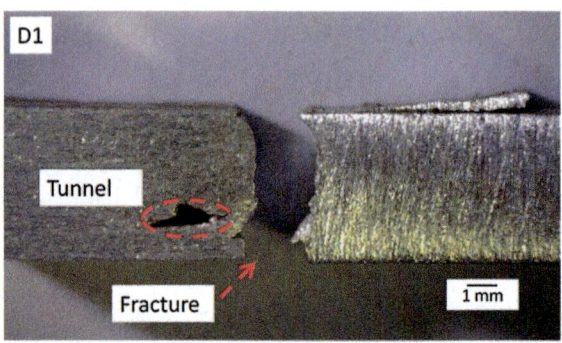

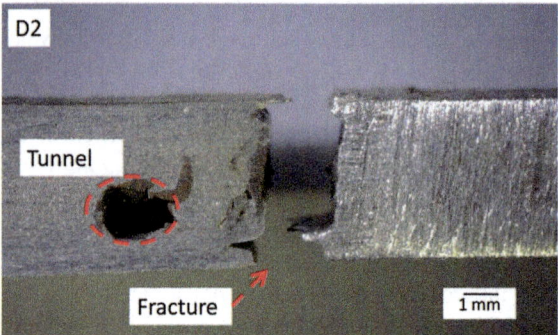

did not occur at the tunnel even for sample D2 with the bigger tunnel. As mentioned previously, Zhang's model of tunnel defect formation hypothesized that increasing tool rotational speed should eliminate or reduce the size of tunnel defects. However, it was found in FSW of dissimilar materials between aluminum alloy 6061 and mild steel, increasing tool rotational speed also increases the tool shoulder interaction with the harder material (mild steel) hence increasing heat to a larger degree than in similar material case. Too much heat input could also lead to improper material flow thus the tunnel defect was formed [14]. It is also important to note that in aluminum alloy and steel welding, the formation of an intermetallic compound (IMC) layer joining a complicated task due to its brittle nature [15]. As a result, in FSW of aluminum and steel, heat input plays a significant role as high process temperature promotes IMC layer growth.

Steel particles were also seen around the tunnel defects in Fig. 6. The steel particles were detached as a result of rotating pin interaction with the steel faying surface and could also disrupt material flow during FSW thus also contribute to the tunnel defect formation. In samples D2, more steel particles were seen in the nugget zone due to the higher tool rotational speed used to perform FSW of the joint. Several previous reports also mentioned the increase in steel particle scattering as well as formation of tunnels and voids in the weld cross section, which was caused by a higher tool offset [16].

Fig. 7 Ultimate tensile strength of tensile specimens

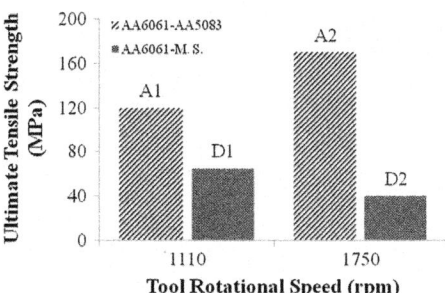

3.2 Tensile Strength

Figure 7 shows the tensile strength of tensile specimens. From these results, it can be seen that increasing tool rotational speed improves the joint strength for FSW of dissimilar AA6061 and AA5083. However, the inverse was true for FSW of dissimilar materials between AA6061 and mild steel. Relating tensile strength results to the macrostructure, it is seen that the improvement of joint strength for the dissimilar aluminum alloy case was due to the reduction of tunnel defect size. The increase in heat input from the faster tool rotation improves material flow. On the other hand, increasing heat input for the dissimilar material FSW between AA6061 and mild steel encouraged IMC layer growth that damaged joint strength [17].

3.3 Hardness

Figures 8 and 9 illustrate the hardness profile of sample A2 and D2, respectively. It was seen that the hardness at the AA5083 side is generally higher than AA6061. The hardness at AA5083 side hit a minimum approximately 10 mm away from the interface. As the hardness profile approached the AA6061 side it started to go down. Similarly, the hardness profile seen in the AA6061 hit a minimum in the region of 10 mm away from the interface. The minimum HV of the joint was found

Fig. 8 Hardness profile for sample A2

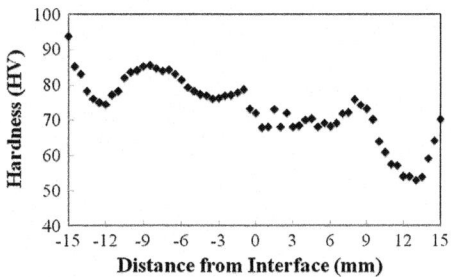

Fig. 9 Hardness profile for sample D2

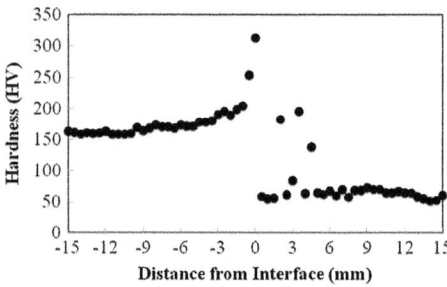

to be at the AA6061 HAZ. This phenomenon explained the failure location of sample A2, due to necking preferring to occur at the point of lowest hardness. For sample D2, the lowest HV was also located at the AA6061 HAZ, however failure did not occur here. This is because the joint failed at the brittle IMC layer located at the interface before necking could occur at the AA6061 HAZ. A similar HV pattern and relationship between failure location and HV minimum was seen in previous studies of dissimilar aluminium alloy FSW [11]. The fluctuating HV seen in D2 from 0 to 5 mm was due to the scattering of steel particles and intermetallics from the rotating tool pin [18].

4 Conclusion

In this paper, FSW between dissimilar aluminum alloys and dissimilar materials were compared. Based on the results, several conclusions were made:

1. For dissimilar aluminum alloy FSW between AA5083 and AA6061, tunnel was formed when the tool rotational speed was 1110 rpm. Failure occurred at the tunnel defect. Increasing the tool rotational speed to 1750 rpm increased the material flow thus reducing tunnel size. Failure location was changed to AA6061 HAZ. The joint created with a tool rotational speed of 1750 rpm also was stronger than the joint made with a tool rotational speed of 1110 rpm.
2. For dissimilar material FSW between AA6061 and S235JR mild steel, tunnel also was seen in the 1110 rpm joint. Increasing the rotational speed made the resulting tunnel defect larger and more separated steel particles were seen in the nugget zone, in the vicinity of the tunnel. Failure for both cases occurred at the interface. Joint strength was reduced when the tool rotational speed was increased from 1110 to 1750 rpm. This may be due to the presence of brittle intermetallic compound layer at the interface that occurs in high process temperature.
3. Minimum hardness for the dissimilar alloy joint was at AA6061 HAZ, which explains why failure occurred here due to necking phenomenon initiating at softer parts of the joint. Minimum hardness for the dissimilar material joint also

was seen at the AA6061 HAZ, but the joint was not strong enough to reach the plastic stage of deformation, therefore necking was not seen. Failure occurred at the interface regardless of hardness profile results due to the brittle nature of IMC layer formed at the interface.
4. For FSW of dissimilar aluminum alloy, increasing tool rotational speed thus heat input solved tunnel defect problem. On the other hand, increasing heat input during FSW of dissimilar material involving aluminum alloy and mild steel deteriorated joint strength due to IMC layer formation that depends on high process temperature.

References

1. Barnes, S.J., Steuwer, A., Mahawish, S., Johnson, R., Withers, P.J.: Residual strains and microstructure development in single and sequential double sided friction stir welds in RQT-701 steel. Mater. Sci. Eng. A **492**(1–2), 35–44 (2008)
2. Chen, J., Ueji, R., Fujii, H.: Double-sided friction-stir welding of magnesium alloy with concave-convex tools for texture control. Mater. Des. **76**, 181–189 (2015)
3. Rohilla, P., Kumar, N.: Experimental investigation of tool geometry on mechanical properties of friction stir welding of AA6061. Int. J. Innov. Technol. Explor. Eng. **3**(3), 56–61 (2013)
4. Kumar, N., Mishra, R.S.: Friction stir welding and processing (2014)
5. Ryabov, V.R.: Aluminizing of steel. Oxonian Press, New Delhi (1985)
6. Lee, W.B., Yeon, Y.M., Kim, D.U., Jung, S.B.: Effect of friction welding parameters on mechanical and metallurgical properties of aluminium alloy 5052–A36 steel joint. Mater. Sci. Technol. **19**(6), 773–778 (2003)
7. Tsujino, J., Hidai, K., Hasegawa, A., Kanai, R., Matsuura, H., Matsushima, K., Ueoka, T.: Ultrasonic butt welding of aluminum, aluminum alloy and stainless steel plate specimens. **40**, 371–374 (2002)
8. Yamamoto, N., Takahashi, M., Aritoshi, M., Ikeuchi, K.: Effect of interfacial layer on bond strength of friction-welded interface of Al-Mg5083 alloy to mild steel. Q. J. Jpn. Weld. Soc. **23**(3), 496–503 (2005)
9. Liu, X., Lan, S., Ni, J.: Analysis of process parameters effects on friction stir welding of dissimilar aluminum alloy to advanced high strength steel. Mater. Des. **59**, 50–62 (2014)
10. Leal, R.M., Leitão, C., Loureiro, A., Rodrigues, D.M., Vilaça, P.: Material flow in heterogeneous friction stir welding of thin aluminium sheets: Effect of shoulder geometry. Mater. Sci. Eng. A **498**(1–2), 384–391 (2008)
11. Rodriguez, R.I., Jordon, J.B., Allison, P.G., Rushing, T., Garcia, L.: Microstructure and mechanical properties of dissimilar friction stir welding of 6061-to-7050 aluminum alloys. Mater. Des. **83**, 60–65 (2015)
12. Zhang, H., Lin, S.B., Wu, L., Feng, J.C., Ma, S.L.: Defects formation procedure and mathematic model for defect free friction stir welding of magnesium alloy. Mater. Des. **27**(9), 805–809 (2006)
13. Wu, A.P., Song, Z.H., Nakata, K., Liao, J.S.: Defects and the properties of the dissimilar materials FSW joints of titanium alloy TC4 with aluminum alloy 6061. In: Proceedings of the 1st International Joint Symposium on Joining and Welding, Elsevier, pp. 243–248 (2013)
14. Kim, Y.G., Fujii, H., Tsumura, T., Komazaki, T., Nakata, K.: Three defect types in friction stir welding of aluminum die casting alloy. Mater. Sci. Eng. A **415**(1–2), 250–254 (2006)
15. Chen, C.M., Kovacevic, R.: Joining of Al 6061 alloy to AISI 1018 steel by combined effects of fusion and solid state welding. Int. J. Mach. Tools Manuf **44**(11), 1205–1214 (2004)

16. Kimapong, K., Watanabe, T.: Friction stir welding of aluminum alloy to steel. Weld. J., pp. 277–282 (2004)
17. Tanaka, T.H.T., Morishige, T.: Comprehensive analysis of joint strength for dissimilar friction stir welds of mild steel to aluminium alloys. Scr. Mater. **61**(7), 756–759 (2009)
18. Chen, W.-B., Lin, T.: A prime study on FSW joint of dissimilar metals. In: Proceedings of the XIth International Congress and Exposition (2008)

Analysis of Drum Brake System for Improvement of Braking Performance

Siti Nor Nadirah Baba, Muhammad Najib Abdul Hamid,
Shahril Nizam Mohamed Soid, Mohd Nurhidayat Zahelem
and Mohd Suyerdi Omar

Abstract The braking performance has become a very important factor for automotive manufacturers and passenger because of the safety requirements. In general, brake squeal occurrences can be reduced by decreasing the friction coefficient; however, the braking performances also decrease. The objectives of this study are obtaining the dynamics properties (natural frequency) of brake shoes and to propose a new modification design of the brake shoe with the aim of improving the performances of the drum brake and stability of squeal. A finite element model of the automotive components has been developed in the analysis to determine the dynamic properties, brake torque measurement and instability of the system. The data of the finite element model is validated by performing an experimental modal analysis. The result showed that the value of natural frequency, brake torque and contact analysis increased when using the new design of brake shoe.

Keywords Drum brake system · Dynamic properties · Brake torque

1 Introduction

The braking system is one of the important systems in vehicle safety. The purpose of the brake system is to prevent or reduce the severe injury during accidents. The function of the brake system is to slow or stop the moving vehicle. Besides, the

---

S. N. N. Baba (✉) · M. N. A. Hamid · S. N. M. Soid · M. N. Zahelem · M. S. Omar
Malaysian Spanish Institute, Universiti Kuala Lumpur, Kulim Hi-Tech Park,
09000 Kulim, Kedah, Malaysia
e-mail: snadirah.baba@s.unikl.edu.my

M. N. A. Hamid
e-mail: mnajib@unikl.edu.my

M. N. Zahelem
e-mail: mnurhidayat@unikl.edu.my

M. S. Omar
e-mail: msuyerdi@unikl.edu.my

© Springer International Publishing AG 2018
A. Öchsner (ed.), *Engineering Applications for New Materials and Technologies*,
Advanced Structured Materials 85, https://doi.org/10.1007/978-3-319-72697-7_28

brake system is also used to maintain the vehicle speed during downhill operation and hold a vehicle stationary on a grade. Recently, many countries presented brake system design criteria and operational requirements for brake systems depending on the vehicle types [1]. Their regulations of brake systems include to slow the vehicle down in a controllable, stable, predictable, and repeatable manner regardless or road, load, weather or partial failure. All this may be achieved when the performance of the brake system retains its stability during vehicle brake operation. A drum brake is widely used in automotive industry for more than forty years. Although disc brakes now are found wisely at the front of almost all cars, the drum brake still can be seen at the rear wheel of the cars and light track due to cheap production price and the fact that the maintenance of drum brakes is cheaper due having all in one design that is easier to replace when repair work is needed. Drum brakes are also self-energizing and can operate as parking brakes while disc brakes require special parking brake mechanisms.

The brake noise is categorized into squeal, groan, and judder dependend, according to the mechanism of generation. Squeal has been the primary subject of past studies of brake systems. Brake squeal is defined as a self-excited friction induced vibration, which occurs at frequencies above 1 kHz and this phenomenon is caused by dynamics instability of the brake owing to the friction material of the drum or disc brake. In general, the brake squeals propensity is reduced by decreasing the friction coefficient, however the braking capability also decreases [2]. To reduce squeal without effect on the brake capability, there is a need to analyse a new design modification of brake shoe that provides high braking force and stability from squeal.

The drum brake problems were studies using various methods including the finite element method, analytical and experimental approaches [3]. The primary problem in drum brake systems is brake squeal. Reducing squeal from the brake system becomes a challenging task for any brake engineer. Previous researchers studied the cause of brake squeal and proposed methods for reducing the brake squeal. Hamabe et al. [4] found the common causes of brake squeal when the friction material has a negative slope in relation to the relative velocity and suggested to reduce the brake that the squeal the friction characteristic of the friction material must be improved. The way to prevent brake squeal is through lining modification [3]. Based on another study, brake drum squeal also can be reduced by modifying the partial of the shoe web [5]. In 2013, researchers found that a way to prevent the brake squeal is to increase the bending stiffness of the cross-sectional area of the drum. While for the shoe, it increases the critical friction coefficient of squeal (CFS) to reduce the bending stiffness and increases the cross-sectional area [2].

Recent investigations of brake noise phenomenons showed that natural frequencies and damping of brake components play an important role regarding the propensity of noise excitation. While works are on-going to include requirements of those parameters into international standardization of disc brakes, there is a lack of knowledge regarding the drum brake noise. Vibrational modes of drum brake shoes are the main contributor to the squeal phenomena [6]. In the different study, Haverkamp et al. [7] investigated the squeal phenomenon of brake shoes using a resonance test that was assembled under an artificial excitation (shaker). The test analysis showed an axial 'hat shape' mode and its first harmonic at squeal frequency that was detected at 3.2 and 4.8 kHz. The corresponding bending modes have been found at the backing plate. The main excitation force occurs during the brake operation in tangential direction in between the friction surface of the shoes and the inner surfaces of the drum. If this leads to excitation of mode shape with maximum deflection in axial direction, it can be assumed that a specific process in involved, transforming the tangential excitation into axial movement. The result indeed clearly showed a torsional vibration of the brake shoes during squeal excitation. The researchers suggest that an alternative way to eliminate squeal without adding weight to the system is a modification of the stiffness of the brake shoes because it can strongly change the coupling between the drum and back plate.

The finite element method for brake analysis has become a preferred method in studying the braking performance such as the analysis of the dynamics properties, thermal distribution analysis, and stability analysis It is because of its flexibility and diversity in providing solutions to problems involving advanced material properties [8]. Temperature distribution analysis is mostly performed using FE method due it powerful tool for numerical solutions for a wide range of engineering problems [9]. Shahril et al. [10] used the finite element software ABAQUS as their simulation none modelling tool because ABAQUS can perform static or dynamics analysis and simulation of the structure. It can deal with bodies with various loads, temperatures, contacts, impacts, and other environmental conditions. Ioannidis et al. [11] investigated the drum brake contact and analysed its influence on the squeal noise prediction by using the FE method. In another study, Zaki et al. [12] used the modal analysis in the scope of the FE method to analyse the dynamic properties of the brake disc to determine the brake dynamic behaviour.

In this study, the brake shoe of a passenger car was investigated. Finite element analysis has been performed based on the dynamic properties of a brake shoe extracted from experimental modal analysis. The brake torque was analysed by using the validated brake shoe and the modification of the brake shoe geometry was proposed for the improvement of drum brake performance.

2 Experimental Setup

2.1 Brake Factor Calculation

2.1.1 Existing Brake Shoe

The brake factor of the leading shoe C_1 is the ratio between the brake force R_1 and the applied force F by evaluating the moment of the point B as seen in Fig. 1:

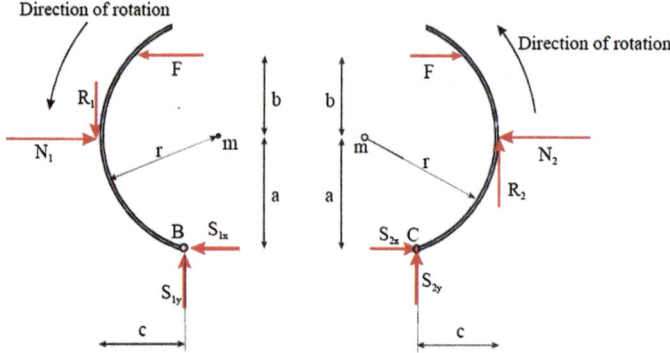

Fig. 1 Simplified free body diagram for simplex drum brake by Mahmoud [13]

$$C_1 = \frac{R_1}{F} = \frac{\mu(a+b)}{a - c\mu} \tag{1}$$

Similarly, for the trailing shoe by evaluating the moment about point C, it will give the shoe factor C_2 as follows:

$$C_2 = \frac{R_2}{F} = \frac{\mu(a+b)}{a + c\mu} \tag{2}$$

An equivalent angle of inclination can be defined as:

$$\alpha = \tan^{-1}\left(\frac{a}{c}\right) \tag{3}$$

The overall shoe factor of the simplex drum brake is the summation of the shoe factor of the primary shoe and the secondary shoe:

$$C^* = C_1 + C_2 = \frac{\mu(a+b)}{a - c\mu} + \frac{\mu(a+b)}{a + c} \tag{4}$$

2.1.2 Modification of Brake Shoe Calculation

Figure 2 shows the free body diagram of the modified brake shoe. The brake factor of the leading shoe C_1 is the ratio between brake force R_1 and the applied force F by evaluating the moment of the point A:

$$C_1 = \frac{R_1}{F_1} = \frac{\mu(a+b)}{(a \cos \alpha - \mu \sin \beta)} \tag{5}$$

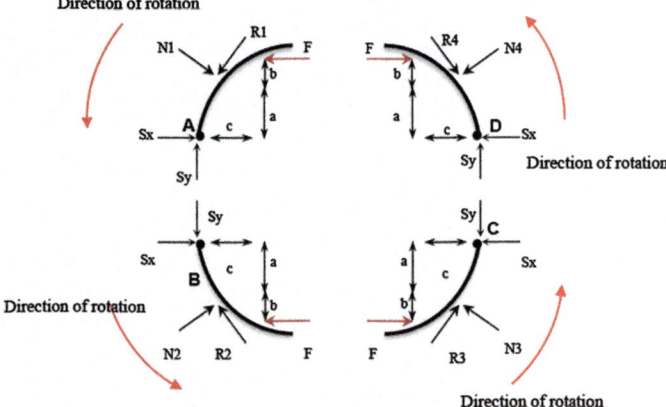

Fig. 2 Simplified free body diagram for modification of simplex drum brake

where $C_1 = C_3$

$$C_2 = C_4 = \frac{R_2}{F_2} = \frac{R_4}{F_4} = \frac{\mu(a+b)}{(a \cos \alpha - c\mu \sin \beta)} \quad (6)$$

Similarly, for the trailing shoe by evaluating the moment about point C, it will give the shoe factor C_2 as follows;

$$C_2 = \frac{R_2}{F_2} = \frac{\mu(a+b)}{a \cos \alpha + c\mu \sin \beta} \quad (7)$$

where $C_2 = C_4$

$$C_2 = C_4 = \frac{R_2}{F_2} = \frac{R_4}{F_4} = \frac{\mu(a+b)}{(a \cos \alpha + c\mu \sin \beta)} \quad (8)$$

The overall shoe factor of the simplex drum brake is the summation of the shoe factor of the two primary shoe and the two secondary shoe:

$$C^* = C_1 + C_2 + C_3 + C_4 \quad (9)$$

2.2 Brake Shoe Modelling

The existing brake shoe model of this study is taken from the rear brake set of the Proton Saga model and the dimension of the drum brake was taken based on the internal diameter for the drum brake which is 182 mm. The CATIA software was

Fig. 3 Existing brake shoe model

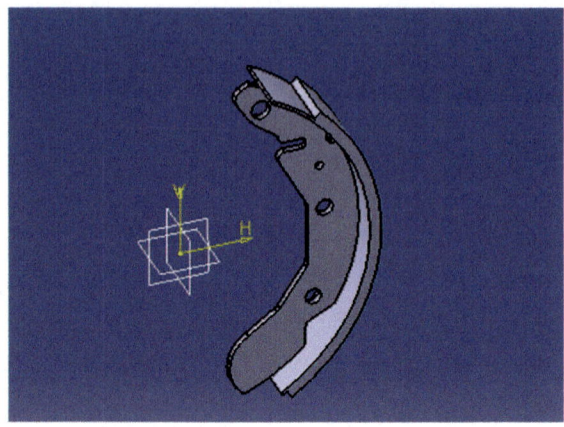

used to create the model of the brake shoe (see Fig. 3) before transfered into the ABAQUS software.

Modal analysis is a method to describe a structure in terms of its natural characteristics which are frequency, damping and mode shapes (dynamics properties) without using a rigorous mathematical treatment. The brake shoe was set in free-free condition by hanging using a rubber-band as shown is Fig. 4. The brake shoe was excited using an impact hammer and the response was measured by an accelerometer. The sensitivities of the transducer and sensor are cited in Table 1. The bandwidth used in this EMA was 5120 Hz. This experiment has been done at the university vibration laboratory.

The FE model of the brake shoe is refined and adjusted to make as close as possible with the experimental modal analysis result. Figure 5 indicates the meshed brake shoe and lining after appropriate simplification to the original parts. The material specification of the brake shoe is showed in Table 2. The validation of the result was done by using randomization design in statistical analysis. The accurate

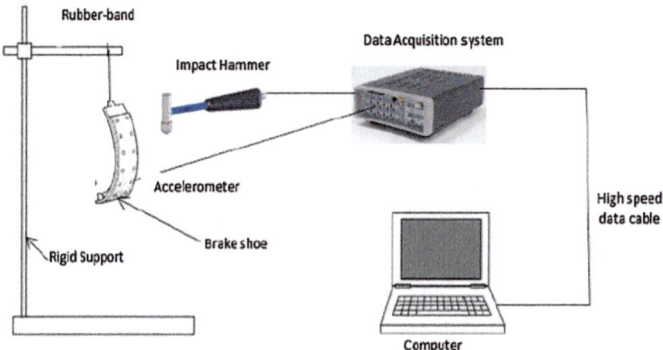

Fig. 4 Experimental modal analysis on the brake shoe

Table 1 EMA set up parameter

Impact hammer	4.91 mV/lbf
Accelerometer	2.83 mV/g
Bandwidth	5120 Hz

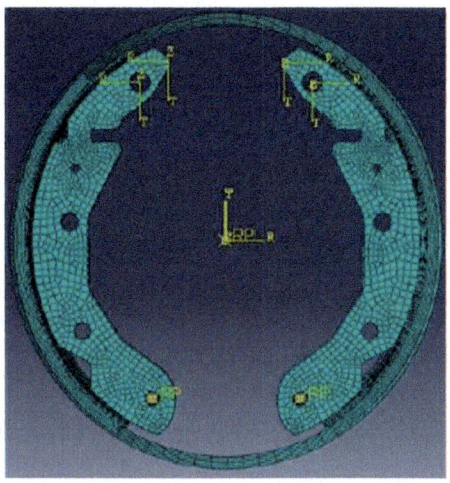

Fig. 5 Existing set up for brake torque analysis

Table 2 Material properties for brake shoes components

Lining	Young's modulus, E	100 GPa
	Poisson's ratio, v	0.23
	Density	1995.40 kg/m³
	Mass	39 g
	Radius	91 mm
	Width	30 mm
	Angle	105°
	Thickness	4 mm
Shoe	Young's modulus, E	210 GPa
	Poisson's ratio, v	0.27
	Density	3102 kg/m³
	Mass	155.1 g
	Radius	87 mm
	Width	30 mm
	Angle	1.5 mm
	Thickness	158°

simulations of the brake shoes model are important for studying the squeal characteristics either experimentally or numerically by FEM. In addition, the FEA model was used to simulate the non-linear contact analysis for the brake torque analysis. The experimental approach used the experimental modal analysis to examine the natural frequencies of the brake shoes.

2.3 Brake Torque Analysis

The non-linear analysis was performed to measure the brake torque and also to determine the contact pressure distribution for a given actuation force. The drum brake assembly consists of three main parts; one drum brake and two brake shoes with linings (see Fig. 5). The drum brake model is created directly by using the discrete rigid type in the ABAQUS software while the brake shoe was imported from the CATIA software. There is a small gap installation between the inner drum and the lining. Three steps were used to analyse the brake torque. Firstly, the contact between lining and drum was established. Then, the force was applied and finally the drum was rotated. The boundary condition has been applied to the drum and brake shoe. There is a small circle used to act as pinned support at the bottom of the brake shoes. The force was applied to approximate the behaviour of the double acting master cylinder and reflecting the pressure distribution. The parameter set up for the brake torque simulation is displayed in Table 3.

2.4 Modified Design of Brake Shoe

The modified shoe (see Fig. 6) was done by taking half of the original brake shoes dimension and drawing it by using the CATIA software. The design was set up by using the same material specification and parameter to run the modal analysis and brake torque simulation by using ABAQUS. For the new design, the brake shoes are added from two shoes into four shoes and the two master cylinders. The assembly of the new design is shown in Fig. 8.

Table 3 Parameter set up for brake torque

Force of each brake shoe	50 N
Friction coefficient	0.3
Speed	60 rpm

Fig. 6 Simulation model for modification brake shoe

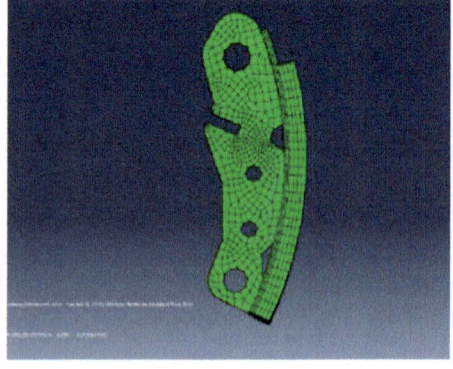

3 Results and Discussion

3.1 Brake Factor Calculation

Based on the brake factor calculation derived by Mahmoud [13] for the simplex drum brake, the value of the brake factor for this existing drum brake is 1.27. A modification of the brake shoe has been made and the value of the brake factor increased to 2.57 as shown in Table 4.

3.2 Modal Analysis

Tables 5 shows the result of the natural frequency of the existing brake shoe by using experimental and numerical modal analysis. The mode shapes are similar to each other. Mode 1 is the first torsional mode and mode 2 is the second torsional mode. Meanwhile mode 3 is the first bending mode and lastly mode 4 is the third torsional mode. Through the statistical test evaluation based on a 5% significance level, there is no significance difference between the experimental and simulation result. The result of the simulation is acceptable to carry on with the model to measure the brake torque and the new design. Data from this table can be compared with the data in Table 6 which shows the natural frequency of the new design.

Table 5 shows the results of the natural frequencies for the modified design brake shoes. From Tables 5 and 6, the values of the natural frequency for mode 1 and mode 2 of the modified design brake shoes are drastically increased compared to mode 1 and mode 2 of the original shoes. The modified design of the brake shoe can produce a high natural frequency than the existing brake shoe. The highest natural frequency is hard to excite and reduces the probability of resonance of the brake shoe and also reduces the brake squeal propensity. Consistent with finding by Rahman et al. [3] that modification of lining of shoes can reduce brake squeal, the study found that the modification of the length of the brake shoes is also a method to reduce brake squeal.

Figure 8 presents the brake torque of the drum brake by using simulation for one and two original brake shoes, and two and four of the new design of brake shoes. Two new design brake shoes can produce 150 Nm torque compared to 100 Nm produced by the existing brake shoes while four new design shoes can produce 300 Nm compared to the existing shoes with 150 Nm. The result found that the new design can produce a higher brake torque than the original design of the brake shoe. Figures 9 and 10 provide the value of the contact pressure for the existing shoe and

Table 4 Overall brake factor of drum brake system

Type of shoe	Total brake factor
Existing shoes	1.27
Modification shoes	2.57

Table 5 Modal analysis of existing shoe

Mode	Experimental	Simulation
1	1192.4 Hz	1151.7 Hz
2	1974.0 Hz	2147.9 Hz
3	2770.7 Hz	2876.8 Hz

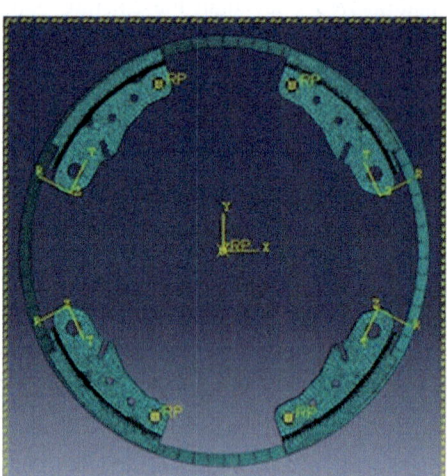

Fig. 7 The new set up for brake torque analysis

Table 6 Modification shoe design

Simulation
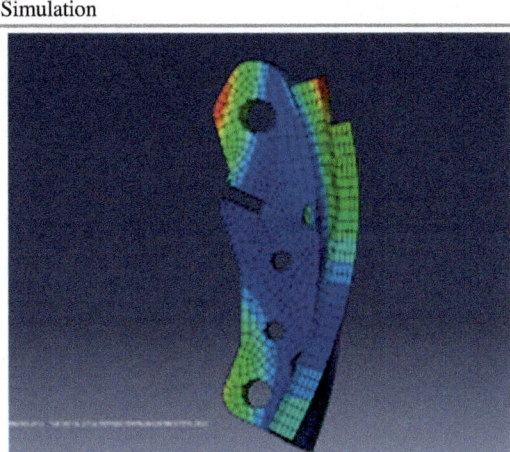 Mode 1: 5047.5 Hz
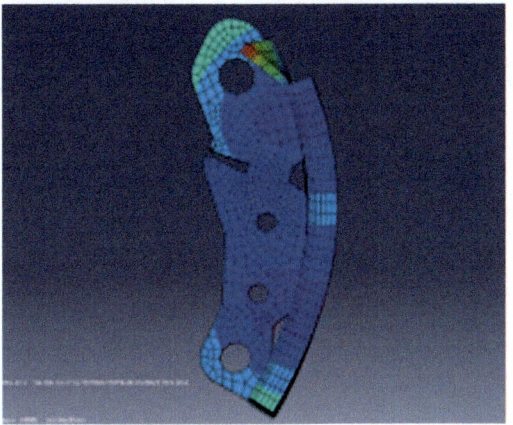 Mode 2: 7337.9 Hz

the new design shoe. The data shoes that the contact value of the new design shoes increases compared to the existing shoes. The brake system with four new design brake shoes produce a higher torque compared to the existing brake shoes because the total contact pressure is higher and has more contact points. By adding the contact pressure and contact point at drum, it can help to improve the stability and braking performance of the drum brake.

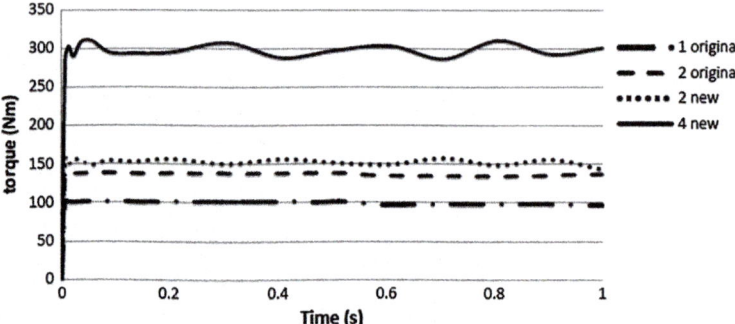

Fig. 8 Brake torque analysis

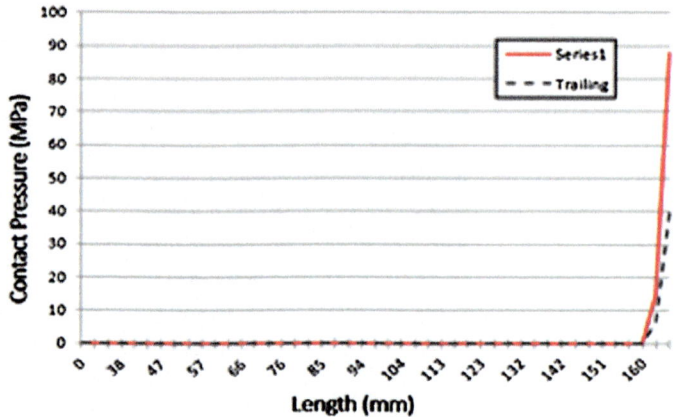

Fig. 9 Contact pressure of existing brake shoe

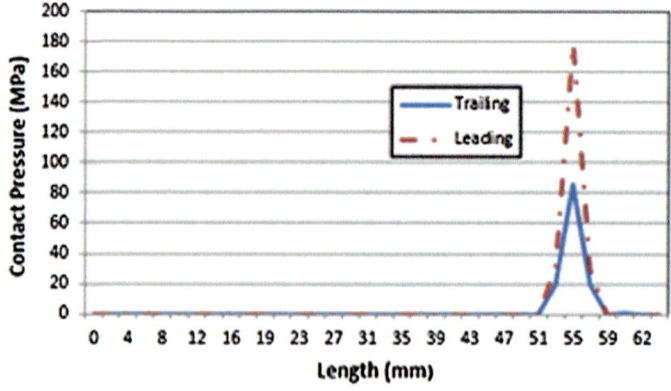

Fig. 10 Contact pressure of modified brake shoe

4 Conclusion

This paper presented a study to improve the existing drum brake by a proposed new design based on dynamics properties and brake torque investigation. The objective of this study was achieved by completing the simulation of the new design using the ABAQUS software to investigate the natural frequency and brake torque measurement. This study found generally that modying the length of the shoe can increase the natural frequency, brake torque and contact pressure of the drum brakes. Future research is recommended to fabricate the new design brake shoe to study the dynamic behaviour by obtained experimental modal analysis so that the analysis results obtained from analysis can be validated.

References

1. E. and E. United Nation: Uniform Provisions Concerning the Approval of Vehicles of Categories M, N and O with Regard to Braking (2014)
2. Chang, S., Cho, S. S. C., Lee, J. M., Yoo, S. W., Park, C. S.: A Study on the Squeal Noise of Drum Brakes, pp. 1369–1375 (2013)
3. Abd Rahman, M.R., Vernin, G., Bakar, A.R.A.: Preventing drum brake squeal through lining modifications. Appl. Mech. Mater. **471**, 20–24 (2013)
4. Hamabe, T., Yamazaki, I., Yamada, K., Matsui, H., Nakagawa, S., Kawamura, M.: Study of a method for reducing drum brake squeal. SAE Tech. Pap. (1999)
5. Lee, J.M., Yoo, S.W., Kim, J.H., Ahn, C.G.: A Study on the squeal of a drum brake which has shoes of non-uniform cross-section. J. Sound Vib. **240**(5), 789–808 (2001)
6. Baillet, L., Errico, S. D., Laulagnet, B.: Article in press Understanding the occurrence of squealing noise using the temporal finite element method. **292**(33), 443–460 (2006)
7. Haverkamp, M., Koopmann, N., Werke, F., Köln, A. G.: Acoustic Quality Control of Drum Brake Shoes, pp. 3–4 (2004)
8. Ali, B., Mostefa, B.: Thermomechanical modelling of disc brake contact phenomena. FME Trans. **41**(1), 59–65 (2013)
9. Adebisi, A.A., Maleque, M.A., Shah, Q.H.: Surface temperature distribution in a composite brake rotor. Int. J. Mech. Mater. Eng. **6**(3), 356–361 (2011)
10. Shahril, K., Nordin, M., Sulaiman, A. S.: Temperature analysis of automotive modeling parts, pp. 285–288 (2012)
11. Ioannidis, P., Brooks, P. C., Barton, D. C.: Drum brake contact analysis and its influence on squeal noise prediction. SAE Tech. Pap. (2003)
12. Zaki, A., Che, B. I. N., Ariff, Z.: Modal Analysis of Car Disc Brake (2010, Dec)
13. Mahmoud, K. R. M.: Theoretical and Experimental Investigations on a New Adaptive Duo Servo Drum Brake with High (2005)

Ultrasonic Based Technique to Measure Residual Stresses in Offshore Structures

Ramesh Ramasamy and Zainah Ibrahim

Abstract The offshore construction environment can be very difficult to be accurately predicted, due to various uncertainties in the geotechnical elements, and highly probabilistic events at which ocean conditions govern the design. The integrity assessment and life extension activities towards end of field life is becoming very common in older fields surrounding Malaysia, and other areas such as the Arabian Gulf and the North Sea. The in-line measurement of the existing residual stresses on these assets is very critical for the remaining life assessments, and also to categorise assets for shutdown to undergo rehabilitation work. The novel idea of using ultrasonic waves is developed and explained in detail in this paper. This method is based on the measurement of the time of propagation of the critically refracted longitudinal wave components travelling along the subsurface region of the specimen. Numerical validation is backed up by experimental work to verify this method to be fit-for-purpose for the intended application, through the development of an acoustoelastic calibration curve for several sectional shape/sizes of carbon steel material. These tests show good correlation between the numerical and experimental work, and this has the potential to be a reliable tool for use in the offshore/marine engineering sector where shutdowns are expensive.

Keywords Residual stresses · Offshore structures · Ultrasonic based technique

1 Introduction

A major problem facing the offshore oil well operators worldwide is ageing and degradations of their wells, specifically their structural components. The mature fields of the North Sea and the Persian Gulf are populated by wells operating

---

R. Ramasamy (✉) · Z. Ibrahim
Faculty of Engineering, University of Malaya, 50603 Kuala Lumpur, Malaysia
e-mail: ramesh.ramasamy@siswa.um.edu.my

Z. Ibrahim
e-mail: zainah@um.edu.my

towards or around their calculated design life of 25 years. This presents a hazard to operational safety, if their integrity is not assessed and verified as being fit-for-purpose. The degradations of the metallic surface due to aqueous corrosion and the internal loss of grouting due to spalling and cracking will introduce major problems in carrying out such arduous task, particularly in fields with hundreds of such wells. The well construction (see Fig. 1) process for a typical platform conductor configuration is carried out in a sequence as listed below:

- Drilling of a 36 in hole, installing the 30 in conductor, and cementing the outside annulus;
- Drilling of a 17 in hole, installing the 13 in surface casing, and cementing the C-annulus;
- Installing the topside equipment (wellhead and blowout preventer);
- Drilling the 14 in hole, installing the 9 in inner casing and cementing the B-annulus;
- Drilling the 5 in hole, installing downhole packers and production tubing, including access to reservoir;
- Removing the blowout preventer and installing the surface tree for production.

The major assumptions during the design phase is that the bottom soil bearing is effectively present, i.e. all casing strings are supported at the bottom by the soil, and that there exists load-sharing between the strings due to the cemented annuli [1].

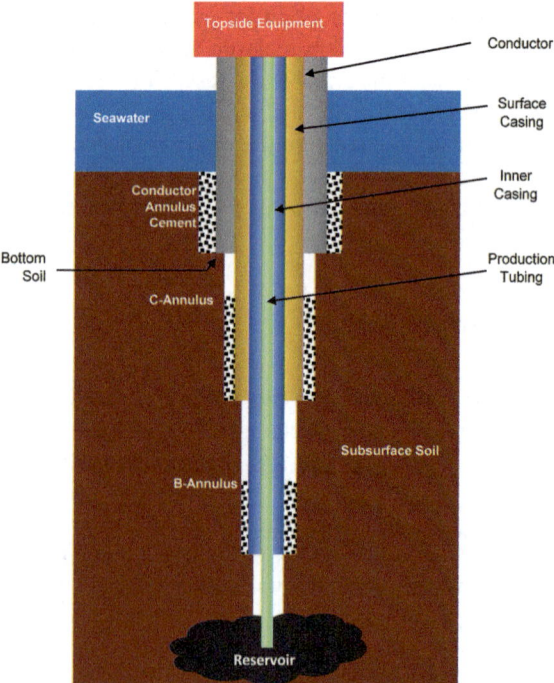

Fig. 1 Well construction schematics

The conductor is designed as an environmental shield to protect the internal strings and cement from the metocean current and waves. Over time, the severe corrosion and pitting formation on the metallic surfaces will reduce the load resistance of the casings and downhole collapse will result in the vertical drop of the topside equipment onto the conductor [2]. The spalling and cracking of the internal cement also will remove any load-sharing ability within the strings, forcing individual string to be overloaded. Since all the strings are connected via spools at the topside, i.e. at the wellhead, this will impart excessive axial load onto the conductor. The conductor, being exposed to the environment, is subjected to severe corrosion and wall loss (up to 70% in some cases, [2, 3]), particularly at the splash zone region (±2 m from the mean sea level). This will increase the axial stresses at this section, and combined with the environmental bending loads, may present the risk of collapse or eventually a catastrophic incident in a producing well.

To evaluate the stresses on the conductor during integrity screening and assessment of ageing wells by using the design loads can be overly conservative, particularly on degraded conductors with reduced capacity and resistance. Therefore, the accurate inspection and measurement of the residual axial load on the conductor is deemed very critical during the life extension planning of ageing platform wells. Standard methods for determining residual stresses on structures are hole drilling and diffraction techniques. The hole drilling method [4] involves drilling of a hole and using strain gauges to detect the strain relieved during material removal at specific regions. The destructive nature of this method, and the spark/fire hazard created on a live well are more than valid reasons to look for a non-destructive and safer alternative. The costly X-ray diffraction technique [5], although very accurate and safe to use is limited to very shallow layers of the material, about 10–25 μm and is unsuitable to adequately detect the axial stresses across the average thickness of the conductor pipe. Other non-destructive techniques such as those based on surface Rayleigh waves and vibration based impacting ball have also been studied and found to be limiting in terms of covered areas and penetration depths of the measurements [6].

The in-service structural components such as the conductor and casings are not easily removable, at least not without the arduous task of shutting down the well and deploying extensive manpower and equipment to the site at such great cost and time. Therefore, the measurement of the residual stresses in such structures need to be carried out during operation, whilst in compliance with the offshore safety protocols. This leads to the investigation and research of manipulating the ultrasonic wave based acoustoelastic technique to inspect and measure the residual stresses on the well structure, and this paper will describe the full extent of this technique.

Theoretically, the ultrasonic waves are categorised into the high frequency sound wave family and are widely used in the fields of non-destructive techniques and examinations (NDT/E) to detect flaws within a structure. The propagation of the ultrasonic compressional and shear wave components in a solid medium is affected by its internal planar stress state, and making this an ideal non-destructive residual stress measurement method. The linear relationship between the time of flight

(TOF) and the residual stress (σ) can be drawn from the measurement of the compressional wave speeds in the initial (unstressed) and stressed states, V_0 and V respectively, by a material dependant acoustoelastic constant, K such that:

$$\frac{V - V_0}{V_0} = K \cdot \sigma \qquad (1)$$

The initial state velocity (V_0) is dependent on the material density (ρ), elastic modulus (E), Poisson's ratio (v) and second-order elastic parameters (λ and μ), and can be evaluated as follows:

$$V_0 = \sqrt{\frac{\lambda + 2\mu}{\rho}} \qquad (2)$$

$$V_0 = \sqrt{\frac{\lambda + 2\mu}{\rho}} \qquad (3)$$

$$\lambda = \frac{vE}{(1+v)(1-v)} \qquad (4)$$

$$\mu = \frac{E}{2(1+v)} \qquad (5)$$

Subsequently, the shear wave speed component can be evaluated as follows:

$$\beta = \sqrt{\frac{\mu}{\rho}} \qquad (6)$$

The longitudinal (or compressional) wave, being the more prominent component of the propagation, is deemed more suitable for the measurement of the residual stresses in the conductor pipe. The compressional wave component excites the medium particles parallel to the direction of the propagation, whilst the shear component displaces the specimen particles in orthogonal direction to its propagation path.

A normal contact with the source will impart the vertically downwards compressional wave propagation, therefore a multi-medium arrangement is considered. This utilises the wave properties which changes its angle via refraction phenomena when entering a new medium/material, and is governed by Snells' Law:

$$\frac{\sin \theta_i}{\sin \theta_r} = \frac{V_i}{V_r} \qquad (7)$$

where the subscripts i and r represent the incidence and refracted components respectively, at the angle θ with speed V.

The longitudinal critically refracted (LCR) wave simply means that the incident angles are controlled to ensure that only the LCR waves are dominant along the path of measurement, as shown in Fig. 2. Poly-methyl-methacrylate (PMMA) wedges are introduced in the path of the wave to produce the dominant LCR wave at the near surface region, typically in the vicinity of 1–2 mm beneath the surface, for a standard 5 MHz probe.

The total time taken for the wave to leave the transmitter probe through the PMMA wedge, into the specimen, travelling along to the receiver wedge and back into the receiver probe can be measured directly from the probes. The delays introduced by the wave travelling through the wedges is then subtracted to determine the net travel time within the specimen and is defined as the TOF, whereby the velocity terms can be evaluated based on the fixed probe distances from each other (x). For various stressed conditions, the TOF can be evaluated and the acoustoelastic constant (K) can be determined from the slope of the linearized curve. This constant will be independent of the section shapes and sizes and depends on the material groups such as steel, aluminium, polymers etc. The material and resulting ultrasonic parameters for PMMA and carbon steel specimen are presented in Table 1, using Eqs. (2)–(6). The PMMA wedge is designed to be 28° form the vertical axis to produce the LCR wave, and the resulting shear wave components into steel will progress at 32°.

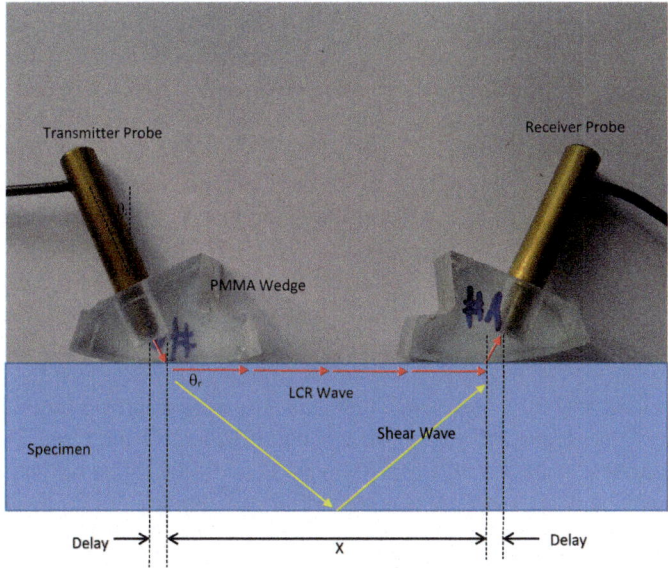

Fig. 2 Ultrasonic measurement concept

Table 1 Material constants

Parameters	Unit	PMMA	Carbon steel
Density	kg/m^3	1100	7950
Elastic modulus	GPa	4	207
Poisson's ratio	–	0.4	0.3
Compression wave speed	m/s	2792	5920
Shear wave speed	m/s	1140	3165
LCR angle (θ_i)	°	28	–
Shear angle (θ_r)	°	–	32

2 Experimental Setup

The laboratory setup is shown in Fig. 3, highlighting the test rig to induce compression stresses on the specimen, the carbon steel test specimen, probes (x = 50 mm), ultrasonic interface box and a laptop computer.

The 200 mm square hollow section specimen with a thickness of 10 mm is selected for this test. Another section size of 150 mm × 6 mm and shapes such as tubular (100 mm × 2 mm and 145 mm × 3 mm) are also used to verify the independence of the sectional shapes and sizes in the acoustoelastic constant evaluation. The loads are applied up to the limiting capacity of the sections, i.e. the buckling or yield limits to ensure consistency and repeatability of the tests. The resulting TOF captured by the in-place measurement system is shown in Fig. 4, and similar plots are obtained for various stressed conditions on the specimen. By using Eq. (1), the acoustoelastic constant can be determined for each stressed condition, and the relationship with the stresses can be derived and are presented in Fig. 5 for all the specimen shapes and sizes being considered. The time delay of approximately 8 μs was incorporated in the TOF evaluations, based on the travel time inside the PMMA wedges at the transmitting and receiving sides of the specimen.

3 Results and Discussion

The finite element (FE) analyses was undertaken to verify and confirm the order of magnitude of the acoustoelastic constant value obtained in the tests. The general purpose nonlinear code ABAQUS [7] was used to model the specimen with S4R shell elements, and the wave propagation within, as outlined in Fig. 6. The 10 mm thick specimen is restrained at its bottom face, and the interactions between the wedges and specimen is modelled by a surface contact. The compressive residual stresses are applied using the initial conditions setting.

The 5 MHz transmitting probe is modelled by the force-time function in Eq. (7), and the resulting perturbation is shown in Fig. 7 [8].

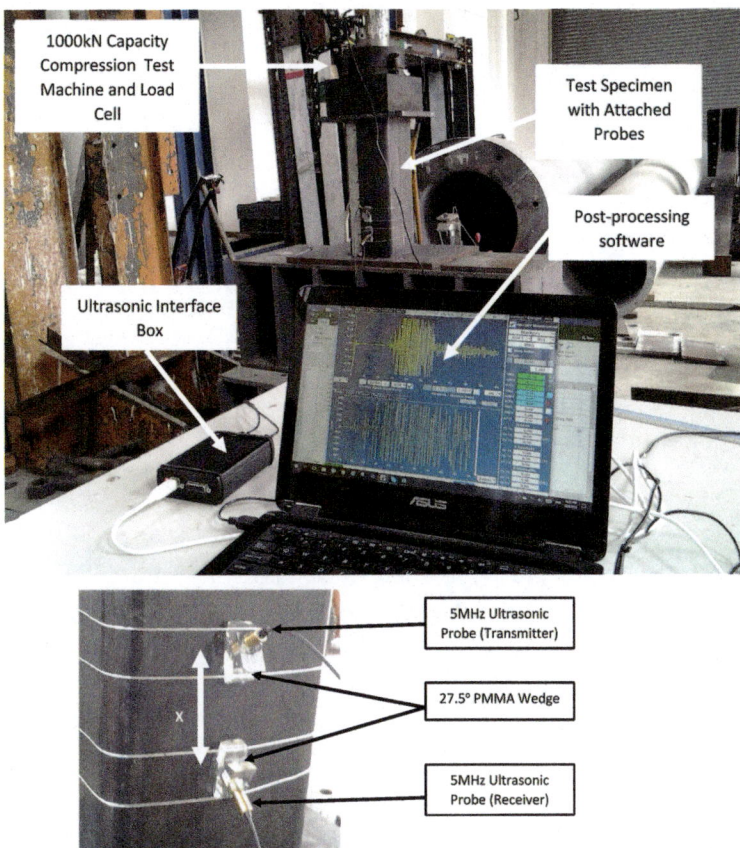

Fig. 3 Laboratory setup

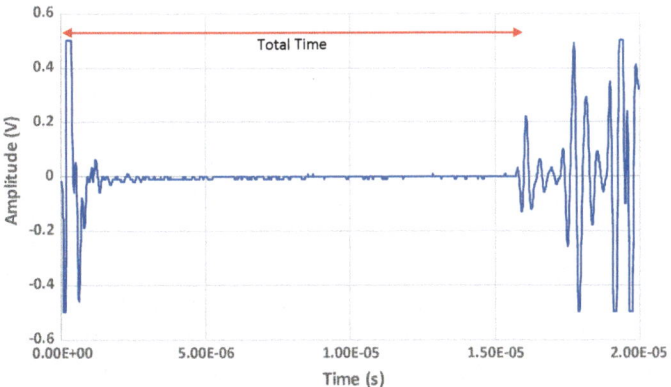

Fig. 4 Recorded ultrasonic wave propagation

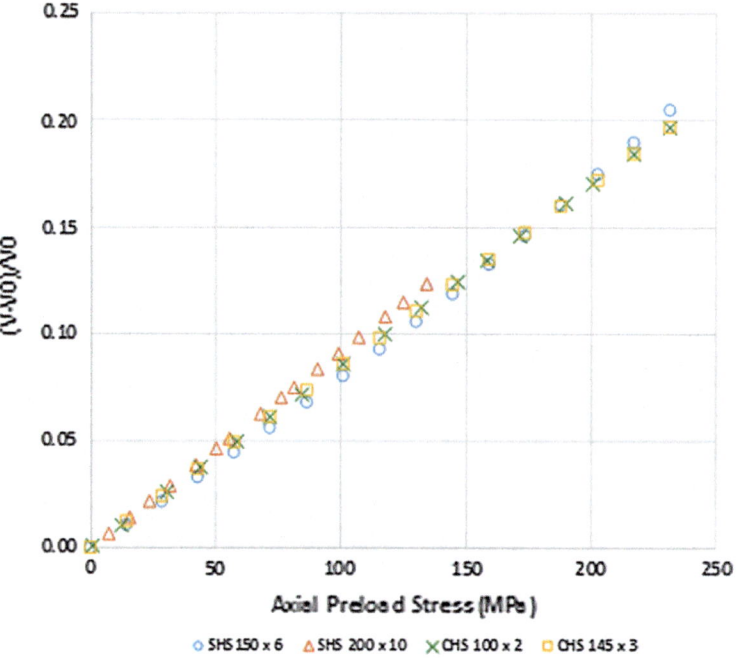

Fig. 5 Acoustoelastic constant calibration curve

Fig. 6 FE model outline

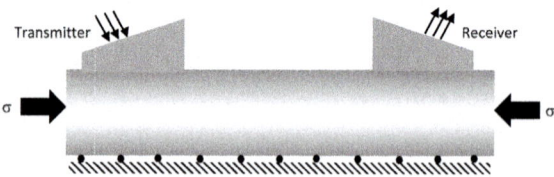

Fig. 7 Ultrasonic perturbation function

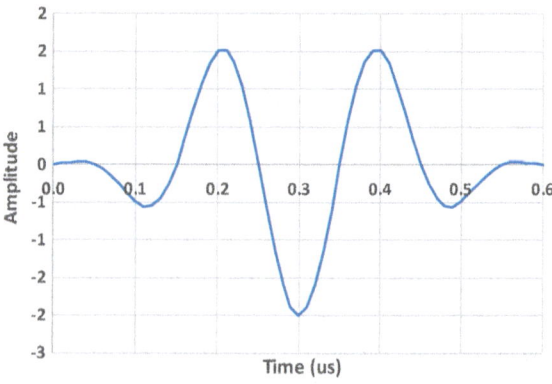

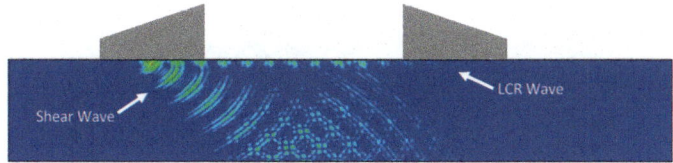

Fig. 8 Final stress wave propagation from FE analysis

$$F(t) = \left[1 - \cos\left(\frac{2\pi ft}{3}\right)\right] \cdot \cos(2\pi ft) \qquad (8)$$

As the model is constructed with linear elastic materials, the simulations at higher residual stresses of up to 1000 MPa is possible, hence allowing for the investigations into variation of the constant at higher stresses, which is limited to section capacity during the laboratory testing phase. The resulting response contour for the stress propagation is shown in Fig. 8, and the propagation stages in Fig. 9a–c, clearly highlighting the LCR wave and the shear wave components.

Detailed post-processing of the FE results also enables the visualisation of the LCR and shear components individually in terms of the displacement of the specimen particles, and are shown in Fig. 10, indicating the parallel particle motion for the LCR component in (a) and the orthogonal particle motion for the shear wave component in (b).

The resulting TOF can be directly extracted from the FE analysis, and presented for the model, as shown in Fig. 11. The probe point for extracting this propagation is set to be at the base of the receiver wedge, thus eliminating the need to subtract the delays inside the PMMA medium.

The resulting wave speed difference for each stressed condition can then be plotted, as shown in Fig. 12. Based on the FE modelling also, the LCR wave speed was computed to be 5986 m/s (diff. = 1.1%), and the shear wave refraction angle

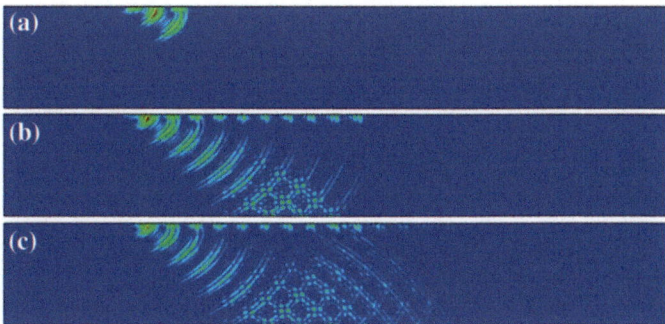

Fig. 9 Wave propagation stages

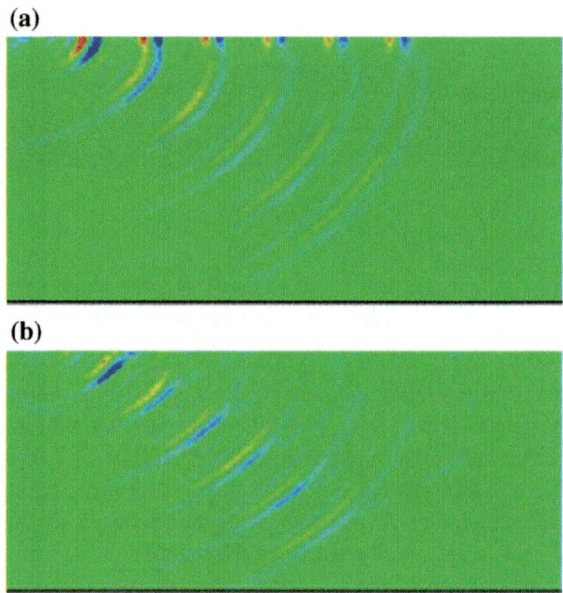

Fig. 10 Individual wave components contour

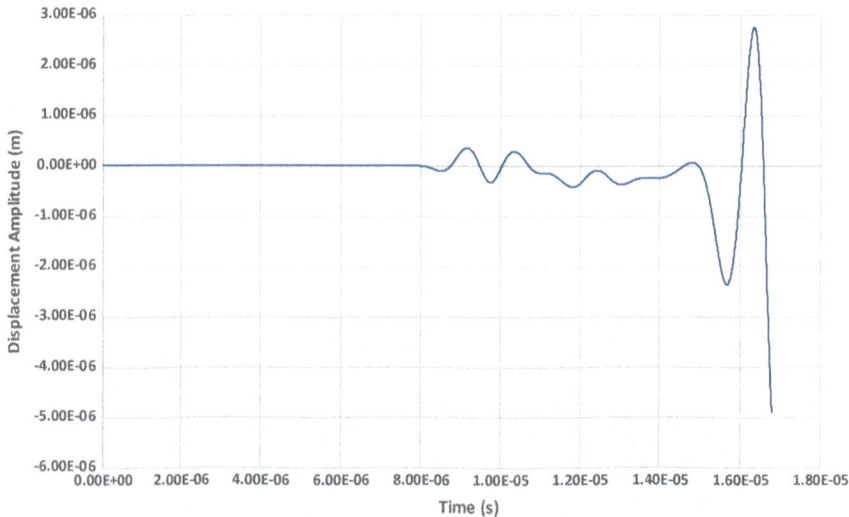

Fig. 11 Extracted wave propagation from FEA

(θ_r) is computed to be 33.7° (diff. = 5%). The depth of penetration of the LCR wave was measured to be approximately 2 mm, and is within the anticipated order of magnitude (~ 2 times the wavelength).

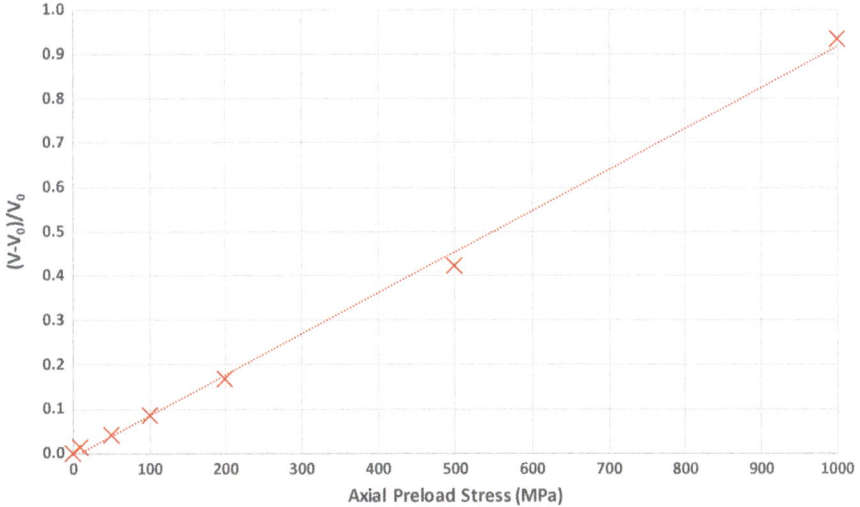

Fig. 12 Acoustoelastic calibration curve from FEA

4 Conclusion

An ultrasonic based non-destructive method to measure the residual stresses of in-service ageing offshore wells structures is being proposed to address the issue of over-conservatisms in predicting the axial loads in conductor pipes. The aqueous corrosion and internal cement shortfalls inside the oil well structural barriers present a severe integrity issue, particularly if the life extension planning is being considered, i.e. to extend the operational life of the well beyond its calculated design life. In order to effectively carry out the integrity assessments of the ageing well conductors, the accurate and safe inspection of the residual axial load is essential to avoid unnecessary and expensive repairs and premature rehabilitations.

The ultrasonic based LCR wave components are generated by carefully designing the PMMA wedges. The LCR wave's sub-surface penetration makes it ideal to measure the axial residual load on the conductor by recording the wave propagation, and extracting the TOF within the specimen length. The acoustoelastic equation is used to evaluate the linear relationship between the stress and change of speed, by monitoring its TOF over a fixed probe spacing. A laboratory setup of the measurement is designed to carry out to experimentally determine the constant to calibrate for the specific carbon steel specimen. Numerical modelling is also carried out to investigate further this linear relationship at higher linear stresses, and the results can be compared, as shown in Fig. 13. The difference of about 7.7% is observed between the tests and numerical based evaluations for the acoustoelastic constant, nevertheless the order of magnitude for carbon steel is within acceptable range.

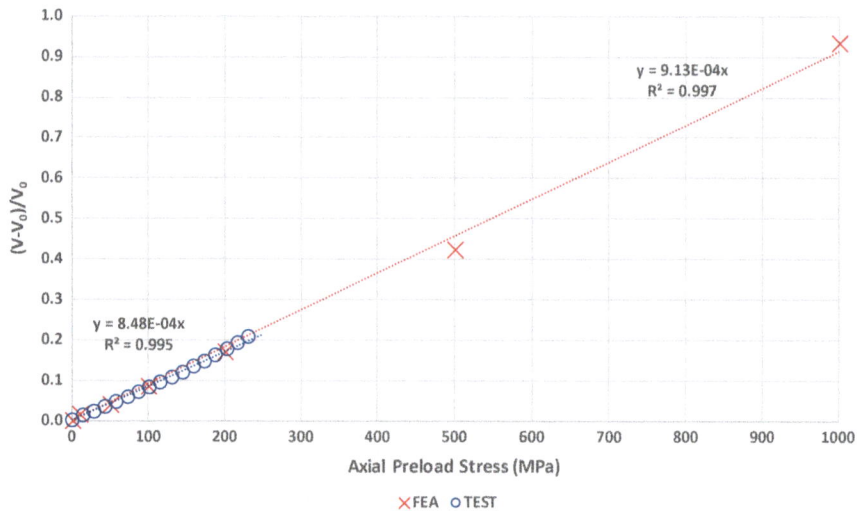

Fig. 13 Test and numerical comparison

The ultrasonic LCR based method to inspect and measure the residual stresses on offshore structures has a great potential to be implemented in a safe and portable manner in the critical aspects of the hydrocarbon industry. The ease of use of this technique/application, along with the level of accuracy it carries, at the reasonable cost of setup makes this technique very convenient and effective to be easily applied on any in-service structural components without having to shut-down the operations at great cost and resources.

Acknowledgements This research was supported by Fundamental Research Grant Scheme, Ministry of Education, Malaysia (FRGS—Project No. FP004/2014AB).

References

1. The Institute of Petroleum, London.: Guidelines for the Analysis of Jackup and Fixed Platform Well Conductor Systems (2001)
2. Ramasamy, R., Aljaberi, M.S., Aljunaibi, H.: Ageing offshore well structural integrity modelling, assessment and rehabilitation. In: Abu Dhabi International Petroleum Exhibition and Conference (2014)
3. Talabani, S., Atlas, B., Al-Khatiri, M.B., Islam, M.R.: An alternate approach to downhole corrosion mitigation. J. Pet. Sci. Eng. **26**, 41–48 (2000)
4. American Society for Testing and Materials.: ASTM E837 Standard Test Method for Determining Residual Stresses by the Hole-Drilling Strain-Gage Method (2001)

5. Prevey, P.S.: X-Ray diffraction residual stress techniques. www.lambdatechs.com
6. Ramasamy, R., Ibrahim, A., Suhatril, M.: Criticality of conductor/casing integrity for ageing offshore well life extension. In: Regional Marine and Mechanical Engineering Conference (ReMME), Malaysia, Oct 2016, Paper-16
7. Abaqus Theory Manual, version 6.11 (2011)
8. Crecraft, D.: Ultrasonic measurement of stresses. Ultrasonics **6**(2), 117–121 (1968)

Influence of Tool Plunge Depth on the Joint Strength and Hardness of Friction Stir Welded AA6061 and Mild Steel

Wan Mohd Syafiq Wan Sulong, Mohd Afendi Rojan and Mohd Noor Mazlee

Abstract The purpose of this paper is to investigate the impact of the tool plunge depth on the mechanical properties of friction stir welding (FSW) of AA6061 and mild steel. In FSW, welding is performed using a non-consumable rotating tool that is allowed to plunge in between two plates in a butt joint and travel along the abutting line. The tool plunge depth determines how deep the pin and shoulder of the tool penetrates the surface of the plates. Tensile testing was performed to determine the influence of plunge depth at different tool travel speeds on joint failure strength. Also, the formation of defects and its relationship with plunge depth were also investigated. Vickers microhardness testing was also done on weld cross sections to study plunge depth's effect on hardness.

Keywords Friction stir welding · Tool plunge depth · Micro hardness

1 Introduction

Friction stir welding (FSW) is a solid-state welding process that is mainly used in joining materials with low melting temperature such as aluminum and copper alloys [1]. Compared to other welding techniques commonly available, it consumes less energy, no filler metals are used during welding and no hazardous conditions such as dangerous fumes and extreme brightness emitted from the welding arc [2–4].

W. M. Syafiq (✉) · M. Afendi
School of Mechatronic Engineering, Pauh Putra Campus, Universiti Malaysia Perlis (UniMAP), 02600 Arau, Perlis, Malaysia
e-mail: wmsyafiq10@gmail.com

M. Afendi
e-mail: afendirojan@unimap.edu.my

M. N. Mazlee
School of Materials Engineering, Universiti Malaysia Perlis (UniMAP), Centre of Excellence for Frontier Materials Research, 01000 Kangar, Perlis, Malaysia
e-mail: mazlee@unimap.edu.my

The process was firstly introduced by The Welding Institute (TWI), United Kingdom in 1991, with initial works mainly on joining aluminum alloy. Currently, the friction stir welding process has now seen developments in joining different types of materials such as copper, magnesium, steel and dissimilar alloys and materials.

One such example of a FSW joint is the welding of dissimilar materials of aluminum alloy and steel. Trying to create aluminum alloy and steel joints using conventional or even advanced fusion welding techniques is a really difficult task. The complicatedness arose from the tendency to form a brittle intermetallic compound (IMC) layer at the interface [5]. These brittle layers severely deteriorate joint strength as it acts as stress concentration points while also reduces the welded area between the two interfaces. Since FSW is a solid-state process where the process temperature does not reach the melting point, IMC layers grow at a lower rate compared to other welding techniques that involve elevated process temperatures which promotes its growth. It is desirable to investigate the welding parameters while carrying out FSW and its effect on mechanical properties of the welded joint. While a lot of studies have been done on parameters such as tool rotational speed and travel speed, investigation on plunge depth has been lacking especially in aluminium-steel FSW.

In this paper, FSW of dissimilar material between the aluminum alloy 6061 and mild steel plates in butt joint configuration is studied. AA6061 is a precipitation hardened Al–Mg–Si alloy with excellent weldability and corrosion resistance, while mild steel is a commonly used material in various manufacturing industries due to its cheap cost and strength. The relationship between tool plunge depth and mechanical properties of the friction stir welded joint is discussed.

2 Experimental Setup

Friction stir welding was the chosen technique to fabricate the joint between the aluminum alloy 6061 and S235JR mild steel plate of 5 mm thickness. The welding was executed by a rotating tool made out of H13 tool steel (Fig. 1). The tool went

Fig. 1 Shoulder and pin of the FSW tool

Table 1 Welding parameters and their respective values

Parameters	Value
Plunge depth, pd (mm)	0.1, 0.5
Rotational speed, ω (rpm)	1110
Travel speed, V (mm/min)	30, 40
Tool offset, TO (mm)	0.2

through a heat treatment process to increase its hardness to 52 HRC. The shoulder diameter of the tool was 20 mm, while the pin diameter was 5 mm. FSW was performed on a conventional milling machine.

Welding parameters such as tool plunge depth, tool rotational speed, tool travel speed and tool offset used to perform FSW throughout this study were tabulated in Table 1. Tool plunge depth is defined as the depth of penetration by the tool shoulder into the weld piece. For better visualization of the plunge depth parameter, Fig. 3 demonstrates how it is defined. The plunge depth parameter was controlled by the ascending of the milling machine table. Figure 2 shows the conventional

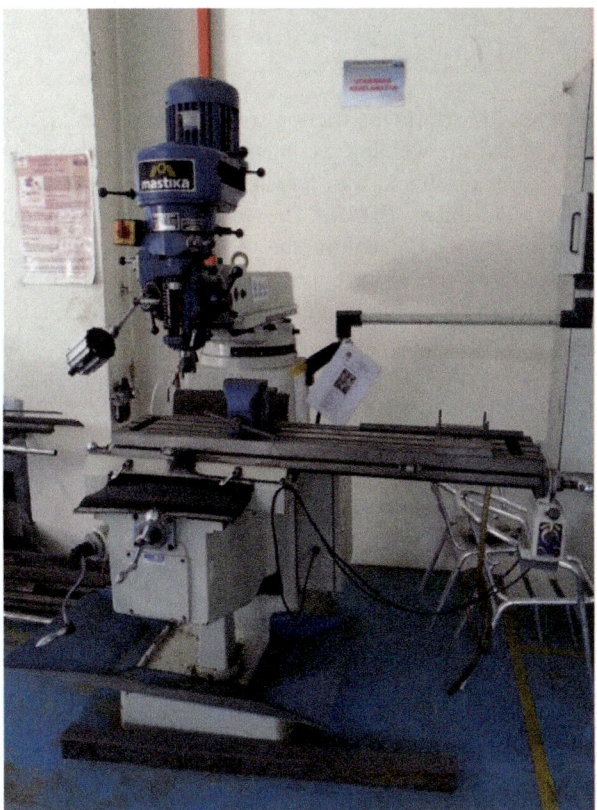

Fig. 2 Milling machine used for FSW

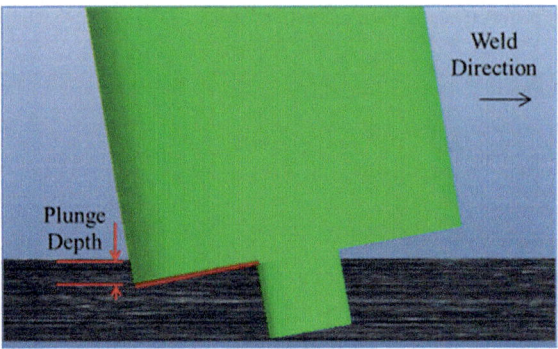

Fig. 3 Definition of tool plunge depth

Table 2 Samples and parameters list

Sample	Rotational speed (rpm)	Travel speed (mm/min)	Plunge depth (mm)
S11	1110	40	0.5
S12			0.1
S21		30	0.5
S22			0.1

milling machine used to perform the friction stir welding process. Table 2 illustrates the specimen labels and the welding parameters used for each specimen.

Welded plates were machined into tensile testing specimens using a wire cut electrical discharge machine (EDM). The tensile testing specimens were cut according to the dimensions stated in the ASTM E8/E8 M-13a for sheet-type standard specimen. Specimens were tested using a universal testing machine (UTM) with crosshead speed of 1 mm/min. Side profiles of the tensile specimens before and after tensile testing were taken via a digital microscope. Vickers microhardness values were also measured at mid thickness of the joint across the joint interface. The indenter, which was a square-based diamond pyramid was applied at a load of 200 N for 15 s.

3 Results and Discussion

3.1 Effect of Tool Plunge Depth on Macrostructure of FSW Joint

3.1.1 Bondline Formation

Figure 4 shows the cross section of the FSW joints, namely S11, S12, S21 and S22. It can be seen that for samples welded with tool plunge depth of 0.5 mm

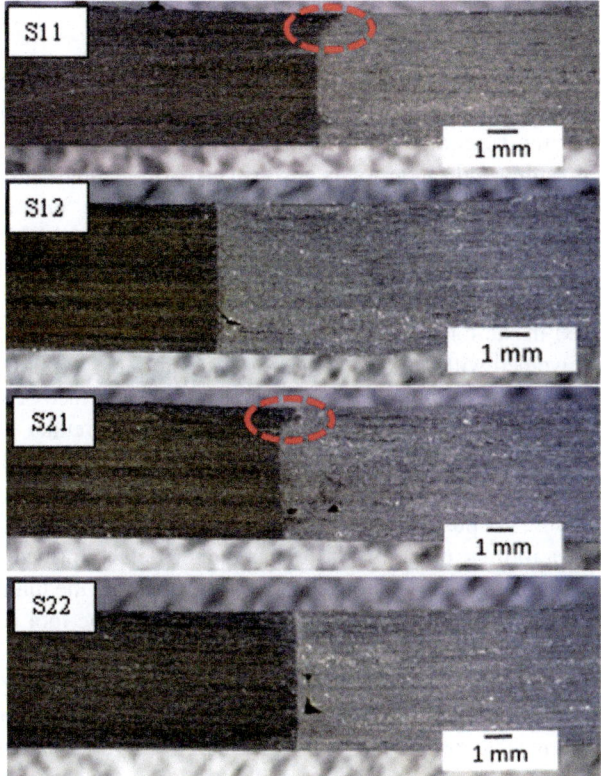

Fig. 4 Cross section of the welded joints

(S11 and S21), there were excessive interaction between the weldpiece's top surface and tool shoulder. This can be seen by the deformed steel at the interface close to the top surface, where the mild steel was seen to deform towards the aluminum matrix. This was caused by the downward pressure exerted by the tool shoulder surface onto the weldpiece.

For samples welded with tool plunge depth of 0.1 mm (S12 and S22), no such deformation was seen on the cross section at the steel interface close to the top surface. The interface was seen as a straight line as opposed to the interface seen in S11 and S12 where the bond line got slightly distorted as it approached the top. Therefore, in terms of macrostructure of the welded joint, from cross sectional diagram it can be said that increasing the tool plunge depth parameter deforms the joint interface close to the tool shoulder due to the increase of tool shoulder/workpiece interaction.

3.1.2 Tunnel Defect Formation

Figure 5 shows the absence and presence of tunnel defects in S11, S12, S21 and S22 respectively. In this study, all welded specimens except S11 exhibited tunnel defects with varying sizes and positions. The tunnels present in the welded joints are highlighted with a red circle for better visualization. Tunnel defect formed in sample S12 was the smallest out of all the tunnel defects seen. Sample 22 also was affected by tunnel defect, which was the largest out of the samples. The tunnel defect was seen close to the interface, and shaped such that it was elongated in the thickness direction. On the other hand, sample S21 contained two small tunnel defects separated in the direction perpendicular to the thickness.

For samples welded with tool plunge depth of 0.1 mm (S12, S22), tunnels were formed approximately at mid thickness, close to the faying surface of the joint. This can be attributed to lower downwards pressure by the tool shoulder which affected material flow [6]. During consolidation, plasticized material was pushed down by the tool shoulder to fill in the void left by the advancing tool pin. Less plunge depth meant less pressure to push the material, thus increasing the likelihood of any tunnel defects to form. Compare this to the tunnel defect formed for the joint welded with a tool plunge depth of 0.5 mm (S21). Notice that the tunnel defect was smaller in size, and was formed further down, away from the tool shoulder. This can be attributed to the improved downwards pressure thus enhanced ability to push plasticized materials into voids. The result is also in agreement with Zhang's model for tunnel defect formation, where an increase in applied pressure, rotational speed and shoulder diameter improves material consolidation to eliminate voids [7]. As sample S11 and S21 demonstrated, it is possible to reduce and even eliminate the tunnel defect formation by increasing the tool plunge depth. Hence, increasing the

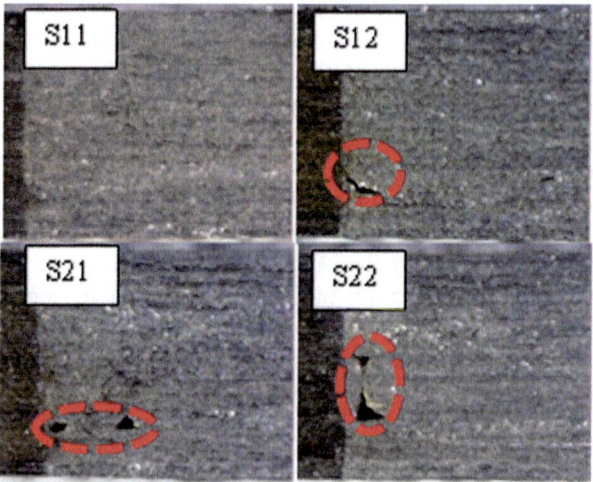

Fig. 5 Absence and presence of tunnel defects in welded joints

tool plunge depth improved downwards pressure onto plasticized material, consequently reducing the tunnel defect formation in FSW joints.

Figure 6 shows the side profile view of tensile specimens following tensile testing. It can be seen that the tunnel defect in S12 and S21 were not big enough to influence the failure location. The failure location in these samples took place at the joint interface, away from the tunnel defects. A similar phenomenon was seen by a previous researcher where failure did not happen at tunnel position, but instead occurring at the interface [8]. However, the failure in sample S22 was seen to happen at the tunnel defect position. The tunnel's big size and elongated shape affected its tensile strength thus made it the weakest point in the welded joint. For this reason, tensile failure occurred at the tunnel defect site.

Fig. 6 Side profile view of welded joints after tensile testing

3.2 Effect of Tool Plunge Depth on Tensile Strength of FSW Joint

Figure 7 shows the ultimate tensile strength of samples recorded during tensile testing. The samples are categorized according to tool plunge depth of 0.1 and 0.5 mm. Sample S11 is labelled in bold because it was the only sample that did not show evidence of tunnel defects. The strongest sample was S12, followed by S22, S11 and S21.

From the results, it can be seen that increasing the tool plunge depth and reducing travel speed had negative effects on the joint strength. This may be caused by the higher contribution towards heat generation by these two parameters, which in turn promoted more growth of brittle IMC layers that compromised the joint strength. The heat input equation proposed by Frigaard to describe the generated heat, Q during the friction stir welding process is as follows [9]:

$$Q = \frac{4}{3}\pi^2 \mu P \omega R_s^3 \tag{1}$$

where μ is the friction coefficient between the tool and workpiece surface, P is the axial load, ω is the tool rotational speed and R_s is the radius of the tool shoulder. To investigate the influence of the tool travel speed, the heat input per unit length of the weld or Q/v can be evaluated where v is th etool travel speed. Consequently, Eq. (1) can be described as follows:

$$\frac{Q}{v} = \frac{4}{3}\pi^2 \mu P \frac{\omega}{v} R_s^3 \tag{2}$$

The axial force was not measured directly in this study, however it was controlled by the tool plunge depth parameter given that increasing plunge depth causes an increase in axial load as shown from a previous study [10].

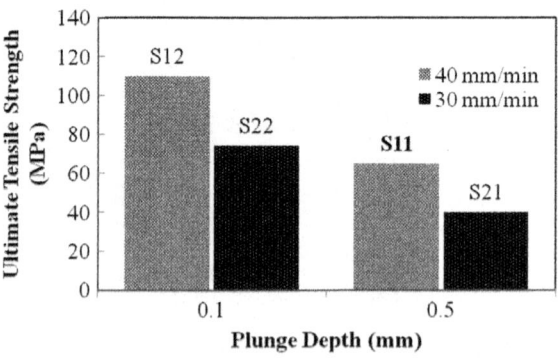

Fig. 7 Ultimate tensile strength of tensile specimens

It is shown from Eq. (2) that reducing the tool travel speed and increasing the tool plunge depth produces more heat generation in FSW, which promoted brittle IMC layer growth at the interface.

Figure 5 shows that sample S11 was the only sample without tunnel defect due to the extra downwards pressure by the 0.5 mm plunge depth. However, based on UTS results this joint was weaker than the joints produced using 0.1 mm plunge depth. Since S11 experienced more heat input, a thicker IMC layers can grow thus deteriorating the joint strength [11].

Samples welded with tool plunge 0.1 mm (S12, S22) had better joint strengths even though tunnel defects were present. The tunnel defect in S22 was formed such that it was elongated in the thickness direction, which reduced the effective bond area of the joint at the interface. Tunnel defect seen in S12 was smaller than the one seen in S22. As can be seen in Fig. 6, S12's tunnel did not influence failure as it happened away from it, whereas failure of S22 took place at tunnel defect. It is seen that the tunnel defect size in S12 was small enough such that it did not serve as the weakest point for the sample during tensile testing. Therefore from the tensile testing results, it can be said that for FSW of aluminum alloy and mild steel, increasing the tool plunge depth may get rid of tunnel defects, but the resulting increase in heat generation worsens the joint strength due to IMC layer formation. Thus, there exists a give and take situation caused by the tool plunge depth parameter.

3.3 Effect of Tool Plunge Depth on Hardness of FSW Joint

The hardness profiles of sample S11 and S12 are demonstrated in Fig. 8 to show the influence of the tool plunge depth on hardness profiles of FSW joints. As can be seen in the figure, no considerable difference was seen between the two hardness profiles. Similar patterns and values were observed in the two profiles, with both S11 and S12 recording maximum HV at the interface due to work hardening of mild steel by the tool shoulder [12].

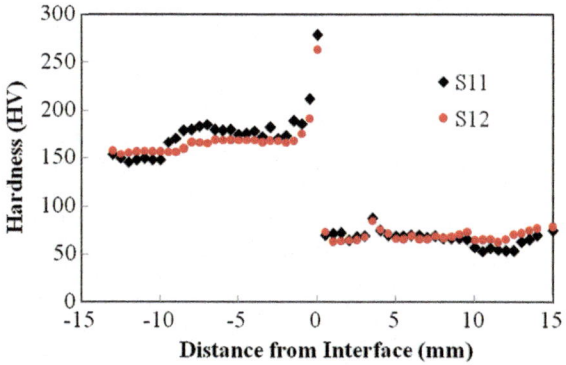

Fig. 8 Hardness profile

Slight spreading of HV was seen in the nugget zone due to scattering of steel particles into the aluminum matrix and dispersion of intermetallics in the nugget zone. The minimum HV for both profiles were seen at the heat affected zone (HAZ). This can be attributed to the dissolution and coarsening of hardening second phase particles in 6xxx aluminum alloys, as well as larger grain size typically seen in HAZ of FSW of 6xxx aluminum alloys [13, 14].

4 Conclusion

In this paper, the effect of tool plunge depth on the joint strength and hardness of dissimilar FSW joint of AA6061 and mild steel was investigated. From the results, the following conclusions were made:

1. Increasing the plunge depth increases the shoulder/workpiece interaction as can be seen by the deformation of the steel at the top surface. This in turn increases the axial force exerted by the shoulder downwards unto the workpiece, increasing heat generation thus promoting IMC layer growth.
2. A low plunge depth means less axial force to push the material into voids thus creating tunnel defect. Increasing the pd increases the force to push the material thus solving this problem.
3. From tensile testing, increasing plunge depth and reducing travel speed have a deteriorating effect on the joint strength. The strongest joint was created at low plunge depth and high travel speed.
4. In terms of hardness, not much difference can be seen caused by the variation of tool plunge depth. Both recorded hardness profiles demonstrated similar patterns and minima/maxima values.
5. A high plunge depth eliminates the tunnel defect problem but produces more frictional heat due to increased shoulder/workpiece interaction, whereas a low plunge depth reduces heat generation but causes tunnel defects to form. Therefore, it can be seen that in FSW of aluminum alloy to steel, the plunge depth is an important parameter.

References

1. McPherson, N.A., Galloway, A.M., Cater, S.R., Osman, M.M.: A comparison between single sided and double sided friction stir welded 8 mm thick DH36 steel plate. Trends In Welding Research 2012: Proceedings of the 9th International Conference (2012)
2. Barnes, S.J., Steuwer, A., Mahawish, S., Johnson, R., Withers, P.J.: Residual strains and microstructure development in single and sequential double sided friction stir welds in RQT-701 steel. Mater. Sci. Eng. A **492**(1–2), 35–44 (2008)
3. Kumar, N., Mishra, R.S.: Friction stir welding and processing. (2014)

4. Othman, N.H., Shah, L.H., Ishak, M.: Mechanical and microstructural characterization of single and double pass Aluminum AA6061 friction stir weld joints. IOP Conf. Ser. Mater. Sci. Eng. **100**, 12016 (2015)
5. Kimapong, K., Watanabe, T.: Friction stir welding of aluminum alloy to steel. Weld. J. 277–282 (2004)
6. Kim, Y.G., Fujii, H., Tsumura, T., Komazaki, T., Nakata, K.: Three defect types in friction stir welding of aluminum die casting alloy. Mater. Sci. Eng. A **415**(1–2), 250–254 (2006)
7. Zhang, H., Lin, S.B., Wu, L., Feng, J.C., Ma, S.L.: Defects formation procedure and mathematic model for defect free friction stir welding of magnesium alloy. Mater. Des. **27**(9), 805–809 (2006)
8. Dehghani, M., Amadeh, A., Mousavi, S. A.: Investigations on the effects of friction stir welding parameters on intermetallic and defect formation in joining aluminum alloy to mild steel. Mater. Des. **49**, 433–441 (Aug 2013)
9. Frigaard, Ø., Grong, Ø., Midling, O. T.: A process model for friction stir welding of age hardening aluminum alloys. **32**, (May 2001)
10. Kumar, K., Kailas, S.V.: On the role of axial load and the effect of interface position on the tensile strength of a friction stir welded aluminium alloy. Mater. Des. **29**(4), 791–797 (2008)
11. Tanaka, T.H.T., Morishige, T.: Comprehensive analysis of joint strength for dissimilar friction stir welds of mild steel to aluminium alloys. Scr. Mater. **61**(7), 756–759 (2009)
12. Kakiuchi, T., Uematsu, Y., Suzuki, K.: Evaluation of fatigue crack propagation in dissimilar Al/steel friction stir welds. Procedia Struct. Integr. **2**, 1007–1014 (2016)
13. Uzun, H., Dalle Donne, C., Argagnotto, A., Ghidini, T., Gambaro, C.: Friction stir welding of dissimilar Al 6013-T4 To X5CrNi18-10 stainless steel. Mater. Des. **26**(1), 41–46 (2005)
14. Krasnowski, K., Hamilton, C., Dymek, S.: Influence of the tool shape and weld configuration on microstructure and mechanical properties of the Al 6082 alloy FSW joints. Arch. Civ. Mech. Eng. **15**(1), 133–141 (2015)

Analysis of Production Layout Model to Improve Production Efficiency

Ahmad Razlee AB Kadir, Baizura Zubir, C. A. Mohd Norzaimi, M. Sabri, M. Zaki, W. Faradiana and Noor Helinahani

Abstract In order to sustain the manufacturing industry under the current global situation of fierce competition, a company needs to reduce or eliminate the idle and or down time of operation in addition to improvement of the current production efficiency and working method. In this case study, the main problem and challenge of oil and gas companies is due to non-optimal operations with inefficiency in the distribution of the work loading. The whole production layout suffers due to the absence of established standard time for activities carried out by workers, unbalanced load carried, non-value added (NVA) activities involved and inefficient methods. The primary goal of this present paper is to focus on the improvement of the production efficiency of current a "Front of Line" production layout by identifying and minimizing unnecessary activities at the bottleneck and to rearrange the allocation of man power in the production line to increment the line balancing. This study is conducted through the Time Study Technique to obtain an estimate for the time allowed for a qualified and well-trained worker in a normal situation to complete a specific task. Result shows that through improving the working method and by rearranging the allocation of manpower, it is possible to well balancing the process flow as well as ensuring economic benefits.

A. R. AB Kadir (✉) · B. Zubir · M. Sabri
Manufacturing Department, Universiti Kuala Lumpur Malaysian
Spanish Institute, Kulim, Kedah, Malaysia
e-mail: ahmadrazlee@unikl.edu.my

B. Zubir
e-mail: baizura@unikl.edu.my

C. A. Mohd Norzaimi · M. Zaki · W. Faradiana · N. Helinahani
Mechanical Department, Universiti Kuala Lumpur Malaysian
Spanish Institute Kulim, Kedah, Malaysia
e-mail: mnorzaimi@unikl.edu.my

W. Faradiana
e-mail: wfaradiana@unikl.edu.my

N. Helinahani
e-mail: noorhelinahani@unikl.edu.my

© Springer International Publishing AG 2018
A. Öchsner (ed.), *Engineering Applications for New Materials and Technologies*, Advanced Structured Materials 85, https://doi.org/10.1007/978-3-319-72697-7_31

Keywords Production efficiency · Bottleneck · Process flow · Time study technique

1 Introduction

In modern manufacturing systems, layout design is the main goal that will maximize the productivity of manufacturing processes. This depends on several factors such as the type of product made, the quality of raw material, the technique of manufacturing process and the arrangement of workstation that contributes to the production line. The challenge in determining the best solution arrangement of the workstation is one of the elements that will make a significant change on the manufacturing system performance. The production layout, also knows as facility layout, includes the planning location, process flow, floor layout and material handling systems. The objective of each manufacturing industry is to ensure a smooth and efficient production process from the first stage of the material until the end of the process based on the production layout design that will reduce the waste activities in the production line and to improve the overall effectiveness in the production.

2 Literature

In manufacturing industry, there are two main problems that need to be considering for the production layout [1]. The first one is the quantitative approach aiming at minimizing the total material handling cost between workstation based on a distance function. Secondly is the approach aiming at minimizing the waste activities or non-value added activities in the production line. Material handling and process flow have a close relationship in the production line [2, 3]. Worker and material have to travel along a distance from the first stage of raw material until the final stage of the finish product in the manufacturing process. A material handling system with low level of automated handling will effect the production efficiency regarding to manual handling among of workers incapable to be controlled compared to automated handling such as conveyors, automated guided vehicles, pallet and others.

This project focuses on the production layout which is the same as the current layout of the cell because of the space available, the machines, the equipment and the process flow, which is suitable for this type of layout. Generally, in production layout the machines and equipment are arranged in a single line, which depends on the sequence of process activity required for the product in the manufacturing system. The material in the production layout typically flows from one workstation directly to the next workstation. The product in the production layout usually occurs in high-volume of production and low variety of product. It means that a manufacturing industry that has a production layout only focusing on minimum variety of products but high volume of production depends on the high demand from the user. Refurbishment services for a liquefied petroleum gas (LPG) cylinder company has been selected to perform the verification and validation of the production layout performance in terms of improved production efficiency. This company provides refurbishment services for liquefied petroleum gas (LPG) cylinder in accordance to the specification and requirement of the customer. Continuously monitoring and improvement of the quality system was applied in the working culture of the organization.

3 Experimental Setup

a. Experiment Preparation.

The methodology was divided into two phases. Phase one includes the research as a literature review about previous studies mentioned in the literature. The judgment condition has been done to analyze the current layout practice at case studies by the applied time study technique to analysis the production output and study the effectiveness of current layout practice. Then the current layout will be modeled for future analysis. Phase 2 is based on the output to design many possible models of production layout as the research main target by applied ALDEP. Once again, the time study technique will be applied again to analyze the effectiveness of the proposed layout with line balancing before running the simulation. Then the simulation will be conducted for each production layout model, and the result will be recorded and analyzed. The verification and validation of the proposed production layout will be done in this phase after completing the simulation as shown in Fig. 1.

b. Experiment Commencement

3.1 Production Layout and Judgement Condition

The production layout was focused on the primary goal which is improvement of the production efficiency in terms of increased production output, elimination of any waste that was added in the production flow that could cause the bottleneck problem. Hence, the identification on the actual layout practice was done for about one and half month at the production plant. Figure 2 demonstrates the summary of the layout problem in FOL.

The critical workflow in phase 1 is the comparison of the actual production output by recording and analyzing of collected data from the previous workflow. Mathematical calculations of the time study in the manufacturing facilities design

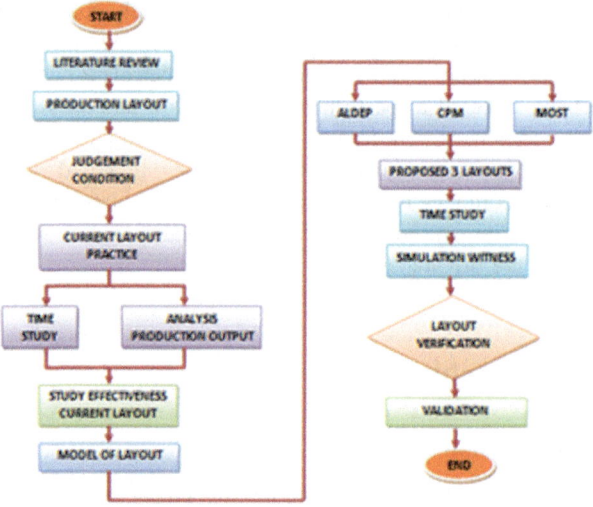

Fig. 1 The methodology flowchart

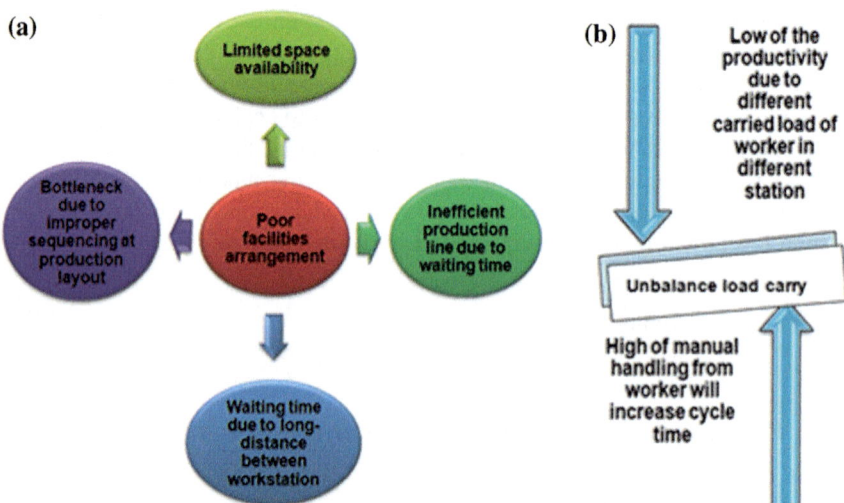

Fig. 2 a Summarize on layout problem. **b** Time study

were done. Before understanding the important use of time study, our research must understand what the term of time standard means [4]. A time standard is defined as "the time required to produce a product at a workstation with the following three conditions: a qualified, well-trained operator; working at a normal pace; and doing a specific task." Table 1 clearly explains the meaning of these three conditions in the time standard that was applied in this methodology.

Table 1 Time standard condition

Condition of time standard	Definition	Example
A qualified, well-trained operator	Experience when usually what makes a qualified, well-trained, and time on the job was the best indication of experience. The time needed to become qualified varies with the job and teh person	In STIC production flow, the less of manual equipment required long learning period. The mistake mode by new time study personnel is time studying someone due to different qualified, well-trained operators in handling their job
Normal pace	The pace at which a trained operator, under normal conditions, performs a task with a normal level of effort. A normal level of effort is one in which the operator can maintain a comfortable pace; not too fast and not too slow. There was only one time standard can be used for each job, even through individual differences in operators cause different results	In STIC production flow, the process involved in the degassing station; 1. Put five gas tester to LPG in 1.25 s 2. Collect 2 LPG in 3.47 s. 3. Movement 2 LPG in 1.73 s In this situation, normal time was done by senior operator, the normal time will cause different results when new operator done same with no experience, it will take time to trained new operator to comfortable that task
Specific task	Detailed description of what must be accomplished. The description of the task must include: 1. The prescribed work method. 2. The material specification. 3. The tools and equipment being used. 4. The position of incoming and outgoing material. 5. Additional requirements such as safety, quality, housekeeping, and maintenance tasks	In STIC production flow, the prescribed work from starting FOL to ship the conveyer station was limited of the tool and equipment due to worker need to move from one station to next station by walking with carried 2 LPG by manually and the distance between unloading station to next station was so far. In this case, the additional requirement such as applied of 5 s housekeeping after done process in each station

3.1.1 Technique of Time Study

This section will present an overview on the time study techniques that were applied in this research activity. Firstly, the further study was done by setting the time standard or by applying any of these techniques. Time study (setting the time standard) covers a wide variety of situations. At one time, a job must be designed, workstation and machine, and a time standard set before the plant is built [5, 6]. In this situation, a predetermined time standards system (PTTS), method time measurement (MTM), or standard data would be the technique used to set the time standard. For example, once the worker or machine has been operating, the stopwatch technique was used to record the time taken of the worker to complete one cycle of operating procedure.

Numerous techniques were used due to the fact that some jobs occurred once or twice a week, whereas others were repeated thousands of times per days, some jobs were very fast, whereas others took hours [7, 8]. A stopwatch time study is the method that most manufacturing employees think of when talking about time standard. The stopwatch technique was introduced around 1880 for studying work [9]. Due to this long history, this technique was part in this research activity to collect the cycle time each worker requires to complete the assigned task in the workstation. The data was recorded by using the stopwatch for 5 times of trial from each process in the workstation. The average of the cycle time was used for the next calculation by adding some mathematical formula.

3.1.2 Time Study Procedure

Step 1: Select the job to study.

In STIC production layout, the job to study refers to the all job that needs to be completed in the FOL production layout. The jobs that need to be studied are degassing, decanting, devalving, shot blasting, QC inspection, hydrostatic, stamping, and welding. The job of unloading LPG and shipment to the conveyer is considered as an offline station.

Step 2: Collect information about the job.

The information regarding the job selection at first step are the working hour per day, working shift, working procedure of each job, cycle time of each procedure, number of operators, daily output target. All this information will relate to each other and will be used for the mathematical calculation in the next step of the time study.

- Working hour = 8 h per day (45 min break)
- Working hour available = 7.25 h per shift
- Working shift = 2 shifts per day
- Available operators = 11 person for 8 workstations
- Working procedure and cycle time
- Daily production target = 1000 units

Step 3: Divide the job into elements.

An element in this step refers to the process or movement of LPG, which is divided into different jobs at different workstations from the collection of information in step 2.

Step 4: Do the actual time study.

This step refers to the essence of the stopwatch time study. The cycle time as shown in Table 2 was record 5 times of the trial for every element stated in step 3 for every standard operating procedure done by the worker in every station. For all the cycle

Table 2 The cycle time at workstation

Workstation	Cycle time (s)	Worker (person)
Degassing	7.42	1
Devalving	23.09	2
Shot blasting	58.63	2
QC inspection	72.3	3
Hydrostatic	69.69	1
Stamping	18.86	2
Total	249.99	11

times recorded in Table 2, their average of the cycle time was calculated and to the average cycle time a fatigue allowance was added to make the time following the standard practice. All calculations in this step were done by using the following formulae:

$$\text{Average Cycle Time per process} = \frac{\text{Sum of Time}}{\text{Number of Trial}} \quad (1)$$

$$\text{Average Cycle Time per station} = \text{Sum of Average Cycle Time per process} \quad (2)$$

$$\text{Cycle Time} = \frac{\text{Average Cycle Time per Process}}{\text{Number of Process}} \times \text{Fatigue Allowance} \, (1.1) \quad (3)$$

The average cycle time of each station was used for the next mathematical calculation in next step of the time study.

Step 5: Selection of workstation to extend the time study

From the data collection of the cycle time in the FOL production layout, there were only 6 workstations selected of a daily production of FOL. The summarize data for the extended time study is shown in Table 2.

Step 6: Analyze Data

The data was analyzed to obtain the time necessary for carrying out the job performance at a defined level of performance in terms of production output and number of workers. In this step of calculation, the longest cycle time was selected for further calculation based on the following formulae:

$$\text{Unit per hour (UPH)} = \frac{3600}{\text{Cycle Time}} \quad (4)$$

$$\text{Daily Output} = \text{Working hour per day} \times \text{UPH Bottleneck} \quad (5)$$

$$\text{Takt Time} = \left(\frac{\text{Working hour per day} \times 3600}{\text{Daily Target}}\right) \quad (6)$$

$$\text{Daily Capacity} = (\text{UPH Bottleneck} \times \text{Working hour per day}) \\ \times \text{Machine Standard Operating}(0.85) \tag{7}$$

$$\text{Number of Operator}: \left(\frac{3600 \times \text{Working Hour per day}}{\text{Total Time}}\right) \times \text{manpower} \\ = \text{daily target} \tag{8}$$

The time study method was used to obtain the time necessary for carrying out the job at different levels of performance to improve the production output and to identify the first objective by applying a time study. The further method will need to complete the time study approach in this research. All the results of time study analysis will be further discussed in the result and discussion section.

3.2 Automated Layout Design Planning (ALDEP)

ALDEP is used as a tool to estimate the space area utilization of the current layout practice. The comparison between the current space utilization and the ALDEP analysis will be compared by using the following basic steps of the ALDEP analysis:

Step 1: Construct a table of the workstation and the number of unit areas according to the current area of FOL production layout.

Step 2: Construct the activity relationship chart by following the weighted value for the relationship activity by referring to ALDEP. If it has an "A" relationship, a value of 64 is added to the rating of the layout. An "E" relationship adds 16, "I" adds 4, and an "O" relationship adds 1 to the rating of the layout. A "U" relationship has no effect on the rating of the layout.

Step 3: Randomly select the first adjacent workstation to enter in the layout as shown in Table 3.

Step 4: Placement of layout by using ALDEP sweep width pattern.

Step 5: Draw the final layout using a block diagram.

Table 3 Workstation area layout

Activity	Area (ft^2)	Number of unit area
Degassing	840	3
Decanting	416	2
Devalving	728	3
Shot blasting	759	3
QC inspection	437	2
Hydrostatic	675	3
Stamping	325	1
Welding	450	1

4 Data Collection and Analysis

The production line was operating in 2 shifts per working day to achieve customer satisfactory based on 1000 units per day. The data collection in this research was done during 6 months. Figure 3 shows the graph of the production efficiency for 6 months of units production of customer demands, company target and actual output. Production efficiency was calculated based on the company target and the actual output. The production efficiency needs to increase by 13.57% to achieve the company target. From the data collection, the current productivity is 900 units per day only. Therefore, the customer demand was not fulfilled by the company. To achieve the productivity needed, the line balancing should be balanced for the entire workstation and the suitability of working method and distribution of work loading are the main issues to be solved.

4.1 Time Study

To assess the improvement in the production rate, the take time of the production line was determined by dividing the total available time with the customer demand. The total available time for the production is obtained by the fixed loss from the total working time per day. The fixed loss consists of meal break, allowed downtime, briefing time and cleaning time. Necessary information for calculating the takt time is given in Table 3.

From the data collection, the graph in Fig. 4 presents the line balancing graph to visualize the cycle time of the current practice production line. Since that, the takt time is 52.2 s (shown by the break line in Fig. 4), whereas the maximum cycle time in current practice is 72.3 s, which is longest than the takt time. Therefore, the bottleneck workstation can be easily identified as the station having the maximum cycle time, i.e. the 4th workstation. The units per hour of the bottleneck workstation are 50 units per hour and this will result the daily capacity is 725 units per day.

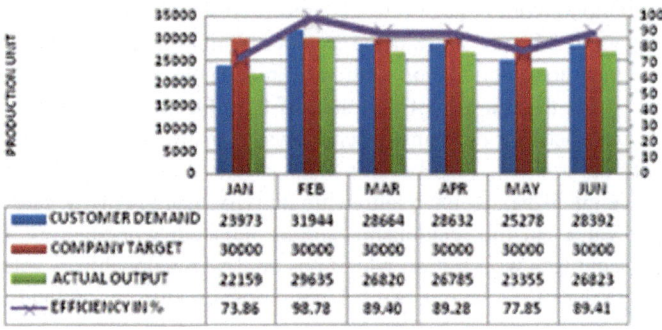

Fig. 3 Production efficiency of current practice

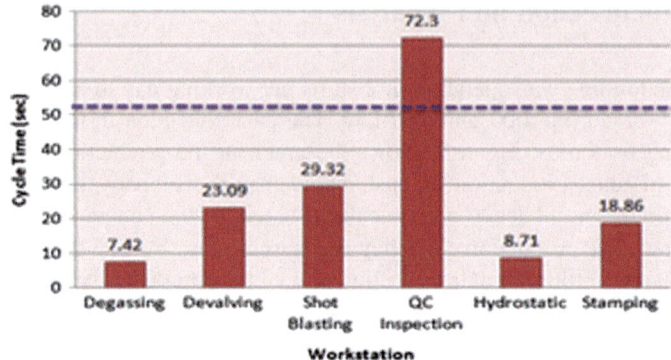

Fig. 4 The line balancing graph of current practice

Nevertheless, the actual daily capacity of the current practice is 614 units. This shows that the capacity of the current practice not equal to the capacity at bottleneck workstation. Therefore, it instigates that the line balancing of the current practice is unbalanced in terms of distribution of the workload and cycle time of man power. The current line balancing is just 36.81% from the worldwide target of 85%.

4.2 Automated Layout Design Program (ALDEP)

This section is representing the analysis of the existing FOL production layout to estimate the space area utilization of each workstation. According to [4], this was evaluated by the adjacency score such as, totaling the numerical values assigned for adjacent department closeness rating. There are four steps to developing a new layout using ALDEP.

The first step is to construct a table of the workstation and the number of unit area according to the current area of FOL production layout. The number of unit area refers to the number of unit to fill up the actual area. In this case, the total area of FOL production layout is 12,500 ft^2. Therefore, researchers decided to estimate 1 unit area equals to 250 ft^2. To find the number of unit area for each workstation, divided the actual area workstation with unit area. Table 3 refers to the actual area of workstation and the number of unit area of the FOL production layout.

Secondly, the activity relationship chart was constructed to identify the relationship score between each workstation. The activity relationship chart in given in Fig. 5. The weighted values for relationship activity using ALDEP rate the layout by assigning values to the relationship among adjacent workstations. If a workstation is adjacent to a workstation which has an "A" relationship, a value of 64 is added to the rating of the layout. An "E" relationship adds 16, "I" adds 4, and an

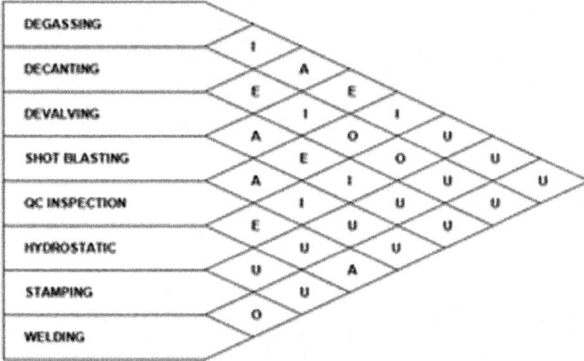

Fig. 5 Activity relationship chart

"O" relationship adds 1 to the rating of the layout. A "U" relationship has no effect on the rating of the layout.

ALDEP requires the specification of the sweep width and the specified level of importance. For the interaction to be described, the sweep width will be assigned a value 2, and the minimum acceptable level of importance will be input as an "E" relationship. As third step, assume that ALDEP randomly selected workstation 3 to be the first workstation to enter the layout for alternative 1 and workstation 5 for alternative 2. The relationship chart is scanned to determine if a workstation has either an "A" or "E" relationship with the first workstation selected. The second workstation has an "E" relationship with the first workstation and is selected to be the second workstation to enter the layout. The unselected workstation has an "A" and "E" relationship with next workstation. Therefore, this workstation will enter the layout next.

After all input data has been completed, the final layout will be constructed by using the ALDEP placement procedure. For this case, the placement routine within ALDEP begins by placing the first workstation in the upper left corner of the layout and extends it downwards. The width of the downward extension of the workstation entering the layout is input by the user and is called the sweep width. The workstation is placed in the layout using the sweep width pattern.

Given that the overall layout is to be 2 units wide and 20 units in length, the placement of workstation 3 in the layout is shown in Fig. 6. Each digit in the layout represents 250 ft^2. To placement the unit number of area by using the input data in Table 4 and the follow adjacent workstation as given in Tables 6 and 7.

After all the input data was analyzed using ALDEP, the actual space requirement for each layout has been identified. The result of the space requirement is shown in Table 5. The total area for the FOL production layout is equal to 4500 ft^2, about 130 ft^2 less than current practice area of 4630 ft^2. The different space requirement

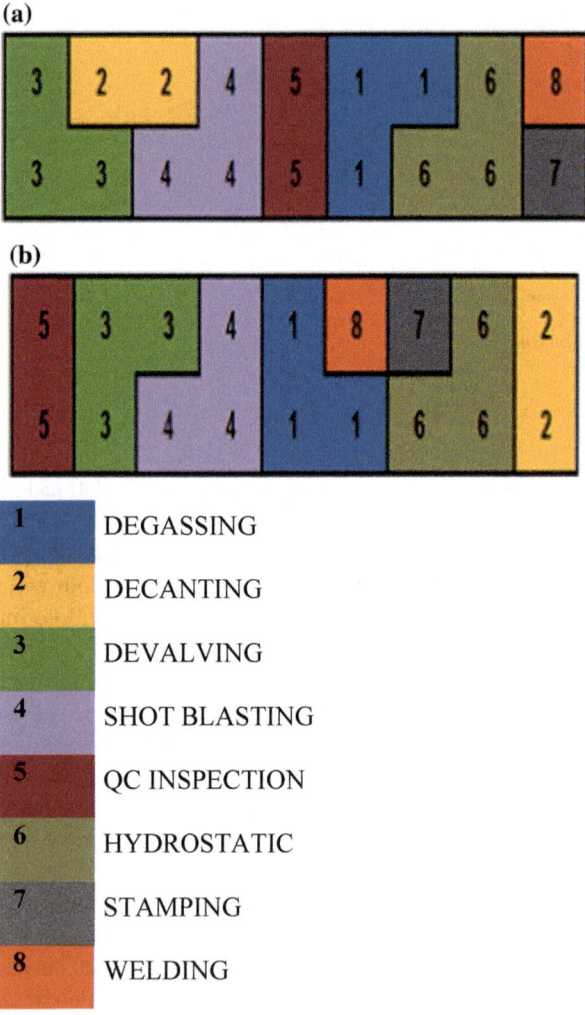

Fig. 6 a Final layout for first alternative, that start sweep pattern with workstation 3. **b** Final layout for second alternative, that start sweep pattern with workstation 3

between the current practice and ALDEP is shows in Table 5. Therefore, the new layout design should be designed based on the space requirement by ALDEP to get more space utilization and to avoid waste in terms of more space utilization and travel time of the workers between each workstation in the FOL production layout.

Analysis of Production Layout Model to Improve Production ...

Table 4 ALDEP specifications

Activity	Area (ft^2)	
	Current layout	ALDEP
Degassing	840	750
Decanting	416	500
Devalving	728	750
Shot blasting	759	750
QC inspection	437	500
Hydrostatic	675	750
Stamping	325	250
Welding	450	250

Table 5 The difference space requirement between current processes ALDEP

No. shift	Total working time	Fixed loss	Available working time	Daily required
2	16 h	1.5 h	14.5 h	1000 units

Table 6 Adjacent workstation

STATION	CURRENT PRACTICE		PROPOSED CHANGE	
	WORKER	CYCLE TIME(SEC)	WORKER	CYCLE TIME(SEC)
1	1	7.42	1	7.42
2	2	8.48	2	14.4
	3	14.61	3	14.4
3	4	14.55	4	15
	5	14.76	5	34.2
4	6	10.07	6	33
	7	38.28	7	9
	8	23.94	8	9.43
5	9	8.71	9	9.43
6	10	9.43		
	11	9.43		Rearrange allocation of manpower

Symbol: Worker

Table 7 Proposed cycle time for second proposal

STATION	CURRENT PRACTICE		PROPOSED CHANGE	
	WORKER	CYCLE TIME(SEC)	WORKER	CYCLE TIME(SEC)
1	1	7.42	1	7.42
2	2	8.48	2	8.48
2	3	14.61	3	14.61
3	4	14.55	4	14.55
3	5	14.76	5	14.76
4	6	10.07	6	34.2
4	7	38.28	7	33
4	8	23.94	8	9
5	9	8.71	9	9.43
6	10	9.43	10	9.43
6	11	9.43		Rearrange allocation of manpower

Symbol: Worker

4.3 Proposal of Improvement

4.3.1 First Proposal

The first proposal is focused to reduce the time at the bottleneck workstation, which is the QC inspection workstation. In this case, the worker does not need to repeat the movement of LPG to the buffers. At the same time, the number of workers will reduce from three to two workers. Table 6 shows the proposed cycle time for the FOL production layout with the rearranged allocation of manpower 6, 7 and 8 at the 4th station.

By combined the weighing process and chipping process, the bottleneck time was reduced about 7.1% from 72.3 to 67.2 s. Therefore, the new output for this proposed layout will also increase due to the shortest time in the bottleneck workstation.

4.3.2 Second Proposal

The second proposal is a continuous improvement of proposal one by eliminating another waste at workstation 1 and workstation 2. The elimination of waste between both workstations is called as buffer. This buffer was caused by the waiting time of the LPG arrival at the next workstation. Table 7 shows the proposed cycle time of the FOL production layout with the proposed suggested manpower allocation on worker 1, 2 and 3.

Table 8 Proposed cycle time for third proposal

STATION	CURRENT PRACTICE		PROPOSED CHANGE		Symbol
	WORKER	CYCLE TIME(SEC)	WORKER	CYCLE TIME(SEC)	
1	1	7.42	1	7.42	Worker
2	2	8.48	2	8.48	
	3	14.61	3	14.61	
3	4	14.55	4	14.55	
	5	14.76	5	14.76	
4	6	10.07	6	10.07	
	7	38.28	7	27.32	
	8	23.94	8	10.39	
5	9	8.71	9	8.71	
6	10	9.43	10	9.43	
	11	9.43	11	9.43	
		Predetermine motion time using MOST			

By elimination of waste at workstation 1 and workstation 2, the total cycle time was reduced by about 3.2%, i.e. from 159.68 to 146.28 s. Nevertheless, in this proposal, the QC inspection workstation still defined the bottleneck. Therefore, the new output for this proposed layout will also increase due to the shorter time in the total cycle time of all workstations.

The new standardized method of checking the series number process at the bottleneck workstation reduced the cycle time by 24.52 s from 72.3 s. The total cycle time will be shorter compared to proposal 1 and proposal 2. Therefore, the new production output will increase more than the current practice and other proposals.

4.3.3 Third Proposal

In the third proposal, a new standardized orientation method of checking the series number process at the bottleneck workstation has been proposed. At the same time, the chipping process also analyzes the predetermined motion time system by applying the MOST technique to estimate the standard time for the chipping process due the longest time during the current practice. Table 8 shows the proposed cycle time for the FOL production layout with new predetermined standard time for worker 7 and 8 at the 4th station.

5 Conclusion

In this case study, the main problem and challenge of an oil and gas company was related to non-optimal operations with an inefficient distribution of the work. The whole production layout suffers due to the absence of an established standard time for activities carried out by workers, unbalanced load carried, the non-value added (NVA) activities involved and the inefficient methods. The primary goal of this research was focused on the improvement of the production efficiency of the current "Front of Line" production layout by identifying and minimizing unnecessary activities at the bottleneck and rearrange the allocation of manpower in the production line to increment the line balancing. It is evident that to sustain in this competitive industrial environment, a company needs to reduce or eliminate idle and or down time, improving the working method, standardize the time as well as enhance the overall capacity planning.

Acknowledgements The authors thanked the Manufacturing Department and Mechanical Department University Kuala Lumpur Malaysian Spanish Institute, Kulim, Kedah, Malaysia.

For research facilities and financial assistance provided for this project.

References

1. Dwijayanti, K., Dawal, S.Z.M., Aoyama, H.: A proposed study on facility planning and design in manufacturing process. In: Proceedings of the International Multi Conference of Engineers and Computer Scientists (2010)
2. Hasan, M.A., Sarkis, J., Shankar, R.: Agility and production flow layouts: an analytical decision analysis. Comput. Ind. Eng. **62**(4), 898–907 (2012)
3. de Carlo, F., Arleo, M.A., Borgia, O., Tucci, M.: Layout design for a low capacity manufacturing line: a case study. Int. J. Eng. Bus. Manage. **5**(Spl.Issue), 1–10 (2013)
4. Khan, A.J., Tidke, D.J., Scholar, M.T.: Designing facilities layout for small and medium enterprises. Int. J. Eng. Res. Gen. Sci. **1**(2) (2013)
5. Shayan, E., Chittilappilly, A.: Genetic algorithm for facilities layout problems based on slicing tree structure. Int. J. Prod. Res. (2004)
6. Zuhdi, A., Taha, Z.: Simulation model of assembly system design. In: Proceeding Asia Pacific Conference on Management of Technology and Technology Entrepreneurship (2008)
7. Jia, Z., Lu, X., Wang, W., Jia, D., Wang, L.: Design and implementation of lean facility layout system of a production line. Int. J. Ind. Eng. Theory Appl. Pract. **18**(5), 260–269 (2011)
8. Hesen, P.M.C., Renders, P.J.J., Rooda, J.E.: Application of a layout/material handling design method to a furnace area in a 300 mm wafer fab. Int. J. Adv. Manufact. Technol. **17**, 216–220 (2011)
9. Meyers, F.E., Stephens, M.P.: Manufacturing Facilities Design and Material Handling (2005)

Effect of Vibration on Occupant Driving Performances: Measurement by Simulated Driving

Mohd Amzar Azizan

Abstract Although the performance of the vehicle driver has been well investigated in many types of environments, however, drowsy driving caused by vibration has received far less attention. Our experimental procedures comprised of two 10 min simulated driving sessions in no-vibration condition and with-vibration condition. In with-vibration condition, volunteers were exposed to a Gaussian random vibration, with 1–15 Hz frequency bandwidth at 0.2 m s^{-2} r.m.s. for 30 min. A deviation of the in lane position and vehicle speed were recorded and analysed. Volunteers have also rated their subjective drowsiness by giving scores using the Karolinska Sleepiness Scale (KSS) in 5 min intervals. Strong evidence of driving impairment following 30 min exposure to vibration was found significant for all volunteers ($p < 0.01$).

Keywords Simulated driving · Vibration · Driving performance

1 Introduction

Drowsy driving is a significant cause of accidents on motorways or major roadways. Drowsy driving has been reported to account for approximately 20% of accidents worldwide [1]. In Australia, there were 251 fatalities (16.6% of total road deaths) caused explicitly by sleep-related accidents in 1998 alone [2]. In addition to that, a new EU regulation about sleepiness in driving with a focus on sleep apnoea patients has been issued. Drivers or driving license applicants with moderate and severe obstructive sleep apnoea shall be referred for further medical advice before a driving license is granted or renewed [3]. This suggests that the effect of drowsy/sleepy driving is comparable to drunk driving. Although some research has demonstrated there is a possible link between short-term exposure to vibration and

M. A. Azizan (✉)
Universiti Kuala Lumpur Malaysian Institute of Aviation Technology, Jalan Jenderam Hulu, Kampung Jenderam Hulu, 43900 Sepang, Selangor, Malaysia
e-mail: mohdamzar@unikl.edu.my

reduction of wakefulness level [4, 5], however drowsiness caused by vibration is not well investigated and characterized in the available literature. Hence, automotive industry standards to limit vibration-induced drowsiness do not yet exist.

It is well established that drowsiness caused by alcohol influence, monotonous driving and night driving have considerable influences on the driver alertness and performance, therefore, can compromise the transportation safety [6–9]. However, the formulation of drowsiness caused by exposure to vibration is not well defined.

The relationship between vibration amplitude and vibration frequency of the vehicle occupant and drowsiness has been assumed without sufficient research. This is because drowsiness is a multifactorial phenomenon, and there is little quantitative data available. Vibration has been found to correlate with a range of physiological reactions of the human body such as lower back pain and heart rate variability [10, 11]. Vibration may also affect muscle and neurological functions, by acting as a stressor [12, 13].

In the automotive industry, the vehicle seat structure is exposed to vibrations from various sources such as vehicle powertrain and road surface. Fundamental vibration modes (resonant frequency and correspondence mode shapes) of the automotive body which can transmit vibrations to the seat structure occur at a frequency below 60 Hz [14]. However, the fundamental resonance of the human body occurs at a frequency below 15 Hz [15]. It is well known that transmitted vibration to the seated human body has a significant influence on human perception and ride comfort [15–17]. The ISO 2631-1 (1997) [18] international standard for evaluation of human exposure to whole-body vibration has been used successfully for several years. Although this international standard (ISO 2631-1) has been developed for the assessment of human body discomfort that is called the "Equivalent Comfort Contour," however, there is no "Equivalent Drowsiness Contour" available.

Hence, there is considerable scope for defining the exact effects of the vehicle and particularly the seat vibration on the driver drowsiness levels. Drowsiness or sleepiness is an intermediate state between being awake and asleep [19]. Various studies have suggested that drowsy driving affects the ability to drive safely [20]. According to several prior investigations on drowsiness and vehicle control, there is a close relationship between drowsiness and lane position variability. The lane position variability is calculated as a standard deviation of the average lateral position. The lane position variability also corresponds to the amount of weaving in the car and the increment of lane variability may ultimately result in the lane crossing into the road shoulder and adjacent traffic lane. Therefore, the primary dependent variables for this investigation were the volunteers' lane position variability measured from the simulated driving vehicle and the speed variability (standard deviation of average vehicle speed).

Although many studies have attempted to demonstrate the links between driving performance and drowsiness, drowsiness caused by vehicle vibration has not been experimentally assessed by simulated driving. Therefore, it was also important to investigate the feasibility and utility of simulated driving in the detection of drowsiness caused by vibration. Hence, it was the primary aim of this study to investigate the effects of vibration on human drowsiness level using both objective (Simulated Driving Test) and subjective (Karolinska Sleepiness Scale) measurement methods.

2 Methodology

2.1 Volunteer Recruitment

Twenty young male ($n = 20$) participated in this investigation with a mean age of 23.0 ± 1.3. They were randomly selected from university students. They had no history of low back pain (LBP) and normal or corrected-to-normal vision. They were 168.2 ± 4.0 cm in height and weighed 64.2 ± 12.2 kg. The average BMI of participants was 22.6 ± 2.54 kg/m^2. All volunteers were screened using the Pittsburgh Sleep Quality Index (PSQI) to measure the sleep quality [21]. Volunteers who showed poor sleep quality index (PSQI > 5) were excluded from the investigation.

2.2 Ethical Consideration

Before the investigation, volunteers were provided with verbal and written explanations of the purpose and contents of the experiment. They were also informed that they have the right to refuse participation in the experiment, and the results of the experiment would remain confidential. Following this, the informed written consent form was obtained from all the volunteers after the procedure of the experiment was explained, and the laboratory facilities were introduced to them. The experimental protocol was reviewed and approved by the Human Research Ethics Committee.

2.3 Experiment Setup

The experimental setup for the drowsiness assessment is illustrated in Fig. 1. The seat used for the experiments is a mid-sized sedan car seat with adjustable headrest. The seat was mounted on a cast aluminum Table (2 m × 1.2 m × 1.2 m), and the table was mounted on four air mountings (regulated to 20 psi). The seat's inclination angle was set at 15° to the vertical direction. The excitation input force for the table was provided by a servo-controlled hydraulic actuator (5 kN) that was placed vertically at the corner of the table. The off-centre excitation provided multi-axial input power in different orientations. It also generated typical vibrations that are usually generated on the vehicle seat mountings. The vibration table below the seat was designed to be dynamically rigid in frequencies below 100 Hz. This is to ensure that there is no interaction with the vehicle seat structural dynamics. Prior to drowsiness measurement, measurement of total transmitted vibration to each volunteer has been done in accordance with ISO 2631-1 (1997). The measurement was carried out to adjust the required hydraulic input force for every volunteer to become 0.2 m s^{-2} r.m.s.

Two tri-axial accelerometer pads (SVANTEK SV-38 V model) were used to measure the transmitted vibration to the human volunteer body at the seat cushion and the seatback [22]. The SV 106 Human Vibration Exposure (HVE) meter (analyser), which was connected to the accelerometer pads, was used to obtain the total frequency weighted transmitted vibration to the seated human body. The HVE analyser used the weighting factors (Wk, Wd, Wc) and multiplication factors to calculate the total frequency-weighted transmitted vibration to the seated human body. The weighting curves (Wk, Wd, Wc) were taken from ISO 2631-1 (1997) [18]. The frequency weighting curves define the values by which the vibration magnitude at each specific frequency is to be multiplied in order to weight the measured vibration in accordance with the human body [15]. The multiplication factors were used to weight the effects of seatback and seat pan vibrations on the ride comfort assessment [15, 18, 22].

2.4 Objective Measures

Volunteers were tested on the York driving simulator software (York Computer Technologies, Kingston, Ontario, Canada) as shown in Fig. 1. The simulator has

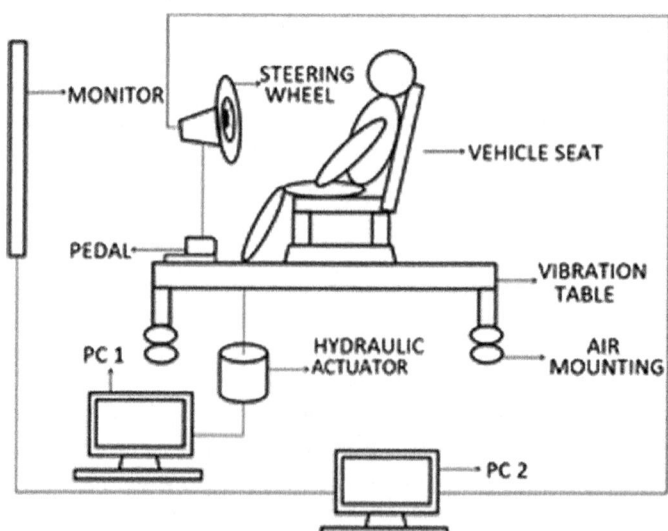

Fig. 1 Configuration of the used driving simulator: An actual vehicle's seat was mounted on a vibration table. A hydraulic actuator located at the corner of the table will provide multi-axial input to the volunteer. A simulator also consists of a personal computer, a 40-inch monitor and peripheral steering wheel, accelerator, and brake accessories

been determined to be an ecologically valid research tool to measure psychomotor performance related to driving [23, 24]. The simulator assembly consists of a personal computer, a 40-inch monitor and peripheral steering wheel, accelerator and brake accessories. A customized driving scenario was developed in which volunteers were presented with a forward view from the driver's seat. The driving simulation showed a cross-country highway, with two lanes in each direction.

The two main instructions for volunteers were:

1. Maintain a steady position within the left traffic lane during the entire test.
2. Maintain a constant speed (usually 100 km/h).

Outcome variables measured by the simulator included the deviation from the center of lane position (SDLP) and deviation from the posted speed limit. The variation in these two outcome measures shows how well the volunteers were able to conduct the test according to this instruction.

2.5 Subjective Measures

Subjective drowsiness level was assessed using the Karolinska Sleepiness Scale (KSS). At every 5-min interval during the simulated driving task, volunteers were prompted by the word "KSS" by the investigator to provide a subjective rating score according to scales visible at all times next to the monitor screen: The use of the scale had been practiced beforehand, and consisted of the following scores: 1 = extremely alert, 2 = very alert, 3 = alert, 4 = rather alert, 5 = neither alert or sleepy, 6 = some sign of sleepiness, 7 = sleepy, but no effort to stay awake, 8 = sleepy, some effort to stay awake, 9 = very sleepy, great effort to stay awake [25] were tested.

2.6 Experimental Procedures

Prior to the experiment, all volunteers were screened using the Epworth Sleepiness Scale (ESS) to detect any abnormalities in sleep. Score > 10 will indicate excessive daytime sleepiness and were excluded from the experiment. Volunteers performed two separate test conditions [baseline (no-vibration) and with-vibration] in the controlled human vibration laboratory in a randomized cross-over design, one week apart. To minimize the learning effect, volunteers underwent a 10-min practice session before baseline and with-vibration conditions to familiarize themselves with the simulator interface. All volunteers were assessed at approximately the same time of day.

During with-vibration condition, volunteers were asked to drive for 10 min with no vibration followed by 30 min sitting with exposure to vibration. Volunteers were exposed to a Gaussian random vibration, with 1–15 Hz frequency bandwidth. The total transmitted acceleration to the human body was kept constant at 0.2 m s^{-2} r.m.s. Volunteers rated their subjective sleepiness scale using KSS before vibration exposure, every 5 min of vibration and after vibration exposure [26]. The rating was initiated by the test leader saying "KSS." Immediately after the 30 min sitting, volunteers were required to drive for another 10 min. Similar procedures and sitting arrangement as in with-vibration condition with the only difference being no vibration exposure were applied for 30 min sitting. The total duration of each condition (no-vibration and with-vibration) was 50 min.

3 Results

3.1 Objective measurement

A total of 20 volunteers completed the investigation. No volunteers reported simulator sickness. There was no significant difference in alertness level measured by the Epworth Sleepiness Scale (ESS) between baseline (no-vibration) and with-vibration condition. Driving performance indexes and subjective sleepiness scale (KSS) in no-vibration and with-vibration condition are presented here. The following two driving performance indexes were assessed and included in the analyses: (1) standard deviation of lane position or lane variability (SDLP), (2) speed deviation from the posted speed limit. The following observations were made in conjunction with this analysis.

Results of the standard deviation of lane position (SDLP) are presented in Table 1. The repeated-measures procedures revealed a significant difference between group variations. As compared with the baseline condition (no-vibration) driving performance index between before exposure and after exposure to vibration showed that 30 min exposure to vibration had significant influences on the volunteer lane keeping performance. We found that following a 30 min exposure to vibration, the deviation from a lateral position or lane variability was significantly increased by 2.6 cm ($p < 0.01$). This indicates poor lane control by all twenty volunteers.

Table 1 Results for lane position variability and speed deviation

Driving index	Baseline		With-vibration	
	Before	After	Before	After
SDLP (cm)	23.5	22.8	23.6	26.2
Speed deviation (kph)	4.68	5.58	5.30	6.28

A significant increase in lane position variability (SDLP) ($p < 0.01$) was observed in vibration condition. No significant changes ($p > 0.05$) in speed deviation were observed in both conditions (baseline and with vibration) following 30 min exposure to vibration and sitting

The analysis of lane position variability showed difficulties in maintaining the vehicle in the middle of the left-hand lane when alertness was lowest after exposure to vibration. A better lane control was observed in the baseline condition (no-vibration) where the lane position variability was reduced by 0.7 cm. It indicates a little improvement of driving performance after 30 min sitting. A significant increase in standard deviation demonstrates the inability of the volunteers to maintain a steady position caused by drowsiness-inducing vibration.

The secondary outcome index measured in this study was the speed deviation. The speed deviation is calculated as the mean sum of the differences of the speed of the vehicle from the posted speed limit. Table 1 shows the average speed deviation measured for all volunteers in no-vibration and with-vibration condition. Although there was a slight increase of speed deviation following the exposure to 30 min vibration (6.28 ± 1.37 kph) and 30 min sitting (5.58 ± 0.85 kph), however, speed deviation failed to exhibit any statistically significant variation within and between groups ($p > 0.05$). The speed deviation did not show significant diurnal variation as compared to lane variability. As expected, the speed deviation was a considerably less sensitive outcome index that can reflex the low level of alertness caused by vibration.

3.2 Subjective measurement

Results of the subjective sleepiness (KSS) for all volunteers in both conditions (no-vibration and with-vibration) were recorded against time and are shown in Table 2. As illustrated here, the initial KSS values did not differ between both

Table 2 Results for subjective sleepiness scale: The table shows an average score of subjective sleepiness scale (KSS) for twenty volunteers for seven interval session in no-vibration compared to with-vibration condition

(KSS)	No-vibration (mean ± SE)	With-vibration (mean ± SE)	P-value
Baseline	2.16 ± 0.19	2.11 ± 0.13	$P > 0.05$
After 5 min.	3.94 ± 0.25	4.37 ± 0.31	$P < 0.01$
After 10 min.	4.53 ± 0.31	5.37 ± 0.30	$P < 0.01$
After 15 min.	4.84 ± 0.32	6.11 ± 0.32	$P < 0.01$
After 20 min.	5.00 ± 0.29	6.79 ± 0.37	$P < 0.01$
After 25 min.	5.10 ± 0.33	7.05 ± 0.43	$P < 0.01$
After 30 min.	5.16 ± 0.29	7.26 ± 0.41	$P < 0.01$

Before the experiment, no significant changes were observed for both conditions ($p > 0.05$). However, the subjective sleepiness scale for all volunteers shows a significant increase following exposure to vibration ($p < 0.01$). This indicates a reduction in alertness level due to vibration exposure. Although, there is an increase of subjective score in no-vibration condition, however, the 30-min sitting is insufficient to induce drowsiness

conditions ($p > 0.05$). Significant increases in KSS score between before vibration exposure and every subsequent 5 min of exposure to vibration were detected using repeated measures-ANOVA test ($p < 0.01$) for all twenty volunteers. As can be seen in Table 2, there is a clear decline in alertness level indicated by a progressive increase in subjective sleepiness score throughout the course of exposure to vibration. Before the experiment, the average KSS score was 2.11 ± 0.13 in with-vibration condition and 2.16 ± 0.19 in no-vibration condition. After 5 min of vibration exposure, KSS scores increased to 4.37 ± 0.31. Drowsiness was pronounced following 15 min exposure to vibration with KSS values of 6.11 ± 0.32 whereas KSS values of 4.84 ± 0.32 were observed in no-vibration condition. After 30 min, KSS score was significantly higher for all conditions and increased more significantly with exposure to vibration than without vibration ($p < 0.01$). To investigate the statistical significance, two-way repeated measures-ANOVA was carried out. It was found out that intra-individual and inter-individual differences in all twenty volunteers were highly significant ($p < 0.001$).

4 Discussion

Driver drowsiness has been one of the primary causes of road accidents [27]. However, drowsiness that is caused by vehicle vibration is not well understood or investigated. This study examined the relationship between human drowsiness levels and exposure to whole body vibration as can be experienced during driving. We demonstrated that the increase of human drowsiness level, measured by simulated driving software and Karolinska Sleepiness Scale (KSS) significantly correlated with exposure to vibration. These data support the hypothesis that exposure to vibration (random vibration with 1–15 Hz frequency band) even for as little as 30 min causes drowsiness and adversely affects the psychomotor performance as measured by lane keeping performance, found in all twenty volunteers.

SDLP is related to the amount of drowsiness in the driver. SDLP greater as the driver becomes drowsier. As drowsiness increases, situational awareness decreases and the driver's ability to predict an upcoming event is lowered. This leads to over-compensation by the driver and the wheel tends to be moved frequently. The continuous variations of the standard deviation of lane position (SDLP) show that the low level of alertness when volunteers were exposed to vibration and this causes difficulties in maintaining the vehicle in the middle of the left-hand lane. It has also been reported that SDLP is the most common and persistent finding in sleep deprivation and drowsiness [28–30]. Another important finding to emerge from this study is that the speed deviation failed to exhibit any statistically significant variation between no-vibration condition and with-vibration condition. This finding is consistent with the literature that longitudinal measures, such as speed deviation, are not found capable of detecting drowsiness [31]. As expected, results from the subjective measurement (KSS) also show a significant declination of the alertness level for all the volunteers following a 30 min exposure to vibration. The increase in subjective sleepiness scores provides important corroborating evidence that exposure to 30 min of vibration level can steadily reduce human alertness levels that are linked to drowsiness [32].

Various methods have been proposed in the past to assess human drowsiness and performance, such as measuring brainwave activity using electroencephalography (EEG) [19, 33–35]. The EEG method has the ability to measure changes in the brainwave power spectrum. However, the implementation of EEG in the real environment is still challenging. Brainwave activity signals measured by the electrode on the human scalp can be easily distorted by movement artefacts such as muscle activity and eye movements [36]. Placement of EEG electrodes may be uncomfortable and, therefore, impractical [10, 19, 37]. An experimental design has been developed to replicate the actual driving condition. Therefore, the vibration perceived by volunteers in this study is similar to the actual vibration felt in a typical vehicle. An actual vehicle seat was selected to ensure good vibration transmissibility to the human body. In many studies, the relationship between vibration and drowsiness has been assumed without supporting research [38]. This study demonstrates a link between exposure to vibration and drowsiness, at least under these experimental conditions.

5 Conclusion

The novel contribution of this study is the characterization of the role of vibration. Our findings have identified vibration as an important source of driver drowsiness. Furthermore, our data clearly demonstrate that exposure to vibration has considerable influence on subjective sleepiness levels, and more importantly, human psychomotor and lapse of attention. Exposure to vibration significantly impacts the lane keeping performance of the driver (SDLP). The findings show a low excitation vibration at 0.2 m s^{-2} r.m.s. increased SDLP by 11%. This line of research can then assist in the development of practical and relevant guidelines for limitation of vibration exposure in the automotive industry, in an effort to reduce the burden of disease of road accidents. This will also complement the existing ISO 2631-1 (effects of vibration on comfort) to extend these guidelines in assessment and establishment of thresholds and safe limits for drowsiness-inducing vibration.

Acknowledgements The authors thanked the University Kuala Lumpur Malaysian Institute of Marine Engineering Technology (UniKL MIMET) and Sciences and Technology Research Institute for Defence (STRIDE), Lumut for research facilities and financial assistance provided for this project.

References

1. Horne, J.A., Reyner, L.A.: BMJ (Clinical research ed.) **310**, 565–567 (1995)
2. Dobbie, K.: Australian transport safety bureau, (2002)
3. EU Commission Directives, Official Journal of European Union, 4 (2014)
4. Satou, Y., Ando, H., Nakiri, M., Nagatomi, K., Yamaguchi, Y., Hoshino, M., Tsuji, Y., Muramoto, J., Mori, M., Hara, K., Ishitake, T.: Ind. Health **45**, 217–223 (2007)
5. Satou, Y., Ishitake, T., Ando, H., Nagatomi, K., Hoshiko, M., Tsuji, Y., Tamaki, H., Shigemoto, A., Kusano, M., Mori, M., Hara, K.: Kurume Med. J. **56**, 17–23 (2009)
6. Larue, G.S.: Predicting effects of monotony on driver's vigilance. Queensland University of Technology (2010)
7. Kamdar, B.B., Kaplan, K.A., Kezirian, E.J., Dement, W.C.: Sleep Med. **5**, 441–448 (2004)
8. Akerstedt, T., Peters, B., Anund, A., Kecklund, G.: J. Sleep Res. **14**, 17–20 (2005)
9. Fairclough, S.H., Graham, R.: Hum. Factors: J. Hum. Factors Ergon. Soc. **41**, 118–128 (1999)
10. Vicente, J., Laguna, P., Bartra, A.: R. Bailón **38**, 89–92 (2011)
11. Callaghan, J.P., McGill, S.M.: Ergonomics **44**, 280–294 (2011)
12. Di Giminiani, R., Masedu, F., Tihanyi, J., Scrimaglio, R., Valenti, M.: Official Journal of the International Society of Electrophysiological Kinesiology. J. Electromyogr. Kinesiol. **23**, 245–251 (2013)
13. de Oliveira, C.G., Nadal, J.: Aviation. Space Environ. Med. **75**, 317–322 (2004)
14. Lo, L., Fard, M., Subic, A., Jazar, R.: J. Sound Vib. **332**, 1141–1152 (2012)
15. Griffin, M.J.: Handbook of Human Vibration. Academic Press, London (1990)
16. Baik, S.: KSCE J. Civ. Eng. **8**, 135–139 (2004)
17. Kitazaki, S., Griffin, M.J.: J. Sound Vib. **200**, 83–103 (1997)
18. ISO 2631-1, Mechanical vibration and shock—evaluation of human exposure to whole-body vibration—Part 1: General requirements. International Organization for Standardization (1997)
19. Lal, S.K., Craig, A.: Biol. Psychol. **55**, 173–194 (2001)
20. Gander, P.H., Nguyen, D., Rosekind, M.R., Connell, L.J.: Aviat. Space Environ. Med. **64**, 189–195 (1993)
21. Buysse, D.J., Reynolds, C.F., Monk, T.H., Berman, S.R., Kupfer, D.J.: Psychiatry Res. **28**, 193–213 (1989)
22. Fard, M., Lo, L., Subic, A., Jazar, R.: Ergonomics **57**, 1549–1561 (2014)
23. Arnedt, J.T., Owens, J., Crouch, M., Stahl, J., Carskadon, M.A.: JAMA: J. Am. Med. Assoc. **294**, 1025–1033 (2005)
24. Chung, F., Kayumov, L., Ph.D. Sinclair, R., Edward, R.: Anesthesiol. **5**, 951–956 (2005)
25. Gillberg, M., Kecklund, G., Akerstedt, T.: Sleep **17**, 236–241 (1994)
26. Akerstedt, T., Gillberg, M.: Int. J. Neurosci. **52**, 29–37 (1990)
27. Hosking, S.: The relationship between driving performance and the Johns Drowsiness scale as measured by the Optalert. **252** (2006)
28. Verster, J.C., Bervoets, A.C., De Klerk, S., Vreman, R.A., Olivier, B., Roth, T., Brookhuis, K.A.: Psychopharmacol. **231**, 2999–3008 (2014)
29. Ramaekers, J.G., Robbe, H.W.J., O'Hanlon, J.F.: Hum. Psychopharmacol. **15**, 551–558 (2000)
30. Verster, J.C., Roth, T.: Int. J. Gen. Med. **4**, 359–371 (2011)
31. Arnedt, J.T., Wilde, G.J., Munt, P.W., MacLean, W.: Accid.; Anal. Prev. **33**, 337–344 (2001)
32. Hallvig, D., Anund, A., Fors, C., Kecklund, G., Karlsson, J.G., Wahde, M., Åkerstedt, T.: Accid. Anal. Prev. **50**, 44–50 (2013)
33. Brown, I.D.: Accid. Anal. Prev. **29**, 525–531 (1997)
34. Cheng, S.Y., Hsu, H.T.: Mental fatigue measurement using EEG. In: Risk Management Trends, (2011)
35. Li, W., He, Q., Fan, X., Fei, Z.: Neurosci. Lett. **506**, 235–239 (2012)
36. Hu, S., Peters, B., Zheng, G.: IET Intel. Transport Syst. **7**, 105–113 (2013)

37. Amin, J., Anopas, D., Horapong, M., Triponyuwasi, P., Yamsa-Ard, T., Iampetch, S., Wongsawat, Y.: Wireless-based portable EEG-EOG monitoring for real time drowsiness detection. In: Engineering in Medicine and Biology Society (EMBC), Annual International Conference of the IEEE, pp. 4977–4980. IEEE, (2013)
38. Mcphee, B.: Heavy vehicle seat vibration and driver fatigue, Australian Transport Safety Bureau (2011)

Simulation Studies of a New Magnetorheological Brake with Difference Gap Size Using Combination of Shear and Squeeze Mode

Lailatul Hamidah Hamdan, Saiful Amri Mazlan, Fitrian Imaduddin, Shamsul Sarip and Ashadi Yusop

Abstract Magnetorheological (MR) brake contains magnetized particles, which are strong, fast and reversible transform in their rheological properties when applied the magnetic field. There are a few types of modes that have been working on in the fluid such as the shear mode, flow mode, squeeze mode and recently a new mode called the magnetic gradient pinch mode. Commonly, shear modes have been widely investigated and used in MR brakes. Nevertheless, limited focus has been given to the combination of shear and squeeze mode due to the design consideration in MR brake. This paper focuses on the design of MR brake with a difference of fluid gap rather than a single gap in one device by using both modes. In this work, a few design criteria are considered to select the basic automotive MR brake configuration such as material selection, MR fluid selection, working surface area, applied current density, and wire size. Then, a Finite Element Method in 2D simulation is performed to analyse the resulting magnetic circuit within the MR brake configuration. Moreover, the simulated results of the magnetic flux density in the MR fluid are used to predict the torque produced by the combination of shear

L. H. Hamdan (✉) · A. Yusop
Politeknik Ungku Omar, Jalan Raja Musa Mahadi, 31400 Ipoh,
Perak, Malaysia
e-mail: Lhhh0110@yahoo.com

A. Yusop
e-mail: ashadi1979@gmail.com

S. A. Mazlan
Malaysia-Japan Institute of Technology, Universiti Teknologi Malaysia,
54100 Jalan Semarak, Kuala Lumpur, Malaysia
e-mail: amri.kl@utm.my

S. Sarip
Universiti Teknologi Malaysia, 54100 Jalan Semarak, Kuala Lumpur, Malaysia
e-mail: shamsul.kl@utm.my

F. Imaduddin
Multimedia University, Jalan Ayer Keroh Lama, 75450 Bukit Beruang,
Melaka, Malaysia
e-mail: fitrian.imaduddin@mmu.edu.my

and squeeze modes. It can be concluded that, the finite element simulation predictions show a good correlation between effect of the current and fluid gap.

Keywords Magnetorheological brake · Automotive · Finite element method 2D simulation

1 Introduction

Magnetorheological (MR) fluid is a smart material which continuously and reversibly from fluid to semi solid within several milliseconds when a magnetic field is applied. There are many studies which have proven that a MR fluid holds great potential in electromechanical interfaces; and one of them are brakes. A brake system is a system, that uses friction to convert kinetic energy into heat by the connection of two surfaces. Effort was made to utilizing the MR fluid in the function of brakes. The fluid transforms the large number of particle to strong parallel chains (fibrils) that consists of thick clusters to build a semi solid fluid when the current is applied to the braking system. Carbon iron particles can easily be peeled by shocks or frictions to thicken irreversibly and thus decrease its performance [1]. The frictions between the particles in the fluid and the rotating rotor, generates the braking torque to stop the rotating wheel from moving vehicles. MR brake with various objective have been investigated in many devices such as in automotive industry [2–4] and in medical application [5–7].

Recently there are four basic operational modes which are the shear mode, flow mode, squeeze mode and the latest is called the magnetic gradient pinch mode which has been proposed by Goncalves and Carlson [8]. Wide ranges of researcher have come out with the MR brake using shear mode as a medium to transform the torque value. Nguyen and Choi [9] designed multiple types of MR brake including the disc-type, drum-type, inverted-type, single disc type, multiple disc type, the hybrid-type, as well as the combination of types such as the T-type. The results have demonstrated that a good correlation between experimental results and Finite Element Analysis (FEA) based optimal results. Although significant improvement has been achieved, there is no usage of bobbin that could cover the wire coil which the MR fluid could run through it and interrupt the current flow. Meanwhile, Song et al. [10] demonstrated a new configuration of MR brake using MRF132DG with coil directly wound to the magnetic bobbin. From the results shown that; by optimized design, even with less turns, the performance of the proposed MR brake could be improve than those from the conventional MR brake. Authors however claimed that; by using the magnetic bobbin configuration, the manufacturing accuracy of the MR brake can be better and the manufacturing cost can be reduced. However, the design showed a small difference of power consumption and it gives a good correlation between them.

All of the researches that were mentioned before are mostly using the shear mode as their main working mode without considering of the combination of modes although they have similarities in using FEA software as a medium to define the magnetic density. The critical disadvantages of the shear mode are connected to the fabrication and assembly under more difficult conditions [11]. In addition, Carlson and Jolly [12] suggested to consider a combination of two or three type of mode to overcome the limitation of each mode and to get the maximum input such as the combination of shear and flow modes or shear and squeeze modes.

In this paper, the design of the MR brake uses a combination of shear and squeeze modes. The basic idea of this combination modes is referring to the findings by Tang et al. [13] and Tao [14] who mentioned that the achieved structure-enhanced static yield stress is up to 800 kPa under magnetic field influence measured after compression. Despite from that, with the same fluid, the yield stress value without compression only can reach until 80 kPa. The intention is the formation of thicker and thus tougher columns of particles that are able to sustain the load. Based on this finding, Kulkarni et al. [15] has defined a possible mixed configuration in the same effective area and named it as the shear-squeeze mode. In this design, the shearing was done by rotating the lower plate (torsional shear) while simultaneously compressing the plate to the other plate. Meanwhile, Zhang et al. [16] designed a device to estimate the effect in linear shear mode of the mechanical compression on MR fluids. However, the apparatus was quite complex and bulky. Their findings revealed that a high compression could enhance τ_y by more than 20 times for a given magnetic field, which was called as the squeeze strengthening effect.

Nevertheless, Sarkar and Hirani [4, 17] have broken the tradition and looked at the interesting modes view in designing MR brakes. The new exposure has been gained in designing the MR brake which included the tensile and compression zone of MR fluid, namely the mode called as compression plus the shear mode. The available torque expressions are assuming the Bingham and Herskel-Bulkley fluid model. Although there has been a large number of studies on the design and application of MR brake using shear mode and rarely research in combination of modes configurations in MR brake have been developed, none have combined shear and squeeze by isolating their effective area. This research proposed a new design of MR brake prototype suitable for medium and small size cars with a combination of the shear and squeeze mode. The proposed MR brake has been modelled and simulated using an equivalent electric circuit to model the magnetic flux which is uniformly distributed over a cross-section of the components in the Finite Element Magnetic Method (FEMM) software package. The design, however, is to achieve the maximum magnitude of the magnetic flux density and to investigate the effect of the input current, adjustable gap size and inner compress disc position on the produced torque. By identify the magnetic flux density and carrier fluid velocity profile, the value of the dynamic shear stress is obtained and at once the torque value is performed.

2 Two-Dimensional Simulation

In this section, the proposed design of the MR brake has been sketched in FEMM in order to predict the amount of torque produced in the shear-squeeze mode. The first consideration in designing the MR brake is to derive the mathematical relation between the braking torque, the parameters of the structure and the magnetic field strength in the effective area [18]. Figure 1a shows a graph of the yield stress versus magnetic field strength and Fig. 1b shows typical magnetic properties from the LORD Corporation technical data sheet for MRF132DG. The magnetic circuit of the designed MR brake is important to confirm the flux flow along the circuit and it can also be used for an initial prediction by the electromagnet that generates the magnetic field density [5]. Therefore, the value of the magnetic field density then can be used to determine the value of the field-induced yield stress using the polynomial approximation for Lord MRF132DG as shown in Eq. (1) [5, 19, 20].

$$\tau_y(B) = 52.962\,B4 - 176.51\,B3 + 158.79\,B2 + 13.708\,B + 0.1442, \quad (1)$$

where B is the magnetic flux density in Tesla which is determined by coil turns and dimensions, current input, and total reluctance of the medium. The coil is made from 250 turns of 22 Standard Wire Gauge (SWG) copper wire with 0.711 mm diameter. Given that the current input was limited to 1 A, where the maximum current power transmission is 1.05 A. The mesh size is a parameter, which has been chosen on the basic convergence of the simulation results to achieve a reasonable accuracy [21]. When a current is applied to the coil, a magnetic circuit is generated and the magnetic flux goes across both the annular duct and the radial ducts.

2.1 Modelling of MR Brake

The modelling of the MR brake is required for the theoretical validation of the torque performance during the experiment. The accumulated torque is expected to be a mixture of the torque produced from both areas of the operational shear and squeeze modes. On the other hand, the shear friction that sandwiched between the rotor and the solidified MR fluid delivers the resultant torque. The performance of the disc type MR brake can be upgraded by looking to the effective area of the shear mode located around the end face gap in the axial direction [22]. The configuration of the prototype MR brake is shown in Fig. 2. The design consists of wire coil which are wound at the end of the flange of the MR brake, rotor, outer housing, inner squeeze plate, body squeeze plate, bobbin, and MR fluid that is filling the gap between the rotor and adjustable inner squeeze plate which can be clearly seen in the cross sectional conceptual design of the MR brake in Fig. 3. The power to be supplied to the coil is district so that the strength of current of the coils can be controlled by adjusting the power supply. Figure 4 shows the 2D model of the MR

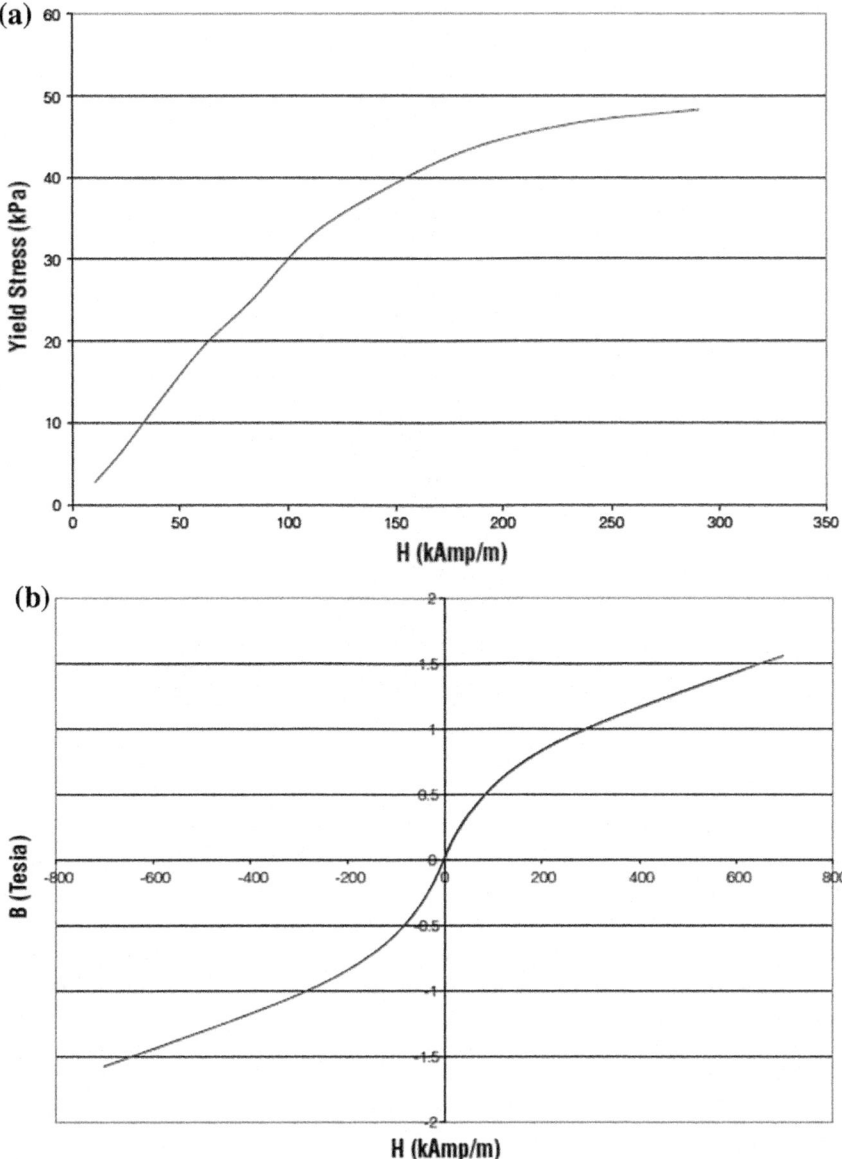

Fig. 1 LORD Corporation technical data for MRF132DG **a** Yield stress versus Magnetic field strength. **b** B-H curve

brake sketched in the FEMM software. The current setting in this simulation has been identified to five different values which are applied to the coil for instance 0.2, 0.4, 0.6, 0.8 and 1.0 A. Wire coils are wound at the end of the flange of the MR brake, which is close with the bobbin to make sure that the fluid will not leak and

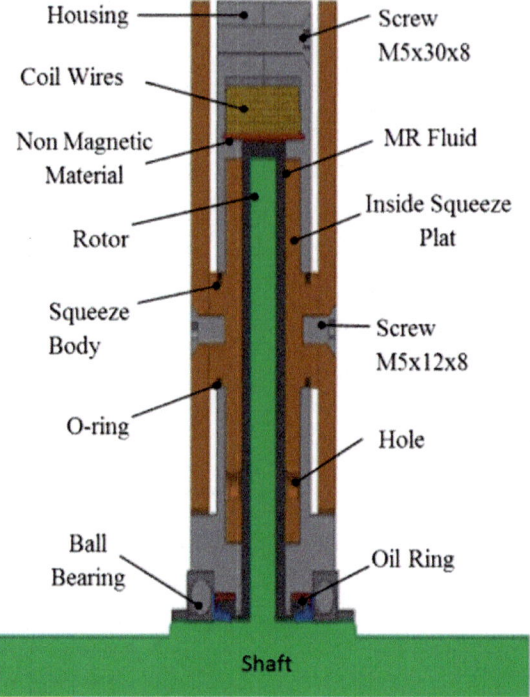

Fig. 2 Conceptual design of MR brake

flow to the coil. The material of each element that included the magnetic and nonmagnetic material (bobbin and rotor) has been defined by looking at the properties of the material which specifically is focusing on the conductivity and permeability in the electromagnetic analysis. The permeability is proportional to the strength of the magnetic field, which can achieved from the B-H curve from the Lord Corporation. The magnetic material is selected with the purpose of directing the flux through the axial gap.

2.2 Magnetostatic Analysis

A simulation based on the FEA is commonly used because of the finding is nearly similar with experimental research especially in the case of FEMM software. Other than that reason, the software can be accessed freely from the internet and also would minimize the calculation error of the proposed design [23]. The analytical braking torque was calculated for each adjustable gap by using the magnetization curve of the MRF132DG in the analyses. The MR fluid is sandwiched between the inside squeeze plate and rotor-disks with an overlapping width of 3 mm from the

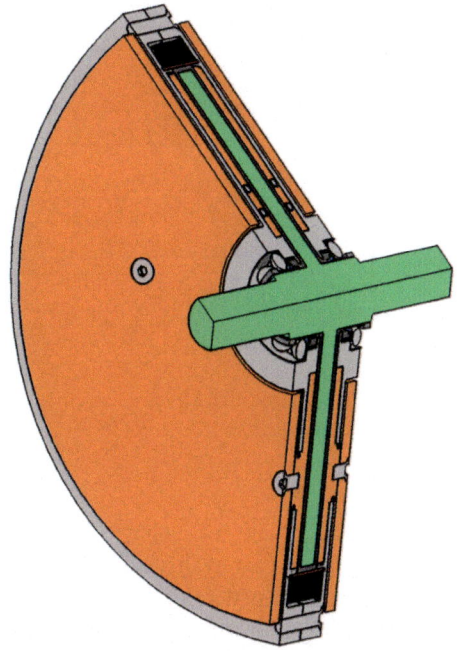

Fig. 3 Cross sectional of conceptual MR brake design

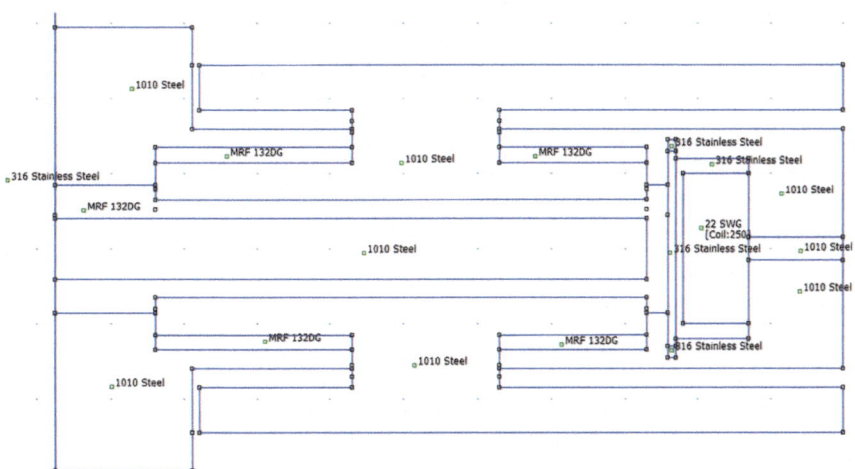

Fig. 4 2D model that sketched in FEMM

outer edge of the rotor disks. The inside squeeze plate can be adjusted to the minimum of 0.5 mm width with help from the blocking plate that slides in between the outer squeeze plate and the housing of the MR brake.

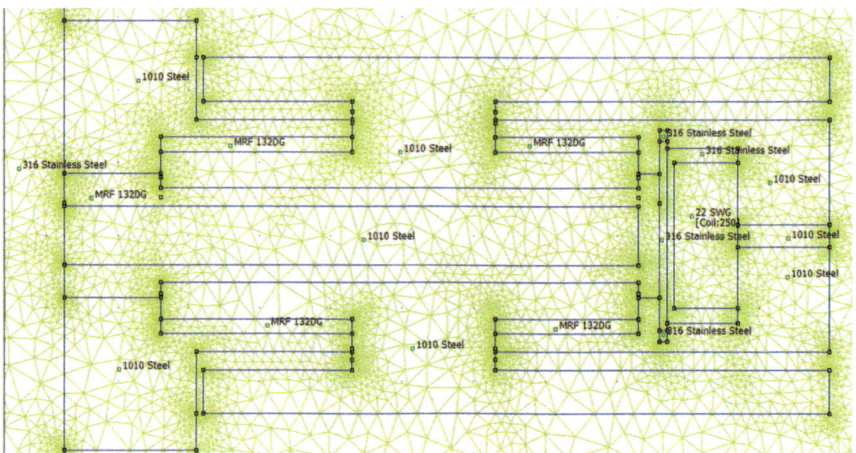

Fig. 5 The finite element mesh

Bear in mind, that the designer can evaluate the magnetic field distribution within the MR brake by looking at the pattern of the magnetic flux density and field intensity. However there is a limitation of the exact value of both requirements based on the design. The value of the flux density obtained from the simulation is to predict the theoretical number of the braking torque. First of all, the design needs to define the brake geometry and the considered material properties, which are based on an axisymmetric problem. The Maxwell's equations have been clearly explained by Mughni et al. [21] until the form of the equations for usage in the magnetic vector potential approach to solve the magnetostatic problems with a non-linear B-H curve. After the geometry is triangulated, the finite element mesh is loaded as shown in Fig. 5, which was created with 14,903 nodes. The "big magnifying glass" icon in the FEMM program had been pressed to run the postprocessor to generate the flux line. To produce a magnetic flux density graph, the red line drawn in between the MR fluid is as in the box shown in Fig. 6.

3 Results and Discussion

The relationship between the magnetic density and the length of the effective area under different fluid gaps and different applied currents is depicted in Fig. 7a–e. The magnetic density increases with increasing the length of the effective area. The usage of magnetic and nonmagnetic material has a significant effect on the flux flow and increases the value of the magnetic density if the selected geometry is focusing into the effective area. The found results are proven by the Kirchhroff's law as below:

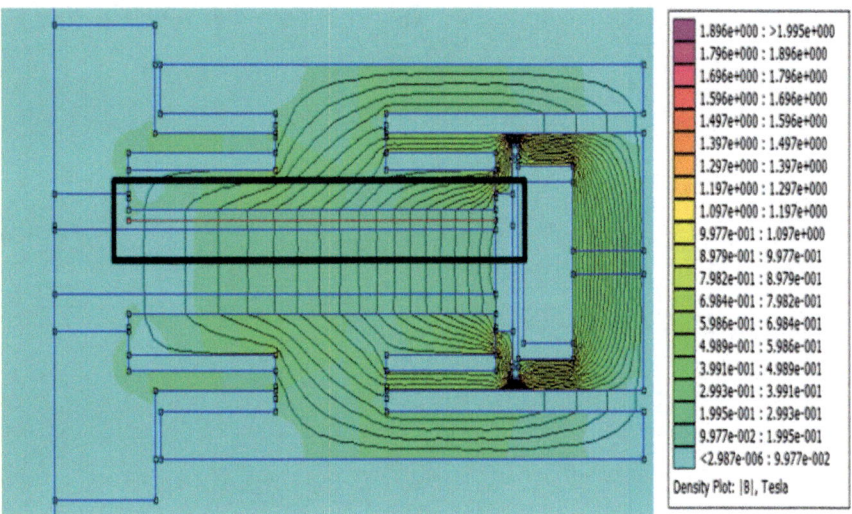

Fig. 6 The flux density flow with colored values

$$\mathrm{NI} = \oint H\, dl \qquad (2)$$

Equation (2) describes the relation for magnetic circuits, the number of turns N and the applied current I which determines the magnitude of the magnetic field strength H [24]. The length l refers to the length of the effective area of the magnetic circuit that passes through the flux lines in the MR fluid. l is larger if the gap size is large. It is shown in Fig. 7a–e that at 0.2 A the magnetic density achieves more than 0.15 T at the smaller fluid gap which is 0.5 mm. A huge difference of the magnetic density value can be seen in 0.5 mm fluid gap and the rest of 1.0, 1.5, 2.0, 2.5 and 3 mm fluid gaps. Due to this, the effect of a minimum gap size is to reduce the magnetic resistance and at the same time reducing the power consumption [25]. Other than that, a minimum gap size is suggested and has been proven as ideal gap to produce a higher torque [17, 26].

In other observations, the trend of every current applied is slightly the same for each fluid gap. When the maximum current is applied (1.0 A) to the proposed MR brake, it gives the highest value of the magnetic density especially at the 0.5 mm fluid gap. Mazlan et al. [24] have mentioned that with the same amount of wire coil turns, a higher electrical current or a small gap size will give a greater magnetic field strength. For the result, the magnetic flux density B values were obtained by simulating the MR brake. Therefore, the values of the magnetic density can be acquired using the proposed design for the shear-squeeze modes.

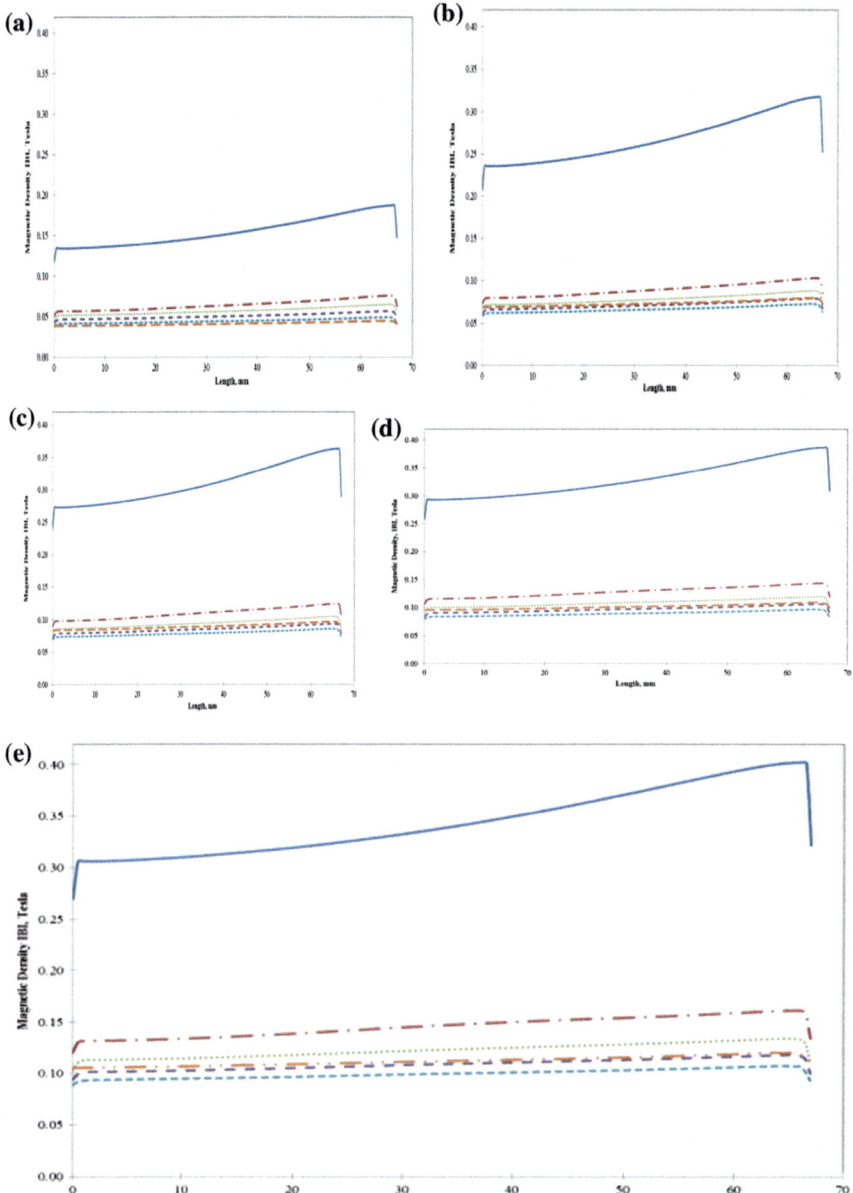

Fig. 7 a Magnetic flux density versus length with difference fluid gap at current 0.2 A.
b Magnetic flux density versus length with difference fluid gap at current 0.4 A.
c Magnetic flux density versus length with difference fluid gap at current 0.6 A.
d Magnetic flux density versus length with difference fluid gap at current 0.8 A.
e Magnetic flux density versus length with difference fluid gap at current 1.0 A

4 Conclusion

This paper aimed at evaluating the performance improvement of MR brakes using simulation outcomes. The results are presented starting with different gap size and with different electric currents to generate the magnetic field. The finite element simulation was performed using FEMM which has been used to study the influence of the electric current on the magnetic flux. In this work, the simulation results of the magnetic density measured by fluid gap have been compared with different electric currents. The magnetic density gives a good correlation to the length of the effective area due to the increment of Tesla as it is near to the location of wire coil. This work estimates the shear squeeze effect to the MR fluid compared to previous findings only based on the shear mode performance. Generally, it can be concluded that on increasing current applied to the MR brake will generate a high braking torque at minimum gap. Meanwhile the adjustable fluid gap that produces the magnetic field strength of the MR fluid, gives benefits to the researcher to increase or decrease the braking torque for multipurpose used neither medium nor small vehicle application.

Acknowledgements This work was supported by Politeknik Ungku Omar (PUO), Malaysia-Japan International Institute of Technology (MJIIT), University Teknologi Malaysia (UTM), and Multimedia University (MMU) Melaka especially in research facilities and financial assistance.

References

1. Claracq, J., Sarrazin, J., Montfort, J.P.: Viscoelastic properties of magnetorheological fluids. Rheol. Acta. **43**(1), 38–49 (2004)
2. Nguyen, Q.H., Choi, S.B.: Optimal design of a novel hybrid MR brake for motorcycles considering axial and radial magnetic flux. Smart Mater. Struct. **21**(5), 55003 (2012)
3. Park, E.J., da Luz, L.F., Suleman, A.: Multidisciplinary design optimization of an automotive magnetorheological brake design. Comput. Struct. **86**(3–5), 207–216 (2008)
4. Sarkar, C., Hirani, H.: Design of a squeeze film magnetorheological brake considering compression enhanced shear yield stress of magnetorheological fluid. J. Phys. Conf. Ser. **412**, 12045 (2013)
5. Ubaidillah, Wibowo, A., Adiputra, D., Tjahjana, D.D.D.P., Rahman, M.A.A., Mazlan, S.A.: Performance prediction of serpentine type compact magnetorheological brake prototype. AIP Conf. Proc. **1788**, 30032 (2017)
6. Nakamura, T., Midorikawa, Y., Tomori, H.: Position and vibration control of variable rheological joints using artificial muscles and magneto-rheological brake. Int. J. Humanoid Robot. **8**(1), 205–222 (2011)
7. Tomori, H., Midorikawa, Y., Nakamura, T.: Vibration control of an artificial muscle manipulator with a magnetorheological fluid brake. J. Phys. Conf. Ser. **412**, 12053 (2013)
8. Goncalves, F.D., Carlson, J.D.: An alternate operation mode for MR fluids—magnetic gradient pinch. J. Phys. Conf. Ser. **149**, 12050 (2009)
9. Nguyen, Q.H., Choi, S.B.: Selection of magnetorheological brake types via optimal design considering maximum torque and constrained volume. Smart Mater. Struct. **21**, 15012 (2011)

10. Song, B.K., Nguyen, Q.H., Choi, S.B., Woo, J.K.: The impact of bobbin material and design on magnetorheological brake performance. Smart Mater. Struct. **22**(10), 105030 (2013)
11. Nam, T.H., Ahn, K.K.: New approach to designing an MR brake using a small steel roller and MR fluid. J. Mech. Sci. Technol. **23**(7), 1911–1923 (2009)
12. Carlson, J.D., Jolly, M.R.: MR fluid, foam and elastomer devices. Mechatronics **10**(4–5), 555–569 (2000)
13. Tang, X., Zhang, X., Tao, R., Rong, Y.: Structure-enhanced yield stress of magnetorheological fluids. J. Appl. Phys. **87**(5), 2634 (2000)
14. Tao, R.: Super-strong magnetorheological fluids. J. Phys. Condens. Matter **13** (S0953-8984-0), 979–999 (2001)
15. Kulkarni, P., Ciocanel, C., Vieira, S.L., Naganathan, N.: Study of the behavior of MR fluids in squeeze, torsional and valve modes. J. Intell. Mater. Syst. Struct. **14**(2), 99–104 (2003)
16. Zhang, X.Z., Gong, X.L., Zhang, P.Q., Wang, Q.M.: Study on the mechanism of the squeeze-strengthen effect in magnetorheological fluids. J. Appl. Phys. **96**(4), 2359 (2004)
17. Sarkar, C., Hirani, H.: Theoretical and experimental studies on a magnetorheological brake operating under compression plus shear mode. Smart Mater. Struct. **22**(11), 115032 (2013)
18. Hung, N.Q., Bok, C.S.: Optimal design of a T-shaped drum-type brake for motorcycle utilizing magnetorheological fluid. Mech. Based Des. Struct. Mach. **40**(2), 153–162 (2012)
19. Imaduddin, F., Mazlan, S.A., Zamzuri, H., Yazid, I.I.M.: Design and performance analysis of a compact magnetorheological valve with multiple annular and radial gaps. J. Intell. Mater. Syst. Struct. **26**(9), 1038–1049 (2015)
20. Nguyen, Q.H., Choi, S.B., Wereley, N.M.: Optimal design of magnetorheological valves via a finite element method considering control energy and a time constant. Smart Mater. Struct. **17**(2), 25024 (2008)
21. Mughni, M.J., Zeinali, M., Mazlan, S.A., Zamzuri, H., Abdul Rahman, M.A.: Experiments and modeling of a new magnetorheological cell under combination of flow and shear-flow modes. J. Non-Newton. Fluid Mech. **215**, 70–79 (2015)
22. Imaduddin, F., Mazlan, S.A., Zamzuri, H.: A design and modelling review of rotary magnetorheological damper. Mater. Des. **51**, 575–591 (2013)
23. Ubaidillah, U., Imaduddin, F., Nizam, M., Mazlan, S.A.: Response of a magnetorheological brake under inertial loads. Int. J. Electr. Eng. Inform. **7**(2), 308–322 (2015)
24. Mazlan, S.A., Ekreem, N.B., Olabi, A.G.: The performance of magnetorheological fluid in squeeze mode. Smart Mater. Struct. **16**(5), 1678–1682 (2007)
25. Kikuchi, T., Kobayashi, K., Inoue, A.: Gap-size effect of compact MR fluid brake. J. Intell. Mater. Syst. Struct. **22**(15), 1677–1683 (2011)
26. Sarkar, C., Hirani, H.: Synthesis and characterization of nano-particles based magnetorheological fluids for brake. Tribol. Online **10**(4), 282–294 (2015)

Optimization of Air-Fuel Ratio and Compression Ratio to Increase the Performance of Hydrogen Port Fuel Injection Engines

Mohd Fazri Shaari, Shahril Nizam Mohamed,
Surenthar Magalinggam, Muhammad Najib Abdul Hamid,
Mohd Nurhidayat Zahelem, Zahelem
and Mohd Farid Muhammad Said

Abstract Climate changes and increased demands for energy, tends researchers in searching for environmental friendly resources of energy that can be used in internal combustion engines (ICEs). Hydrogen fuel is one of alternative fuels that produces zero emission. It has been extensively studied by various researchers in internal combustion engines. However, optimum performance could not be achieved due to some limitations on the configurations of air fuel ratio (AFR) and compression ratio. In this study, hydrogen fuelled engine performance and the effect of different air fuel ratios and compression ratios were analyzed. The engine performance characteristics are investigated based on the brake torque, brake power, brake mean effective pressure (BMEP), and peak pressure. Modenas Kriss 110 cc petrol engine was selected for experimentation and engine modelling. A three-dimensional engine model was created using CATIA V5R21 and the engine performance simulation was performed using GT-Suite v7.4.3. The optimum configuration was

---

M. F. Shaari (✉) · S. N. Mohamed · S. Magalinggam · M. N. A. Hamid · M. N. Zahelem
Malaysian Spanish Institute, Universiti Kuala Lumpur, 09000 Kulim Hi-TechPark,
Kulim, Kedah, Malaysia
e-mail: mofazuni10@gmail.com

S. N. Mohamed
e-mail: shahrilnizam@unikl.edu.my

S. Magalinggam
e-mail: surenthar1437@gmail.com

M. N. A. Hamid
e-mail: mnajib@unikl.edu.my

M. N. Zahelem
e-mail: mnurhidayat@unikl.edu.my

Zahelem · M. F. M. Said
Faculty of Mechanical Engineering, Universiti Teknologi Malaysia,
81310 Johor Bahru, Johor, Malaysia
e-mail: mfarid@mail.fkm.utm.my

© Springer International Publishing AG 2018
A. Öchsner (ed.), *Engineering Applications for New Materials and Technologies*,
Advanced Structured Materials 85, https://doi.org/10.1007/978-3-319-72697-7_34

obtained using a design of experiments (DOE) software. From the analysis, it was found that the performance of the hydrogen fueled engine was improved by changing to a leaner AFR and higher compression ratio. The improvement was found around 7.6, 9.3, 8.0, 25.9, 30.6% for brake torque, brake power, BMEP, peak pressure and brake specific fuel consumption, respectively.

Keywords Hydrogen · Engine performance · GT-suite

1 Introduction

The unstable market price for crude oil and electricity are pushing the government and non-government bodies to find an alternative or renewable green energies to fulfill the users need. Although there are many options of renewable and alternative energies, it is not much help in preserving the quality of environment and some of them contribute to higher pollutions. Many studies have been conducted by researchers on various types of alternative fuel to power an engine and it turned up that hydrogen is one of the non-carbonaceous fuels that exists on earth [1]. It can be used as a fuel source for powering an engine. There were many experimental studies conducted using hydrogen as a fuel in an engine [2–4]. Results show that there are some limitations in configuring the air-fuel ratio, injection timing and also the compression ratio that can affect the engine performances [4, 6]. The air-fuel ratio has a significant effect on the performance of hydrogen fueled engines, where the brake mean effective pressure (BMEP) and brake thermal efficiency was found decreasing with the increment of the air fuel ratio while the brake specific fuel consumption (BSFC) increased with increases of the air-fuel ratio [4].

Studies on hydrogen addition in internal combustion engines also had found that hydrogen fueled engines suffer from power output degradation due to its low heating value (volume basis) which results in lean mixture operation [5]. In a natural-gas direct injection spark ignition engine, the compression ratio has a significant influence on the combustion duration at lean combustion. The exhaust hydrocarbon (HC) and carbon monoxide emissions decreased with an increase in the compression ratio, while the exhaust nitrogen oxide emission increased with an increase in the compression ratio [6]. Hydrogen compressed natural gas engines, on the other hand, show that the performance and emissions characteristics can be improved respectively with a hydrogen content of 30% by volume. At higher compression ratio, the specified torque value could be achieved with spark timing retard; an excess air ratio of 1.8 was found to be preferable, and the nitrogen oxide emissions are also satisfying the standard of Euro VI [7]. In the performance of hydrogen fueled engines, it was found that the peak power output was reduced up to 20% [8].

The air fuel ratio and compression ratio play a significant role on the performance of hydrogen fueled engines. This research will analyze these effects and optimize the air-fuel ratio and compression ratio in a simulated hydrogen port fuel injection engine. The air to fuel ratios are studied from rich to lean mixture, as well as increasing the compression ratios from low to high. The optimum configurations achieved from the analysis is studied and the reasons behind will be explained in detail for future improvements.

2 Methodology

The initial preparation taken to achieve the objectives of this study is by collecting the overall engine performance results from the experimental setup. The engine performance testing was conducted using a dynamometer along with a gas analyzer coupled to the Modenas Kriss 110 cc engine, in the Engine Performance Lab located at UniKL MSI. The Wide Open Throttle (WOT) method was used during the petrol fueled engine performance test. Furthermore, different engine speeds were tested from 2000 to 8000 rpm and the results obtained from it were used for the purpose of validation and comparisons of the graphically developed engine model.

2.1 Experimental Setup

Figure 1 shows the experimental engine setup. The apparatus involved during the experimentation are the engine dynamometer (Focus Applied-Model G7.5LC), dynamometer controller, controller software, gas analyzer (SPTC 085241), PCB Piezotronics Pressure Transducer (Model 112A05) and Dewetron Data Acquisition (DAQ) (Model DEWE-30-8). The Modenas Kriss 110 cc engine is fitted alongside with the engine dynamometer to obtain engine performance results. Wide Open Throttle (WOT) mode [8] was performed to gather the overall engine performances. The pre-process of engine parameters set via dynamometer controller which controls the dynamometer and the post-process are shown in the controller software.

A gas analyzer has been used to measure the value of various air fuel ratios during the performance test. Besides that, the peak pressure in the cylinder is measured by the pressure transducer fitted on the cylinder head. The output signal from the sensor transferred to the Dewetron Data Acquisition and the values of in-cylinder pressure were acquired through it. Performance parameters involved in this study are the brake torque, brake power, brake mean effective pressure (Bmep) and peak pressure. Equation (1) shows the formula to calculate the engine torque:

$$T = F \cdot R \tag{1}$$

where T is the torque in (N m), F is the load applied in Newton (N) and R is the distance of load cell from dynamometer in meter (m).

Fig. 1 Experimental apparatus involved

$$\text{Brake torque} = \frac{\text{WOT Torque (N m)}}{\text{Engine Ratio}} \quad (2)$$

The brake torque can be calculated using Eq. (2). Where the WOT Torque in (N m) achieved from the computer software and the engine ratio is set to be 3.31.

$$b.p = \frac{2\pi NT}{60} \quad (3)$$

where, b.p is the brake power in (kW), N is the engine speed in (rpm) and T is the torque in (N m). Finally, the brake mean effective pressure (Bmep) can be calculated using Eq. (4):

$$\text{Bmep} = \frac{2\,b.p}{ALNn} \quad (4)$$

where A is the area of bore in square meter (m^2), L is the stroke in meter (m) and n is the number of cylinders.

2.2 Experimental Procedure

The experiments were carried out at engine speeds of 2000–8000 rpm in increment of 1000 rpm. The engine started while the dynamometer was set to speed control mode. Then, the gear shifted to top gear and the throttle was slowly opened until it reached the maximum position. The engine and dynamometer were allowed to stabilize for a few seconds before collecting data and move on to the next engine speed. For every run the air-fuel ratio was collected from the exhaust gas analyzer. Other data such as torque, brake power and fuel consumption also was collected from the engine dynamometer setup.

2.3 Engine Modelling and Simulation

The engine modelling construction starts with creating a three-dimensional model of the engine parts by referring to the actual geometry of Modenas Kriss 110 cc engine. The CATIA V5R21 CAD software was used to model the air box, intake runner 1 and 2, intake port, exhaust port and exhaust runner. Precision of parts dimensions in terms of its volume, diameters of runner's throat, radius of bending on intake and exhaust port and also the length of parts were taken into consideration during modeling. Every part was created separately in order to be exported in GEM-3D which is a module in GT-Suite v7.4.3 software. Once the parts were successfully modelled in 3D version, they were imported in GEM-3D to be discretized which transforms every part into a 1D diagram. During the discretization process in GEM-3D, the diameter of every pipe section and the wall temperature were set accordingly.

Once the discretization process was completed, every part was exported to the GT-Power module in GT-Suite v7.4.3 software. In this module, the 1D converted parts were arranged accordingly, starting from the air box until the tail pipe. Every part was connected using a string called junction as shown in Fig. 2. The arrangements of the engine are different according to the engine specification and can be found in various previous works [4, 9]. Then, using the engine technical specifications as shown in Table 1 and temperature for engine parts in GT-Suite v7.4.3 software as shown in Table 2, detailed configurations were set for every parameter and engine components are involved. Valves diameters, wall temperatures, flow arrays, valves opening and closing durations were taken into serious consideration. The next thing was setting different air fuel ratios for different engine speeds and the separation to several cases. The simulation runs according to the cases set so that, the output performance can be compared to the actual experimental results which runs with different configurations of air fuel ratios for each and every different engine speed. The output from the successfully performed simulation were shown in GT-Post. The results were collected and tabulated to be used for the validation [10], comparison and optimization process.

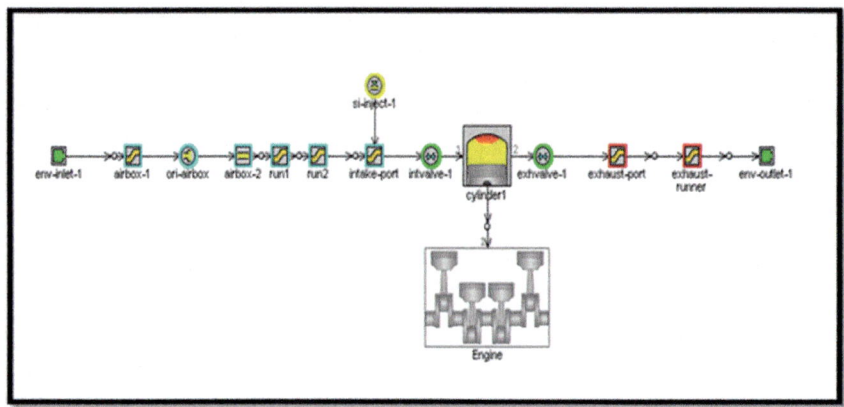

Fig. 2 1D engine model arrangements

Table 1 Engine technical specifications

Engine parameter	Value
Bore × Stroke	53 mm × 50.6 mm
Total displacement	111 mL
Maximum torque	9.3 N m @ 4000 rpm
Maximum horsepower	6.6 kW @ 8500 rpm
Compression ratio	9:3:1
Intake Valve Open (IVO)	20° BTDC
Intake Valve Close (IVC)	60° ABDC
Exhaust Valve Open (EVO)	55° BBDC
Exhaust Valve Close (EVC)	25° ATDC
Total duration intake/exhaust	260°

Table 2 Engine parts temperature in GT-suite software

Engine parts	Temperature (K)
Cylinder head	550
Piston	590
Cylinder block wall	450

3 Results and Discussion

To analyze the effect of different air-fuel ratios and compression ratios on the performance of the hydrogen port fuel injection engine, three different studies have been used.

Phase 1: Comparison and validation of gasoline simulation engine model with real gasoline engine based on the percentage difference in engine performance results.

Phase 2: Conversion of gasoline fuel to hydrogen in the engine model and comparison between results with gasoline and hydrogen.

Phase 3: Optimization of the hydrogen fueled model with different altered parameters obtained from the Design of Experiments module and rerun the simulation with those parameters. Confirmation test performed to validate those results.

3.1 Model Validation

Validation is the process of confirming the results accuracy of the developed model with the results from experiments. Both engine performance results were compared to ensure that the differences are in the range of 5%. At the initial stage, the results from the model shows a quite significant amount of differences compared to the experimental results. The final outcome from the engine performance results comparison between experiment and model are shown in Fig. 3.

During the model development process, the initial parameter sets of the simulation model are rechecked and some minor alterations are performed. There are some increments shown from the alteration of cam timing angle, angle multiplier, lift multiplier and flow array. The cam timing angle for the intake valve is set to 231° and for the exhaust to 123.9°. The angle multiplier is 0.83 and the lift array is 0.5 for both intake and exhaust valves. The flow array is configured using an excel

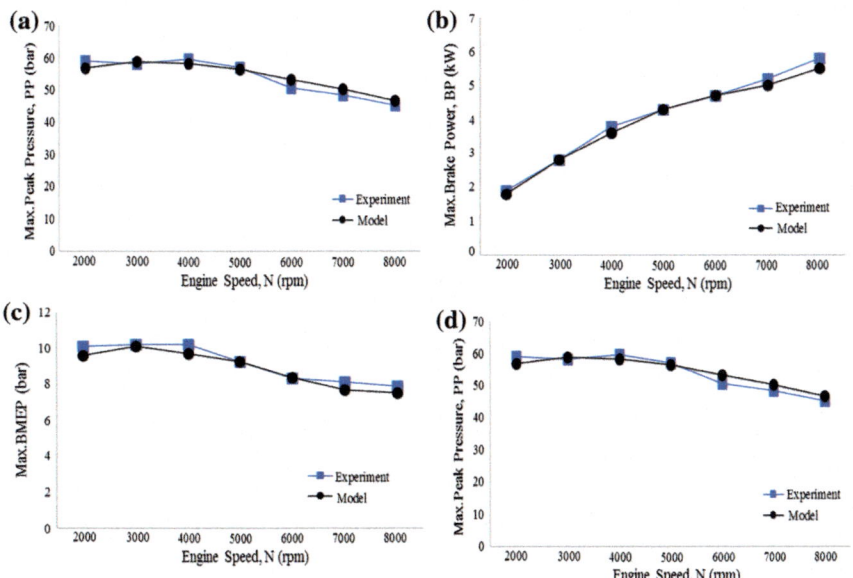

Fig. 3 a Max. brake torque. b Max. brake power. c Max. Bmep. d Max. peak pressure

self-calculating file in GT-Suite. The forward coefficient is adjusted to 0.95 to obtain the flow array values.

With the ten cycle runs of simulation, the engine performance results were plotted in GT-Post. Differentiations between average value of experiment and model are in the range of 0.6–3% for brake torque, brake power, BMEP and peak pressure for all number of engine speed. The accuracy and precision of the developed model is validated with the output results achieved and no more than 5% difference was obtained compared to the experimental engine values for all brake torque, brake power, Bmep and peak pressure.

3.2 Hydrogen Fueled Model

Before the gasoline fueled model is converted into a hydrogen fuel injected engine, the equivalence ratio for gasoline is calculated to suit with the hydrogen model. The stoichiometric air-fuel ratio for hydrogen is 34.3:1 while for gasoline is 14.7:1. To provide a fair comparison, the same equivalence ratios are used for both of the engine systems. The equivalence ratio from gasoline fuel is obtained by dividing the stoichiometric air-fuel ratio of gasoline with air-fuel ratio measured during the experiment [11]. Then, the stoichiometric air-fuel ratio of hydrogen is divided with the same equivalence ratio to obtain the value of air-fuel ratios needed for the hydrogen model as shown in Table 3.

An increase in the equivalent ratio of hydrogen fuel leads to a lower value of air-fuel ratio; this means that the engine operates at richer condition. From Fig. 4, it was found that the engine performance degrades at about 29% for the brake torque, brake power and also BMEP for hydrogen fueled engine. This can be considered as power loss. The loss in brake torque and brake power is mostly due to the low volumetric efficiency and backfiring. Buildup of a small amount of hydrogen in intake manifold when the injection duration is longer than the intake valve opening duration creates backfire by ignites the hydrogen leftover in it with hot exhaust gases released at exhaust stroke [8, 12]. Meanwhile, drops in Bmep are due to

Table 3 AFR configurations of gasoline and hydrogen using same equivalence ratio

Engine speed (rpm)	AFR (Gasoline)	Equivalence ratio	AFR (Hydrogen)
2000	13.41	1.09	31.46
3000	13.05	1.12	30.62
4000	11.16	1.32	25.9
5000	10.66	1.38	24.8
6000	10.54	1.39	24.6
7000	11.34	1.29	26.5
8000	10.68	1.37	25.0

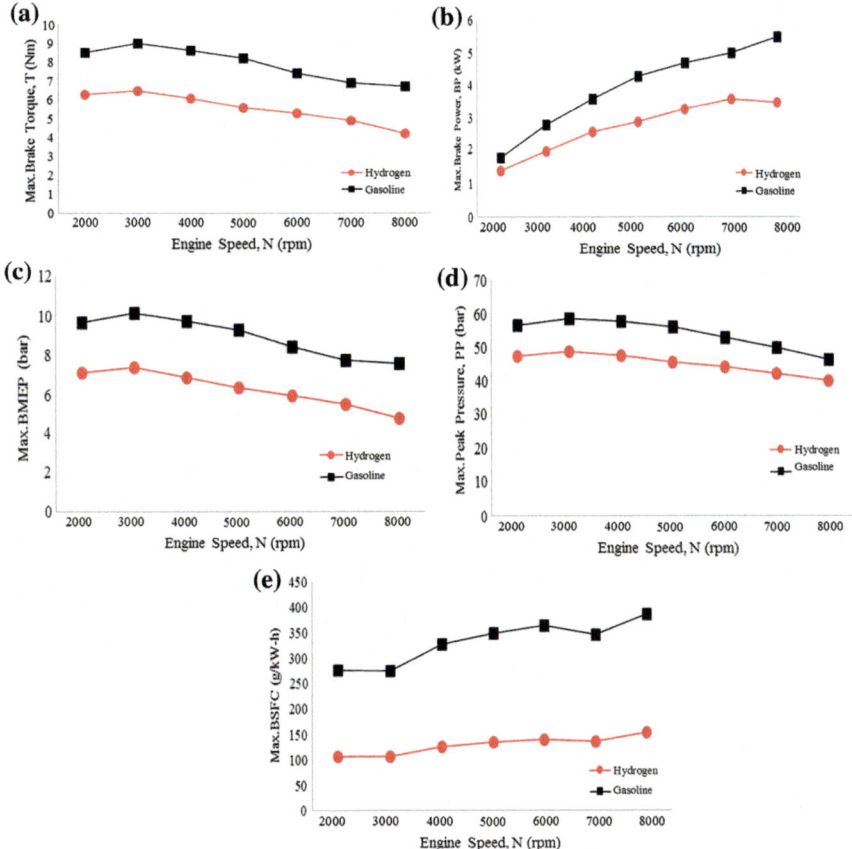

Fig. 4 a Max. brake torque. **b** Max. brake power. **c** Max. Bmep. **d** Max. peak pressure. **e** Max. BSFC

improper combustion. This is due to increasing of the hydrogen fraction in the overall fuel intake which reduces the air intake [6].

For the peak pressure, an average value of 16% difference is achieved from the comparisons. A lower peak pressure is produced due to improper combustion mostly due to lower amount of oxygen induced during the intake process. However, the Bsfc for hydrogen is better than the gasoline. The fuel consumptions of hydrogen fueled model are better at an average value of 61% compared to the gasoline fueled engine. Although the overall engine performance of hydrogen is low compared to gasoline, the Bsfc for hydrogen fueled is better, probably due to the lower mass flow rate of hydrogen compared to gasoline. Therefore, optimization of air-fuel ratio and compression ratio could lead to an improvement in engine performances.

3.3 Hydrogen Model Optimizations

Table 4 shows the design configurations for optimization from the Design of Experiments (DoE) module. The DoE module has been used to obtain various setup values for the air-fuel ratio and compression ratio to study their effects on the performance of the engine. Thirty set of tests were performed using different configurations for hydrogen engine with the engine speed of 3000, 4000, 5000 and 6000 rpm. The inlet valve open timing also was advanced to 40 degree and retarded to 0° before TDC (standard timing is 40° BTDC). While for exhaust valve open timing, it was advanced to 75° and retarded to 35° before BDC. Outputs from the model (brake torque, brake power, Bmep, peak pressure and BSFC) are used as the respond in the software, for optimization process. Table 5 shows the selected optimum configurations which resulted from the Design Expert software.

Figure 5 shows the analysis of engine performance based on the suggested configuration by the Design Expert software. It was found that all of the engine performances had improved due to the optimization of the air-fuel ratio, intake/exhaust valve opening timing and compression ratio. The brake torque increased by an average value of 7.6% compared to before optimization. For the brake power, it had achieved 9.3% improvement with the optimized configuration. The Bmep also had achieved some improvement although still lower compared to gasoline. The peak pressure had achieved the major improvement for the hydrogen fueled engine, and even exceeded the gasoline engine. The improvement on the overall engine performance is due to the increased compression ratio. When the compression ratio is increased, it increases the in-cylinder pressure and temperature, thus increases the in-cylinder pressure during the compression stroke [13]. The higher temperature during the compression stroke resulted in a better fuel conversion efficiency. The air-fuel ratio also plays a significant factor as the combustion is suggested to occur at slightly rich condition. Moreover intake valve open timings are suggested to be advanced to provide longer duration for air and fuel mixing process. Advancing the IVO will ensure a better preparation of fuel and air mixture thus produces a complete and fast combustion process [14].

The brake specific fuel consumption also had shown improvements with higher compression ratios. The reason behind this is the enhancement of fuel conversion efficiency at high engine pressure and temperature. Moreover, the increasing of Bsfc is due to the improvement of power output compared to low amount (mass) of fuel used for combustion process.

Table 4 Optimization values from design expert and respond

No	Air-fuel ratio	Compression ratio	IVO (BTDC)	EVO (BBDC)	Respond
1	29.5	15	20	55	1. Brake torque
2	29.5	11	0	55	2. Brake power
3	24	7	40	35	3. BMEP
4	35	11	20	55	4. Peak pressure
5	24	11	20	55	5. BSFC
6	29.5	11	20	55	
7	29.5	11	40	55	
8	29.5	11	20	55	
9	24	7	0	35	
10	29.5	11	20	55	
11	29.5	11	20	35	
12	35	7	40	35	
13	29.5	11	20	75	
14	29.5	11	20	55	
15	35	15	40	75	
16	35	7	0	35	
17	24	15	0	75	
18	35	7	40	75	
19	35	15	0	35	
20	24	15	0	35	
21	29.5	7	20	55	
22	24	15	40	75	
23	35	15	40	35	
24	29.5	11	20	55	
25	35	7	0	75	
26	35	15	0	75	
27	24	7	40	75	
28	24	15	40	35	
29	29.5	11	20	55	
30	24	7	0	75	

Table 5 Hydrogen engine performance at optimum configurations

Engine speed (rpm)	Air-fuel ratio	Compression ratio	IVO	EVO
3000	35.00	14.95	39.14	49.95
4000	34.99	12.12	30.74	35.00
5000	35.00	12.07	17.43	35.71
6000	34.61	11.74	25.52	35.98

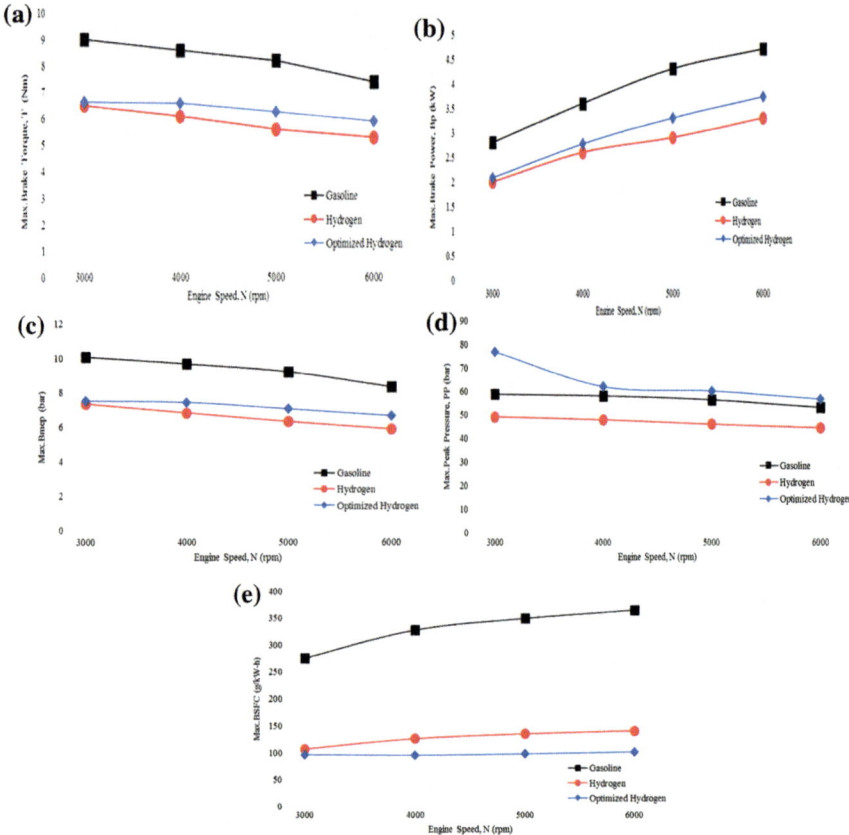

Fig. 5 **a** Max. brake torque. **b** Max. brake power. **c** Max. Bmep. **d** Max. peak pressure. **e** Max. BSFC

4 Conclusion

Based on the simulation study the findings can be summarized as follows;

(a) The analysis with different air-fuel ratios and compression ratios gives significant improvements in the overall engine performance of the hydrogen fueled engine at average values of 7–30%, respectively.
(b) With higher compression ratio; brake torque, brake mean effective pressure and brake power increases throughout every engine speed been tested. Higher compression ratios also produced high peak pressure and thus improved the brake specific fuel consumption.
(c) From the optimization, the total average value for valve opening duration shows that advance the IVO and retard the EVO would improve the overall engine performance.

As the overall conclusion, higher compression ratio and advance IVO would improve the performance of the engine. Balanced configuration in between air to fuel ratio and compression ratio should be handled with the actual hydrogen engine and simulated model in the future.

Acknowledgements The authors thanked the University Kuala Lumpur Malaysian Spanish Institute (UniKL MSI) for research facilities and financial assistance provided for this project.

References

1. Rahman, M.M., Mohammed, M.K., Bakar, R. A.: Effects of Air Fuel Ratio and Engine Speed on Performance of Hydrogen Fueled Port Injection Engine (2009b)
2. Xu, J., Zhang, X., Liu, J., Fan, L.: Experimental study of a single-cylinder engine fueled with natural gas-hydrogen mixtures. Int. J. Hydrogen Energy **35**(7), 2909–2914 (2010)
3. Verhelst, S., Wallner, T.: Hydrogen-fueled internal combustion engines. Prog. Energy Combust. Sci. **35**(6), 490–527 (2009)
4. Rahman, M.M., Mohammed, M.K., Bakar, R.A.: Effect of air fuel ratio on engine performance of single cylinder port injection hydrogen fueled engine: a numerical study. Computer, II (2009a)
5. Karim, A.G.: Hydrogen as a Spark Ignition Engine Fuel, 5 (n.d.)
6. Zheng, J.-J., Wang, J.-H., Wang, B., Huang, Z.-H.: Effect of the compression ratio on the performance and combustion of a natural-gas direct-injection engine. Proc. Inst. Mech. Eng. Part D J. Automobile Eng. **223**(1), 85–98 (2009)
7. Park, C., Lim, G., Lee, S., Kim, C., Choi, Y.: Effects of the ignition timing retard and the compression ratio on the full-load performance and the emissions characteristics of a heavy-duty engine fuelled by hydrogen-natural-gas blends. Proc. Inst. Mech. Eng. Part D J. Automobile Eng. **227**(9), 1295–1302 (2013)
8. Hari Ganesh, R., Subramanian, V., Balasubramanian, V., Mallikarjuna, J.M., Ramesh, A., Sharma, R.P.: Hydrogen fueled spark ignition engine with electronically controlled manifold injection: an experimental study. Renew. Energy **33**(6), 1324–1333 (2008)
9. Said, M.F.M., Aziz, A.A., Latiff, Z.A., Andwari, A.M., Soid, S.N.M.: Investigation of Cylinder Deactivation (CDA) strategies on part load conditions. SAE Technical Paper 2014-01-2549 (2014)
10. Bos, M.: Validation Gt-Power model cyclops heavy duty diesel engine. Master Thesis, pp. 1–110 (2007)
11. Sonthalia, A., Rameshkumar, C., Sharma, U., Punganur, A., Abbas, S.: Combustion and performance characteristics of a small spark ignition engine fuelled with HCNG. J. Eng. Sci. Technol. **10**(4), 404–419 (2015)
12. Verhelst, S., Maesschalck, P., Rombaut, N., Sierens, R.: Efficiency comparison between hydrogen and gasoline, on a bi-fuel hydrogen/gasoline engine. Int. J. Hydrogen Energy **34**(5), 2504–2510 (2009)
13. Patil, S.: Investigation on effect of variation in compression ratio on performance and combustion characteristics of C.I engine fuelled With Palm Oil Methyl Ester (POME) and Its blends by simulation. Glob. J. Res. Eng. **12**(3), 1–8 (2012)
14. Soid, S.N., Zainal, Z.A., Iqbal, M.A., Miskam, M.A.: Macroscopic spray characteristics of palm oil-diesel blends in a contant volume combustion chamber. J. Sci. Ind. Res. **71**, 740–747 (2012)

Evaluation of the Hardness Distribution and Fracture Location in Friction Stir Welded AA6063 Pipe Butt Joints

Azman Ismail, Mokhtar Awang, Fauziah Ab Rahman, Bakhtiar Ariff Baharudin, Puteri Zarina Megat Khalid and Darulihsan Abdul Hamid

Abstract Frictional heat generated during the friction stir welding (FSW) process will somehow affect the mechanical properties of friction stir welded joints. In this present study, the FSW of aluminium alloy (AA) 6063 pipe butt joints with 5 mm wall thickness has been performed, and the evaluation of fracture and hardness distribution across the weld cross section is primarily discussed. Several selected rotational and travel speeds are used for specimen preparation. It was found that the fracture location closely matched with the lowest hardness region either on the advancing or retreating side of the welded joints.

Keywords Friction stir welding · Fracture · Hardness · Aluminium Pipe

---

A. Ismail (✉) · F. A. Rahman · B. A. Baharudin · P. Z. M. Khalid
Universiti Kuala Lumpur Malaysian Institute of Marine Engineering Technology,
Jalan Pantai Remis, 32200 Lumut, Perak, Malaysia
e-mail: azman@unikl.edu.my

F. A. Rahman
e-mail: fauziahabra@unikl.edu.my

B. A. Baharudin
e-mail: bakhtiarab@unikl.edu.my

P. Z. M. Khalid
e-mail: puterizarina@unikl.edu.my

A. Ismail · M. Awang
Department of Mechanical Engineering, Universiti Teknologi PETRONAS,
Bandar Seri Iskandar, 32610 Teronoh, Perak, Malaysia
e-mail: mokhtar_awang@utp.edu.my

D. A. Hamid
Universiti Kuala Lumpur Institute of Product Design and Manufacturing,
No.9, Jalan Perdana 7/3, Taman Shamelin Perkasa, 56100 Kuala Lumpur, Selangor, Malaysia
e-mail: darulihsan@unikl.edu.my

© Springer International Publishing AG 2018
A. Öchsner (ed.), *Engineering Applications for New Materials and Technologies*,
Advanced Structured Materials 85, https://doi.org/10.1007/978-3-319-72697-7_35

1 Introduction

Friction stir welding (FSW) is the best option available in order to perform well on difficult-to-weld materials such as aluminium. This happens without distortion and is free from defects. Besides butt joints, several other joint configurations can be made such as tee joints and lap joints for a wide range of material thickness which is up to 100 inches. This green technology of joining process has become the choice of many industries nowadays in aerospace, shipbuilding and automotive fields [1]. The FSW technology can be used to join similar and dissimilar metal alloys with the utilization of frictional heat which softens and allows the pin tool to stir/join along the weld line to form a soundly joint. FSW could eliminate various weld defects that are associated with fusion arc welding.

Most studies are done on microstructure changes, temperature profiles, tensile properties and hardness distribution of the friction stir welded (FSWed) joint, while a few are focused on the fracture behavior [2–4]. The origin of the tensile fracture is due to the entrapment of an oxide layer in the weld zone and is only dependent on the microhardness distributions across the FSWed joint [1, 5]. The best combination of welding parameters can be applied in order to improve the properties of these joints. The present work focuses on the evaluation of fracture and hardness distribution in friction stir welded aluminium alloy (AA) 6063 pipe butt joints. The experimental results will be discussed accordingly.

2 Experimental Setup

The FSW of aluminium alloy (AA) 6063 of 89 mm outside diameter pipe butt joint with 5 mm wall thickness was performed. The length of each specimen was set to 100 mm each. Then, each two sections were joined together by the FSW process. A special jigs was utilized to perform this joining process on the Bridgeport 2216 CNC milling machine. Figure 1 shows the FSW experimental setup on the Bridgeport 2216 CNC milling machine. The tool was made available by dimensioning and the setting from [6] with 2° tilt and 6 mm offset forward the centerline.

The welding parameters were set into two categories, with the increment of rotational speed at constant travel speed and with the increment of travel speed at constant rotational speed. The selected welding parameters were based on previous studies done by other researchers [7, 8]. Tables 1 and 2 show the welding parameters for this experiment setting which were able to produce quality joints.

The hardness for each specimen was measured by the hardness tester model AFFRI 206 RTD with a maximum load of 100 kg based on the ASTM E18 requirement [9]. The Rockwell B scale (HRB) was used for this hardness measurement unit. Several locations were identified as point of measurement covering the stirred zone (SZ), thermo-mechanically affected zone (TMAZ), heat affected zone (HAZ) and base metal (BM) as shown in Fig. 2. The visual inspection

Fig. 1 The FSW jig setup on Bridgeport 2216 CNC milling machine

Table 1 FSW parameters with increment of rotational speed and constant travel speed

FSW samples	Welding parameters	
	Rotation speed (rpm)	Travel speed (mm/min)
FSW#1	900	72
FSW#2	1200	72
FSW#3	1500	72

Table 2 FSW parameters with increment of travel speed and constant rotational speed

FSW samples	Welding parameters	
	Rotation speed (rpm)	Travel speed (mm/min)
FSW#3	1500	72
FSW#4	1500	108
FSW#5	1500	144

Fig. 2 Hardness measurement area

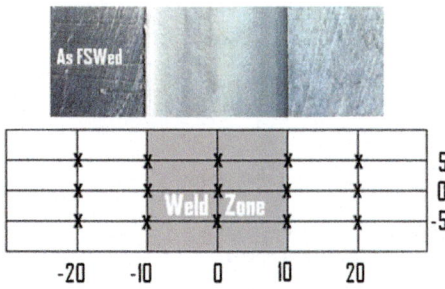

was based on the ISO 25239 standard acceptance level [10]. Three tensile specimens were prepared for each parameter setting. These tensile tests were performed using a servo controlled universal testing machine based on the ASTM E8 M-04 standard [11].

3 Results and Discussion

The hardness measurement is shown in Figs. 3 and 4. Figure 3 shows the hardness measurement for the increment of rotation speed at constant travel speed meanwhile Fig. 4 shows the hardness measurement for the increment of travel speed at constant rotational speed.

Based on these figures, the hardness dropped to the lowest at 33.8 HRB with the increment of rotational speed. Meanwhile, the hardness stayed at a high value of 40.6 HRB with the increment of the travel speed. The frictional heat generated with the increment of rotational and travel speed was responsible for the drop in hardness. The stirred zone (SZ) gave the lowest range of hardness due to the mechanical stirring process heavily occurred in this zone which caused the occurrence of the high dislocation density. The base metal (BM) shows a high hardness value in this zone. The less heat affected zone will produce higher reading of hardness value except for the less rotational speeds parameter setting. The variation in hardness presence was due to variation of welding parameters.

The fracture location for each friction stir welded specimen is shown in Table 3 below. The friction between the tool shoulder and pipe surface generated heat. This heat generated mainly depended on the tool rotation and travel speed set for this experiment. Either compatibility or incompatibility of these welding parameters (rotational and travel speeds) will determine the result of these joining quality.

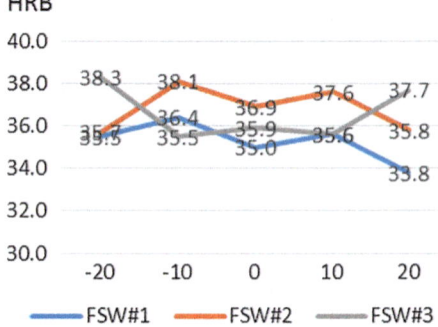

Fig. 3 Hardness value along the cross section for increment of rotational speed at constant travel speed

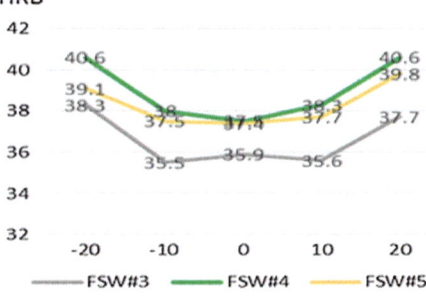

Fig. 4 Hardness value along the cross section for increment of travel speed at constant rotational speed

Table 3 Fracture locations of FSW specimens

FSW Id No.	Advancing	Retreating
FSW#1		
FSW#2		
FSW#3		
FSW#4		
FSW#5		

In this present study, the lowest hardness value caused the fracture to occur at the center line especially in the stirred zone (SZ) for FSW#1 and FSW#5. The compatibility of welding parameters used gave a better mixture of materials thus causing the fracture to occur either on the advancing or retreating side of the specimens as shown in FSW#2, FSW#3 and FSW#4.

4 Conclusion

In this study, the hardness distributions and fracture locations across the friction stir welded joints were evaluated experimentally. The compatibility of welding parameters can be determined from the relation between the characteristic of these hardness distribution and fracture location. The fracture locations of the joints are mainly dependent on the weld defects (if any) and hardness distributions of the joints, which are usually related to the welding parameters.

Acknowledgements The authors would like to acknowledge the facilities provided by Universiti Teknologi PETRONAS and Universiti Kuala Lumpur. This project was financially funded by MARA Product and Innovation Research Grant.

References

1. El-Nasr, A.B.A.A.: Mechanical properties and fracture behavior of friction stir welded 7075-T6 aluminum alloy, J. Eng. Comput. Sci. Qassim University, **3**(2), 147–161

2. Khodir, S.A., Shibayanagi, T., Naka, M.: Microstructure and mechanical properties of friction stir welded AA2024-T3 aluminum alloy, Mater. Trans. The Japan Institute of Metals, **47**(1), 185–193
3. Khodir, S.A., Shibayanagi, T., Naka, M.: Control of hardness distribution in friction stir welded AA2024-T3 aluminum alloy. Mater. Trans. **47**(6), 1560–1567 (2006)
4. Ismail, A., Awang, M.: Surface hardness of friction stir welded AA6063 pipe. MATEC Web Conf. **13**(04025), 1–5 (2014)
5. Liu, H.J., Fujii, H., Maeda, M., Nogi, K.: Tensile properties and fracture locations of friction-stir-welded joints of 2017-T351 aluminum alloy. J. Mater. Process. Technol. **142** (2003), 692–696 (2003)
6. Ismail, A., Awang, M., Fawad, H., Ahmad, K.: Friction stir welding on aluminum alloy 6063 pipe, In: Proceedings of the 7th Asia Pacific IIW International Congress 2013, pp. 78–81. (2013)
7. Lammlein, D.H., Gibson, B.T., DeLapp, D.R., Cox, C., Strauss, A.M., Cook, G.E.: The friction stir welding of small-diameter pipe: an experimental and numerical proof of concept for automation and manufacturing, Proc. Inst. Mech. Eng. Part B: J. Engineering Manufacture, pp 1–16. (2011)
8. Gercekcioglu, E., Eren, T., Yildizh, K., Kahraman, N., Salamci, E: The friction behavior on the external surface of the friction stir welding of AA6063-T6 tubes, 5th Int Conf on Tribology, pp. 225–228. (2005)
9. ISO 25239, Friction stir welding—aluminium, International Standard Organization, (2011)
10. ASTM E18, standard test method for Rockwell hardness and Rockwell superficial hardness of metallic materials, ASTM International
11. ASTM E8 M-04, Standard test methods for tension testing of metallic materials, ASTM International

Study on Gas Emission of Saline Water from a Hydrogen System

Zakiman Zali, Norsheila Buyamin, Nazihah Mohd Noor, Rawaida Muhammad, Normah Ishak and Rohimi Yusof

Abstract The crisis of global pollution is increasing dramatically. Thus, energy generating processes through the use of natural resources such as water, sunlight, air or minerals from the earth with low gas emission are becoming more and more important. The objective of this study is to produce hydrogen through electrolysis of saline water on a vehicle capable of delivering positive impact on emission reductions. The hydrogen produced would support the petrol system used in combustion engines. The prototype model of the hydrogen engine is fabricated to become a hydrogen generation system, which is reliable in meeting the basic needs of the internal continuously burning combustion engine. This process occurs continuously to help the petrol engine system and to reduce the emissions. Then, by using the prototype, the exhaust gas from the combustion engine is analysed to evaluate the engine performance and also to diagnose the gas emission. Four different combination modes of hydrogen and petrol have been tested in order to get the best combination that may lead to lowest fuel consumption and low level of pollution. Based on the analysis done, a combination of 75% petrol and 25% hydrogen gas shows a low level of harmful gases pollution and this combination of 75% is good to minimize the effect on the ozone layer. Thus, it can be concluded that by using this hydrogen system, the level of pollution can be decreased.

Keywords Gas emission · Saline water · Hydrogen system · Pollution

---

Z. Zali (✉)
Department of Marine Engineering, Politeknik Ungku Omar,
Jalan Raja Musa Mahadi, 31400 Ipoh, Perak, Malaysia
e-mail: namikazzali@gmail.com

N. Buyamin · N. M. Noor · R. Muhammad · N. Ishak · R. Yusof
Department of Mechanical Engineering, Politeknik Ungku Omar,
Jalan Raja Musa Mahadi, 31400 Ipoh, Perak, Malaysia
e-mail: nsheila@puo.edu.my

N. M. Noor
e-mail: nazihah@puo.edu.my

© Springer International Publishing AG 2018
A. Öchsner (ed.), *Engineering Applications for New Materials and Technologies*,
Advanced Structured Materials 85, https://doi.org/10.1007/978-3-319-72697-7_36

1 Introduction

In recent years, concerns about exhaust emission from vehicles are drastically increasing. Apart from that, the issue of improving the emission performance of internal combustion (IC) is very important in the era of globalization and many researchers are studying to find clean alternative fuels for IC. Vehicles are greatly contribution to gas emissions in the socio-economic development with high environmental impact [1]. Normally, IC engines are using petrol fuel and are emitting gasses to the environment such as carbon dioxide (CO_2), nitrogen oxides (NO_x), sulfur dioxide (SO_2) and carbon monoxide (CO_2).

In fact, potential alternative fuels have been proposed such as methanol, biodiesel, ethanol, hydrogen, boron, natural gas, liquefied petroleum gas (LPG), Fischer-Tropsch fuel, p-series, electricity and solar fuels [2]. One of these alternative fuels, i.e., hydrogen, has unique properties as a fuel use in IC engines where it gives the potential of near zero emissions of all emission gases. Moreover, hydrogen has special properties such as colourless, odourless, tasteless and non-toxic. Hydrogen can be produced from natural gas, oil, coal and from electrolysis sources. As shown in Table 1 [2], approximately 48% of the hydrogen was produced from natural gas while 4% of the hydrogen was produced from electrolysis source.

Table 1 Annual global hydrogen production share by source

Source	Bcm^a/y	Share
Natural gas	240	48
Oil	150	30
Coal	90	18
Electrolysis	20	4
Total	500	100

^{a}Bcm: Billion cubic meters

This study is focused on an electrolysis source where hydrogen will be produced from saline water by using a variety of energy sources such as wind, solar, geothermal or nuclear energy. Since saline water is similar to sea water in terms of composition, hydrogen can be produced from this aqueous solution. In the electrolysis of sea water as a hydrogen producer, two options exist for the performance of

---

R. Muhammad
e-mail: rawaida@puo.edu.my

N. Ishak
e-mail: normahishak@puo.edu.my

R. Yusof
e-mail: rohimi@puo.edu.my

the electrolysis process. The first option is to remove all dissolved salts and to produce essentially distilled water while the second option is to design an electrolyze system capable of utilizing sea water for direct electrolysis [3]. The main advantage of using sea water is the natural elimination of the waste brine which is only slightly enriched with salts. Furthermore, saline water is a more economical capital cost and environmental friendly where this production is directly produced for marine application. Apart from that, the gas emission from gasoline engines should be determined to show the positive impact emission reduction from hydrogen ignition. In comparison with a burning petrol fuel, the NO_x emission is far less for hydrogen fuel [2]. While previous studies have shown that using hydrogen has decreased the CO and HC gas emission, the NO_x gas emission was slightly increased [4–6].

Nowadays, the crisis of global pollution is increasing dramatically and this gives an effect to the ecosystem of the earth. Apart from that, energy will be produced through the use of natural resources such as water, sunlight, air or minerals from the earth with low gas emission. Therefore, the objective of this study is to produce hydrogen gas (H_2) by an electrolysis system using saline water as a supporter of the fuel that is capable of delivering a positive impact emission reduction.

2 Experimental Setup and Procedure

2.1 Experimental Setup

The experiment was carried out in a gasoline engine with the following specifications: 169 cc, air-cooled, 4-cycle single head cylinder, and manufactured by SUBARU Fuji Heavy Industries, Japan. However, this engine was modified to enter hydrogen gas and to mix it together with petrol fuel into the combustion chamber.

Apart from that, the hydrogen system was built up to produce hydrogen gas as shown in Fig. 1. This study used saline water and stainless steel as a solution and an electrode in the electrolysis process. The chemical formulation for the reduction at the cathode reads:

$$2H_2O(aq) + 2e^- \rightarrow H_2(g) + 2OH^-$$

In salt water at the negatively charged cathode, a reduction reaction takes place, with electrons (e^-) from the cathode being given to the hydrogen cations to form hydrogen gas. The oxidation at the anode is:

$$2Cl^-(l) \rightarrow Cl_2(g) + 2e^-$$

At the positively charged anode, an oxidation reaction occurs, generating chlorine gas and giving electrons to the anode to complete the circuit. The overall reaction is:

$$2Cl^-(l) + 2H_2O(aq) \rightarrow Cl_2(g) + H_2(g) + 2OH^-$$

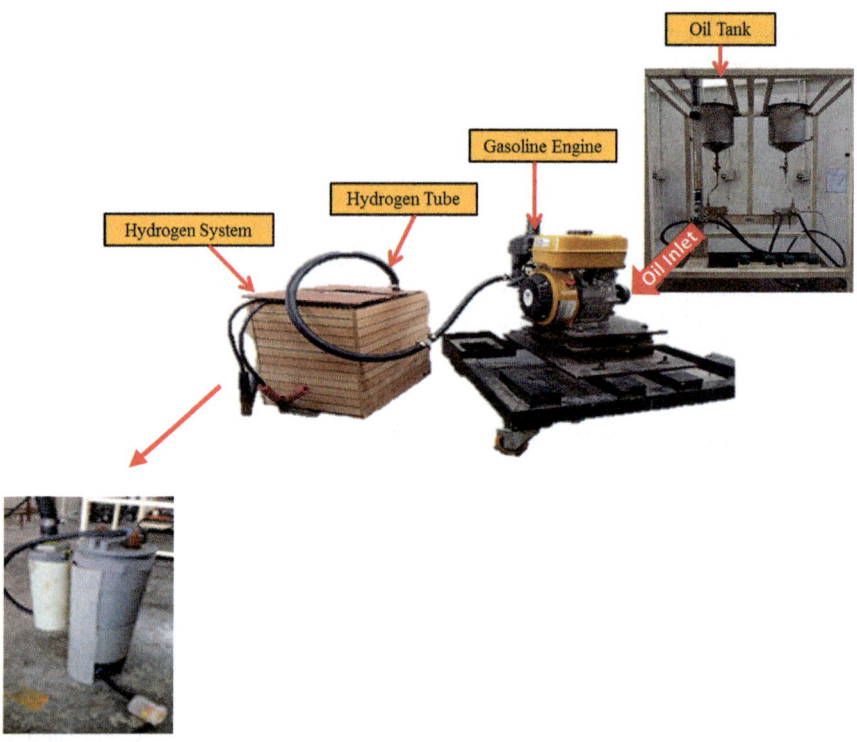

Fig. 1 Design of hydrogen system for experimental setup

The number of hydrogen molecules produced is twice the number of chlorine molecules. Assuming equal temperature and pressure for both gases, the produced hydrogen gas has twice the volume of the produced chlorine gas. The number of electrons pushed through the saline water which is twice number of generated hydrogen molecules with four times the number of generated chlorine molecules.

Thus, this has been suggested as a way of shifting the society towards using hydrogen as an energy carrier for powering electric motors and internal combustion engines. Electrolysis of an aqueous solution of table salt (NaCl or sodium chloride) produces aqueous sodium hydroxide and chlorine, although usually only in minute amounts. NaCl (aq) can be reliably electrolysed to produce hydrogen. All the hydrogen produced will be stored into a generator tank before entering to the gasoline engine. The exhaust emission of CO, NO_x, SO_2 and H_2S are measured by the OPTIMA 7 emission monitoring system analyser.

2.2 Experimental Procedure

In this study, four experimental configurations have been tested with different combinations of hydrogen and petrol in order to get the best result in regards to lower fuel consumption and low level of pollution to the environment. The experiment started when the engine was fully warmed with the idle engine speed and fully petrol. The control valve of hydrogen was closed to ensure the hydrogen gas cannot flow to the engine. During the experiment, the amount of hydrogen gas in the total intake was uniformly increased due to the combination mixture between hydrogen and petrol. For a specified hydrogen quantity, the amount of petrol fuel was reduced and controlled by the control valve. Hydrogen and petrol were injected into each intake port and mixed together into the IC with air. The data of gas emission has been taken when the engine speed was increased at the specified value.

3 Results and Discussions

Figure 2 illustrates the effect of hydrogen on the carbon monoxide emission between idle speed and 3500 rpm engine speed for the gasoline engine. This experiment was performed for four combinations ratios which are 100% petrol, 75% petrol with 25% hydrogen, 50% petrol with 50% hydrogen and 25% petrol with 75% hydrogen. The figure shows that the emission gas from 100% of petrol was increased drastically followed by the combination of 50% hydrogen and 25% hydrogen. On the other hand, a combination of 75% hydrogen was declined dramatically at an engine speed of 2000 rpm and continuously decreased slightly from 4393 to 4153 ppm. Currently, Malaysia is the second largest per capita greenhouse gas (GHG) emitter among the ASEAN countries [1]. Therefore, a combination of 75% hydrogen is good for the environment to minimize the effect on the ozone layer and also to control the global warming potential (GWP).

Figure 3 depicts the hydrogen effect on the amount of NO_x emission due to additional engine speed. For lower engine speeds, the data on gas contamination NO_x for 100% petrol is lower than for others combination tests. Normally, the NO_x emission is generally increased in the engine start processes [6]. However, the emission of NO_x for 100% petrol was increased drastically after engine speed is 2000 rpm and the others combination were decreased slightly at the specified engine speed. As is shown in Fig. 3, NO_x was increased after hydrogen addition from 25 to 50% because the fast burning velocity and high flame temperature of hydrogen. The addition of 25% hydrogen is a good formation supporter for the fuel to reduce the level of pollution and harmful gas emission.

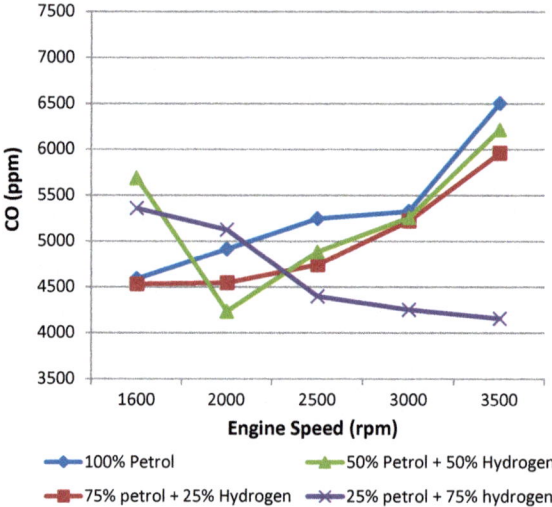

Fig. 2 Effect of addition engine speed (rpm) on CO emission at different combination hydrogen

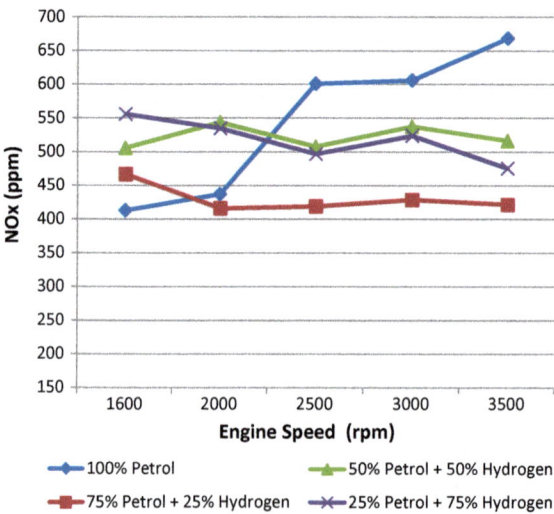

Fig. 3 Effect of addition engine speed (rpm) on NO_x emission at different combination hydrogen

Figures 4 and 5 show the effect on other gas emission by SO_2 according to the addition of engine speed at different combination ratios of hydrogen. As shown in Fig. 4, the 75% petrol mixed with 25% hydrogen emits the lowest SO_2 at any engine speeds (1600–3500 rpm). Figure 5 shows a decreasing trend of H_2S emission for the red-line with increasing of engine speed even though at 1600 rpm it is marginally higher compared to 100% petrol.

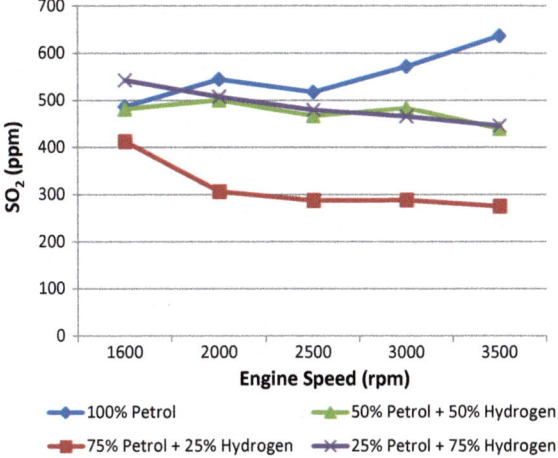

Fig. 4 Effect of addition engine speed (rpm) on SO_2 emission at different combination hydrogen

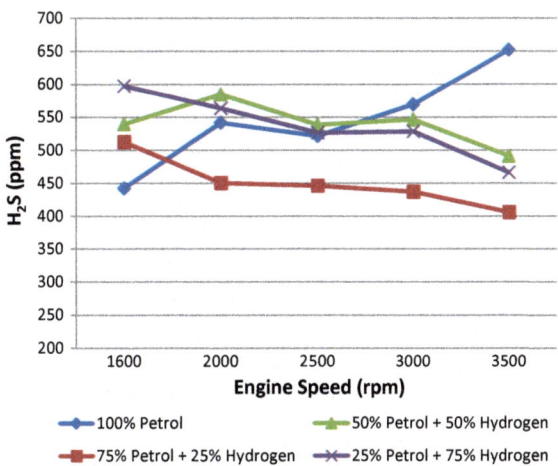

Fig. 5 Effect of addition engine speed (rpm) on H_2S emission at different combination hydrogen

For both figures, by using only petrol generated higher SO_2 and H_2S at higher engine speeds. While mixtures of 50% petrol and hydrogen and also 25% petrol and 75% hydrogen showed almost identical results of SO_2 and H_2S emission. In general, mixed 75% petrol with 25% hydrogen shows the lowest results of SO_2 and H_2S emission. The burn duration for hydrogen gas and petrol were found to be increasing respectively which infers that hydrogen has a very short combustion duration rather than petrol fuel [7]. So, harmful gasses emissions like SO_2 and H_2S were reduced for that hydrogen gas rather than for petrol.

4 Conclusions

This study introduced an experiment aiming at investigating the effect of engine speed on emission gases at different combinations of hydrogen. The test was carried out on a modified single head cylinder gasoline engine with a hydrogen system. The main conclusions are listed below:

- By accomplishing this design, at lower engine speed with lower petrol use, gas the contamination from this modified gasoline engine is higher than for full petrol but these emission gases become lower than for full petrol when the engine speed is increased.
- The level of pollution for a combination of 75% petrol and 25% hydrogen gas is lower than using full petrol, even at the same engine speed.
- But, a combination of 25% petrol and 75% hydrogen is good to minimize the effect of the ozone layer and GWP.
- Hydrogen that is produced by electrolysis of saline water is unlimited, thus the energy that is produced by a combination of both hydrogen gas and petrol can be considered as renewable energy. Hopefully, by implement this kind of system into modern gasoline engines, the rate of gas pollution would be lower and possibly can replace the fossil fuel in the future.

Acknowledgements The authors would like to acknowledge Ministry of Education Malaysia and Ungku Omar Polytechnic (PUO) to support this research.

References

1. Shahid, S., Minhans, A., Puan, O.C.: Assessment of greenhouse gas emission reduction measures in transportation sector of Malaysia. Jurnal teknologi **70**, 1–8 (2014)
2. Balat, M.: Potential importance of hydrogen as a future solution to environmental and transportation problems. Int. J. Hydrogen Energy **33**, 4013–4029 (2008)
3. Abdel-Aal, H., Zohdy, K., Kareem, M.A.: Hydrogen production using sea water electrolysis. Open Fuel Cells J. **3**, 1–7 (2010)
4. Du, Y., Yu, X., Wang, J., Wu, H., Dong, W., Gu, J.: Research on combustion and emission characteristics of a lean burn gasoline engine with hydrogen direct-injection. Int. J. Hydrogen Energy **41**, 3240–3248 (2016)
5. Ji, C., Wang, S.: Effect of hydrogen addition on lean burn performance of a spark-ignited gasoline engine at 800 rpm and low loads. Fuel **90**, 1301–1304 (2011)
6. Ji, C., Yang, J., Liu, X., Wang, S., Zhang, B., Wang, D.: Enhancing the fuel economy and emissions performance of a gasoline engine-powered vehicle with idle elimination and hydrogen start. Appl. Energy **182**, 135–144 (2016)
7. Chitragar, P., Shivaprasad, K., Nayak, V., Bedar, P., Kumar, G.: An Experimental Study on Combustion and Emission Analysis of Four Cylinder 4-Stroke Gasoline Engine Using Pure Hydrogen and LPG at Idle Condition. Energy Procedia **90**, 525–534 (2016)

Analysis of Human Behavior During Braking for Autonomous Electric Vehicles

Khairul Ikram, Wan Khairunizam, A. B. Shahriman, D. Hazry, Zuradzman M. Razlan, Hasri Haris, Hafiz Halin and Chin S. Zhe

Abstract Nowadays, the development of modern technologies in the transportation area has increased rapidly. The latest technology is focused on autonomous vehicles. The most important aspect in autonomous vehicle control systems is safety, smooth operation and comfort. It can be provided by the natural element which is from human behavior as a reference to the autonomous control system. This paper is briefly describing the human behavior during braking for autonomous electric vehicle (AEV) control system development. To obtain the human behaviour during accelerating and braking, two units of angle sensors were installed to the accelerator and brake pedal of an electric car to measure the angle of both pedals. Real-time speed is recorded by using a GPS unit. The experiment is focused on braking characteristic of the electric vehicle. It was carried out based on four different distances with fixed initial speed before braking action was recorded. A software interface was designed to display and control real-time speed and pedal angle value during operation.

Keywords Autonomous electric vehicle · Transportation · Autonomous control system · Human behaviour

---

K. Ikram (✉) · W. Khairunizam · A. B. Shahriman · D. Hazry · Z. M. Razlan · H. Haris · H. Halin · C. S. Zhe
School of Mechatronic Engineering, Universiti Malaysia Perlis UniMAP, Kampus Pauh Putra, 02600 Arau, Perlis, Malaysia
e-mail: iqram832@yahoo.com

W. Khairunizam
e-mail: khairunizam@unimap.edu.my

A. B. Shahriman
e-mail: shariman@unimap.edu.my

H. Haris
e-mail: hasri@unimap.edu.my

H. Halin
e-mail: hafiz@unimap.edu.my

© Springer International Publishing AG 2018
A. Öchsner (ed.), *Engineering Applications for New Materials and Technologies*,
Advanced Structured Materials 85, https://doi.org/10.1007/978-3-319-72697-7_37

1 Introduction

In the recent time, the technology in transportation has become more advanced day by day [1]. It has transformed from old internal combustion engines to hybrid systems before the autonomous electric vehicle (AEV) was invented. Longitudinal control becomes the fundamental component in AEV technologies to keep the vehicle moving at the desired speed by controlling throttle and brake precisely. This is to ensure that the driver is safe and comfortable during accelerating and braking. In acceleration control, Bjomberg [2] has proposed an autonomous intelligent cruise control technique to aid the driver to keep the desired speed on the highway.

For braking control, many new research and inventions were developed day by day. In 2003, Emami [3] was proposing an antilock braking system (ABS) using a discrete-time adaptive fuzzy sliding mode controller. The non-linear function of the plant was controlled using two fuzzy approximations. The aim of the project was to reduce the dependence on mathematical models, just assuming on certain upper and lower bound of uncertainties. The development of autonomous control continued using the robotic driver by Nicholas [4] in 2008. PID controllers were used to control both steering and pedal actuator. Input is from a combination between a 6-axis inertial measurement unit and GPS. In 2011, Chang [5] developed an electronic braking system (EBS) to increase the braking performance in standard cars. This system focused on passenger comfort. The main point was to obtain the maximum deceleration based on different road condition.

In 2012, Jonas [6] has introduced a system to avoid collision by automatic control on brake and steering during emergency events. Closed-form expressions were derived based on longitudinal prediction and measurement errors. For autonomous braking, Kopert [7] highlighted the autonomous emergency braking system in 2013 for high safety standard in commercial cars. In 2014, Sachin [8] highlighted the automatic car braking system using a combination between fuzzy logic and PID controller. Responses on both controllers were compared to achieve the best result. By latest, Frederik [9] has introduced driver intention recognition to automatically give warning and braking in emergency situation on safe behavior planning on 2015. The automatic system could be cancelled in case of existing braking intention by the driver.

This paper is organized as follows. Section 1 describes the literature review in development of brake and acceleration control during recent time. The following section introduces hardware configurations, design of experiment and all the methods used to extract data in this paper. Section 3 shows the test result of the experiment and the conclusion.

2 Methodology

2.1 Flow Chart

The overall process to obtain braking characteristic of human behavior is shown in Fig. 1. The speed control configuration of the electric cars (golf buggy car) is determined before deciding the methods for measuring human characteristic behavior on pedal control. Based on the speed change characteristic, it is proportional to the movement of the accelerator pedal, so we decided to use the pedal angle as the primary measure for the position of the pedal.

As to measure the pedal angle, a pair of the potentiometers (see Fig. 2) was installed into the car for both brake and accelerator pedals. Global Positioning System (GPS) was used to indicate the real time speed (see Fig. 3).

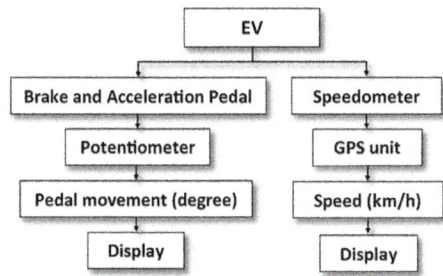

Fig. 1 Flowchart of the buggy car speed control

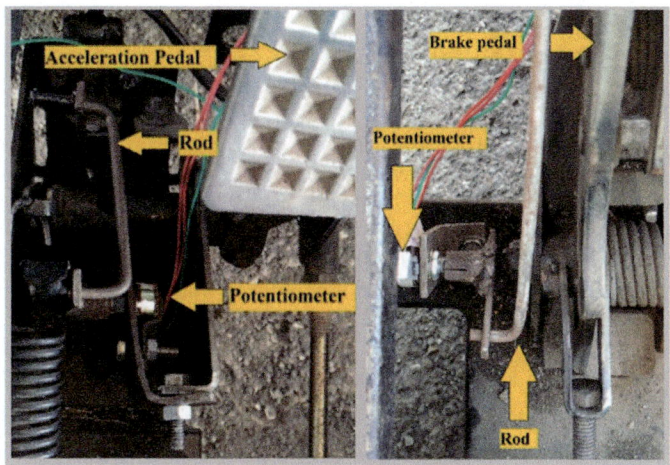

Fig. 2 Potentiometer installation on brake and accelerator pedal

Fig. 3 The GPS unit for speed measurement

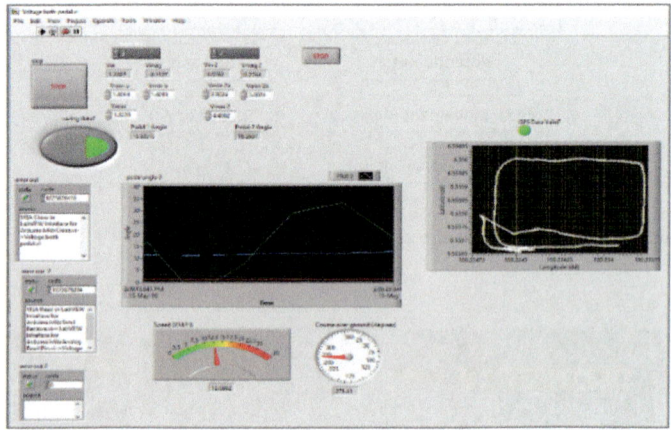

Fig. 4 User interface by using LabVIEW software

During the experiment, all input data will be displayed on a computer by using a pre-designed user interface based on the LabVIEW software as shown in Fig. 4. The real-time data will be recorded before analyzed after the experiment.

2.2 Design of Experiment

The experiment is carried out at the Institut Kemahiran MARA Perlis (IKM). Based on the design of experiment (DOE) as shown in Fig. 5, to obtain appropriate data for braking characteristic, the test is performed in terms of braking distance. Four different distances were chosen which is 5, 8, 10 and 17 m by considering the speed of the car before stopping action. The hallway has 50 m in distance to maintain the car speed at 15 km/h before entering the braking zone. There were five selected subjects as the respondents in this trial of braking [4]. The respondents are males and ages between 19 and 26 years old with driving experience.

2.3 Experimental Result

In this section we present the experimental result based the DOE in Fig. 5. An analysis of the experimental graph in Fig. 6 shows the relationship between the accelerator and brake pedal angle with the speed change, especially the effect of braking action in terms of speed drop. The first and second graph from the top, shows the acceleration and braking characteristic based on pedal angle (degree). Maximum pedal angle for accelerator and brake pedal are respectively 40° and 22°. The bottom graph indicates the corresponding speed (km/h). All the graph's x-axis represents the time in second (s). Four different color's lines in the graph indicate the four different distances that were tested.

Refer to Fig. 6, the initial condition is maintaining speed at 17 km/h for all distances. When entering the braking zone, volunteers must stop the car before the stop line (based on the given braking range) which is 5, 8, 10 and 15 m. Volunteers were advised to operate the pedals naturally as they drive a normal car. We can observe two main aspects from the graph which is the time taken to brake from zero-degree angle to fully pressed and the time taken for the vehicle start to decelerate until it fully stops. The different distance ranges have different results. During 5 m operation, braking action started on 20th second and the pedal was fully pressed on 22th second. So it takes 2 s to fully press the pedal. The 8 and 10 m braking graph is equally consuming 3 s to before the brake pedal is fully

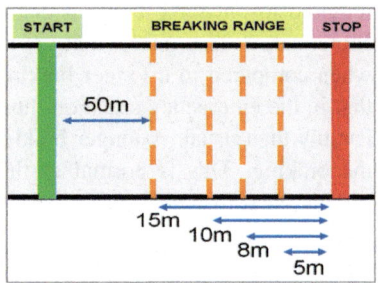

Fig. 5 Design of experiment (DOE)

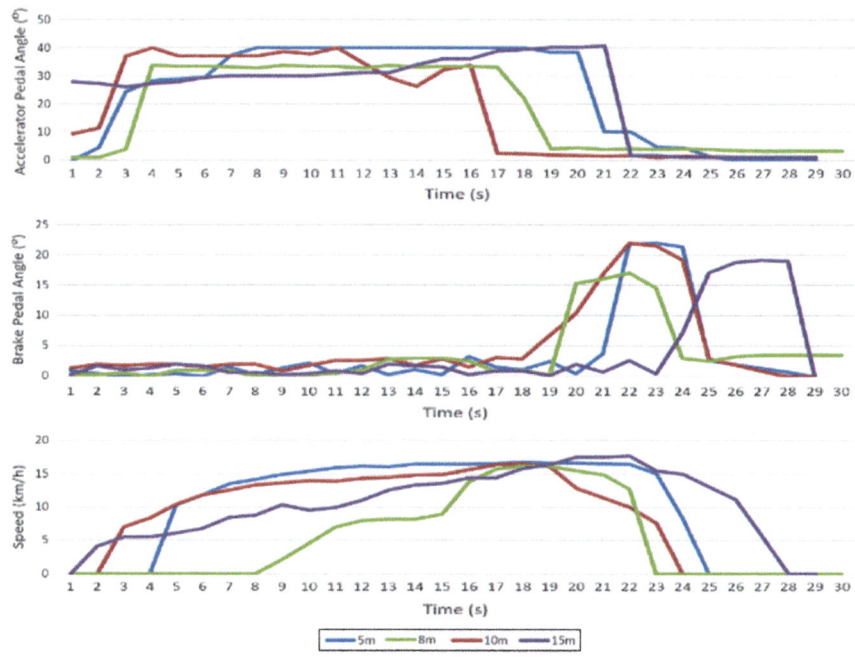

Fig. 6 Experimental results for 5, 8, 10 and 15 m braking range

pressed. The 15 m braking graph is taking brake action from 23rd second to 20th second, which is used the 5 s to perform the braking.

On the speed changes, for 5 m stop range, the speed starts to decrease on 22th second and the vehicle fully stops at 25th second. So its only take 3 s to reduce the speed from 17 to 0 km/h. On 8 and 10 m distance, the speed starts dropping equally at 19th second, but both are fully stopped at one second time interval, which is 23rd second and 24th second. So it takes 4 s in 8 m, and 5 s in 10 m to stop the vehicle. For 15 m, it is required 6 s to fully stop which is the speed start to drop at 22th second before fully stop at the 28th second. The graph pattern is similar generally, the only difference is the time taken for braking. So, the 15 m graph has a bigger curve than the 5 m graph.

As the conclusion, the closer the distance for braking, the shorter the time required to press the brake pedal from zero-degree angle until maximum degree. Moreover, its also means that the vehicle speed is falling faster from 17 to 0 km/h when compared to a longer braking distance. The shorter braking distance shows that in the emergency situation, human is in panic and will press the braking more harshly than usual. A longer braking distance gives more space to slowly perform the braking. This is normal to the human behavior, as we are not a computer programmed machine.

Acknowledgements I would like to thank all team members of the UniMAP Advanced Intelligent Computing and Sustainability Research Group (AICoS), School of Mechatronics Engineering, University Malaysia Perlis (UniMAP), Unit Penyelidikan Teknologi Sukan Permotoran UniMAP (MOTECH) and Centre of Excellence For Unmanned Aerial Systems (COEUAS) for their work and their help in this research.

References

1. Li, L., Wang, F.: Advanced Motion Control and Sensing for Intelligent Vehicles, Tucson AZ85721. Springer, Berlin (2007)
2. Bjomberg, A.: Autonomous Intelligent Cruise Control, Gothenburg (1994)
3. Akbarzadeh, R.: Adaptive Discrete-Time Fuzzy Sliding Mode Control for Anti-Lock Braking Systems, pp. 335–340 (2003)
4. Wong, N., Chambers, C., Stol, K., Halkyard, R.: Autonomous Vehicle Following Using a Robotic Driver, pp. 2–4 (2008)
5. Chang-fu, Z.: A Control Strategy of Electronic Braking System based on Brake Comfort, pp. 1265–1268 (2011)
6. Nilsson, J., Odblom, A.C.E., Fredriksson, J.: Worst Case Analysis of Automotive Collision Avoidance Systems, vol. 11, no. 4 (2015)
7. Kopetz, H., Poledna, S.: Autonomous Emergency Braking: a System-of-Systems Perspective, Vienna (2013)
8. Hirulkar, S., Damle, M., Rathee, V., Hardas, B.: Design of Automatic Car Braking System Using Fuzzy Logic and PID controller, pp. 413–418 (2014)
9. Diederichs, F., Spath, D.: Driver Intention Algorithm for Pedestrian Protection and Automated Emergency Braking Systems, pp. 1049–1054 (2015)
10. Schmitz, M., Hanig, M., Jagiellowicz, M., Maag, C.: Impact of a combined accelerator–brake pedal solution on efficient driving. IET Intell. Transp. Syst. **7**(2), 203–209 (2013)

The Mapping of Full Weld Cycle Heat Profile for Friction Stir Welding Pipe Butt Joints

Azman Ismail, Mokhtar Awang, Fauziah Ab Rahman,
Bakhtiar Ariff Baharudin, Puteri Zarina Megat Khalid,
Darulihsan Abdul Hamid and Kamal Ahmad

Abstract Aluminium alloy 6063 pipe sections were friction stir welded to produce a sound joining. In this present study, the full weld cycle of heat input will be gathered for rotational speed ranging between 1000 and 1600 rpm and travel speed between 3 and 5 mm/s. There will be 5 specimens to be friction stir welded. It was found that, the heat input will be increased with the increment of rotational speed up to 396 °C and decreased with the increment of travel speed with a minimum of 221 °C. These conditions were recorded by utilizing K-type thermocouples and the Signal Express software with a National Instrument model cDAQ-9171 data logger.

Keywords Friction stir welding · Pipe section · Heat input · Full weld cycle
Aluminium

A. Ismail (✉) · F. A. Rahman · B. A. Baharudin · P. Z. M. Khalid · K. Ahmad
Universiti Kuala Lumpur Malaysian Institute of Marine Engineering Technology,
Jalan Pantai Remis, 32200 Lumut, Perak, Malaysia
e-mail: azman@unikl.edu.my

F. A. Rahman
e-mail: fauziahabra@unikl.edu.my

B. A. Baharudin
e-mail: bakhtiarab@unikl.edu.my

P. Z. M. Khalid
e-mail: puterizarina@unikl.edu.my

K. Ahmad
e-mail: kamalmat@unikl.edu.my

A. Ismail · M. Awang
Department of Mechanical Engineering, Universiti Teknologi PETRONAS,
Bandar Seri Iskandar, 32610 Teronoh, Perak, Malaysia
e-mail: mokhtar_awang@utp.edu.my

D. A. Hamid
Institute of Product Design and Manufacturing, Universiti Kuala Lumpur Institute of Product Design and Manufacturing, No.9, Jalan Perdana 7/3, Taman Shamelin Perkasa, 56100 Kuala Lumpur, Selangor, Malaysia
e-mail: darulihsan@unikl.edu.my

1 Introduction

The friction stir welding (FSW) was developed in 1991 at TWI, United Kingdom [1]. This process was initially developed to cater for soft materials such as aluminium and then expanding its application to a wide variety of hard materials such as copper, steel, stainless steel, titanium and iconel [2, 3]. However, most of the research was done on aluminium especially on plates instead of pipe sections. The heat supplied for FSW process was generated by friction between the pipe surface-shoulder and plastic deformation during stirring process of a high rotation tool. Figure 1 shows the basic concept of friction stir welding on how the heat is supplied and the joint occurs.

Previous studies showed that the heat input could cause a detrimental effect on the mechanical properties especially on hardness and tensile strength of aluminium specimens [3–5]. This heat input could cause annealing and reduction in hardness and strength of the specimen. The control of heat input imposed towards these materials could prevent the reduction on mechanical properties in the weld joint region. A suitable combination of rotational and travel speed will produce sound friction stir weld and incompatibility of these welding parameters could cause weld defects due to insufficient heat input for appropriate material flow in the friction stir welded zone. Besides that, a tubular shape in pipe joining posed a unique challenge. In order to complete a full weld cycle in friction stir welding of pipe joining, the rotating tool will start and stop at the same point thus will also introduce secondary heating [6, 7].

In this present study, a butted section joining of AA6063 pipe was performed and the temperatures were gathered for a full complete weld cycle in order to determine suitable heat input generated based on selected rotational speeds (i.e. 1000, 1300 and 1600 rpm) and travel speeds (i.e. 3, 4 and 5 mm/s). This joining necessitates the use of special holding equipment which is known as the orbital clamping unit as shown in Fig. 2.

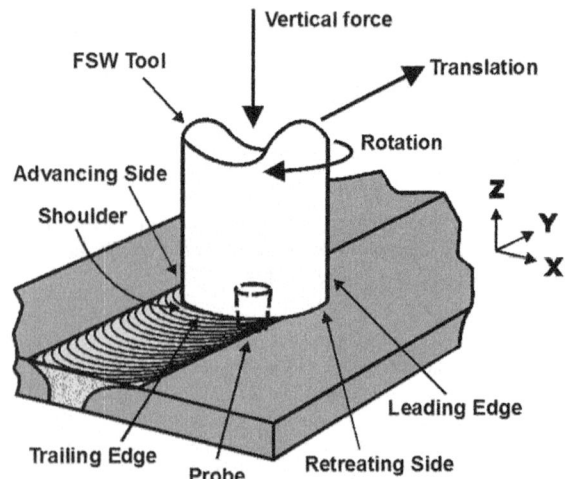

Fig. 1 Basic concept of friction stir welding process

Fig. 2 Orbital clamping unit on FSW machine

2 Experimental Setup

A set of two AA6063 pipe sections were joined together by the FSW process. The 89 mm outside diameter of the pipe with 5 mm wall thickness was utilized as weld specimen. These 2 pieces of 100 mm length of pipe were joined together to produce full penetration friction stir welds. The FSW tools used for this experiment setting was made from surface-hardened H13 high carbon steel with a 20 mm diameter shoulder and 5 mm diameter of pin. The length of the pin was set for 3.8 mm. The FSW tool was offset at 6 mm from the centerline in order to supply sufficient plowing effect for a sound weld joint.

Eight units of k-type thermocouples were attached onto the pipe circumferentially on the advancing side based on previous research findings [8, 9]. The first 4 units were put at a distance of 20 mm between each other and the last 4 units were put at a distance of 44 mm each, 15 mm gap towards advancing side, represented by A0, B, C, D, E, F, G, H, and A1 where is A0 and A1 were the same thermocouples at the same start and stop position. This is shown in Fig. 3. Figures 4 and 5 show the setting of thermocouples onto pipe sections and tool positioning respectively. The welding parameters of these experiments are shown in Table 1. The plunge depth and dwell time were set to 4 mm and 25 s, respectively. The temperature was measured by K-type thermocouples and recorded using a National Instrument (NI) Signal Express application on a laptop. A NI CompactRIO, model NI cDAQ-9171 was used as data logger.

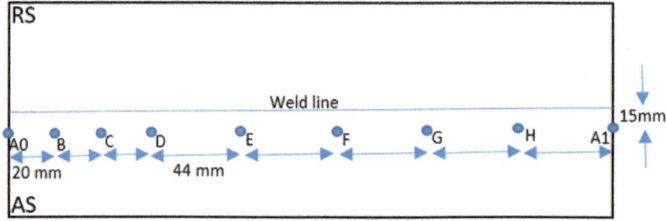

Fig. 3 Thermocouples positioning based on pipe circumferences

Fig. 4 Pipe settings on FSW machine

3 Results and Discussion

The temperature profiles measured at various welding parameters are shown in Figs. 6 and 7, respectively. Figure 6 shows the temperature profile at constant travel speed with the increment of rotational speed. Meanwhile Fig. 7 shows the same data at constant rotational speed with the increment of travel speed. Heat generation in FSW is vital. Sufficient heat will produce a sound friction stir welded joint. In this case, the welding parameters such as rotational and travel speed will affect the generation of heat in FSW.

Based on Fig. 6, the heat input increased with the increment of rotational speed at constant travel speed. Meanwhile, the heat input decreased with the increment of travel speed at constant rotational speed as shown in Fig. 7. The highest heat measured

Fig. 5 Tool positioning (side view)

Table 1 Friction stir welding parameters

FSW Specimen	Welding parameters	
	Rotational speed (rpm)	Travel speed (mm/s)
FSW-1	1000	3
FSW-2	1300	3
FSW-3	1600	3
FSW-4	1600	4
FSW-5	1600	5

at both conditions are 396 °C. The combination of these parameters play a major role as the heat is generated solely caused by friction. Some condition gave fluctuate reading due to vibration and pipe eccentricity but still within acceptable range.

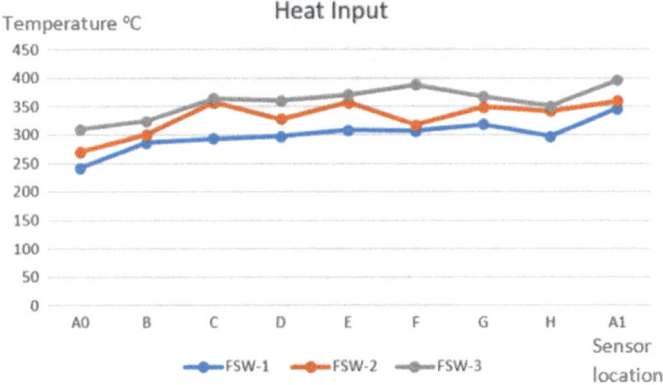

Fig. 6 Heat input with increment of rotational speed

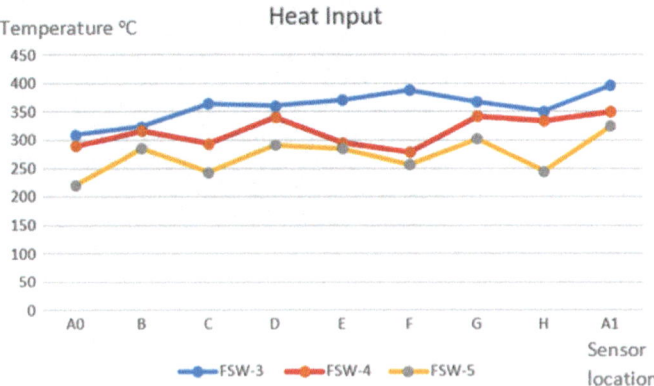

Fig. 7 Heat input with increment of travel speed

4 Conclusion

Based on the results of the present study, these conclusions could be made as follows;

a. The heat increased with the increment of rotational speed but vice versa with the increment of travel speed.
b. The maximum heat generated for both conditions is 396 °C.
c. The combination of these parameters such as rotational and travel speed are important in order to control the heat generated towards the weld region in FSW process.

Acknowledgements The authors thanked Majlis Amanah Rakyat (MARA) for fund provided. The research facilities provided by Universiti Teknologi PETRONAS and Universiti Kuala Lumpur (MIMET) were greatly acknowledged.

References

1. Abnar, B., Kazeminezhad, M., Kokabi, A.H.: Effects of heat input in friction stir welding on microstructure and mechanical properties of AA3003-H18 plates. Trans. Nonferrous Met. Soc. China **25**, 2147–2155 (2015)
2. Threadgill, P.L., Leonard, A.J., Shercliff, H.R., Withers, P.J.: Friction stir welding of aluminium alloys [J]. Int. Mater. Rev. **54**(13–14), 49–93 (2009)
3. Mishra, R.S., Mahoney, M.W.: Friction stir welding and processing. ASM International (2007)
4. Mishra, R.S., Ma, Z.Y.: Friction stir welding and processing. Mater. Sci. Eng. R **50**, 1–78 (2005)
5. Liu, H.J., Fujii, H., Nogi, K.: Tensile properties of a friction stir welded thin-sheet of 1050-H24 aluminum alloy. Charact. Control of Interfaces for High Qual. Adv. Mater. **146**, 129–136 (2006)
6. Lammlein, D.H., Gibson, B.T., DeLapp, D.R., Cox, C., Strauss, A.M., Cook, G.E.: The friction stir welding of small-diameter pipe: an experimental and numerical proof of concept for automation and manufacturing. Proc. Inst. Mech. Eng. Part B J. Eng. Manuf. **226**, 1–16 (2012)
7. Ismail, A., Awang, M., Samsudin, S.H., Rojan, M.A., Jasri, M.A.H.M.: The hardness variation due to secondary heating in friction stir welding of small diameter aluminium alloy 6063 pipe. Mach. Joining and Modifications of Adv. Mater. Adv. Struct. Mater. **61**, 87–94 (2016)
8. Khourshid, A.M., Sabry, I.: Friction stir welding study on aluminum pipe. Int. J. Mech. Eng. Rob. Res. **2**(3), 31–339 (2013)
9. Ismail, A., Awang, M., Rojan, M.A., Samsudin, S.H.: The characteristic of temperature curves for friction stir welding of aluminium alloy 6063-t6 pipe during tool plunging stage. ARPN J. Eng. Appl. Sci. **11**(1), 277–280 (2016)

Assessment of Thermal Comfort in a Car Cabin Under Sun Radiation Exposure

Mohamad Asyraf Othoman, Mohd Sahril Mohd Fouzi and Adzuieen Nordin

Abstract The number of vehicles in Malaysia increases rapidly and these cars are usually parked at open or un-shaded parking areas due to limited roof parking facilities. Most of the car users need a few minutes to cool down the car cabin temperature because of the sun radiation exposure. The increase of car cabin temperature results in a degradation of the material and the interior quality of the car cabin and is also connected to the danger of heat strokes of passengers. Therefore, the objective of this study is to identify the temperature and relative humidity of the car cabin when it was exposed to hot and sunny days for the case of tropical climate. The data was collected in five different modes. It was observed for three days in order to reduce uncertainty. The results show that the car cabin temperature could be exceed to 60% from the ambient temperature when it was exposed to the sun within two hours and the relative humidity decreased from 25 to 54%.

Keywords Thermal · Car cabin · Sun radiation · Vehicle

1 Introduction

Due to the economic stability and income growth among Malaysians, majority of Malaysians afford to purchase their own vehicle. Hence, the numbers of vehicles rapidly increased. Regarding this situation, Malaysian drivers tend to park their vehicles at uncover or un-shaded parking areas due to the lack of roof parking accommodation. As people have to park their cars at uncovered areas for several

---

M. A. Othoman (✉) · M. S. M. Fouzi · A. Nordin
Mechanical Engineering Department, Politeknik Ungku Omar,
Jalan Raja Musa Mahadi, 31400 Ipoh, Perak, Malaysia
e-mail: masyraf@puo.edu.my

M. S. M. Fouzi
e-mail: mechanical86@puo.edu.my

A. Nordin
e-mail: adzuieen@puo.edu.my

© Springer International Publishing AG 2018
A. Öchsner (ed.), *Engineering Applications for New Materials and Technologies*,
Advanced Structured Materials 85, https://doi.org/10.1007/978-3-319-72697-7_39

minutes even hours under hot weather to go for shopping or working purposes, the sun radiation will heat up the temperature in the vehicle's cabin. Affected to this condition, the car owner is forced to wait for a certain times to cool down the temperature in the vehicle's cabin by either lowering the window or switch on the air-conditioner system at full level which leads to high fuel consumption and interior component degradation due to rapid cooling.

Many researchers have done research on indoor air quality in buildings [1]. Guidelines for indoor air quality for office buildings have been implemented from [2, 3]. References [4, 5] have discussed a fundamental study on the ventilation rates and volatile organic compounds concentration in the car compartment and particles, ultra-fine particles and ozone in the car compartment, respectively.

According to Al-Kayiem et al. [6], solar irradiation increases with time to maximum values between 12.00 and 14.00 h. The mean value of 500 ± 50 Wm^{-2} was dependent on the cloud conditions. They have also done experimental and numerical analyses on the effect of a 20 mm window gap and sunshade on thermal accumulation inside a parked car cabin. The use of the sunshade and lowering windows on both sides reduced the heat accumulation due to fresh air exchange with the exterior environment.

The temperature level in the car interior can be more than 20 °C and above the ambient temperature when a car is parked and exposed to the sun radiation as noted by Dadour et al. [7]. They built a simple 'greenhouse' model for predicting the daily interior vehicle temperatures of black and white vehicles. They also demonstrated that the car interior temperature can be reduced typically 3 °C by lowering the driver's window by 25 mm.

Kaynakli et al. [8] measured the temperature and humidity at various points inside a sedan car during heating period. The results showed that the heat losses were very high at the beginning of heating period. Mezrhab et al. [9], demonstrated a numerical model to study the performance of thermal comfort inside the car interior according to climatic conditions and materials that compile the vehicle. According to the American Heritage Science Dictionary [10], relative humidity is the ratio of the actual amount of water vapour present in a volume of air at a given temperature to the maximum amount that the air could hold at that temperature, expressed as a percentage. Warm air can hold more water vapour than cool air, so a particular amount of water vapour will yield a lower relative humidity in warm air than it does in cool air.

The relative humidity influenced the comfort that occupants feel in a certain space. The ASHRAE [11] advises a relative humidity range of 30–60% in occupied spaces. The occupied spaces refer to office or educational buildings, entertainment venues and medical facilities. Some medical facilities have different relative humidity requirements, which are based on the condition and purpose.

Low relative humidity can have negative effects on human wellbeing and health. Arundel et al. [12] stated that the low relative humidity caused an eye irritation and increased the occurrence of infective aerosols produced by coughing or exhaling. People who stay indoors for a long time period during winter were easy to get respiratory infection by weakening the defences provided by the mucous membranes. They also state that a relative humidity between 40 and 70% can minimize the infectivity of bacterial and viral organisms.

High relative humidity is considered when it is above 60%. Too high percentage of humidity can cause occupant discomfort, building and paper-based materials damage, mold and fungi growth, the existence of dust mites, and develop infectivity of bacteria and viruses. When humidity is increased, the human body has more complexity in cooling itself through perspiration. If the perspiration cannot disappear from the skin due to the increased relative humidity, then the people will become uncomfortable. Asthma can be triggered by allergens produced by dust mites.

High relative humidity can contribute to problems in building materials. Moisture formed from air absorbed by acoustical ceiling tiles can cause sagging. Wallboard can absorb enough moisture in a high humidity space to the extent that it supports mold or fungi will growth, which itself can damage the wallboard. High relative humidity also can contribute to the degradation of books and other paper-based materials as they become an attractive place for mold and for house dust mites. House dust mites can increase in space where the relative humidity exceeds 50% and fine organic particles are readily available. These dust mites can be found in all types of indoor environments besides residential properties. Arundel et al.'s [12] has summarized the effect of relative humidity on chemical and biological factors.

Thermal comfort is a term that generally is regarded as a desirable or positive state of a person. The feeling of comfort is a result of mixture sensations that varies depending on the person, lifestyle, and habits. For example, during winter, when you are from the outside region and coming in (from temperature of −5 °C to temperature 20 °C), warm sensation will be experienced and the same goes to cool sensation, from hot to cold area. Thermal comfort mainly concerns the interior temperature of rooms, maintaining and distributing it evenly and the quality of air.

There are many levels of discussion regarding the meaning and nature of thermal comfort and there was much activity and debate in the 1960s and 1970s on this topic. Much has been achieved since 1970 in understanding the conditions that create thermal comfort. However, because of the need to take the understanding to a higher level in order to meet increased and new requirements, these discussions are again coming to the fore. Based on ASHRAE standard 55, a good thermal comfort is achieved where eighty percent or more people in a room feel comfortable. A less neutral and more positive concept related to thermal comfort is thermal pleasure. Thermal pleasure is found in cool temperatures compensated for by the neat of the sun. Such conditions are usually beyond thermal comfort.

The rapid rise of car the interior temperature inside static vehicles without natural or auxiliary ventilation in direct sun radiation exposure will damage property and harm human or pets left inside. Excessive heat inside fully closed cabin temperature will increase the level of air borne and chemical substances in car interiors. Exposure to these substances will increase the potential to get allergy and asthma symptoms and cause eye, nose and throat irritation [13]. Temperature and relative humidity is a part of indoor air quality components that is used to determine how our air is in acceptance level due confined space.

The climate in Malaysia is hot and humid. The data obtained by the Malaysian Meteorological Department [14] for a ten-year period records relatively uniform outdoor temperatures with an average of between 23.7 and 31.3 °C throughout a day with the highest maximum recorded temperature as 36.9 °C with an average relative humidity of between 67 and 95%. Therefore, the objective of this study is to assess the car cabin thermal comfort level in direct sun radiation during a hot and sunny day in tropical country weather.

2 Methodology

Meteorological data was used for this research. Meteorological data from a fixed air quality monitoring station nearby to the study area was provided by the Malaysian Meteorological Department. The data consist of temperature and relative humidity. The data was used to compare with the actual ambient data taken in the testing area. On the average, Malaysia receives about 6 h of sunshine per day due the cloud cover cuts off a substantial amount of solar radiation. This study was conducted in a selected parking area (latitude 4.57, longitude 101.1) and altitude: 53 m above mean sea level in Ipoh, the capital city of Perak state. The testing area was surrounded by residential flats, terrace houses, schools, and government buildings. The area also was secluded by limestone hills; i.e. to the northeast, east and southeast of the testing area.

In order to collect data for this study, a white Proton Persona (Model 2011) and car sunshade are used in the experiments. The overall dimensions of the car are 4477 mm in length, 1725 mm in width and 1438 mm in height. All side windows are made of white transparent glass. The front and rear windscreen are also made of white glass. The car sunshade is shown in Fig. 1 has a silver foil front and back. The car was parked and exposed to direct sun at free air in the period of time at Ipoh. The test area was the car park of the learning institution at Ipoh, Perak, Malaysia which is a typical bitumen surfaced open car park. Vacant spaces were left between and either side of the test vehicles to ensure full sun exposure. The test vehicles oriented South-East to ensure full sun exposure for the whole period of the test. The selected location ensured that there was no shading of the vehicles during the period of the test. A TPI 597C1 digital hygrometer, also shown on Fig. 1, was used to monitor the indoor dry bulb temperature and relative humidity. The instrument was mounted on the middle of the rear seat about 30 cm from the floor. The instrument was not directly exposed to sun radiation, therefore, only the air temperature was measured.

Temperature and relative humidity were measured from 12.00 to 14.00 h for the duration of three days for each mode. Each measurement was at constant atmospheric temperatures between March and May. All experiments were made almost at the same conditions (car direction and daily hours). The starting temperature of

Fig. 1 Position of car sunshade and digital hygrometer inside the car interior

the vehicles was normalized by parking in an under-roof car park for several hours' priors to each data measurement sequence. Air conditioning was off and set at low fan speed, immediately prior to parking the car in its test and commencing the test, in order to replicate typical driving condition.

Five different parking conditions were investigated consisting of all side glass windows openings, front shield shading and normal parking condition. The modes are described in Table 1. For all modes except mode 1, the car was tested at an open space (un-shaded) parking area in an under-roof parking area and all windows were totally closed. Mode 2 is measured while the car was parked at an open space (un-shaded) parking area and windows were totally closed. Mode 3 when the car windows were totally closed with applied car sunshade in front and rear windscreen. Mode 4 when the car is lowering the four windows by 20 mm and mode 5 when the car is lowering the four windows by 20 mm with applied car sunshade at

Table 1 Experiment modes description

Mode	Setup				
	Roof parking area	Open parking area	All windows closed	Sunshade	Windows gab 20 mm
1	√		√		
2		√	√		
3		√	√	√	
4		√			√
5		√		√	√

front and rear windscreen. Indoor air quality parameter data for control (fully shaded) condition for each mode were measured to make a comparison study.

3 Results and Discussion

The results are divided into two sections, which look into the dry bulb air temperature and relative humidity.

3.1 Temperatures Variation Assessment

The temperature variations of the car cabin in different modes are plotted in Fig. 2. The data was recorded at fifteen minutes intervals at all modes. Deviation temperature for ambient and mode 1 is small compared to others modes when the car was fully exposed to sun radiation.

Table 2 summarizes the maximum, minimum and average car interior temperature values in all modes. The minimum ambient temperature recorded was 33.7 °C

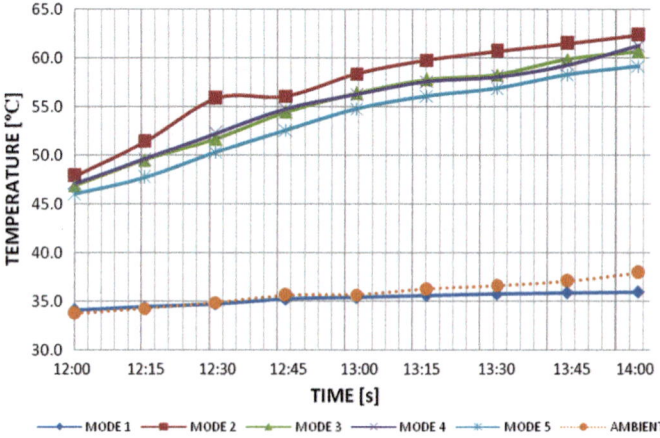

Fig. 2 Temperature of car cabin in different modes

Table 2 Maximum, minimum and average interior temperature values in deferent modes

	Mode 1	Mode 2	Mode 3	Mode 4	Mode 5	Ambient
Maximum (°C)	35.9	62.4	60.7	61.3	59.2	38.0
Minimum (°C)	34.1	47.9	46.9	47.1	46.0	33.7
Average (°C)	35.2	57.1	55.1	55.2	53.5	35.8
% Reduction	38.4	0.0	3.5	3.3	63	37.3

at the beginning of the test and maximum was 38 °C at 14.00 h. The average data recorded for ambient temperature is 35.8 °C. The interior temperature difference for mode 1 is slightly the same as the ambient temperature but has a difference 2.1 °C below the ambient peak temperature.

In mode 2 as a closed interior recorded the highest reading among the five modes. The minimum mode 2 interior temperature recorded was 47.9 °C at the beginning of the test and increased to the peak temperature 62.4 °C. This was reached within 120 min testing period. The average data recorded for mode 2 temperature is 57.1 °C.

For mode 3, the recorded interior temperature was reduced 3.5% by reducing the solar projection area with an installed aluminium foil sunshade underneath of the front and rear windscreen. The minimum mode 3 interior temperature recorded was 46.9 °C at the beginning of the test and the peak temperature was 60.7 °C. The average data recorded for mode 3 temperature is 55.1 °C.

When the car window is lowered by 20 mm in mode 4 the interior temperature reduction is 3.3%. This small amount reduction values is caused by small the gap window to let sufficient fresh air flows from outside and circulates inside the car. The minimum mode 4 interior temperature recorded was 47.1 °C at the beginning of the test and the peak temperature was 61.3 °C. By averaging the data recorded an average mode 4 car interior temperature of 55.2 °C was calculated. A combination of sunshade and lowering windows in mode 5 gave a better amount of heat reduction. Almost 6.3% of the car interior temperature was reduced.

The minimum mode 5 interior temperature recorded was 46.0 °C at the beginning of the test and the peak temperature was 59.2 °C. The average data recorded for mode 4 temperature is 53.5 °C.

3.2 Relative Humidity Levels Measurement

As show in Fig. 3, results of relative humidity (RH) of air in the car interior with time in fifteen minutes interval throughout the test under various modes.

Table 3 summarizes the value of maximum, minimum and average interior relative humidity as well as the percent of reduction in various modes.

Mode 1 interior recorded high an average relative humidity (57.7%) compared to others modes. The maximum mode 1 interior relative humidity recorded was 59.7% at the beginning and slightly decreased to a minimum value of 56.4% after 120 min of the test. In ambient mode the humidity average value is 50.2%, lower than mode 1 but still in the comfort range recommended by DOSH [2]. The humidity in ambient mode begins with 54.8% and gradually decreases to 47.3%. Mode 2 as closed car and exposed to sun radiation recorded the worst humidity reduction from others modes. Almost 61% reduction compared to mode 1 when the car parked under roof parking area.

The average humidity value in mode 2 is 22%. It begun with 29% and continuously decreases to the lowest humidity value of 18.6% after 120 min. When the

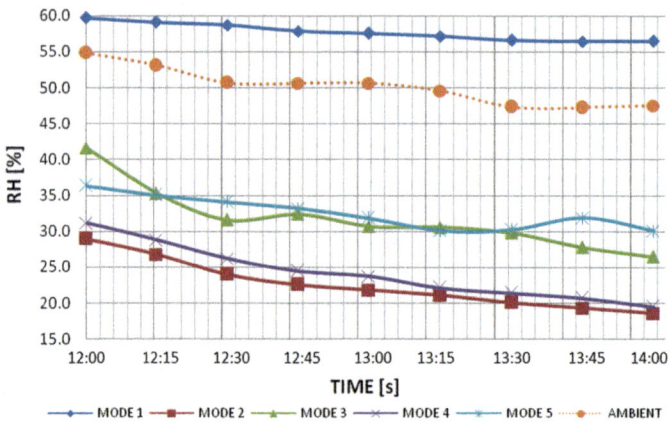

Fig. 3 Relative humidity of car interior in different modes

Table 3 Maximum, minimum and average interior relative humidity in various modes

	Mode 1	Mode 2	Mode 3	Mode 4	Mode 5	Ambient
Maximum	59.7	29	41.6	31.2	36.3	54.8
Minimum	56.4	18.6	26.5	19.5	30.1	47.3
Average	57.7	22.6	31.8	24.2	32.5	50.2
% Reduction	0	61	45	58	44	13

car was equipped with the sunshade in mode 3, the humidity value started with 41.6% and decreased to 26.5% at the end of the test. The average humidity value in mode 3 was 31.8%.

Humidity reduction in mode 3 is 45% compared to mode 1. When the car is just lowering all windows under sun radiation in mode 4, the average humidity value was 24.2% and the humidity reduction was about 58%.

At start the humidity interior value in mode 4 was 31.2% and constantly decreased to 19.5%. When sun exposed car applied a sunshade and lowering all window in mode 5, the interior humidity started at 36.3% and decreased to 30.1% and the average humidity value was 32.5%. The humidity reduction in mode 5 was 44% compared to mode 1.

Figure 4 shows the relationship between the temperature and relative humidity in the car cabin interior. Increasing the temperature inside the car interior will decrease the relative humidity value. The humidity level achieved the comfort range (40–70%) when the temperature in the car interior was below 40 °C.

Without applied sunshade underneath the front and rear windscreen in mode 2 and mode 4, the relative humidity depletion was very high. With applied sunshade in mode 3 and mode 5, the solar projection area were reflected, hence it resulted in positive improvement of maintaining relative humidity in the car interior with 44–45% humidity reduction compared to mode 1.

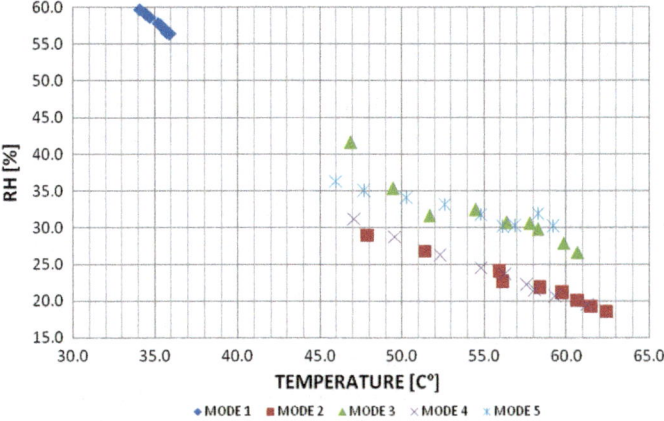

Fig. 4 Effect of temperature on relative humidity in car cabin

While comparing with mode 2, the improvement is about 41–44%. Lowering the windows by 20 mm in mode 4 and 5 gave 7.1% improvement when compared with mode 2 although caused by the circulation between the warm air inside the cabin and the ambient air hence retain relative humidity from rapidly decreasing.

Table 4 shows the percentage of result difference of IAQ compared to the recommendation limit set by the Department of Occupational Safety and Health (DOSH), Malaysia and Institute of Environmental Epidemiology (IAE), Singapore. The yellow highlight colour indicates results that exceed from the recommended limit and the blue highlight colour indicates results below from the recommended limit.

The temperature result for all modes include ambient exceed the comfort limit (23–26 °C) by DOSH and (22.5–25.5 °C) by IAE with assumed 26 °C as comfort temperature. Mode 2 recorded the highest value, 140% exceeding from the limit. When the car is exposed to sun radiation, the relative humidity of the car interior recorded exceeds the limit especially for the enclosed car where the recorded maximum exceeded the limit by 54%.

By lowering all windows 20 mm in mode 4, the recorded data showed 51% of relative humidity which is slightly improved but still exceeded the limit value.

Table 4 Comparison between result obtained and recommendation limit by DOSH, Malaysia and IAE, Singapore

Parameter	Mode 1	Mode 2	Mode 3	Mode 4	Mode 5	Ambient	DOSH	IAE
Temperature (°C)	35.9	62.4	60.7	61.3	59.2	38.0	23–26	22.5–25.5
	38%	140%	133%	136%	128%	46%		
Relative humidity (%)	56.4	18.6	26.5	19.5	30.1	47.3	40–70	70
	41%	54%	34%	51%	25%	18%		

Sunshade installation underneath front and rear wind screen at mode 3 recorded 34% exceeding the limit. A combination of gap of window and sunshade installation recorded 25% exceeding the limit. Therefore roof parking or shaded area will maintain the relative humidity 41% below the minimum recommended limit.

4 Conclusion

Study on the five modes indoor air quality of car interior had been carried out experimentally. The measurements were carried out during a sunny day at 12.00–14.00 h under un-shaded parking and roof parking environment. Temperature level in car interior can be more 20 °C and 60% above the ambient temperature and exceed comfort range (23–26 °C) [15]. Although sunshade installation underneath the front and rear windscreen and lowered all windows by 20 mm, the interior temperature reduction is in between 3 and 7%.

The relative humidity decreased when the car interior temperature increased. For the case when the car parked at the under roof parking area, the relative humidity of the car interior was in the range of 40–70%. When the car is parked at the un-shaded parking area, the relative humidity was in the range of 18–40%. The use of sunshades at the front and rear windscreens resulted in an increase of 40% relative humidity of the car interior. By lowering all windows by 20 mm, it will increase 8% of interior relative humidity.

References

1. Sulaiman, S.A., Isa, N., Raskan, N.I., Harun, C., Farhane, N.: Study of indoor air quality in academic buildings of a university. In: Applied Mechanics and Materials, pp. 389–393 (2013)
2. D. o. O. S. a. H. (DOSH). Industry Code of Practice on Indoor Air Quality, vol. 127/379, pp. 4–39 (2010)
3. Institute of Environmental Epidemiology and M.O.T. Environment. Guidelines for good indoor air quality in office premises (2009)
4. Yokoyama, Y., Iwashita, G., Yoshinami, Y., Nagayama, H., Nakagawa, J.: Fundamental study on particles, ultra-fine particles and ozone in the car compartment. In: Proceedings of the Sixth International Conference on Indoor Air Quality, Ventilation and Energy Conservation in Buildings, pp. 28–31 (2007)
5. Nakagawa, J., Iwashita, G., Yoshinami, Y., Nagayama, H., Yokoyama, Y.: Fundamental study on the ventilation rate and VOCS concentration in the car compartment. In: 6th International Conference on Indoor Air Quality, Ventilation & Energy Conservation in Buildings IAQVEC, Sendai, Japan (2007)
6. Al-Kayiem, H.H., Sidik, M.F., Munusammy, Y.R.: Study on the thermal accumulation and distribution inside a parked car cabin. Am. J. Appl. Sci. **7**(6), 784–789 (2010)
7. Dadour, I., Almanjahie, I., Fowkes, N., Keady, G., Vijayan, K.: Temperature variations in a parked vehicle. Forensic Sci. Int. **207**, 205–211 (2011)

8. Kaynakli, O., Kilic, M.: An investigation of thermal comfort inside an automobile during the heating period. Appl. Ergon. **36**, 301–312 (2005)
9. Mezrhab, A., Bouzidi, M.: Computation of thermal comfort inside a passenger car compartment. Appl. Therm. Eng. **26**, 1697–1704 (2006)
10. Dictionary.com. Relative-humidity. In: The American Heritage® Science Dictionary (2014)
11. The American Society of Heating, and Air-Conditioning Engineers (ASHRAE). Standard, vol. 62 (1999)
12. Arundel, A.V., Sterling, E.M., Biggin, J.H., Sterling, T.D.: Indirect health effects of relative humidity in indoor environments. Environ. Health Perspect. **65**, 351 (1986)
13. McLaren, C., Null, J., Quinn, J.: Heat stress from enclosed vehicles: moderate ambient temperatures cause significant temperature rise in enclosed vehicles. Pediatrics **116**, e109–e112 (2005)
14. N. W. P. D. Section, T. D. Division, M. M. Department, and T. A. I. Ministry of Science. Climate Change Scenarios for Malaysia 2001–2009. Malaysian Meteorological Department Scientific Report (2009)
15. M. O. H. R. Department of Occupational Safety and Health, Malaysia. Industry Code of Practice on Indoor Air Quality 2010, JKKP DP(S) 127/379/4-39 (2010)

Defects of Post Weld Heat Treatment on A36 Carbon Steel Welded by Shielded Metal Arc Welding

Norfadhlina Khalid, Zaherrudin Yusof and M. A. Mun'aim Mohd Idrus

Abstract This study is conducted to assess the influence of post weld heat treatment on weld defects. The purpose of this study is to identify and determine the defect type before and after post weld heat treatment (PWHT) with three different soaking temperatures and three ASTM A36 carbon steel specimens. The specimen is welded by using the shielded metal arc welding (SMAW) process and an AWS E6013 electrode is used with welding current range from 90 to 100 A. All the specimens are inspected using ultrasonic testing (UT) before and after the PWHT process. The specimens in the PWHT process are subjected to different soaking temperatures which are 490, 540 and 610 °C. The result is observed from abnormality criteria to the obtained defects and it can be concluded that by increasing the soaking temperature values from the PWHT, certain types of current defects will be elongated.

Keywords Post weld heat treatment · Ultrasonic testing · Carbon steel shielded metal arc welding

1 Introduction

Post weld heat treatment (PWHT) is a method for reducing and redistributing the residual stresses in the material that have been introduced by welding. The extent of relaxation of the residual stresses depends on the material type and composition, the temperature of PWHT and the soaking time at a certain temperature [1].

N. Khalid (✉) · Z. Yusof · M. A. Mun'aim Mohd Idrus
Malaysian Institute of Marine Engineering Technology, Universiti Kuala Lumpur,
Jalan Pantai Remis, 32200 Lumut, Perak, Malaysia
e-mail: norfadhlina@unikl.edu.my

Z. Yusof
e-mail: zaherrudin@gmail.com

M. A. Mun'aim Mohd Idrus
e-mail: mamunaim@unikl.edu.my

© Springer International Publishing AG 2018
A. Öchsner (ed.), *Engineering Applications for New Materials and Technologies*,
Advanced Structured Materials 85, https://doi.org/10.1007/978-3-319-72697-7_40

The welding process generally involves melting and subsequent cooling and the result of this thermal cycle is distortion if the welded item is free to move or cause residual stresses if the item is securely held [2]. Ultrasonic testing has frequently been used to assess the reliability of structural components which have been applied to the inspection of welding. This study covers the ultrasonic testing of welds to determine any defects. An ultrasonic technique uses the property of imperfections, then reflects and diffracts a wave in order to locate and size the defects. So far with ultrasonic techniques, defects in a weld can be localized but it is very difficult to determine its orientation and exact size. In this study, there are three specimens of carbon steel (ASTM A36) which are joined by butt joint which uses the shielded metal arc welding process (SMAW) and each of the specimen is marked as Specimen 1, Specimen 2 and Specimen 3, respectively [3]. The purpose of the PWHT is to find some abnormality criteria to the defects that were found before, so that it can be a point form that needs to be elaborated. PWHT process can also be a role as a study perimeter with different soaking temperatures to each of the three specimens which are high, medium and low temperature.

2 Experimental Methodology

2.1 Tested Material

ASTM A36 plate is a carbon steel that exhibits good strength coupled with formability [3]. It is easy to machine and fabricate and is weldable. The steel can be galvanized to provide increased corrosion resistance (ASTM A36) [3]. A36 plate can be used for a wide range of applications, depending on the thickness and corrosion resistance of the alloy. The carbon content in the carbon steel is 0.25–0.3%. The material used in this study is carbon steel by using code ASTM A36 flat bars and the size adheres to the American Welding Institute code, AWS D1.1 with composition as given in Table 1 and the rectangular bars size of 180 mm × 75 mm × 10 mm thickness are sectioned from the bars [4].

Table 1 Carbon steel chemical composition

Element	Content (%)
Carbon, C	0.25–0.30
Copper, Cu	0.20
Iron, Fe	98.0
Manganese, Mn	1.03
Phosphorus, P	0.040
Silicon, Si	0.280
Sulfur, S	0.050

2.2 SMAW Welding Process

The specimens were welded together using the shielded metal-arc welding (SMAW) process. AWS E6013 electrodes were used with direct current arc welding process using a 2.4-mm welding rod diameter [5]. Table 2 shows the specification of welding through SMAW.

2.3 Ultrasonic Testing (UT)

Ultrasonic testing (UT) is a non-destructive inspection method that uses high frequency sound waves that are above the range of human hearing, to detect the presence of internal cracks, inclusions, segregations, porosity, lack of fusion, and similar discontinuities in all types of metals [6]. In ultrasonic testing, very high frequency sound waves are transmitted through the part to be tested. The sound waves then return to the sender and are displayed as a graph on a monitoring screen for interpretation [7]. When there is a discontinuity such as a crack in the wave path, part of the energy will be reflected back from the flawed surface.

The reflected wave signal is transformed into an electrical signal by the transducer and is displayed on a screen [8]. In general, ultrasonic testing is based on the capture and quantification of either the reflected waves or the transmitted waves [6]. Each of the two types is used in certain applications, but generally, pulse echo systems are more useful since they require one-sided access to the object being inspected. Figure 1 shows the basic diagram of ultrasonic testing.

2.4 Post Weld Heat Treatment (PWHT)

The machined samples were subjected to stress relieving heat treatment. Three different soaking temperatures were used. The process of the heat treatment started from the heating with 222 °C/25 mm/h, followed by soaking or holding temperature/25 mm thickness/h as shown in Table 3 and then cooling with 280 °C/h/25 mm thickness and lastly cooling by air. The heat-treatment procedure adopted was a soaking temperature range of 490, 540, and 610 °C [9]. The weld

Table 2 Welding specification in SMAW process

Welding method	Shielded metal arc welding (SMAW)
Welding type	Butt joint
Rod/filler	E6013 (∅2.4 mm)
Current range	90–100A
Standard code	AWS A5.1

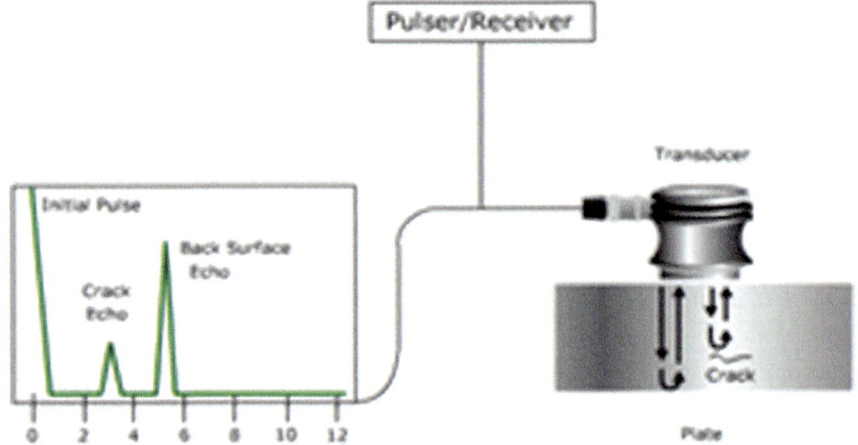

Fig. 1 Ultrasonic testing basic diagram

Table 3 PWHT soaking temperature

Specimen	Soaking temperature (°C)
SPC-01	490
SPC-02	540
SPC-03	610

samples were soaked for 20 min in accordance with the ASTM soaking time requirement of 3–4 min/mm thickness at each annealing temperature [3]. Three samples were cooled by using air.

3 Experimental Results

3.1 Data and Result

Figure 2 shows the temperature versus time graph. It shows the time taken for every specimen to complete the post weld heat treatment (PWHT) process. For Specimen 1 (black line), it takes 4 h and 53 min to complete the PWHT process. This was followed by the second specimen (red line), it takes 4 h and 22 min to complete the PWHT process. Last, the third specimen (green line), it takes 3 h and 57 min [see Fig. 2b] to complete the PWHT process. Post weld heat treatment was accomplished by heating a metal below the critical temperature and then cooling uniformly [1].

Figure 3 shows the scan plan gained from the ultrasonic testing for the Specimen 1 before post weld heat treatment which was exposed to the highest soaking temperature. The other specimen, i.e. 2 and 3, are not included because there was no

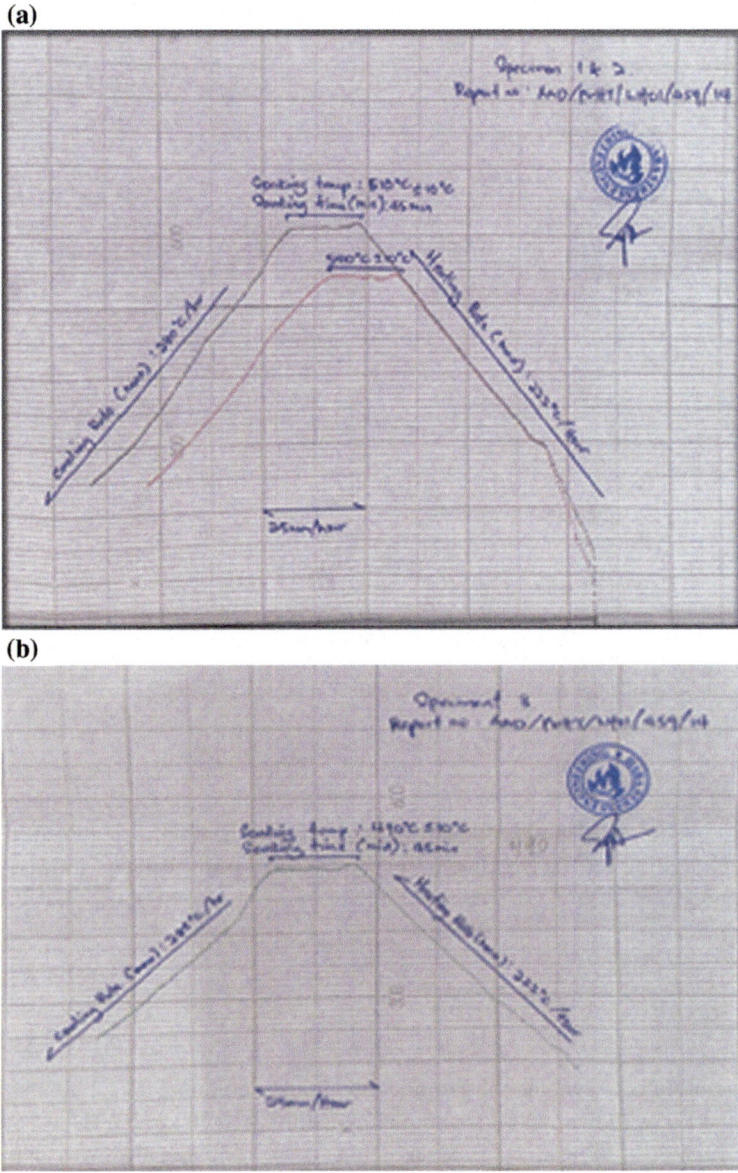

Fig. 2 a The temperature versus time for 1st and 2nd specimens in PWHT process. b The temperature versus time for 3rd specimen in PWHT process

reaction to the current defects after PWHT process within the medium and low temperature.

The crack defect shows some growth for its structure or size after being soaked by the highest temperature of PWHT with the depth changes from 12 to 13 mm and

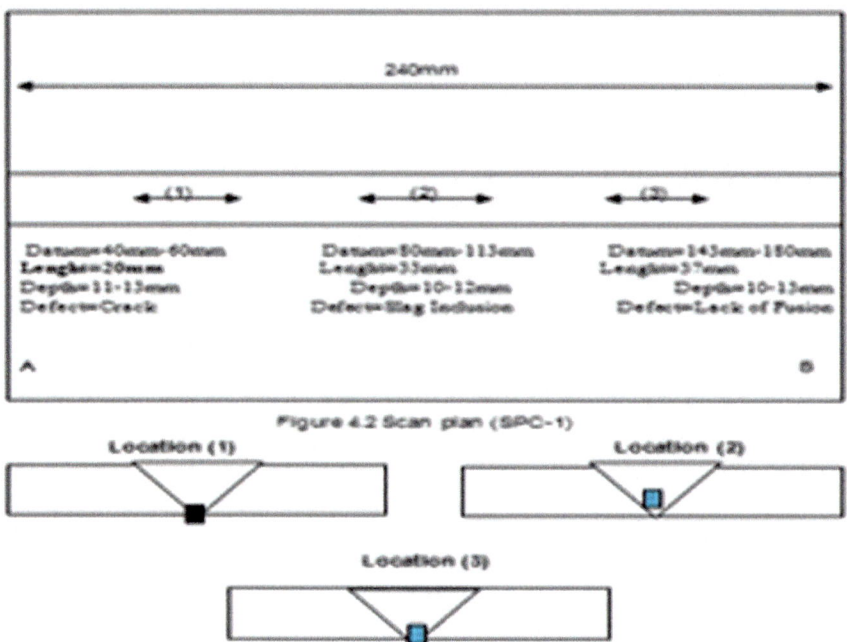

Fig. 3 The scan plan for Specimen 1 before PWHT process [13]

the length elongated from 20 to 22 mm. Other defects from Specimen 1 which are slag inclusion and lack of fusion still maintained their size by not having any changes due to their length or depth. Figure 4 shows the scan plan gained from the ultrasonic testing for the Specimen 1 after post weld heat treatment process which was exposed to the highest soaking temperature.

3.2 Results and Discussion

All type of defects remained the same without any additional defect although after doing PWHT with the highest soaking temperature (610 °C) of PWHT was applied to Specimen 1 followed by medium temperature (540 °C) that was applied to Specimen 2 and then the lowest soaking temperature (490 °C) was applied to Specimen 3. Table 3 shows that there are only three types of defects which were obtained before and after the PWHT process from this study which are crack, lack of fusion and slag inclusion but a little bit of changes in the length and depth of specimens (Table 4).

In Specimen 3, there are two types of defects detected from the ultrasonic testing which are crack and slag inclusion. These two defects do not change their size after

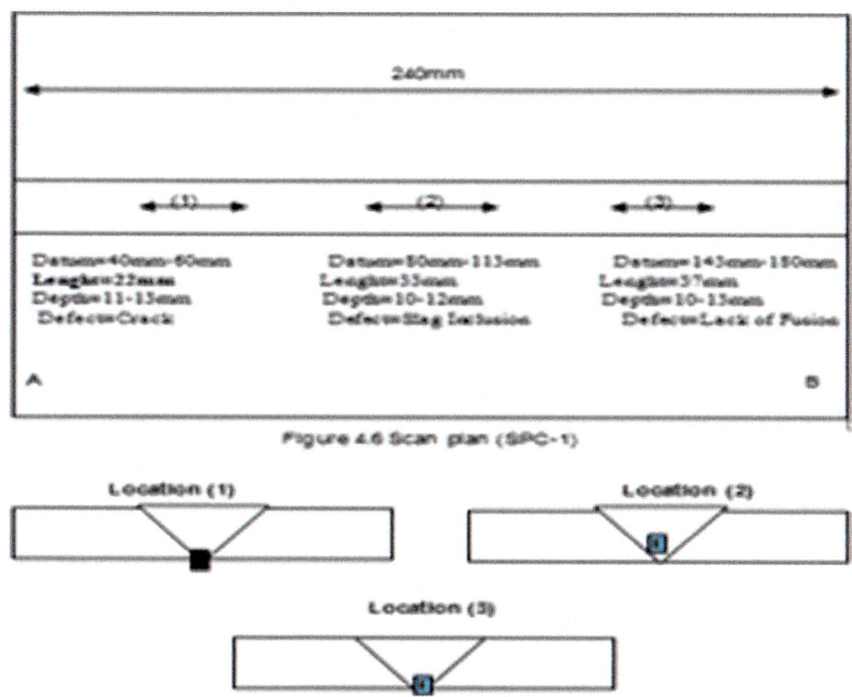

Fig. 4 The scan plan for Specimen 1 after PWHT process. *Source* Ultrasonic Examination Report by Kejuruteraan Escon Sdn. Bhd

Table 4 Types of defects detected before and after PWHT process. *Source* Ultrasonic Examination Report by Kejuruteraan Escon Sdn. Bhd

Specimen	Defect before PWHT			Defect forming after PWHT		
	Type	Length (mm)	Depth (mm)	Type	Length (mm)	Depth (mm)
SPC-1	Crack	20	12	Crack	22	13
	Slag isolation	33	11	Slag isolation	33	11
	Lack of fusion	37	12	Lack of fusion	37	12
SPC-2	Crack	22	9	Crack	22	9
	Slag isolation	25	9	Slag isolation	25	9
SPC-3	Crack	25	8	Crack	25	8
	Slag isolation	25	11	Slag isolation	25	11

being heated with low temperature of PWHT with crack defect still remaining at 25 mm in length and 8 mm in depth.

The slag inclusion defects are also similar i.e. 25 mm in length and 11 mm in depth. There are also no changes for Specimen 2 after being burned by medium temperature of PWHT with only two defects obtained from the ultrasonic testing. These defects occurred during the welding process, not due to the PWHT process. The crack and slag inclusion defect from Specimen 2 still remain their size in length and depth by 22 mm × 9 mm (crack) and 25 mm × 9 mm (slag inclusion).

Figure 5 shows three types of defects detected from the welding process before and after doing PWHT which are crack, lack of fusion and slag inclusion. It is also shown that lack of fusion is the longest defect obtained in this study followed by slag inclusion and crack as the smallest defect.

After doing some investigation, these three defects were identified to have come from poor welding technique. Welding technique has an important role to play in preventing defects. Probably the defects like lack of fusion and slag inclusion do not affect the physique because their defective conditions are not related to any melting microstructure deformation although heated by high temperature [10]. Therefore, the elongated crack defect in Specimen 1 might be caused by high temperature from heat treatment that makes an internal weld structure problem such as structural stress. It is well known that crack is a life defect and it will always increase if exposed to any stress. This is the reasons why there is not any tolerance or acceptance level for this type of major defect [6].

Another probable factor that causes crack elongation is the extremely too high heating temperature that can create hot tears [11]. Hot tears disturb the surface or near surface cracks in the material due to sudden different heating rates [12]. This is evidenced when another crack defect in Specimen 2 and 3 do not show any changes because they were soaked or heated with lower temperature rates.

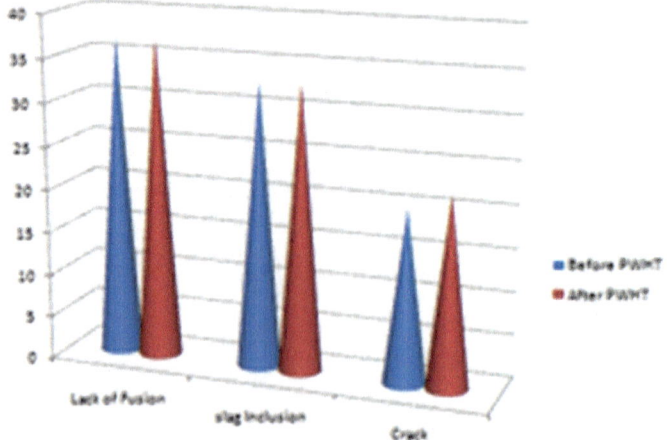

Fig. 5 The defects detected from welding process before and after PWHT process

4 Conclusion

In this study, the defects of post weld heat treatment on carbon steel welded joints by shielded metal arc welding are explained in order to identify and determine the defect type occurring through ultrasonic testing before and after PWHT process. The findings defined only three types of defects found in all specimens such as crack, slag inclusion and lack of fusion. The findings also identified that these three defects came from poor welding technique. For example, an existing crack will grow in size if exposed to any stress. Re-melting weld microstructure of the base metal from high temperature of heat treatment will also create a crack if it is located in the base metal before spreading [11]. However, some defects do not cause alterations to the structures after doing PWHT process when it is not related to microstructure deformation.

Acknowledgements The authors thanked the University Kuala Lumpur Malaysian Institute of Marine Engineering Technology (UniKL MIMET) Lumut for research facilities and financial assistance provided for this project.

References

1. Bipin, K.S., Tewari, S.P., Jyoti, P.: A review on effects of preheating and/or post weld heat treatment (PWHT) on mechanical behavior of ferrous metals. Int. J. Eng. Sci. Technol. **2**(4), 625–626 (2010)
2. Griffin, I.H.: Basic welding techniques, Van Nostrand Reinhold Co. (1977)
3. American Society for Testing and Materials (ASTM) -A36-low carbon steel Standard booklet (2007)
4. American Welding Society, AWS D1.1, American National Standard-Structural Welding Code Steel (2000)
5. American Welding Society, AWS A5.1/A5.1 M, American National Standard-Specification for Carbon Steel Electrodes for Shielded Metal Arc Welding (2004)
6. Kapranos, P.A., Whittaker, V.N.: Ultrasonic inspection of fatigue cracks in Austenitic 316 and 347 weldments. Brit. J. NDT **24**(3), 129–134 (1982)
7. Honeycombe, T., Gooch, G.: The effect of compositional and process variables on microcracking in fully austenitic stainless weld metal, The Welding Institute Research Report (1973)
8. SA-577/SA-577 M (ASTM A 577/A577 M-86) Standard specification for ultrasonic angle-beam examination of steel plates
9. Bray, D.E., Stanley, R.K.: Nondestructive evaluation: a tool for design, manufacturing and service (1989)
10. Jeffus, L.: Welding Principle and Application (Fourth edition), Cengage Learning (1999)
11. Hammar, O., Svensson, U.: Influence of steel composition on segregation and microstructure during solidification of austenitic stainless steels (1979)
12. Masumoto, I., Kutsuma, M.: Hot cracking of austenitic steel weld metal (1972)
13. Salam Abdullah, A.: Ultrasonic examination report. In: Seri Alam Workshop, Kejuruteraan Escon Sdn. Bhd., Masai, Johor, Malaysia. Report No.: UT-UNI-MARINE-14-004 (2014)

Automatic Tug Assistance

Yasseen Adnan Ahmed

Abstract Usually in the harbour area, a big ship with single rudder-single propeller often requires a set of adequate thrust device or tug assistance which exactly takes into account the surge, sway and yaw rate to execute such crabbing motion. The number of tugs involved in each operation depends on the size of the ship as well as existing wind disturbances. The effect of wind varies with its relative direction and the speed of the ship. Therefore, the tug operators must be aware of the disturbances effect and adjust the tow or pulling forces by seeing the response of the ship. The same task, i.e. the adjustment the thrust values, could be done in an automated way which will eliminate the possibilities of human errors and thus ensure proper safety in tug operation. This paper discusses about a controller that can calculate the required tug forces by observing the ship response and then adjust the forces to finally align the ship with the pier with a minimum sway velocity under reasonable wind disturbances.

Keywords Tug assistance · Harbour area · Single rudder-single propeller

1 Introduction

Ship berthing requires precise and gentle control. Everyday ship operators demonstrate such control in ports all over the world. Depending on the experience of the operators, most of the time ships dock safety but the outcome of such sophisticated manoeuvring is not always successful especially in the presence of environmental disturbances. Therefore, bringing automation in the ship berthing sector would be a great relief in terms of safety assessment. Ship berthing is a combination of systematic work done to achieve a 100% successful result. For an example, the operator must slow down the ship speed, the approach must be

Y. A. Ahmed (✉)
Universiti Kuala Lumpur, Malaysian Institute of Marine Engineering Technology,
Jln Pantai Remis, 32200 Lumut, Perak, Malaysia
e-mail: yaseen.ahmed@unikl.edu.my

controllable, he must plan the total process well before and of course proper team work. By doing so, when the ship enters into a narrower zone close to the pier by reducing its speed, then the tugs are requested to assist the ship to navigate further and finally align it with the pier. Therefore, this paper will focus on the last stage of the whole berthing process, i.e. the tug assistance.

It is clear that a berthing phenomenon must incorporate a reduced ship speed, i.e. reduced propeller and rudder inflow velocity. This will drastically reduce the ship manoeuvrability and increases the effect of environmental disturbances. In controlling aspect, if a ship motion is considered as a signal and disturbances as noises, then in low speed running the signal-noise ratio becomes low enough for any controller to separate the noises from the actual ship motion. Thus, selection of the proper controller for such system is very important to ensure its effectiveness. Initially, to develop a controller for side thrusts/tugs assistance, ANN has been investigated as explained by Tran and Im [1] under no wind condition. However, considering the wind that is mostly unpredictable, there is no other easy way to maintain consistency in teaching data that is very important to ensure the effective ANN controller. As a result, the ANN controller trained with inconsistent teaching data may provide good results in few limited cases but in most of the cases, it will fail due to the high noises in low speed running.

In this paper as a feedback controller, a PD controller is given preference over other intelligent controllers, like ANN. It is assumed that two tugs/thrusters will be used to control the sway velocity of a tanker ESSO OSAKA, 3-m model ship and to control the forward motion, longitudinal thrust is also involved. Instead of uniform wind disturbances, gust wind is considered to generate the realistic disturbing environment in the simulation. Real time simulation is carried out to find the actual ship response while operating with the proposed controllers.

2 Wind and Its Effect

It is very important to consider the significance of wind while handling any high sided ships such as car carriers, container ships, bulk and tankers in ballast. Although the wind force and direction can be predicted from information obtained from a variety of sources, such as weather forecasts or the ship's own wind instrumentation, local conditions may change rapidly with little warning. Therefore, the control of a ship can be easily lost during the passage of a high gust wind. There is an obvious need to understand how wind will affect your ship, and how this effect can be difficult to predict. For better understand, let us consider a stopped ship. To understand the behaviour of a stopped ship under wind disturbances, it is necessary to have an idea about the centre of lateral resistance and the point of influence of wind. A brief description of these can be found in master's guide to berthing [2] and also given as follows:

The centre of lateral resistance: The point of influence of underwater forces acting on hull to resist the wind-induced motion is known as the centre of lateral resistance (CLR). Therefore, CLR is the point on the underwater hull at which the total hydrodynamic force can be considered to act. In case of ship with motion, it is usual to consider the pivot point (P) rather than CLR when discussing the effects of wind. On the other hand, a stopped ship does not have a pivot point. Therefore, in such cases CLR should always be used.

The point of influence of wind: This is the point (W) on the above-water structure of the hip upon which the total wind force can be considered to act. This point is not fixed like ship's centre of gravity (CG). Moreover, the point of influence of wind moves depending on the profile of the ship exposed to the wind. Thus, W will be close to the mid-length when a ship's beam is facing to the wind. On the other hand, it may move slightly forward or backward depending on the superstructure position of the ship.

In order to consider the effect of wind while executing the crabbing motion, W must be viewed in relation to CLR. A ship under wind disturbances, always wants to settle into an equilibrium position where the pivot point and the point of influence of wind are in alignment. If a stopped ship faces the wind on its beam, as defined in Fig. 1, W will be close to the mid-length of the ship. Similarly, the CLR will also be at its mid-length. The difference between the two points produces a small moment, and the ship will turn towards the wind with its head facing to it. As the ship continues to turn, W also starts to move until it is close to the CLR. Therefore, the couple reduces gradually to zero and the ship settles on its heading. Figure 2 illustrates such phenomenon.

In this paper, while starting the crabbing motion with automatic tug assistance, the ship might have some forward speed. Therefore, the ship Esso Osaka with its pivot (P) forward of the midship will experience a large lever with the point W at midship. The resultant force will cause the ship's head to turn to the wind as shown in Fig. 3. Therefore, allocation of side thrusts under gust wind disturbances is very difficult. Bui et al. [3] solved such thrust allocation problem by using the redistributed pseudo inverse approach to determine the thrust and direction of each individual tugboat. The main goal of that approach was to minimise the power supplied to the tugboat. However, this paper deals with the side thrusts that act perpendicular to the ship hull and for simplicity the pulsating nature of thrust output is ignored.

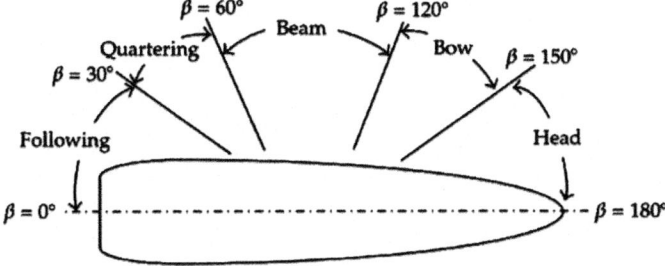

Fig. 1 Definition of different wind direction

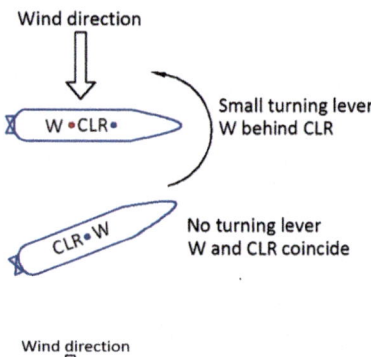

Fig. 2 Wind effect on a stopping ship

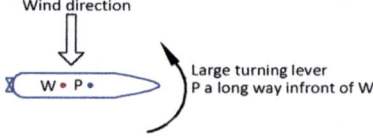

Fig. 3 Wind effect with forward motion

3 Subject Ship and Mathematical Model

3.1 Subject Ship

First implementation of a newly proposed controller often involves risks. Therefore, in this research the Esso Osaka' 3-m tanker model is chosen as subject ship. The model is scaled as 1:108.33. Its details are given in Table 1.

3.2 Mathematical Model—Manoeuvring

To use a precise and accurate mathematical model is very important in this research. Based on the model's predictability, the controller is tuned for the desired action. Therefore, any inappropriate prediction will directly hamper the effectiveness of the proposed controller. In this work, a modified version of the mathematical model based on MMG is used for describing the ship hydrodynamics in three degrees of freedoms. In the MMG model, not only hull, propeller and rudder forces are

Table 1 Principal particulars and parameters of model ship

Hull		Propeller		Rudder	
L (m)	3	Dp (m)	0.084	b (m)	0.083
B (m)	0.48	P (m)	0.06	h (m)	0.1279
D (m)	0.2	Pitch ratio	0.7151	AR (m^2)	0.0106
Cb	0.831	Z	5	Λ	1.539

considered separately, but their interactions are also taken into account. This MMG model can predict both the forward and astern motion of the ship for any particular rudder angle and propeller revolution. Details of such mathematical model can be found in the 23rd ITTC meeting report [4] and Ueda and Ueno's [5] paper on Esso Osaka. The corresponding equations of motions at CG (centre of gravity) of the ship are expressed in Eq. (1).

$$\begin{aligned}
(m+m_x)\dot{u} - (m+m_y)vr &= X_H + X_P + X_R + X_W + X_{tug} \\
(m+m_y)\dot{v} + (m+m_x)ur &= Y_H + Y_P + Y_R + Y_W + Y_{tug} \\
(I_{ZZ} + J_{ZZ})\dot{r} &= N_H + N_P + N_R + N_W + N_{tug}
\end{aligned} \quad (1)$$

where, m is the mass of ship, m_x and m_y are the added mass in x and y direction, I_{zz} is the moment of inertia, J_{zz} is the polar moment of inertia, u is the surge velocity, v is the sway velocity, r is the yaw rate and the right-hand side includes the hydrodynamic forces and moment term due to hull, propeller, rudder, wind disturbances and tug assistance, respectively. Figure 4 shows the coordinate system considered in the formation of equations of motions.

3.3 Mathematical Model—Wind

To consider the influence of wind disturbances during ship manoeuvring, the famous Fujiwara wind model [6] is used for calculating the wind forces and moment. Equation (2) is used for such calculation.

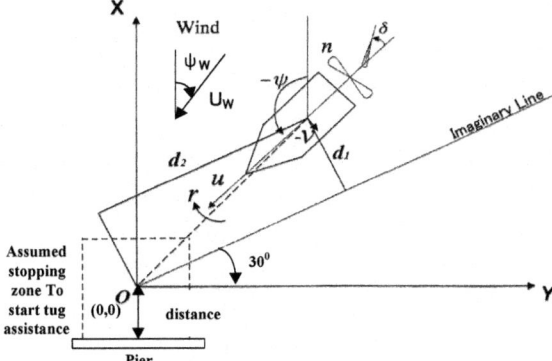

Fig. 4 Coordinate system

$$X_W = \frac{1}{2}C_X \rho V_R^2 A_T$$
$$Y_W = C_Y \rho V_R^2 A_L \qquad (2)$$
$$N_W = \frac{1}{2}C_N \rho V_R^2 A_L L_{OA}$$

where, L_{OA} is the length overall of the ship, A_T is the transverse projected area of the ship, A_L is the lateral projected area of the ship, V_R is the relative wind speed, X_W is the fore-aft component of wind force, Y_W is the lateral component of wind force, N_W is the yawing moment and C_X, C_Y, C_N are the coefficients calculated using Fujiwara's model. To create a realistic environment in simulation, fluctuating wind pattern, i.e. gust wind is considered using the Davenport [7] power spectrum.

4 Controller and Controlling Scheme

It is usual that a ship gradually reduces its speed and stops before starting its approach to berth with tug assistance. The final step into align it with the actual pier. Usually, a big ship with single rudder single propeller often needs tug assistance for executing such crabbing motion. The number of tugs involves in such operation depends on the size of the ship as well as existing environmental disturbances. The plant to be controller here is similar to the ship motion control and given in Fig. 5.

At first, to develop a controller for side thrusts, ANN has been tried in a similar way as explained by Tran and Im [1] under no wind condition. However, considering wind that is mostly unpredictable, there is no other easy way to maintain

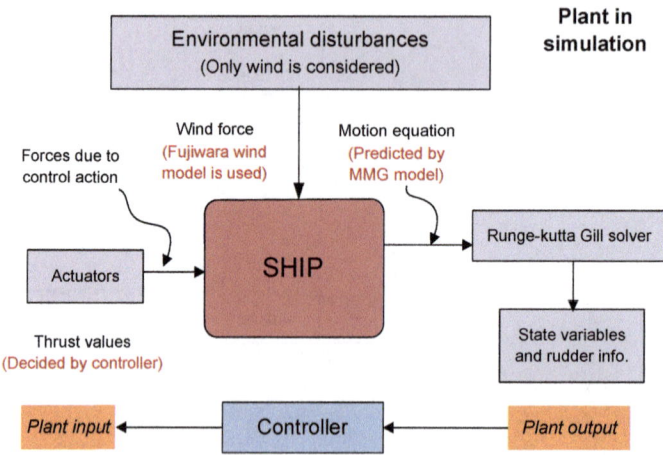

Fig. 5 Plant for tug assistance

consistency in teaching data that is very important to ensure the effectiveness of the trained ANN controller. As a result, a simple but effective PD controller has been chosen over ANN in such cases under wind disturbances. Moreover, to control the forward motion, especially in wind, longitudinal trust is also involved. The methodologies considered while designing the PD controllers are heading angle correction in terms of minimising the difference between the X-coordinate value of fore and aft peak of ship, surge and sway velocity control, ship position control and reverse thrust when almost reaching the destination i.e. making the sway velocity minimum as possible. The following expressions describe the PD controllers used for automatic thrust generation in lateral and longitudinal direction.

if $\Psi < 270°$ and $dis\_fore > dis\_rev$

$$\begin{aligned} T_{fore} &= C_1 * (X_{fore} - 1.5 - X_{fore}) + C_2 * \text{sway} \\ T_{aft} &= C_1 * (X_{fore} - 1.5 - X_{fore}) + C_2 * \text{sway} + C_3 * \text{diff} \end{aligned} \quad (3.1)$$

if $\Psi > 270°$ and $dis\_aft > dis\_rev$

$$\begin{aligned} T_{fore} &= C_1 * (X_{aft} - 1.5 - X_{aft}) + C_2 * \text{sway} + C_3 * \text{diff} \\ T_{aft} &= C_1 * (X_{aft} - 1.5 - x_{aft}) + C_2 * \text{sway} \end{aligned} \quad (3.2)$$

if $\Psi < 270°$ and $dis\_fore < dis\_rev$

$$\begin{aligned} T_{fore} &= C_1 * (-1.5 - X_{fore}) + C_2 * \text{sway} \\ T_{aft} &= C_1 * (-1.5 - X_{fore}) + C_2 * \text{sway} + C_3 * \text{diff} \end{aligned} \quad (3.3)$$

if $\Psi > 270°$ and $dis\_aft < dis\_rev$

$$\begin{aligned} T_{fore} &= C_1 * (-1.5 - X_{aft}) + C_2 * \text{sway} + C_3 * \text{diff} \\ T_{aft} &= C_1 * (-1.5 - X_{aft}) + C_2 * \text{sway} \end{aligned} \quad (3.4)$$

Longitudinal thrust

$$T_{long} = C_4 * \text{surge} + C_5 * \text{Ypos} + C_6 * \text{distance} \quad (3.5)$$

where, Ψ is the ship heading, X_{fore} and X_{aft} are the x-coordinate of ship's fore and aft peak respectively, *diff* is $abs(X_{fore}-X_{aft})$, *distance* is the perpendicular distance of ship's CG from the actual pier, *dis_fore* and *dis_aft* are the perpendicular distance of ship's fore and aft peak respectively from the actual pier, *dis_rev* is the perpendicular distance from the actual pier to start reverse thrust, Y_{pos} is the y-coordinate of the ship's CG in the earth fixed coordinate, $C_1 \sim C_6$ are the coefficients.

Considering Eqs. (3.1) and (3.2) for providing side thrusts, the first part belongs to a constant value irrespective of the ship position to withstand the wind force up to 1.5 m/s (15 m/s for a full scale ship). The second part is for controlling the sway velocity and the third part activates if a correction for ship heading is needed. On the other hand, if the ship reaches the zone to provide reverse side thrusts as given by Eqs. (3.3) and (3.4), the first part is no longer constant and rather increases the thrust value gradually with the decrement of the distance value to minimise the sway velocity upon reaching the pier. Other parts remain the same. Here, the value of *dis_rev* depends on the steady sway velocity while approaching to the pier using side thrusters in presence of wind disturbances form different direction. Considering the longitudinal thrust given in Eq. (3.5), the first part is for controlling the forward velocity. The second part is for controlling the ship position in longitudinal direction and the third part is for controlling the thrust value with respect to the ship's distance from the actual pier.

5 Simulation Results

In this research, simulations are carried out on real time basis to know the actual response of the model ship. Before starting the simulation for automatic tug assistance, several inputs to the system must be considered. These are: ship's initial position, heading, surge velocity, sway velocity, yaw rate and wind information. Therefore, there are hundreds of possible sets of initial conditions from which the simulation can be done. This paper analysis the results based on the following three conditions.

5.1 Starting from One Point but Different Heading

These simulations are done for stationary ship, i.e. ship's surge, sway velocities and yaw rate are considered as zero. With that assumption, the ship starts from a given point under wind disturbances with different initial headings. The wind disturbances considered here is 1.5 m/s average gust which is 15 m/s for a full scale ship from 45°. Three different initial headings are considered as 270°, 240° and 300°. The results are illustrated in the following Figs. 6, 7 and 8. In these figures, the left-hand side shows the resulting trajectory due to the controlling action and the right-hand side shows the controlling action itself, i.e. force acting on fore, aft and longitudinal direction. The last row of the right-hand side also shows the gust wind distribution throughout the operation.

Considering Fig. 6, the ship was initially set parallel to the pier. However, when starting the crabbing motion, the ship tends to generate a yaw rate and takes the stable position so that W and CLR coincide to each other (discussed in Sect. 2). With that stable heading it executes the rest of crabbing motion. Later on, when the ship approaches close to the pier, the controller finally corrects its heading and provides the reverse thrusts to minimise the sway value as much as possible.

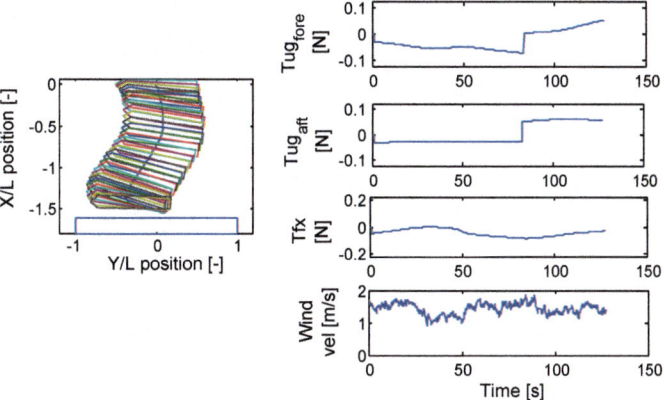

Fig. 6 Ship heading 270°, wind 1.5 m/s from 45°

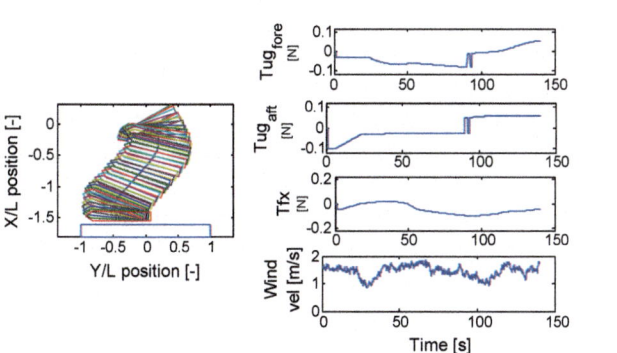

Fig. 7 Ship heading 240°, wind 1.5 m/s from 45°

Figure 7 demonstrates the result for ship's initial heading of 240°. In such cases, the controller first corrects the heading so as to make it parallel to the pier. Then, it starts the crabbing motion. The rest is the same as described for Fig. 6. For Fig. 8, the ship starts with heading 300° i.e. the bow is already towards the direction of wind. Therefore, the heading is just adjusted for the stable value while executing the crabbing motion. Here, it is noted that the action of the controller highly depends on the direction on wind.

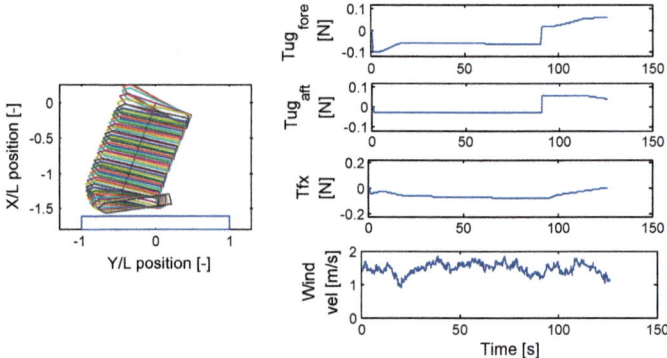

Fig. 8 Ship heading 300°, wind 1.5 m/s from 45°

5.2 Starting from Different Points But Same Heading

These simulations are also done for stationary ship. Two different positions are selected to simulate the scenario considering the ship is parallel to the pier under wind disturbances. The wind disturbances considered here is the same as before, i.e. average gust of 1.5 m/s from 45°. The following Figs. 9 and 10 illustrate the results.

Considering Fig. 9, the longitudinal thrust actives from the beginning with relatively big value to push the stationary ship forward and at the same time adjusts its value as it appears closer to the pier. The fore and aft thrusts do similar types of action as compared with Fig. 6. Finally the ship appears to the pier with almost parallel to it and minimum sway velocity. Figure 10 shows, negative longitudinal thrust at the

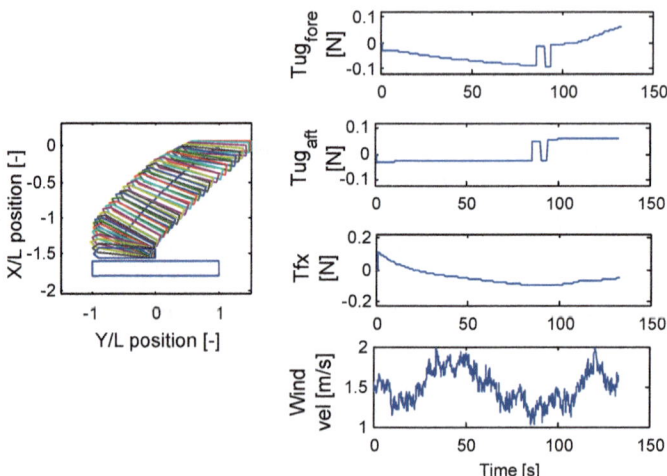

Fig. 9 Ship heading 270°, starting from (3, 0)

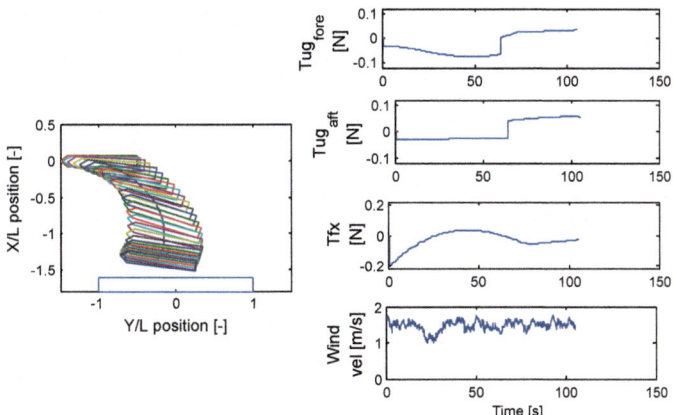

Fig. 10 Ship heading 270°, starting from (−3, 0)

beginning to push the ship backward and at the same time adjusts its value as the ship appears closer to the pier. Together with it, the crabbing motion is executed by the two thrusters at fore and aft to finally make the ship align with the pier.

5.3 Ship with Initial Surge, Sway and Yaw Rate

These simulations are done with the ship in motion, i.e. it may have relatively small surge, sway velocity and yaw rate at the beginning of tug assistance. The following Figs. 11 and 12 illustrate the results for ship with surge velocity 0.2 m/s, sway velocity −0.1 m/s and yaw rate 0.2 deg/s. Considering Fig. 11, the ship starts from (0, 0) point under average gust of 1.5 from 315°. In is noted that the controller immediately activates the negative longitudinal thrust as the ship starts to pass the

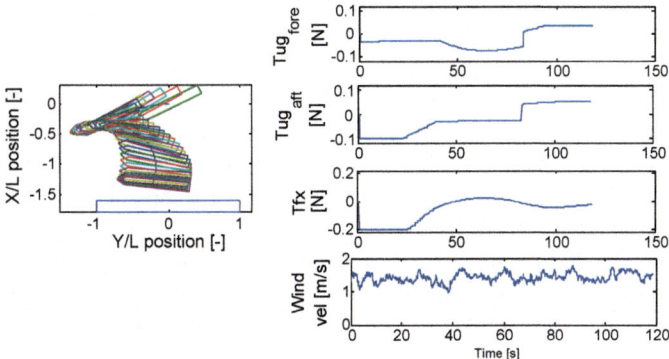

Fig. 11 Ship heading 240°, starting from (0, 0)

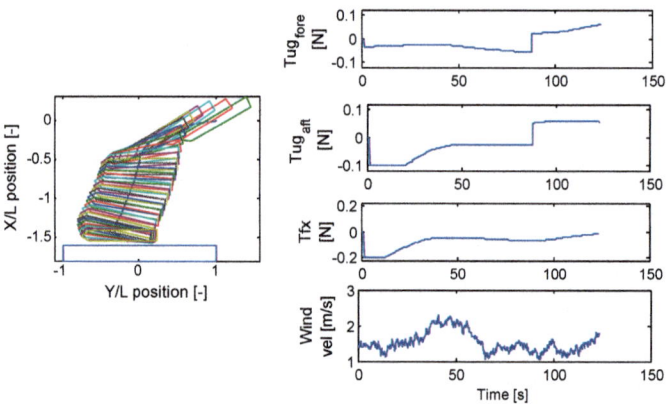

Fig. 12 Ship heading 240°, starting from (3, 0)

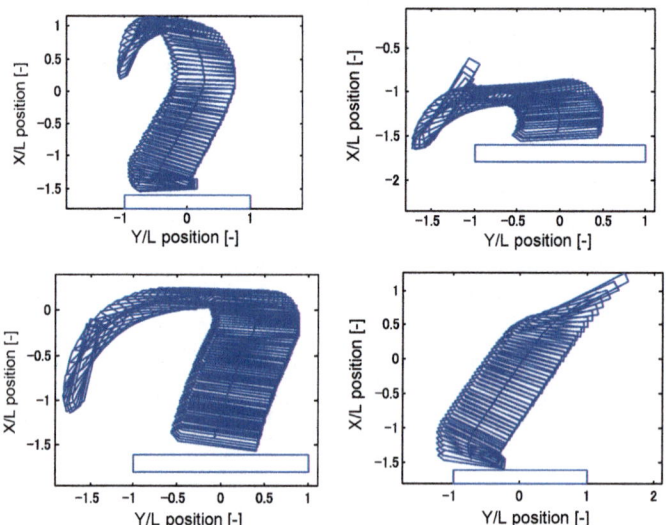

Fig. 13 Ship with arbitrary initial condition

pier. Later on it adjusts the value to make sure the final approach remains within the given pier. The side thrusters also adjust the ship heading first and then push the ship towards the pier for crabbing motion. In Fig. 12, the ship starts from (3, 0) under wind of 1.5 m/s from 0°.

Here, the longitudinal thrust starts with a negative value to reduce the initial ship speed. However, the values gradually modify as the ship tends to pass the projected area of the pier. It also remains zero for some time when the ship's CG perfectly align with the (0, 0) point. Together with it, the two thrusters adjust the ship heading and finally accomplish a successful berthing (Fig. 13).

6 Conclusions

The proposed controller works effectively under the wind disturbances and also with the ship in motion with reduced velocity. The simulations are verified for different initial conditions. Here the simulations are carried out on real time basis. Therefore, the time interval to get the controller's feedback can be tuned to match with actual ship scenario. However, the PD controlled thrusters mentioned in this paper are tested in simulations only. Using air fans attached on board, experiments for the developed PD controller would be possible in the future to validate its application in real ship cases.

References

1. Tran, V.L., Im, N.K.: A study on automatic berthing with assistance of auxiliary devices. J. Nav. Archit. Ocean Eng. **4**, 199–210 (2012) (Korea)
2. A Master's Guide to Berthing, Berthing in Wind. Source: http://www.shipinspection.eu/index.php/themariner-s-handbook/a-master-s-guide-toberthing/item/462-berthing-in-wind
3. Bui, V.P., et al.: Modelling and control allocation for ship berthing system design. In: International Conference on Control, Automation and Systems, pp. 195–200, Oct 2010
4. The Specialist Committee on Esso Osaka. Final Report and Recommendations to the 23rd ITTC. In: Proceedings of the 23rd ITTC **2**, 573–609 (2002)
5. Ueda, N., Ueno, M.A.: A comparative study on experimental results for manoeuvring hydrodynamic coefficients (in Japanese). Bachelor Thesis paper, Osaka University (1982)
6. Fujiwara, T., et al.: Estimation of wind forces and moment acting on ships. J. Soc. Nav. Archit. Jpn. **183**, 77–90 (1998)
7. Davenport, A.G.: The dependence wind loads on meteorological parameters. In: Proceedings of Conference on Wind Effects on Buildings and Structures (1967)

Excitation Force Between Two Ship Models in Waves

Faizul Amri Adnan

Abstract In the presence of incident waves and their interaction between two ships, the waves may scatter and excite the adjacent moving ship. On the assumption of a slender ship and linearized boundary condition, the velocity potentials are expanded by taking into consideration the radiation and diffraction potential due to the motions and scattering waves of the adjacent ship, respectively. By combining this technique with the strip method all the motion equations are derived except for surge. The excitation forces are numerically calculated and evaluated with results from the present experiment of two ships in waves. In this research, the interaction effects of two ships have been found significant in views of heave and sway force. The change of the incident wave angle also affects the increases/decreases of force/moment acting on the hull form of the ship.

Keywords Strip method · Hydrodynamic interaction · Wave forces
Diffraction potential

1 Introduction

The study of seakeeping performance for two bodies of ships usually is related with hydrodynamic interaction effects. In theory, a freely floating body may generate waves and may induce motions to another body. The interaction forces between two ships where one of the ships is fixed freely floating and the other ship is passing nearby has been explored by Islam [1] where the additional forces created by the scattered wave from a moving ship will excite the nearby moored ship. As elaborated by Skejic et al. [2] these interaction effects exist due to the sudden and strong unpredictable motion which is resulting on the ship to attract or repulse with each other.

F. A. Adnan (✉)
Malaysian Institute of Marine Engineering Technology, Universiti Kuala Lumpur,
Dataran Teknologi Kejuruteraan Marin, Bandar Teknologi Maritim, Jalan Pantai Remis,
32200 Lumut, Perak, Malaysia
e-mail: faizul.amri@unikl.edu.my

The interaction effect between two ships may vary dependent on the involved interaction loads. As elaborated by Skejic et al. [2] the maneuvering of ships is affected by the hydrodynamic interaction loads between the ships. The interaction loads are creating unpredictable forces and moments depending on the ships hullform, size, their lateral and longitudinal separation distance, speeds, wetted hull shapes, water depth, and transverse distance from a channel bank. In addition, Wang [3] mentioned that both ships will experience a sequence of repulsion, attraction and repulsion through meeting/passing period where the peak attraction force on each body increases relatively faster with their size, proximity and the period of time as well. Based on the study conducted, he concluded that the increasing of speed, size and proximity distance contributes to the bigger forces and moment acting on the ship bodies.

The two ships motion response studied which considered the interaction problem is paid particularly on the distance between two ships. The distance separation is considered as the most interconnecting variable being discussed by previous researchers. Such cases were studied by Wang [3] who focused on the interaction between two slender bodies at angle of yaw translating in parallel path in very close proximity. He conclude that the forces and moments on both bodies increase with their sizes and proximity distance where the peak attraction force on each body increases relatively faster with their size, proximity and the period of time as well. A similar case was analyzed by Skejic and Faltinsen [4] when two ships advancing in abeam, indicated that each ship experienced a peak of attraction force and attracted transversely to each other as the yaw moment tend to swing the bow of both ships from each other.

However, when environmental factors were included, the interaction load may be induced by the strong lateral force of wind and waves. Chen and Fang [5] mentioned that when two ships are advancing in waves, the interaction load is strongly depending on the position of the ships, the wave frequency and wave headings. The interaction problem is analyzed by applying the strip theory developed by Fang and Kim [6] and the three-dimensional approach proposed by Fang and Chen [7]. Chen and Fang [5] performed the combination of the numerical solution based on the three-dimensional potential flow theory and compared the results with the experimental data and 2D strip theory. For zero speed effect, they had summarized that the interaction effect takes place in sway force at the heading wave and the sway amplitude is higher at the oblique wave. However, the effect of clearance distance between the ships is not studied in their research and the forward speed effect results are not validated with experiment data.

In addition, Fang and Chen [7] also proposed for the interaction of ships in waves where the clearances of gap and speed should be accounted with the wave heading consideration. They concluded that in the effect for relative motions of ships with gap distance may vary depending to the speed and ship position toward the waves. The use of potential theory is widely being explored for predicting the seakeeping performance. Faizul [8] developed a new version of strip method

(NSM) in predicting the ship motion in waves with small lateral drift. The 3D hydrodynamic coefficients were derived by integration of two-dimensional NSM along the longitudinal section.

In the present research, the hydrodynamic interaction effect between two ships in waves was developed based on the new strip method platform. By combining the method used by Fang and Chen [7] and the strip method technique [8], the 3D hydrodynamic coefficients for the two ships were derived. The objective of this research was to identify the pattern of force and moment acting on the ship model due to the interaction effect between two ship models in waves. All the computed forces and moments were validated with the experiment result from the present model testing on two ship models in waves conducted at the NAHRIM facility located at Seri Kembangan Selangor.

2 Mathematical Formulation

2.1 Coordinate System

Consider ship A and B advance in wave at close proximity side by side to each other. Both ships are assumed advancing in waves with constant forward speed U_a and U_b, where the average ship advancing direction is taken in x_a-axis for ship A and x_b-axis for ship B. Initially both ships are assumed to be experiencing the same drift effect with the same forward speed. The hull size and dimension for both ships is considered in similar geometric properties where the ship container type SR108 was used in this research. The coordinate systems for ship A and B are defined in o_a x_a y_a z_a and o_b x_b y_b z_b. While the O-X-Y-Z coordinate system is the fixed coordinate in space. Based on Fig. 1, it is defined that χ is the angle of incident wave,

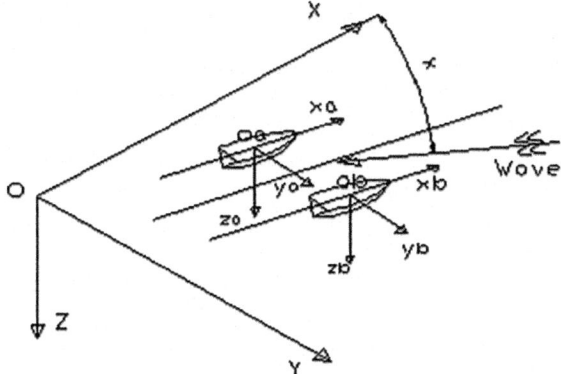

Fig. 1 Coordinate system for ship A and ship B

For ship A,

$$X_a = x_a \cos \chi + y_a \sin \chi$$
$$Y_a = -x_a \sin \chi + y_a \cos \chi \qquad (1)$$
$$Z_a = z_a$$

For ship B,

$$X_b = x_b \cos \chi - y_b \sin \chi$$
$$Y_b = -x_b \sin \chi + y_b \cos \chi \qquad (2)$$
$$Z_b = z_b$$

It is assumed that both ships A and B are at the same speed and induced by the same incident wave. Hence the velocity potential of the incident wave is for both ships the same as follows:

$$\phi_I = \Re\left\{\frac{gA}{i\omega} \exp[-vz - iv\{x \cos \chi + y \sin \chi\} + i\omega t]\right\} \qquad (3)$$

As mentioned above, the variables A and v denote the wave amplitude and the wave number of the incident waves. The circular frequency of the wave, ω for both ships A and B are as shown below:

$$\omega_a = \omega_b = \omega \qquad (4)$$

The wave frequency ω of the incident wave for both ships A and B is derived from differentiating the ϕ_I in $o_a x_a y_a z_a$ and $o_b x_b y_b z_b$ with respect to time.

$$\frac{\partial}{\partial t} = i\omega + iv(U_0 \cos \chi + U_0 \sin \chi) \qquad (5)$$

Hence, the encounter frequency of the incident wave can be written as:

$$\omega_e = \omega + v(U_0 \cos \chi + U_0 \sin \chi) \qquad (6)$$

The flow of fluid around the ships is to be assumed inviscid, irrotational and incompressible so that the potential theory can be applied. Neglecting all steady disturbance terms, the perturbation potential $\Phi(x, y, z, t)$ which is applicable for both ships is defined in the ships coordinate system as:

$$\phi(x, y, z, t) = -U_0 x + \Re[\varphi_w(x, y, z)e^{i\omega t}] + \Re[\varphi(x, y, z)e^{i\omega t}] \qquad (7)$$

Where $U_{ao}x$ and $U_{bo}x$ are the ambient uniform current potential defined in x and y directions as stated in both ships coordinate system. While $\varphi_w(x, y, z)$ is referred to as the time independent incident wave velocity potential on ship A and B which is defined as:

$$\phi_w = \Re\left\{\frac{gA}{i\omega}\exp[-vz - iv\{x\cos\chi + y\sin\chi\}]\right\} \quad (8)$$

In Eq. (9), φ is the velocity potential in complex form, which is composed of the scattering and radiation on ship A and ship B. This complex potential for both ships must satisfy the Laplace equation and the boundary conditions with respect to the free surface and the ship hull.

2.2 Boundary Condition

In Eq. (12), $n = (n_1, n_2, n_3)$ denotes the outward normal unit vector on the hull surface. Variable α is the vector of motion displacement amplitude at the hull surface and given by:

$$\frac{\partial^2 \varphi}{\partial x^2} + \frac{\partial^2 \varphi}{\partial y^2} + \frac{\partial^2 \varphi}{\partial z^2} = 0 \quad (9)$$

$$\left(i\omega_e - U_o\frac{\partial}{\partial x} - V_o\frac{\partial}{\partial y}\right)^2 \varphi - g\frac{\partial \varphi}{\partial z} = 0 \quad \text{On } z = 0 \quad (10)$$

$$\frac{\partial \varphi}{\partial n} = \left(i\omega_e \alpha - U_o\frac{\partial \alpha}{\partial x} - V_o\frac{\partial \alpha}{\partial y}\right) \cdot n - \frac{\partial \varphi_w}{\partial n} \quad \text{On } S_H \quad (11)$$

$$\alpha_a = \begin{Bmatrix} \xi_{1a} + (z-z_G)\xi_{5a} - y\xi_{6a} \\ \xi_{2a} - (z-z_G)\xi_{4a} + x\xi_{6a} \\ \xi_{3a} + y\xi_{4a} - x\xi_{5a} \end{Bmatrix} \quad (12)$$

$$\alpha_b = \begin{Bmatrix} \xi_{1b} + (z-z_G)\xi_{5b} - y\xi_{6b} \\ \xi_{2b} - (z-z_G)\xi_{4b} + x\xi_{6b} \\ \xi_{3b} + y\xi_{4b} - x\xi_{5b} \end{Bmatrix} \quad (13)$$

Where $\xi_i = (i = 1, 2, \ldots 6)$ describe the ship oscillation for surge, sway, heave, roll, pitch and yaw, respectively and are expressed as complex numbers. Using the slender ship assumption, gives the simplification to the relation of the ship's hull normal vector.

$$n_1 \ll n_2, n_3 \quad (14)$$

Hence it implies that the change in flow in x-direction of the ship hull is moderate than those in y-and z-direction. Thus,

$$\frac{\partial}{\partial x} \ll \frac{\partial}{\partial y}, \frac{\partial}{\partial z} \quad (15)$$

The derivatives with respect to the x-direction become smaller than the change of flow in y- and z-direction which is implied in the Laplace equation stated in Eq. (9).

$$\frac{\partial^2 \varphi}{\partial y^2} + \frac{\partial^2 \varphi}{\partial z^2} = 0 \qquad (16)$$

The simplification to he slender theory also applicable for the boundary condition stated in Eqs. (10) and (11), where the term $\partial/\partial x$ has been neglected.

Free surface condition,

$$\left(i\omega_e - V_o \frac{\partial}{\partial y}\right)^2 \varphi - g\frac{\partial \varphi}{\partial z} = 0 \quad \text{on } z = 0 \qquad (17)$$

Hull surface condition,

$$\frac{\partial \varphi}{\partial N} = (i\omega_e \xi_3 - i\omega_e x \xi_5 + U_o \xi_5)n_3$$
$$+ (i\omega_e \xi_2 + i\omega_e x \xi_6 - U_o \xi_6 + i\omega_e z_G \xi_4)n_2 \qquad (18)$$
$$+ i\omega_{ae} \xi_{4a}(yn_3 - zn_2) - \frac{\partial \varphi_w}{\partial N}$$

where,

$$\frac{\partial}{\partial N} = n_2 \frac{\partial}{\partial y} + n_3 \frac{\partial}{\partial z} \qquad (19)$$

At this stage, it is convenient to change the notation for the outward normal unit vector from 3D into 2D as follows:

$$N_2 = n_3, \ N_1 = n_2, \ N_3 = yn_3 - zn_2 \qquad (20)$$

Hence the hull surface condition for 2D for ship A and B is

$$\frac{\partial \varphi_a}{\partial N} = [i\omega_{ea}(\xi_{3a} - x\xi_{5a}) + U_{a0}\xi_{3a}]N_2$$
$$+ [i\omega_{ea}(\xi_{2a} + x\xi_{6a} + z_G\xi_{4a}) - U_{a0}\xi_{6a}]N_1 \qquad (21)$$
$$+ i\omega_{ea}\xi_{4a}N_3 - \frac{\partial \varphi_w}{\partial N}$$

$$\frac{\partial \varphi_b}{\partial N} = [i\omega_{eb}(\xi_{3b} - x\xi_{5b}) + U_{a0}\xi_{3b}]N_2$$
$$+ [i\omega_{eb}(\xi_{2b} + x\xi_{6b} + z_G\xi_{4b}) - U_{b0}\xi_{6}]N_1 + i\omega_{eb}\xi_{4b}N_3 - \frac{\partial \varphi_w}{\partial N} \qquad (22)$$

Referring to the resultant velocity potential applied by Fang (2000) for Ship A and B when advancing in waves, the velocity potential for N.S.M can be derived by expanding the Eqs. (21) and (22).

Velocity potential for ship A and B,

$$\varphi_a = \sum_{j=1}^{3} \beta_{jaa} \varphi_{jaa} + \beta_{jba} \varphi_{jba} + \varphi_{4aa} + \varphi_{4ba} \tag{23}$$

$$\varphi_b = \sum_{j=1}^{3} \beta_{jbb} \varphi_{jbb} + \beta_{jab} \varphi_{jab} + \varphi_{4bb} + \varphi_{4ab} \tag{24}$$

φ_{jaa}, φ_{jba}, ($j = 1, 2, 3$) represent the radiation potential with respect to sway, heave and roll motions for 2D ship section on ship A due to ship A or ship B respectively. While φ_{jbb}, φ_{jab}, represent the radiation potential with respect to sway, heave and roll motions for 2D ship section on ship B due to ship B or ship A respectively. φ_{4aa}, φ_{4ab} represent the scattering potential in 2D resulting from disturbance of the incident wave by the hull on ship A due ship A or ship B, respectively. While φ_{4bb}, φ_{4ba} is defined as the scattering potential in 2D resulting from disturbance of the incident wave by the hull on ship B due ship B or ship A, respectively. Hence, the hull surface condition for φ_{jaa}, φ_{jbb} can be written as:

$$\begin{aligned} \frac{\partial \varphi_{jaa}}{\partial N} &= N_j \quad \text{on } S_H \ (j = 1, 2, 3); \\ \frac{\partial \varphi_{4aa}}{\partial N} &= -\frac{\partial \varphi_w}{\partial N} \\ &= -v[iN_1 \sin(\chi) - N_2]\varphi_w \quad \text{on } S_H \end{aligned} \tag{25}$$

The free surface conditions can be written as:

$$\frac{\omega_{ea}^2}{g} \varphi_{jaa} + \frac{\partial \varphi_{jaa}}{\partial z} = 0 \quad \text{on} \quad z = 0 \text{ with } (j = 1, 2, 3, 4) \tag{26}$$

2.3 Special Treatment for the Diffraction Problem

The incident wave potential can be expressed as:

$$\varphi_w = \frac{gA}{i\omega} \varphi_I \exp(ilx) \tag{27}$$

where

$$\varphi_I = \exp[-\nu\{z - iy \sin \chi\}] \qquad (28)$$

where $l = \nu \cos \chi$.

Since the flow is periodic in the x-direction, scattering potential φ_{4aa}, φ_{4bb} can be written as follows:

$$\varphi_4 = \frac{gA}{i\omega} \varphi_s \exp(ilx) \qquad (29)$$

The boundary condition for φ_s, which here is the solution, rather than φ_4, can be rewritten as:

$$\frac{\partial \varphi_s}{\partial N} = -\nu[iN_1 \sin(\chi) - N_2]\varphi_I \quad S_H \qquad (30)$$

$$\frac{\omega_e^2}{g} \varphi_s^{(1)} + \frac{\partial \varphi_s}{\partial z} = 0 \quad \text{on} \quad z = 0 \qquad (31)$$

In summary, φ_j ($j = 1, 2, 3, S$) for a given ship cross section, are unknown variables, and can be obtained by solving the 2D boundary value problems shown in this section under the proper radiation condition. Using φ_j, the pressure acting on the ship's hull can be calculated.

2.4 Pressure and Forces

By using the linearized Bernoulli equation, the pressure action on both ship's hull is expressed as:

$$P_H = \rho g z - \rho \left(\frac{\partial}{\partial x} - U_o \frac{\partial}{\partial x} - V_o \frac{\partial}{\partial y} \right) \Phi \qquad (32)$$

Time independent motion displacement for ship A and B is expressed in term Ξ_i ($i = 1, 2, 3$), Φ_1, θ_1, ψ_1 as shown below:

$$\begin{aligned}
\Xi_{ia} &= \Re[\xi_{ia} e^{i\omega_a t}] & \Xi_{ib} &= \Re[\xi_{ib} e^{i\omega_b t}] \\
\phi_{1a} &= \Re[\xi_{4a} e^{i\omega_a t}] & \phi_{1b} &= \Re[\xi_{4b} e^{i\omega_b t}] \\
\theta_{1a} &= \Re[\xi_{5a} e^{i\omega_a t}] & \theta_{1b} &= \Re[\xi_{5b} e^{i\omega_b t}] \\
\psi_{1a} &= \Re[\xi_{6a} e^{i\omega_a t}] & \psi_{1b} &= \Re[\xi_{6b} e^{i\omega_b t}]
\end{aligned} \qquad (33)$$

The static pressure is defined from the first term in the linearized Bernoulli equation. Static pressure change due to the ship motion P_s and can be expressed as:

$$P_s = \rho g(\Xi_3 + y\phi_1 - x\theta_1) \qquad (34)$$

From Eq. (7) the radiation potential is expressed as:

$$\Phi_{rj} = \varphi(x, y, z)e^{i\omega t} \qquad (35)$$

$$\sum_{j=1}^{3} \Phi_{rj} = \Phi_{r1} + \Phi_{r2} + \Phi_{r3} \qquad (36)$$

Where Φ_{rj} ($j = 1, 2, 3$) is expressed as the summation of radiation potential for lateral, vertical and rotational motion. Hence the summation for pressure due radiation component can be written as:

$$P_{rj} = P_{r1} + P_{r2} + P_{r3} \qquad (37)$$

where,

$$\frac{P_{rj}}{\rho} = -\left(\frac{\partial}{\partial t} - U_o \frac{\partial}{\partial x}\right)\Phi_{rj} \qquad (38)$$

By using the velocity potential in the complex form $\varphi_j = \varphi_{jc} + i\varphi_{js}$, the radiation potential components for ship A and B can be written as:

$$\begin{aligned}\Phi_{r1a} =& (\varphi_{1Caa})\dot{\Xi}_{2a} - (\varphi_{1Saa}\omega_{ea})\Xi_{2a} + \left[x(\varphi_{1Caa}) - \frac{U_{a0}}{\omega_{ea}}(\varphi_{1Saa})\right]\dot{\psi}_{1a} \\ & - [U_{a0}(\varphi_{1Caa}) + x(\varphi_{1Saa}\omega_{ea})]\psi_{1a} + z_G\dot{\phi}_{1a}(\varphi_{1Caa}) - z_G\phi_{1a}(\varphi_{1Saa}\omega_{ea}) \\ & + \dot{\Xi}_{2b}(\varphi_{1Cba}) - \Xi_{2b}(\varphi_{1Sba}\omega_{eb}) + \left[x(\varphi_{1Cba}) - \frac{U_{b0}}{\omega_{eb}}(\varphi_{1Sba})\right]\dot{\psi}_{1b} \\ & - [U_{b0}(\varphi_{1Cba}) + x(\varphi_{1Sba}\omega_{eb})]\psi_{1b} + z_G\dot{\phi}_{1b}(\varphi_{1Cba}) - z_G\phi_{1b}(\varphi_{1Sba}\omega_{eb})\end{aligned} \qquad (39)$$

$$\begin{aligned}\Phi_{r2a} =& (\varphi_{2Caa})\dot{\Xi}_{3a} - (\varphi_{2Saa}\omega_{ea})\Xi_{3a} + \left[x(\varphi_{2Caa}) - \frac{U_{a0}}{\omega_{ea}}(\varphi_{2Saa})\right]\dot{\theta}_{1a} \\ & - [x(\varphi_{2Saa}\omega_{ea}) + U_{a0}(\varphi_{2Caa})]\theta_{1a} + (\varphi_{2Cba})\dot{\Xi}_{3b} - (\varphi_{2Sba}\omega_{eb})\Xi_{3b} \\ & + \left[x(\varphi_{2Cba}) - \frac{U_{b0}}{\omega_{eb}}(\varphi_{2Sba})\right]\dot{\theta}_{1b} - [x(\varphi_{2Sba}\omega_{eb}) + U_{b0}(\varphi_{2Cba})]\theta_{1b}\end{aligned} \qquad (40)$$

$$\Phi_{r3a} = (\varphi_{3Caa})\dot{\phi}_{1a} - (\varphi_{3Saa}\omega_{ea})\phi_{1a} + (\varphi_{3Cba})\dot{\phi}_{1b} - (\varphi_{3Sba}\omega_{eb})\phi_{1b} \qquad (41)$$

$$\Phi_{r1b} = \dot{\Xi}_{2b}(\varphi_{1Cbb}) - \Xi_{2b}(\varphi_{1Sbb}\omega_{ea}) + \left[x(\varphi_{1Cbb}) - \frac{U_{b0}}{\omega_{eb}}(\varphi_{1Sbb})\right]\dot{\psi}_{1b}$$
$$- [U_{b0}(\varphi_{1Cbb}) + x(\varphi_{1Sbb}\omega_{ea})]\psi_{1b} + z_G\dot{\phi}_{1b}^{(1)}(\varphi_{1Cbb}) - z_G\phi_{1b}(\varphi_{1Sbb}\omega_{ea})$$
$$+ (\varphi_{1Cab})\dot{\Xi}_{2a} - (\varphi_{1Sab}\omega_{ea})\Xi_{2a}^{(1)} + \left[x(\varphi_{1Cab}) - \frac{U_{a0}}{\omega_{ae}}(\varphi_{1Sab})\right]\dot{\psi}_{1a}$$
$$- [U_{a0}(\varphi_{1Cab}) + x(\varphi_{1Sab}\omega_{ea})]\psi_{1a} + z_G\dot{\phi}_{1a}(\varphi_{1Cab}) - z_G\phi_{1a}(\varphi_{1Sab}\omega_{ea}) \tag{42}$$

$$\Phi_{r2b} = (\varphi_{2Cbb})\dot{\Xi}_{3b} - (\varphi_{2Sbb}\omega_{eb})\Xi_{3b} + \left[x(\varphi_{2Cbb}) - \frac{U_{b0}}{\omega_{eb}}(\varphi_{2Sbb})\right]\dot{\theta}_{1b}$$
$$- [x(\varphi_{2Sbb}\omega_{eb}) + U_{b0}(\varphi_{2Cbb})]\theta_{1b} + (\varphi_{2Cab})\dot{\Xi}_{3a} - (\varphi_{2Sab}\omega_{ea})\Xi_{3a} \tag{43}$$
$$+ \left[x(\varphi_{2Cab}) - \frac{U_{a0}}{\omega_{ea}}(\varphi_{2Sab})\right]\dot{\theta}_{1a} - [x(\varphi_{2Sab}\omega_{ea}) + U_{a0}(\varphi_{2Cab})]\theta_{1a}$$

$$\Phi_{r3b} = (\varphi_{3Cbb})\dot{\phi}_{1b} - (\varphi_{3Sbb}\omega_{eb})\varphi_{1b} + (\varphi_{3Cab})\dot{\phi}_{1a} - (\varphi_{3Sab}\omega_{ea})\varphi_{1a} \tag{44}$$

The diffraction potential is composed by the incident wave and scattering wave due to the disturbance of incident waves as they strike the ship hull. Since the incident wave velocity potential is present in complex velocity. Hence, the pressure due to incident wave P_I can be expressed as:

$$\varphi_I \exp(ilx) = \varphi_{Ic} + i\varphi_{Is} \tag{45}$$

$$-\frac{P_I}{\rho} = gA(\varphi_{Ic}\cos\omega_e t - \varphi_{Is}\sin\omega_e t) \tag{46}$$

From the above section, we adopted a similar treatment so that the pressure due to scattering potential can be expressed as:

$$\varphi_S \exp(ilx) = \varphi_{Sc} + i\varphi_{Ss} \tag{47}$$

$$-\frac{P_{Sa}}{\rho} = \frac{gA}{\omega_a}\left[\begin{array}{c}\omega_{ae}(\varphi_{Scaa}\cos\omega_{ea}t - \varphi_{Ssaa}\sin\omega_{ea}t)\\ -U_{ao}\left(\frac{\partial\varphi_{Scaa}}{\partial x}\sin\omega_{ea}t + \frac{\partial\varphi_{Ssaa}}{\partial x}\cos\omega_{ea}t\right)\end{array}\right]$$
$$+ \frac{gA}{\omega_b}\left[\begin{array}{c}\omega_{ae}(\varphi_{Scba}\cos\omega_{eb}t - \varphi_{Ssba}\sin\omega_{eb}t)\\ -U_{ao}\left(\frac{\partial\varphi_{Scba}}{\partial x}\sin\omega_{eb}t + \frac{\partial\varphi_{Ssba}}{\partial x}\cos\omega_{eb}t\right)\end{array}\right] \tag{48}$$

$$-\frac{P_{Sb}}{\rho} = +\frac{gA}{\omega_b}\left[\begin{array}{c}\omega_{eb}(\varphi_{Scbb}\cos\omega_{eb}t - \varphi_{Ssbb}\sin\omega_{eb}t)\\-U_{bo}\left(\frac{\partial\varphi_{Scbb}}{\partial x}\sin\omega_{eb}t + \frac{\partial\varphi_{Ssbb}}{\partial x}\cos\omega_{eb}t\right)\end{array}\right]$$
$$+\frac{gA}{\omega_a}\left[\begin{array}{c}\omega_{eb}(\varphi_{Scab}\cos\omega_{ea}t - \varphi_{Ssab}\sin\omega_{ea}t)\\-U_{bo}\left(\frac{\partial\varphi_{Scab}}{\partial x}\sin\omega_{ea}t + \frac{\partial\varphi_{Ssab}}{\partial x}\cos\omega_{ea}t\right)\end{array}\right] \quad (49)$$

By integrating the wave pressure along the ship's cross section, the hydrodynamic forces acting on the section are obtained. From the hydrodynamic coefficient in two-dimensional derived, the added mass and damping are denoted as M_{lj} and N_{lj}. While the exciting forces in 2D are denoted as E_{Cj} and E_{Sj}.

$$M_{\ell j} - i\frac{N_{\ell j}}{\omega_e} = -\rho\int_{S_H}\varphi_j N_\ell ds \quad (50)$$

$$E_{cj} + iE_{sj} = \rho g\frac{\omega_{ea}}{\omega_a}\int_{S_H}\varphi_S e^{ilx}N_j ds \quad (51)$$

In this section, the equation of motion will be presented in two set of equations for ship A and Ship B. Based on the equation derived, variable m denotes the mass of the ship, A_{ij} and B_{ij} is defined as added mass and damping with respect to the i-th as mode induced and j-th as motion. Where the suffixes $i, j = 2, 3, 4, 5, 6$ represent as sway, heave, pitch, yaw, and roll modes. While C_{ij} denotes the hydrostatic restoring force and F_{jc} and F_{js} are denotes to the exciting forces component for j-th motion.

Heaving Motion

$$\begin{aligned}&\left(m_{a,b} + A_{33aa,bb}\right)\ddot{\Xi}_{3a,b} + B_{33aa,bb}\dot{\Xi}_{3a,b} + C_{33aa,bb}\Xi_{3a,b} + A_{35aa,bb}\ddot{\theta}_{1a,b}\\&+ B_{35aa,bb}\dot{\theta}_{1a,b} + C_{35aa,bb}\theta_{1a,b} + A_{33ba,ab}\ddot{\Xi}_{3b,a} + B_{33ba,ab}\dot{\Xi}_{3b,a}\\&+ C_{33ba,ab}\Xi_{3b,a} + A_{35ba,ab}\ddot{\theta}_{1b,a} + B_{35ba,ab}\dot{\theta}_{1b,a} + C_{35ba,ab}\theta_{1b,a}\\&= A\left(\begin{array}{c}F_{3Caa,ab}\cos\omega_{ea}t - F_{3Saa,ab}\sin\omega_{ea}t\\+F_{3Cba,bb}\cos\omega_{eb}t - F_{3Sbb,ba}\sin\omega_{eb}t\end{array}\right)\end{aligned} \quad (52)$$

Pitching Motion

$$\left(I_{ya,b} + A_{55aa,bb}\right)\ddot{\theta}_{1a,b} + B_{55aa,bb}\dot{\theta}_{1a,b} + C_{55aa,bb}\theta_{1a,b} + A_{53aa,bb}\ddot{\Xi}_{3a,b}$$
$$+ B_{53aa,bb}\dot{\Xi}_{3a,b} + C_{53aa,bb}\Xi_{3a,b} + A_{53ba,ab}\ddot{\Xi}_{3b,a} + B_{53ba,ab}\dot{\Xi}_{3b,a}$$
$$+ C_{53ba,ab}\Xi_{3b,a} + A_{55ba,ab}\ddot{\theta}_{1b,a} + B_{55ba,ab}\dot{\theta}_{1b,a} + C_{55ba,ab}\theta_{1b,a} =$$
$$A\begin{pmatrix} F_{5Caa,ab}\cos\omega_{ea}t - F_{5Saa,ab}\sin\omega_{ea}t \\ + F_{5Cba,bb}\cos\omega_{eb}t - F_{5Sba,bb}\sin\omega_{eb}t \end{pmatrix} \tag{53}$$

Swaying Motion

$$\left(m_{a,b} + A_{22aa,bb}\right)\ddot{\Xi}_{2a,b} + B_{22aa,bb}\dot{\Xi}_{2a,b} + C_{22aa,bb}\Xi_{2a,b}$$
$$+ A_{24aa,bb}\ddot{\phi}_{1a,b} + B_{24aa,bb}\dot{\phi}_{1a,b} + C_{24aa,bb}\phi_{1a,b}$$
$$+ A_{26aa,bb}\ddot{\psi}_{1a,b} + B_{26aa,bb}\dot{\psi}_{1a,b} + C_{26aa,bb}\psi_{1a,b}$$
$$+ A_{22ba,ab}\ddot{\Xi}_{2b,a} + B_{22ba,ab}\dot{\Xi}_{2b,a} + C_{22ba,ab}\Xi_{2b,a}$$
$$+ A_{24ba,ab}\ddot{\phi}_{1b,a} + B_{24ba,ab}\dot{\phi}_{1b,a} + C_{24ba,ab}\phi_{1b,a}$$
$$+ A_{26ba,ab}\ddot{\psi}_{1b,a} + B_{26ba,ab}\dot{\psi}_{1b,a} + C_{26ba,ab}\psi_{1b,a}$$
$$= A\begin{pmatrix} F_{2Caa,ab}\cos\omega_{ea}t - F_{2Saa,ab}\sin\omega_{ea}t \\ + F_{2Cba,bb}\cos\omega_{eb}t - F_{2Sba,bb}\sin\omega_{eb}t \end{pmatrix} \tag{54}$$

Yawing Motion

$$\left(I_{zza,b} + A_{66aa,bb}\right)\ddot{\psi}_{1a,b} + B_{66aa,bb}\dot{\psi}_{1a,b} + C_{66aa,bb}\psi_{1a,b}$$
$$+ A_{64aa,bb}\ddot{\phi}_{1a,b} + B_{64aa,bb}\dot{\phi}_{1a,b} + C_{64aa,bb}\phi_{1a,b} + A_{62aa,bb}\ddot{\Xi}_{2a,b}$$
$$+ B_{62aa,bb}\dot{\Xi}_{2a,b} + C_{62aa,bb}\Xi_{2a,b} + A_{66ba,ab}\ddot{\psi}_{1b,a} + B_{66ba,ab}\dot{\psi}_{1b,a}$$
$$+ C_{66ba,ab}\psi_{1b,a} + A_{64ba,ab}\ddot{\phi}_{1b,a} + B_{64ba,ab}\dot{\phi}_{1b,a} + C_{64ba,ab}\phi_{1b,a}$$
$$+ A_{62ba,ab}\ddot{\Xi}_{2b,a} + B_{62ba,ab}\dot{\Xi}_{2b,a} + C_{62ba,ab}\Xi_{2b,a}$$
$$= A\left(F_{6Caa,ab}\cos\omega_{ea}t - F_{6Saa,ab}\sin\omega_{ea}t\right)$$
$$+ A\left(F_{6Cba,bb}\cos\omega_{eb}t - F_{6Sba,bb}\sin\omega_{eb}t\right) \tag{55}$$

Rolling Motion

$$\left(I_{xxa,\,b} + A_{44aa,\,bb}\right)\ddot{\phi}_{1a,\,b} + B_{44aa,\,bb}\dot{\phi}_{1a,\,b} + C_{44aa,\,bb}\phi_{1a,\,b}$$
$$+ A_{42aa,\,bb}\ddot{\Xi}_{2a,\,b} + B_{42aa,\,bb}\dot{\Xi}_{2a,\,b} + C_{42aa,\,bb}\Xi_{2a,\,b} + A_{46aa,\,bb}\ddot{\psi}_{1a,\,b}$$
$$+ B_{46aa,\,bb}\dot{\psi}_{1a,\,b} + C_{46aa,\,bb}\psi_{1a,\,b} + A_{44ba,\,ab}\ddot{\phi}_{1b,\,a} + B_{44ba,\,ab}\dot{\phi}_{1b,\,a}$$
$$+ C_{44ba,\,ab}\phi_{1b,\,a} + A_{42ba,\,ab}\ddot{\Xi}_{2b,\,a} + B_{42ba,\,ab}\dot{\Xi}_{2b,\,a} + C_{42ba,\,ab}\Xi_{2b,\,a}$$
$$+ A_{46ba,\,ab}\ddot{\psi}_{1b,\,a} + B_{46ba,\,ab}\dot{\psi}_{1b,\,a} + C_{46ba,\,ab}\psi_{1b,\,a}$$
$$= A\left(F_{4Caa,\,ab}\cos\omega_{ea}t - F_{4Saa,\,ab}\sin\omega_{ea}t\right)$$
$$+ A\left(F_{4Cba,\,bb}\cos\omega_{eb}t - F_{4Sba,\,bb}\sin\omega_{eb}t\right) \tag{56}$$

3 Model Experiment

In this experiment, the models used are the same as used in the Fortran simulation where SR108 container type-MTL032 as ship A and Tenaga oil LNG—MTL 057 represent as Ship B. Detail dimensions of ships' model are shown in Table 1. The measurement of forces will take place at ship A (container type) by installing two transducers at 20 cm of aft from the mid ship and 20 cm of fwd from the midship. The experiment is designed under diffraction conditions where both models are fixed or not freely floating in waves. Due to the diffraction effect, the measured force can be obtained in pure sway and heave without any radiation force. For this experiment both ships are positioned side by side with separation distance as shown in Fig. 2. The separation distance between the two models is 35 cm (25 m full scale) from side to side of each other [4]. The wave properties are set up from 0.6, 0.8, 0.9 and 1.0 s wave period with a wave height 5 cm. During all times of the experiment, both models are fixed and not allowed to move in heave, sway and surge. Therefore, the support structure is designed to hold both models in position without forward speed.

Table 1 Dimension of ships' model

	Model ship A	Ship (cont.)	Model ship B	Ship (LNG)
Ship length L_{pp} (m)	2.50	175.00	2.375	268.41
Breadth, B (m)	0.3629	25.40	0.371	41.6
Draft, d (m)	0.1215	9.50	0.0994	11.13
Displacement\\nabla (t)	0.0618	24742	0.0663	94280

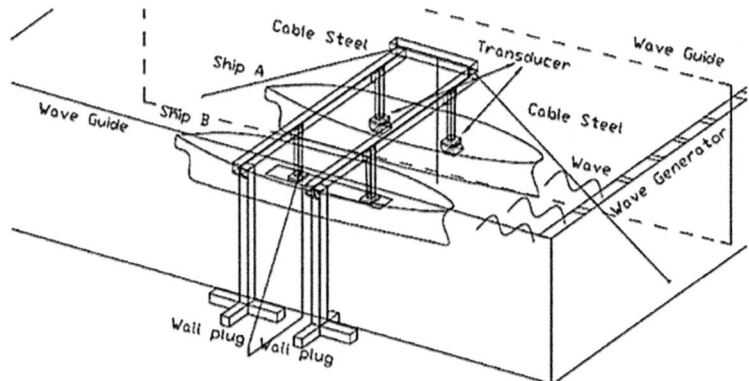

Fig. 2 Schematic diagram of ship model for side by side experiment

4 Discussion

Based on the mathematical formulation derived, all the excitation forces were calculated in 6 DOF except for surge by assuming both ships are slender. However, for validation purpose the results were only presented and discussed on sway and heave forces as well as yaw and pitch moments. Roll motion was not included as the results require further treatment due to the nonlinearity effects. Besides that, the measurement and calculation of the exciting force were performed only for ship A which considers the interaction effect. The related result had been non-dimensional in terms of the following variable EF (excitation force), G (gravity), Δ (displacement), and λ/L_{pp} (wavelength).

In the heading wave, the sway forces experienced by ship model A are shown in Fig. 3 where the force amplitudes calculated are consistent at 0.14 at high and low frequency. However, the experiment result shows a different pattern graph and slightly change at every wavelength when compared to the computed result. For the two body case, the amplitude sway force from computed result manages to capture the pattern of the graph projected from the experiment result. Besides that, the sway force from computation and experiment results shows an average difference in amplitude of 41 and 12.8% at heading and oblique wave. The experment result for the sway forces were not zero as expected similar with the computed result. Usually for a single body case, the sway force acting on ship A is supposed to be zero however the result for two bodies is not because of the effect of interaction between the two ships. Both of sway graphs as shown in Fig. 4, illustrate that the amplitude of force are significantly increased at the change of wave angle toward the ships from zero angle. Therefore, it shows that the sway force is significant at oblique wave and even higher in the two-body case.

Even though the experiment and computational result in Fig. 4 are not giving the same graph characteristic however the result from both methods does show that heave forces experienced by the ship model are not effected by the increasing wave

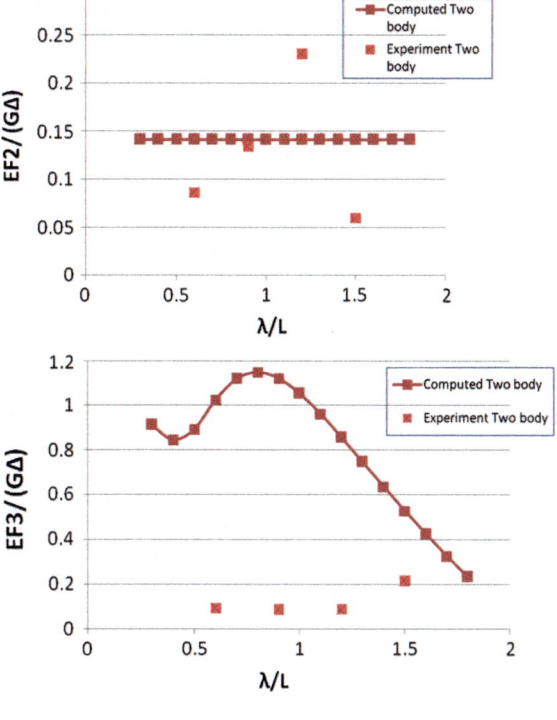

Fig. 3 Sway (EF2/GΔ) and heave force (EF3/GΔ) at heading wave

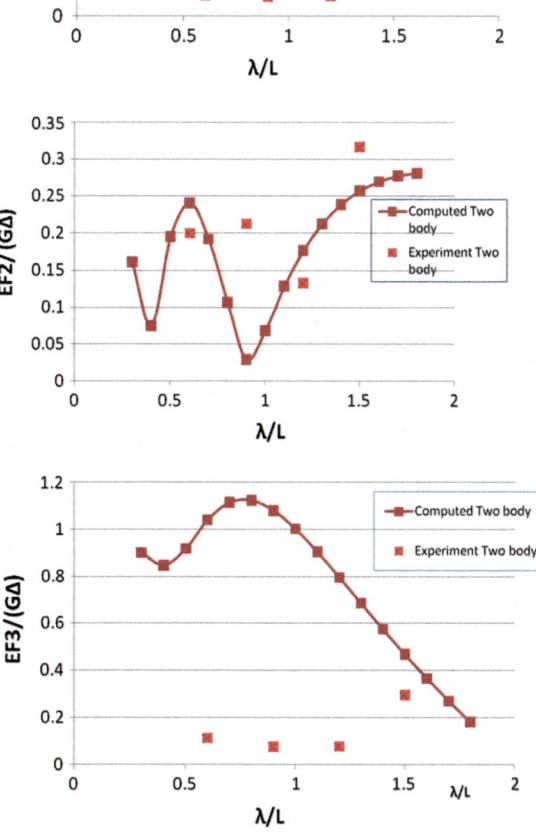

Fig. 4 Sway (EF2/GΔ) and heave force (EF3/GΔ) at oblique wave

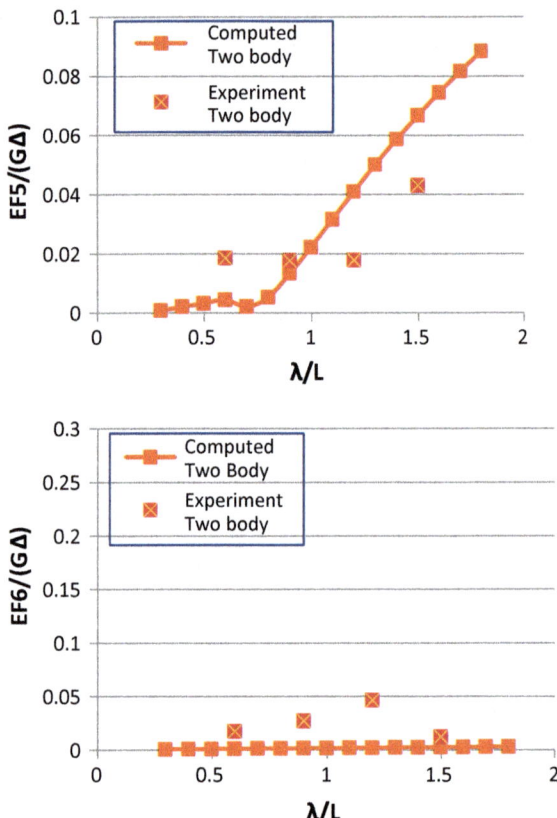

Fig. 5 Pitch moment (EF5/GΔ) and yaw moment (EF5/GΔ) at heading wave

angle. Besides that, the heave force does slightly decrease at the oblique wave which has proven that for the two-body case, the interaction effect is not significantly influenced by the change of wave angles. The differences in amplitude between both methods are big and may result from the effect of distance separation as the experiment defined the distance in 25 m (full scale) while the numerical method di not specifically include the separation distance in the mathematical formulation. However, the interaction effect was considered and numerically solved by taking into account both the highest amplitude as the highest amplitude occurs at a distance of closest separation between the ships.

From pitch moment results presented in Fig. 5, the graph from computed result managed to capture the same characteristic as shown by the graph from the experimental result. The pitch moments computed for two ships are generally agreeable with the experiment result where the average amplitude for both methods are small with 20.8% for heading wave and 11.9% at oblique wave. Based on the

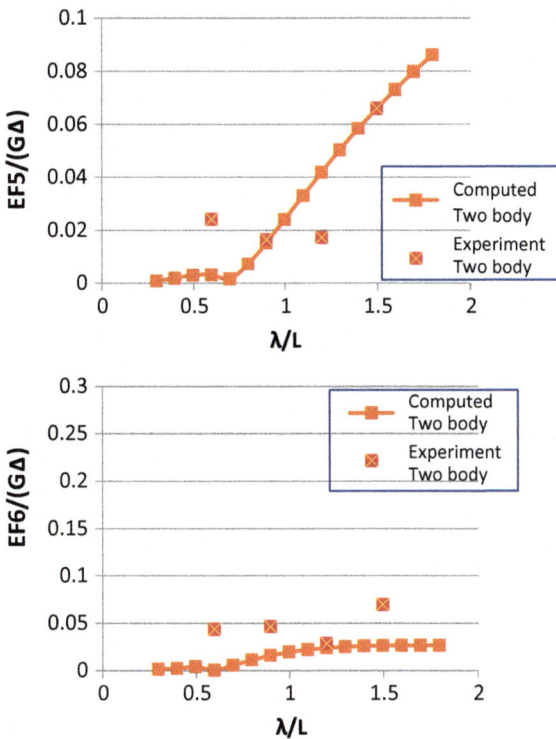

Fig. 6 Pitch moment (EF5/GΔ) and yaw moment (EF5/GΔ) at oblique wave

graphs pattern, the amplitudes of pitch increase with the increasing of wavelength where it illustrated that at short wave the amplitude is lower compared at long waves. However, the amplitude of the pitch moment for both cases was too small where the highest amplitude is below than 0.1. Furthermore, the pitch amplitude shows the same graph characteristic as the heave force where the amplitude slightly decreases with the increasing of wave angle.

The amplitude of yaw force from experiment and computed methods are small for both conditions in heading and oblique wave. Based on the graph plotted in Fig. 6, the yaw moment is almost zero for both wave angles. The yaw moments acting for both methods measured are below from 0.07 where the average differences in amplitude between the computation and experimental result are two times higher at the heading wave and even highest at oblique wave. However, at oblique wave, the yaw moment does show a small increase of amplitude which illustrates that yaw moment is significantly affected by the change of wave angle.

5 Conclusion

All the hydrodynamic coefficients have been derived by using the new strip method for the case when two ships operate in waves with considered hydrodynamic interaction between the ships. Besides that, an experiment was conducted and able to meet the requirement for interaction effect due to diffraction potential. Based on the results obtained, it can be concluded that:

1. When two ships operate in side-by-side position, the interaction effects are highly significant in sway and heave force.
2. The interaction effects are also significant in pitch and yaw moments. However, the amplitudes are too small compared with the other forces.
3. For case study of two ships in waves, the positions of ships toward wave direction are an important interaction variable, which needs to be considered. As the amplitudes of force/moment especially sway and yaw are significantly affected by the change of the wave angle.
4. The danger zone has been predicted as for the heave force are higher at short wave (high wave frequency) and sway force are higher at oblique wave.

References

1. Islam, M.R., Yaakob, Omar, Maimun, Adi: Forces on a Freely Floating Ship While Other Ship Passing Nearby. Universiti Teknologi Malaysia, Malaysia (2008)
2. Skejic, R., Faltinsen, O.M.: A Unified Seakeeping and Maneuvering Analysis of Two Interacting Ships. Norwegian University of Science and Technology, Norway (2007)
3. Wang, Q. X.: An analytical solution for two slender bodies of revolution translating in very close proximity. J. Fluid Mech. **582**, 223–251 (2007)
4. Skejic, R., Faltinsen, O.M.: A unified seakeeping and maneuvering analysis of ships in regular waves. J. Mar. Sci. Technol. **13**, 371–394 (2008)
5. Chen, G.R., Fang, M.C.: Hydrodynamic Interactions Between Two Ships Advancing in Waves. National Cheng Kung University, Taiwan (2000)
6. Fang, M.C., Kim, C.H.: Hydrodynamically coupled motions of two ships advancing in obliquewaves. J. Ship Res. **30**(3), 159–171 (1986)
7. Fang, M. C., Chen, G. R.: The effect of clearance and speed on the relative motions between two ships advancing in waves. In: Proceedings of the Thirteenth (2003) International Offshore and Polar Engineering Conference, Honolulu, Hawaii, USA (2003)
8. Faizul, A. A.: A Strip Method for a laterally drifting ship in waves. Ph.D. Thesis, Hiroshima University, Japan (2006)

A Simplified Computational Fluid Dynamics Approach for a Self-propelled Ship Using the Actuator Disc Theory

Iwan Mustaffa Kamal, Muhamad Sufri Shamsuddin and Jonathan Binns

Abstract This paper describes a simplified approach in modelling the propeller in a CFD computation. Instead of using the full geometry to represent the propeller, the propeller was represented with a simple disc applying a theory called the actuator disc theory developed by Rankine (Trans. RINA 6, 1865) and Froude (Trans. RINA 30, 1889). The computation using this simplified approach would reduce the computation time required if it is compared to a finite volume approach. This paper investigates the ability of the commercial PF solver, FS-Flow®, to predict the interaction between the hull and the propeller through a case study using a training vessel, i.e. the TV Bluefin. To validate the computed results, the results were compared with the results from a full scale speed trial of the TV Bluefin at sea and results from a towing tank experiment using the TV Bluefin's ship model. The computed results show a good agreement with the full scale sea trial results, whereas when compared to the towing tank results, the CFD computation seems to over predict the delivered power. In conclusion, the CFD computation using the actuator disc theory was able to provide a reasonable prediction of the full scale delivered power required by the ship.

Keywords Propeller modelling · CFD computation · Actuator disc theory

---

I. Mustaffa Kamal (✉)
Malaysian Institute of Marine Engineering Technology,
Universiti Kuala Lumpur, Dataran Teknologi Kejuruteraan Marin,
Bandar Teknologi Maritim, Jalan Pantai Remis, 32200 Lumut, Perak, Malaysia
e-mail: iwanzamil@unikl.edu.my

M. Sufri Shamsuddin
Labuan Shipyard and Engineering Sdn Bhd, KM 9 Jalan Rancha-Rancha,
P.O. Box 81210, 87031 Wilayah Persekutuan Labuan, Malaysia
e-mail: sufri.shamsudin@labuanship.com

J. Binns
Australian Maritime College, University of Tasmania,
100 Newnham Dr. Newnham TAS, Launceston 7248, Australia
e-mail: jonathan.binns@utas.edu.au

Nomenclature

a	Inflow factor at the actuator disc
b	Inflow factor after the actuator disc
C_T	Thrust coefficient
C_{TS}	Ship's total resistance coefficient
dA_O	Annular elements of the propeller disc area
dT	Developed thrust at any element at the blade
dQ	Absorbed torque at any element at the blade
D	Diameter of the propeller
D_{HUB}	Diameter of the propeller hub
D_S	Ship's propeller diameter
F_M	Towing force for the model
F_R	Froude number
g	Gravitational acceleration
J	Advance coefficient of propeller
J_O	Advance coefficient corresponding to K_{TO} at the thrust 'identity'
J_P	Advance coefficient at the model self-propulsion point
J_S	Advance coefficient of ship propeller
J_{TS}	Operating value of the advance coefficient of the ship propeller
K_{FD}	Towing force coefficient
K_{QO}	Open water propeller torque coefficient
K_{QS}	Full-scale propeller torque coefficient
K_{QOS}	Full-scale open water propeller torque coefficient
K_{QP}	Propeller torque coefficient in the behind condition
K_{TO}	Open water propeller thrust coefficient
K_{TOS}	Full-scale open water propeller thrust coefficient
K_{TP}	Propeller thrust coefficient in the behind condition
K_{TS}	Operating value of the thrust coefficient of the ship propeller
n_M	Model propeller shaft speed in rev/sec
n_S	Ship propeller shaft speed in rev/sec
P_D	Delivered power
P_{DS}	Delivered power for the ship
Q_M	Model propeller torque
r	Radius of an annular element in a propeller blade
S_S	Wetted surface area for ship's hull
t	Thrust deduction fraction
T	Thrust of the propeller
T_M	Model propeller thrust
V	Ship speed
V_A	Advance speed
w_{TS}	Wake fraction of the ship
π	Pi
ρ	Water density
ω_2	The rotational velocity behind the propeller disc

1 Introduction

In recent years there has been steady progress in the use of computational fluid dynamics (CFD) to simulate the self-propulsion test with the influence of propellers. Accurate prediction of a self-propelled ship is still a challenging task for CFD computations. Much of the research nowadays in the CFD computations of a self-propelled ship simulates the real geometry of the propeller operating behind the hull. Such computations are complicated and require a huge amount of computation time [3]. By modelling a disk of a finite thickness replacing the real geometry propeller as shown in Figs. 1 and 2, the computations are simplified and therefore offer significant savings in computational time and resources [4]. This actuator disk theory was developed by Rankine [1] and Froude [2]. The detail explanation of the actuator disc theory can be found in Manen and Oossanen [5] and Carlton [6]. Basically this simple theory of propeller action was based on the axial motion of the water passing through the propeller disc.

With this simplification, the theory did not consider the geometry of the propeller which was producing the thrust. This propeller disc, which was located at the propeller plane, is considered to absorb all of the power of the engine and dissipate this power by causing a pressure jump, therefore increasing the total head of the fluid, across the two faces of the disc. It is assumed in the theory developed by Rankine that:

1. The propeller works in an ideal fluid in other word the flow is frictionless.
2. The propeller can be replaced by an actuator disc with an infinite number of blades.
3. The propeller disc imparts a uniform acceleration to all the fluid passing through the disc, therefore the thrust generated is uniformly distributed over the disc.
4. The propeller disc can produce thrust without causing rotation in the slipstream.

Fig. 1 The propeller modelled in real geometry

Further development by Froude in his subsequent work [2] removed the fourth assumption and allowed the propeller to impart a rotational velocity to the slipstream which is a more realistic action of a propeller. To summarise, the developed thrust dT by any element in the propeller disc can be written as:

$$dT = \rho dA_O V_A^2 (1+a) b \tag{1}$$

The torque dQ absorbed by the element is written as:

$$dQ = \rho dA_O V_A (1+a) r^2 \omega_2 \tag{2}$$

where ρ is the water density, dA_O is the annular elements of the propeller disc area, V_A is the axial velocity just forward of the propeller disc, a is the inflow factor at the disc, b is the inflow factor after the disc, r is the radius of annular element and ω_2 is the rotational velocity behind the propeller disc. Previous investigations using the actuator disc theory can be found in Villa et al. [3] and Krasilnikov [7]. But most of the approach mentioned in Villa et al. [3] and Krasilnikov [7] were computed using the RANSE solver. In this paper, a different approach was tested using a potential flow (PF) solver or also known as the panel method. The work presented in this paper is then primarily focused to address the accuracy of the PF solver coupled with the actuator disk placed at the propeller plane to predict the interaction between the hull and the propeller through a case study. In this case study the training vessel TV Bluefin was used as there are numerous available data of the vessel from powering trials at sea and experiments conducted in a towing tank.

A comparison was made with the CFD simulated results against results from sea-trials and experiments. The values compared in this case study are the thrust deduction fraction and the delivered power to the propeller. A mesh independence study was made as to find the best resolution of the mesh panel prior to computation. All details on how the panel mesh was created in FS-Flow® are explained in this paper. A brief explanation was also made in this paper on how the delivered power was measured during sea-trials and towing tank experiments.

2 Ship and Propeller Particulars

A 1:20 scaled ship model of the Australian Maritime College's (AMC) 35 m training vessel TV Bluefin was used in this study as shown in Fig. 3. The main particulars of the training vessel are shown in Table 1. The particulars of the ship are shown in full scale and model scale. The actual geometry of the propeller was laser scanned from full scale during the ship docking period, in order to develop and fabricate the physical model propeller. This is necessary as to replicate the real propeller of the TV Bluefin, as the original drawing of the propeller was not available.

Fig. 2 The propeller being replaced with a propeller disc

Table 1 Main particulars of *TV Bluefin* in full scale and model scale

Parameter	Full scale	Model scale
Length overall (m)	34.50	1.725
Waterline length (m)	31.38	1.569
Beam (m)	10.0	0.5
Draught (m)	3.91	0.196
Wetted surface area (m^2)	373.52	0.934
Longitudinal centre of gravity from transom (m)	16.08	0.804
Vertical centre of gravity above baseline (m)	5.1	0.255
Displacement (m^3)	529.3	0.065
Displacement	540.5 t	65.9 kg
No. of propeller	1	1
No. of propeller blades	4	4
Propeller diameter (m)	2.2	0.11

3 Computational Works

A numerical analysis using the commercial CFD software FS-Flow® was carried out to determine the full-scale and model scale self-propelled powering performances. This CFD code is a Rankine-Source panel code with the capability to solve the boundary value problem using the potential theory including nonlinear free-surface [8]. The potential flow approach assumes that the fluid is inviscid and the flow is irrotational around the body of the hull. The numerical analysis using this code consisted of a model replicating the experimental work done in the AMC's towing tank and the vessel at full scale. The hull in Fig. 4 represents a full scale model of the TV Bluefin, where the three-dimensional hull geometry was developed using the Bentley Maxsurf® and Rhinoceros 3D® V5.0 modelling

Fig. 3 Sections of *TV Bluefin*

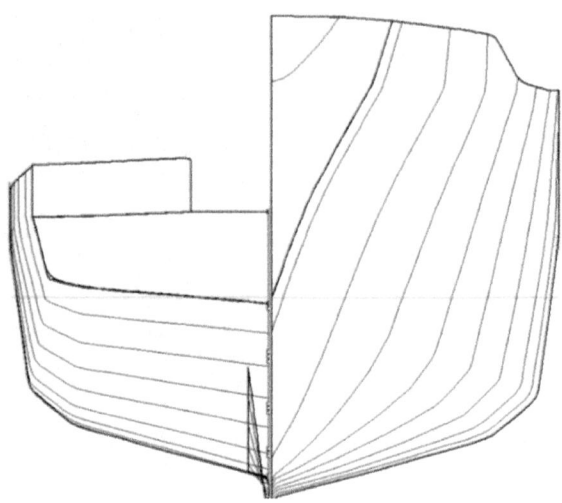

software. The model itself was divided into 3 sections (nose, bow and stern region) allowing the hull to be fair for a smooth transition, see Fig. 4. This allows for group meshing and meshing size control once it is exported to FS-Flow®.

3.1 Mesh Independence Study

As the hull panel distribution is important for the computation of the wave resistance to be accurate enough, the panel size selected must be small enough to resolve all essential aspects of both the hull geometry and the flow field. Local refinement of the panel was needed in the bow region of the Bluefin hull to resolve the strong complex curvature and the expectations that a large pressure gradient will occur in this region, see Fig. 4 for the local refinement region and the half of the hull wetted surface meshes. Basically a too coarse hull panelling will primarily affect the accuracy of the wave resistance by pressure integration over the hull, while the wave pattern is somewhat less sensitive. According to Larsson et al. [9], a few thousand of panels for each symmetric half of the hull is considered to be acceptable in terms or results accuracy but this is not the case for the FS-Flow® code. Therefore, a few computational mesh settings were chosen in order to find the best resolution for the panel mesh number.

There were four sets of mesh settings chosen for this study, from a coarse mesh setting as in Case 1 with 27 hull panels and 110 free surface panels to a fine mesh setting as in Case 4 with 729 hull panels and 1110 free surface panels. In this mesh settings study, it was found that the results were consistent when the hull panels are more than 100 and the free surface were more than 1000 panels. However, the simulation encountered errors, if the mesh panels sizing were more

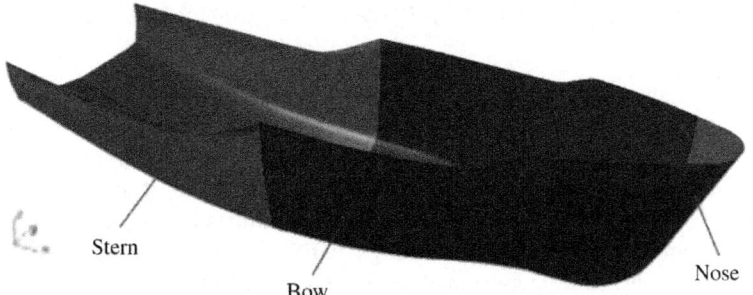

Fig. 4 The 3 sections of TV Bluefin in full scale modelled in Bentley Maxsurf®

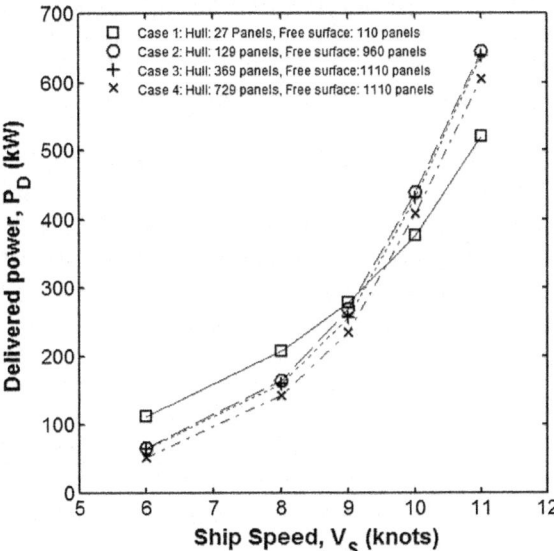

Fig. 5 Delivered power, P_D with respect to ship speed V_S

than the number of panels as in Case 4, where it was found that the simulated wave height was unrealistic. Figure 5 shows that the Case 2 and Case 3 results converged, where the delivered power was consistent with each other. It was clearly shown from the plots that the delivered power for Case 1 and Case 4 decreases as the ship speed increases and not converging to a common result. Therefore, for the CFD simulation, it was decided to use Case 3 with 369 hull panels and 1110 free surface panels for further simulation work (Fig. 6).

3.2 Domain and Meshes in FS-Flow®

In FS-Flow® the mesh generation was done by using the most practical way which is using the panel generation tool, which is based on the imported surface from the CAD software as described earlier. In order to make geometric entities available to the meshing algorithms, a geometry manager which is readily available in FS-Flow® was used to import and link geometric entities like surfaces, curves and lines from another CAD program through IGES files. Once the geometries were imported, the meshes were generated using built-in mesh generators such as BODY mesh, HULL mesh, FREE mesh, TRANSOM mesh and PROP mesh which are featured in the FS-Flow® code. The body fitted surface meshes on the hull were created using the BODY mesh module which generates a patch with Rankine source panels on the hull geometry. The meshes were mapped onto the geometric surface which were linked and stored in the computer input file. The BODY mesh module also allows for updates of elevated intersection with the free surface which were generated using the FREE mesh module, and this intersection were updated during the iteration.

The free surface of the water surrounding the Bluefin hull was created using the FREE mesh module, where it automatically generates a patch with Rankine source panels to represent the free surface of the water. The free surfaces that were created in this computation started in front of the hull, and extended behind the hull at about the half length of the vessel. The free surface panels generated must be small enough to be able to accurately model the wave elevation and to resolve the principal waves. In this case, 1110 panels half the water surface were used. Figure 7 illustrates the final free surface panelling which was used for the simulation, 1110 panels per symmetric half, in this case.

The flow off a transom stern has some particular properties that requires a separate treatment in the simulation. A TRANSOM mesh was used to generate a free surface mesh behind the transom of the hull. Therefore, this mesh was attached

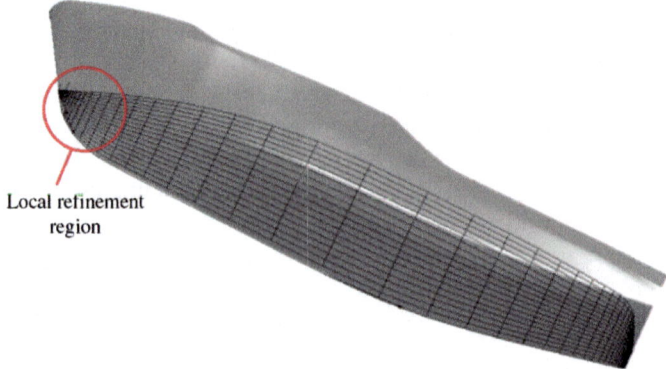

Fig. 6 The half wetted surface meshes with local mesh refinement region at the bow area

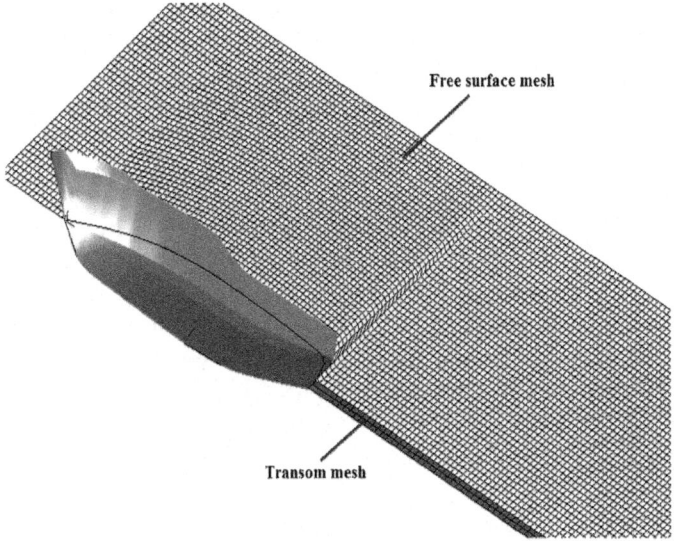

Fig. 7 The free surface mesh and transom mesh panels

to the aft edge at the stern of the model. For a dry transom, the Kutta [10] condition requires that the speed in front of the first panel or mesh corresponds according to Bernoulli law to the height of the transom edge. It is important to note that only a flow regime in which the free surface detaches smoothly from the transom edge (dry-transom flow) can be adequately modelled in this method which was represented in the simulation setup as a transom factor. In this case, a factor of 1 was used which assumed such a dry transom condition. The actual geometry of the Bluefin propeller is represented with the use of actuator disks in FS-Flow® based on the actuator disk theory. The PROP mesh module generates a circular mesh to represent the effect of a propeller under defined loading conditions. This representation is illustrated in Figs. 8 and 9. During the computation the source strength on the propeller was computed according to an ideal fluid condition. To simulate the self-propulsion test, the "free thrust" function was applied into the computations allowing the program to modify the thrust in order to achieve a balance between drag and thrust, which is called the self-propulsion point.

3.3 FS-Flow® Numerical Simulations

The computation was conducted using Bluefin hull form in model scale as well in full scale. In the initial first iteration computation of the source strength on the propeller, it was based on an initial ideal thrust, which will be valid without the hull interaction. During the iterations the FS-Flow® program modified the thrust so as the delivered thrust including the interaction with the hull equals the resistance. The initial thrust coefficient C_T is in a non-dimensional form of thrust and is defined by the following Eq. (3):

$$C_T = \frac{8T}{\rho V^2 (D^2 - D_{HUB}^2)\pi} \qquad (3)$$

where T is the thrust of the propeller in N, ρ is the water density in kg/m³, V is the ship speed in m/s, D is the diameter of the propeller in meter and D_{HUB} is the diameter of the propeller hub in meter. The 'Free Thrust' function was allowed in order for the propeller thrust to adjust freely as to equalize the resistance force. Therefore, by doing this in full scale, the delivered power, P_D, can be obtained. Thus the delivered power, P_D obtained by the CFD simulation then can be compared to the sea trial results.

4 Validations with Experimental Test and Sea Trials Results

In order to evaluate the accuracy of the results of the CFD simulation done in full scale and model scale, the simulation results for these two different scales were compared with towing tank experiment results and data from sea-trial result available from the work of DePaoli [11].

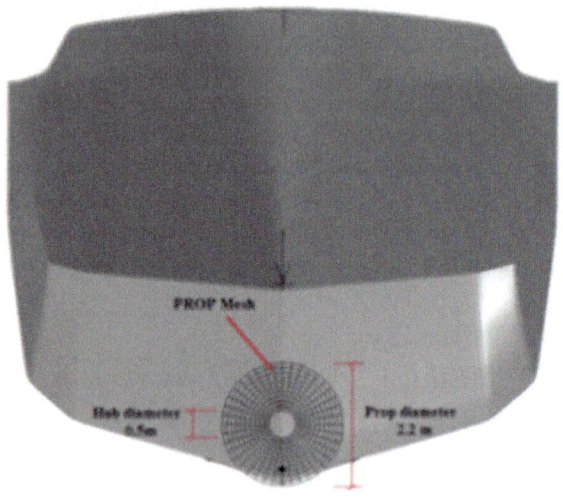

Fig. 8 Actuator disk representation in the FS-Flow

4.1 Towing Tank Experiments

The self-propulsion test experiments were conducted at the Australian Maritime College's towing tank. The towing tank principal dimensions are 100 m in length, 3.55 m wide and 1.5 m deep. The self-propulsion test was conducted using the TV Bluefin's model and propeller at a scale of 1:20 as shown in Figs. 10 and 11. Apart from the self-propulsion test, an open water test is also necessary in order to obtain the open water characteristics of the Bluefin propeller that were being tested. The open water tests were also conducted in the AMC's towing tank using a typical open water test set-up. Due to the limitation in the length of the AMC's towing tank, the self-propulsion test was conducted as load-varying tests or sometimes referred as the "British" method. In this method, the propeller revolution speed n_M was varied at a number of propeller revolution speeds and the towing tank carriage speed V_M was kept constant. The steady-state values of the model towing force F_M, propeller thrust T_M and propeller torque Q_M at each setting of the propeller revolution speed n_M was then recorded.

4.2 Extrapolations to Full Scale

The towing tank test results were extrapolated using ITTC 1978 procedure to obtain the full-scale results. This step was necessary as the model results cannot be directly scaled to full-scale using the scale ratio. In the extrapolation procedure, all the data from the three physical model tests, namely, the self-propulsion test, the open water test, and the calm water bare hull resistance test were combined. The outline of the ITTC1978 'Powering performance predictions for single screw ship' method is described in detail by Bose [12], where once the advance coefficient, J_P, at the model self-propulsion point is obtained using the curve of the non-dimensional form of towing force K_{FD}, and K_{FD}/J_P^2, the values of the propeller coefficients, in the behind condition, K_{TP} and K_{QP}, can be found from the results of the self-propulsion test. Using the "thrust identity" method, the value of propeller thrust in the behind condition K_{TP} is used to find the value of the advance coefficient, J_O in the results from the open water test of the propeller. Some corrections have to be made to the model open water thrust and torque coefficients, K_{TO} and K_{QO}, to obtain the full-scale open water propeller thrust and torque coefficients, K_{TOS} and K_{QOS} and the requirement for thrust are given in the form of $K_T/J^2 = S_S \cdot C_{TS}/2D_S^2(1-t)(1-w_{TS})^2$ [12]. This intersection leads to the operating values of K_{TS}, K_{QS} and J_{TS} of the ship propeller. Then it is possible to calculate the delivered power, P_{DS}.

4.3 Sea-Trial Results

The sea-trial results were obtained from the work of DePaoli [11], where he conducted measurements on-board the TV Bluefin measuring the torque on the propeller shaft with the use of strain gauges linked to a data acquisition system (DAQ) using a wireless telemetry system. Apart from the torque, the shaft speed was measured using a proximity sensor with a photo-reflective strip placed on the propeller shaft. These arrangements are shown in Figs. 12 and 13. Once the torque and the shaft speed were known, the shaft power was then calculated. All the sea-trial results were corrected to the relative wind velocity and wind direction.

5 Comparison of the Results

From the plot of towing force F_M with respect to the propeller thrust T_M using the values measured in the self-propulsion test conducted in the towing tank as shown in Fig. 11, the thrust deduction fraction t, can be obtained from the linear regression equation. These lines of towing force with respect to the thrust are normally linear as commented by Bose [12], where this is normally true for most normal ship forms and are positive values of the towing force. A linear regression line fitted through the points at each Froude number yields a straight line with a slope of $t - 1$, where t is the thrust deduction fraction and the y-axis intersection represents the resistance of the model in the self-propulsion condition. For example the regression line of $y = 0.8789x + 2.9647$, has a slope of $t - 1$ which is equal to -0.8789. Therefore, the value of t can be calculated using the equation, $t - 1 = -0.8789$, giving the value of t equal to 0.121.

The values of t are plotted with respect to the Froude number as shown in Fig. 14, where the thrust deduction fraction of model scale and full scale obtained from CFD simulation were compared against the thrust deduction fraction, t obtained using the slope equation of the curve of towing force versus thrust from the towing tank experiments results. There are some variations observed in the thrust deduction fraction, t, between CFD in full scale, CFD in model scale and experiments. It was found that the thrust deduction fraction in the CFD model scale has a

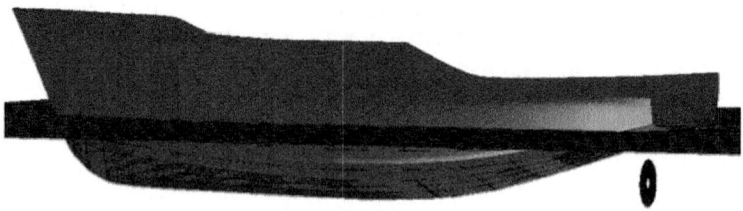

Fig. 9 Perspective view of the hull (from starboard side) and the prop mesh

Fig. 10 The Bluefin ship model in a self-propulsion test in the towing tank

Fig. 11 The set-up of the self-propulsion test showing the load cell

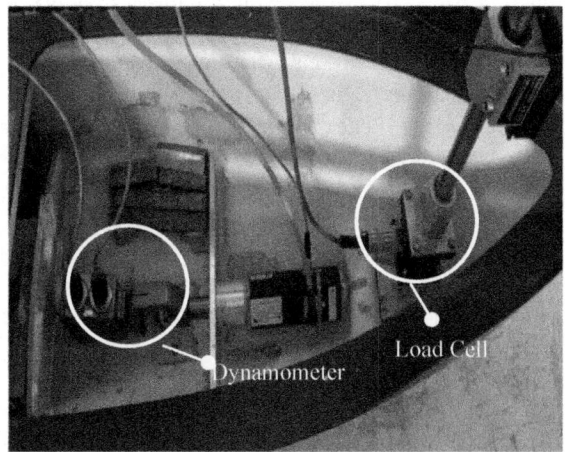

small difference to the experiment results compared to the CFD in full scale. The discrepancies of the CFD model scale to the experiment tend to be large at low Froude number and the difference kept decreasing as the Froude number increases, where at the corresponding speed of 11 knots which is equivalent to Froude number 0.33, the difference is only at 1.8%. The reason of the difference between the CFD simulation with the experimental results are due to the differences of the slope gradient of the linear plotted towing force versus propeller thrust. This may due to the non-correctness of the CFD simulation in capturing the reality of the transom flow, where the dry transom was applied throughout the whole range of speed.

In order to compute the delivered power, P_D, the CFD simulation has been carried out only in full scale at the ship self-propulsion point for ship speed of 6, 8, 9, 10 and 11 knots. To achieve the ship self-propulsion point, the 'Free Thrust'

Fig. 12 The strain gauge arrangement on the *TV Bluefin* output shaft

function was activated allowing the thrust to adjust itself freely as to equalise with the resistance force. No CFD simulation was done in model scale as one needs to extrapolate it to full scale for this comparison purpose. The extrapolation was impossible as the values of torque Q_M, and the shaft speed, n_M in the model scale were unknown. The comparison between the CFD simulation, towing tank experiment and the sea-trial results were plotted as shown in Figs. 15 and 16.

It was found that the differences are large at the lower speed, i.e. 6–9 knots, and the differences decreased as the ship speed increases. The difference at 11 knots is only at 5.6%, whereas at low speed of 6 knots the difference is at 64.7%. Then, if we compare the delivered power obtained from the CFD simulation in full scale with the sea-trial results [11, 13], it was observed that the CFD simulation is in

Fig. 13 The set-up of the telemetry system on the output shaft

Fig. 14 The plot of towing force F_M with respect to the propeller thrust T_M

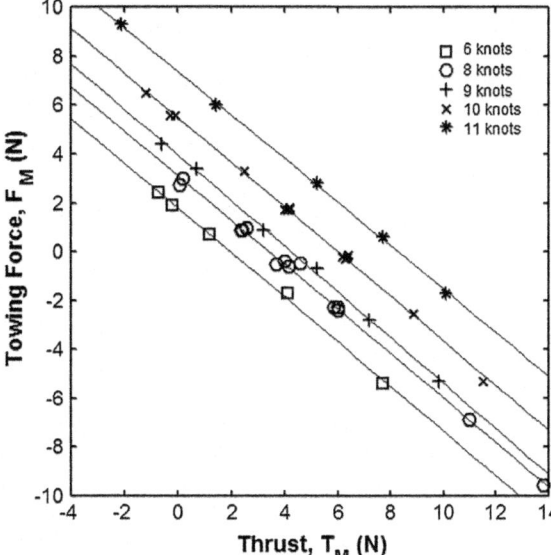

Fig. 15 The thrust deduction fraction with respect to the Froude number

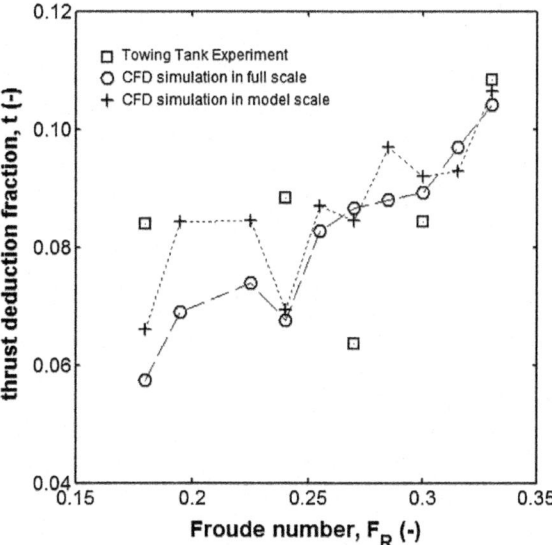

agreement with the sea-trial results at ship speed 5.9, 7.9 and 11.1 knots. The results for 9.7 and 10.4 knots were slightly varied at 17.9 and 23.6 percent respectively. Therefore, it can be concluded that the CFD simulation correlates better with the full-scale results than the towing tank experiments results (Figs. 15 and 16).

Fig. 16 The comparison in terms of delivered power

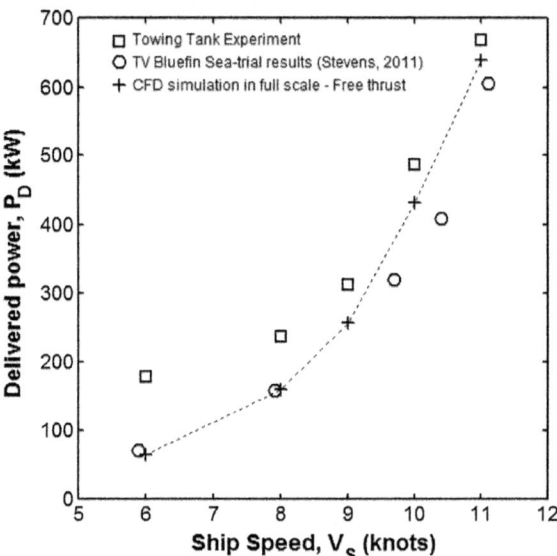

6 Conclusions

The self-propulsion test of a training vessel has been simulated using the potential flow theory with a combination of actuator disk theory to represent the action of a propeller at the stern of the vessel. The FS-Flow® code was used for the simulation. A mesh setting analysis reveals that the optimum mesh panels number of the hull body is from 129 to 369 numbers of panels, while for the free surface the optimum number of panels is from 960 to 1110 numbers of panels. The towing force versus the propeller thrusts simulated in CFD full scale has a larger difference to the experiment results compared to the CFD in model scale. However, the delivered power in CFD full scale simulation shows good agreement with the sea-trial results even though the results varied at 17.9 and 23.6% for a ship speed of 9.7 and 10.4 knots, respectively. This initial work has demonstrated that this CFD method is a promising method in evaluating the self-propulsion characteristics of ships providing important information to the powering design process although there are some issues of inaccuracy in the method.

Acknowledgements The author acknowledges the support from the Australian Research Council Linkage Project "Powering Optimisation of Large Energy-Efficient Medium Speed Multi-Hulls", Project ID: LP1101000080. The research has been conducted as part of a collaborative project between Incat Tasmania Pty Ltd, Revolution Design Pty Ltd, Wärtsilä Corporation, Maritime Research Institute of Netherlands (MARIN), and the Australian Maritime College at the University of Tasmania. This study was conducted in the early stage of the authors' Ph.D. research project, researching for a suitable CFD code to predict the propeller-hull interaction of a vessel.

References

1. Rankine, W.J.: On the mechanical principles of the action of propellers. Trans. RINA **6** (1865)
2. Froude, R.E.: On the part played in the operation of propulsion differences in fluid pressure. Trans. RINA **30** (1889)
3. Villa, D., Gaggero, S., Brizzolara, S.: Ship self-propulsion with different CFD methods: from actuator disk to viscous inviscid unsteady coupled solvers. In: 10th International Conference on Hydrodynamics, St. Petersburg, Russia (2012)
4. Mustaffa Kamal, I., Shamsuddin, M.S., Binns, J.R.: Simulating the self-propulsion of FTV bluefin using potential theory CFD. In: 9th International Conference on Marine Technology, MARTEC 2014, Surabaya (2014)
5. Manen, J.D.V., Oossanen, P.V.: Principles of Naval Architecture, Volume 2. Society of Naval Architects and Marine Engineers. Jersey City, New Jersey (1988)
6. Carlton, J.: Marine Propellers and Propulsion. Butterworth-Heinemann, Oxford, UK (2007)
7. Krasilnikov, V.I.: Self-propulsion RANS computations with a single-screw container ship. In: 3rd International Symposium on Marine Propulsors smp'13. Launceston, Tasmania (2013)
8. DNV GL Maritime: Use Manual of FS-Flow. Version 14.0228 (2014)
9. Larsson, L., Raven, H.C., Paulling, J.R.: Ship Resistance and Flow. Society of Naval Architects and Marine Engineers, Jersey City, New Jersey (2010)
10. Prandtl, L., Tietjens, O.G.: Applied Hydro- and Aero-mechanics. McGraw-Hill, New York (1934)
11. DePaoli, R.J.: Measurement of Power and Propulsion of High-Speed Marine Vehicles. University of Tasmania, Hobart (2011)
12. Bose, N.: Marine Powering Prediction and Propulsors. Society of Naval Architects and Marine Engineers, Jersey City, New Jersey (2008)
13. Stevens, H.: Powering Performance of FTV Bluefin: a Full and Model Scale Investigation. Australian Maritime College. Launceston, Tasmania (2011)

Study of MSI300 Propeller Characteristics Using Computational Fluid Dynamics Analysis

Mohamad Sabri Mohamad Sidik, Ruziah Bolhassan,
Mohd Nurhidayat Zahelem and Muhamad-Husaini Abu-Bakar

Abstract Autonomous Underwater Vehicles (AUVs) are becoming increasingly popular for ocean exploration, industrial and military application. In particular, AUVs are becoming the attractive option for underwater search and survey operation as they are inexpensive compared to manned vehicles. Most of an AUV power is utilized by its propulsion system. Since the AUV power only depends on the onboard battery, the power consumption becomes a crucial issue in optimizing the AUV performance. In this study, the MSI300 propeller was analyzed in regards to its performance by using computational fluid dynamics (CFD). At first, a 3D scanner was used to capture the surface of the propeller and the geometry was converted into a nonuniform rational B-Spline Curve (NURBS) model. Then, CFD simulation was performed to analyze the propeller in the turbulent and Bollard pull condition. It was found that the thrust and torque coefficients are descending with the increasing value of advance ratio. The propeller efficiency is at 61.05%, which occurred at a 0.8 advance ratio. It shows that the CFD simulation has successfully represented the characteristics of the propeller.

Keywords AUV · CFD simulation · Propeller characteristics

M. S. Mohamad Sidik · R. Bolhassan · M. Nurhidayat Zahelem · M.-H. Abu-Bakar
Mechanical Section, Malaysia Spanish Institute, Universiti Kuala Lumpur,
Kulim Hi-Tech Park, 09000 Kulim, Kedah, Malaysia
e-mail: ruziah.bolhassan@unikl.edu.my

M. N. Zahelem
e-mail: mnurhidayat@unikl.edu.my

M.-H. Abu-Bakar
e-mail: muhamadhusaini@unikl.edu.my

M. S. Mohamad Sidik (✉) · M. Nurhidayat Zahelem · M.-H. Abu-Bakar
System Engineering and Energy Laboratory, Malaysia Spanish Institute,
Universiti Kuala Lumpur, Kulim, Kedah, Malaysia
e-mail: msabri@unikl.edu.my

1 Introduction

Deep-sea exploration was considered as relatively new to the recent human activity compared to other areas of geophysical research at the depths of the sea. The ocean depths are still under-explorated areas of the planet and still an entirely undiscovered domain due to the limited abilities of the human being. According to Alam et al. [1], AUVs are robotic mobile instruments which carry self-contained propulsion systems, sensors, and embedded with an artificial intelligent in decision making process. AUV also can take sampling and survey tasks with little or no human intervention. Military uses AUVs for search and rescue missions where high speed, high mobility, and acoustic stealth performance is required. However, the AUV technology still has not achieved its maturity despite the fact that there was a lot of improvement in performance, energy savings, navigation control, and communication [2].

A group of researcher from the UniKL MSI has designed and developed an AUV for sea bed mapping and underwater sampling. This newly develop AUV was named as the MSI300 in regards to UniKL MSI and designed to operate at 300 m depth or at pressure of 3 MPa. The AUV system performance in terms of power consumption depends on the thruster design because the thruster system drains out most of the AUV power. Most thrusters that are available in the market do not fit with the MSI300 in terms of size, holder, power requirement, efficiency and costs. Thus, a study must be conducted to design a thruster specifically for the MSI300. However, it is difficult to analyse the thruster analytically due the hydro dynamical force, which hits the blade surface. The propulsion system is crucial in underwater operations to ensure the manoeuvrability. For example, the AUV should have a strong manoeuvrability in the vertical plane, to escape from colliding with seamounts quickly by moving upward. Therefore, many sectors demanding high driving capability and extensive manoeuvring range are required.

CFD analysis was run to study the characteristic of the thruster in the water with turbulent condition on the thruster and propeller CAD model. Barros et al. [3] claimed that propeller ducts can help the AUV to accelerate from zero to cruising speed. However, the propeller duct drag may effect to the propeller performance. In addition, the propeller duct may contribute to increase the damping effect during operation. According to Rauber et al. [4], it is essential to understand the physical phenomena and concern for the dynamics of the vehicle. Thus, study of the thruster behaviour in common work condition needs to be conducted. The time response and some stress or motor's saturation must be predicted, and analysed based on the physics equations. Yi et al. [5] claimed that the manoeuvrability of an AUV can be increased by designing as a fish-like underwater vehicle. The simulation must be done to verify the hydrodynamic performance of the AUV manoeuvrability. The lift force and pitching moment will become bigger when the velocity is increased. The hull line must be optimized to minimize the lift force and pitching moment.

2 Methodology

The purpose of the CFD simulation was to analyze the propeller in the turbulent condition. By using CFD, the thrust and torque of the propeller will be measured. CFD was used to overcome the difficulty in testing the propeller under axial water condition in order to get the performance of the propeller. However, to confirm the CFD simulation, a Bollard pull test was run to validate the numerical tests. The efficiency of the propeller varies at different advanced ratios. Even though the performance of the propeller was calculated using the vortex lattice method, the value was purely theoretical and predicted that the efficiency of the propeller at 98% of the actuator disc efficiency. This assumption was untrue in the real situation, where the fluid flow does not necessarily move straight and hits the propeller blade. Therefore, CFD analysis was required to resolve this problem. In this study, the speed of the propeller ranges from 500 to 1500 rpm. 1500 rpm was set to be the maximum speed in order to match the propeller vortex lattice (PVL) of propeller speed design. The advance ratio ranged from 0.1 to 1.2. The advance ratio J_S can be calculated using Eq. (1):

$$J_S = \frac{V_S}{nD} \quad (1)$$

V_S, represents the velocity in meter per second, n is the revolution per second and D is the propeller's diameter. Therefore, in the commercial software ANSYS Fluent, the value of the inlet velocity must be based on the motor velocity (V^S) where calculated by using Eq. (1). The speed of the propeller rotation was set to 500–2000 rpm. For the simulation process, the STEP file of the thruster was imported into the ANSYS. Figure 1 shows the CAD drawing of the thruster which was saved as a STEP file.

Fig. 1 CAD drawing of thruster and propeller

2.1 Numerical Method

CFD has been successfully applied in many areas of fluid mechanics, including aerodynamics of cars and aircraft; hydrodynamics of ships; pumps and turbines; combustion and heat transfer; chemical engineering. Applications in civil engineering include wind loading and dynamic response of structures; wind, wave and tidal energy; ventilation; fire and explosion hazards; dispersion of pollutants and effluent; wave loading on coastal and oil and gas industry. The mass and momentum conservation were written in the governing equations:

$$\frac{\partial \rho}{\partial t} + \nabla \cdot (\rho \vec{V}) = 0 \qquad (2)$$

$$\rho \frac{\partial \vec{V}}{\partial t} + \rho (\vec{V} \cdot \nabla) \vec{V} = -\nabla p + \nabla \cdot (\bar{\bar{\tau}}) \qquad (3)$$

where $\vec{v}$ is the velocity vector in the Cartesian coordinate system, p is the static pressure and $\bar{\bar{\tau}}$ is the stress tensor given by;

$$\bar{\bar{\tau}} \equiv \mu \left[\nabla \vec{V} + \nabla \vec{V}^T - \frac{2}{3} \nabla \cdot \vec{V} \right] \qquad (4)$$

where μ are the molecular viscosity, and the second term, on the ride-hand side, is the effect of volume dilation. Once the Reynolds averaging approach for turbulence modeling is applied, the Navier–Stokes equations can be written in Cartesian tensor form as follows;

$$\frac{\partial \rho}{\partial t} + \frac{\partial}{\partial x_i}(\rho u_i) = 0 \qquad (5)$$

$$\frac{\partial}{\partial t}(\rho u_i) + \frac{\partial}{\partial x_j}(\rho u_i u_j) = -\frac{\partial p}{\partial x_i} + \frac{\partial}{\partial x_j}\left[\mu\left(\frac{\partial u_i}{\partial x_j} + \frac{\partial u_j}{\partial x_i} - \frac{2}{3}\delta_{ij}\frac{\partial u_l}{\partial x_l}\right)\right] + \frac{\partial}{\partial x_j}\left(-\rho \overline{u_i' u_j'}\right) \qquad (6)$$

where δ_{ij} is the Kronecker delta, and $-\rho \overline{u_i' u_j'}$ are the Reynolds stresses. The Reynolds stress term is related to the mean velocity gradients, i.e., turbulence closure, by the Boussinesq hypothesis as;

$$-\rho \overline{u_i' u_j'} = \mu_t \left(\frac{\partial u_i}{\partial x_j} + \frac{\partial u_j}{\partial x_i}\right) - \frac{2}{3}\left(\rho k + \mu_t \frac{\partial u_i}{\partial x_i}\right)\delta_{ij} \qquad (7)$$

The k–omega model is one of the most widely used turbulence models for external aerodynamics and hydrodynamics and has shown a better potential to predict the key features of vertical separated flows than other models. In this study, the CFD code employs a cell-centered finite-volume method, which allows the use of computational elements with arbitrary triangulation shapes.

2.2 CFD Control Data Setup

Due to the fact that the thruster consumes a lot of energies, the relationship of speed versus propeller thrust must be determined. For this reasons, the CFD simulation was required to get the efficiency of AUV by getting the speed and thrust relationship. In this research, the propeller speed was varied from 200 to 1200 RPM, but the axial flow magnitude was constant at 2 m/s. This experiment is also known as the Bollard pull test where the water was at static while the propeller was moving. To validate the CFD simulation, it was crucial to study the relationship of thrust and speed of the thruster.

2.3 Simulation Procedure

There were several constraints to consider when completing this simulation. The 3D scan of the propeller was made to get the CAD drawing. There was a problem with the surface of the propeller due to some damages; the Geomagic software was used. The Geomagic software helps to repair the holes, bad scratches, pitch and the point cloud. The model was saved to NURBS condition and saved as a STEP file. The propeller CAD then was imported into the SolidWorks. The thruster body was drawn by using real size and the CAD was saved into a STEP file. The CAD modeling was ready for simulation stage. An accurate setup for boundary conditions was critical in CFD simulation solutions.

2.4 Geometry Setup

Because the simulation was run using a multiple reference frame method, two computational domains were required to represent a full fluid and propeller. A cylinder with dimension 1500 mm in length and 500 mm in diameter was set as a fluid domain. A smaller cylinder with dimension 40 mm in length and 50 mm in diameter was set as the propeller domain. The smaller domain was rotated regarding to the propeller speed value and full fluid domain velocity inlet was set to 2 m/s. The geometry then was meshed and the simulation was run in the ANSYS software. The mesh area at the rotate part must be smaller compared to the domain mesh.

It was a crucial step in order to reduce discretization faults, due to the propeller's complex geometry.

2.5 Boundary Conditions

During the validation process, the velocity inlet was set to zero. Bollard pulls condition occurs, where the water was static, and the MSI300 AUV's propeller was without movement. After the validation process, the variation speed of MSI300 AUV was set at the velocity inlet to investigate the thruster's performance. Figure 2 shows the boundary condition of the MSI300 AUV simulation. Pressure outlet was set to zero. Backflow direction specification method was set to normal to the boundary. The water pill box must be set as an interface in order to ensure that the water in the pill box and the main domain interacts. The blade surface was set as a non-slip wall. The surface must be set symmetrically in order the domain was divided into 3 sections. This set up can be utilized to avoid redundant calculations in other identical parts.

2.6 Grid Dependency Setup

In the CFD simulation, the accuracy of the analysis will impact the grid setup. There are three types of simulation error that were recognized, namely; round off, iterative convergence and discretization errors. Husaini et al. [6] explained that

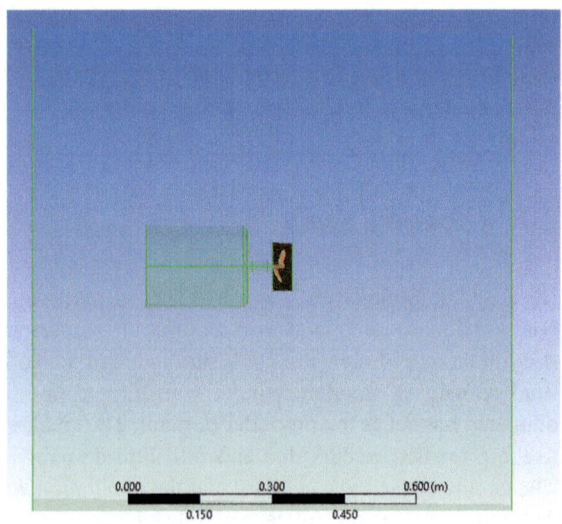

Fig. 2 Boundary condition of propeller CFD simulation

discretization errors can be minimized by a reduction of the timing step and mesh size. This will cause the computing time and the amount of memory required to increase. Grid of sensitivity takes place in order to get the optimum setup for memory and computational time beside the accuracy of the result. For the grid sensitivity study, the same domain geometry was kept with three different types of mesh. The meshed grids were named as; coarse mesh, the medium mesh and fine mesh.

3 Results and Discussion

3.1 Thrust Coefficient

Figure 3 shows the relationship of thrust coefficient versus advance ratio. From the graph, the value of K_T decreased when the advance ratio increased. Since the relation between propeller speed and advance ratio is inversely correlate, the thrust value is decrease when the advance ratio increase when the axial speed was constant. This phenomenon shows a good agreement with momentum theory where the thrust proportionally increase with the speed of propeller.

3.2 Torque Coefficient

Figure 4 shows the torque coefficient against the advance ratio. The pattern of this graph is similar to Fig. 3. The graph shows the propeller torque will decrease when the advance ratio is increased. This occurs because the drag force on the surface of

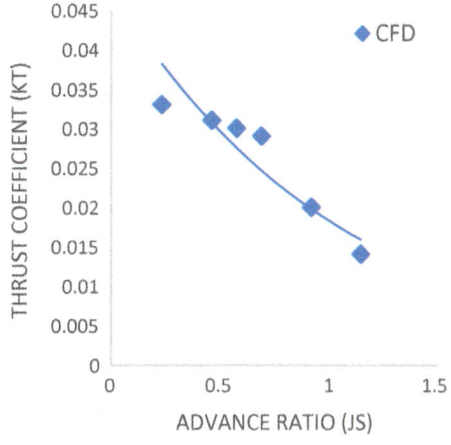

Fig. 3 Graph of thrust coefficient versus advance ratio

Fig. 4 Graph of torque coefficient versus advanced ratio

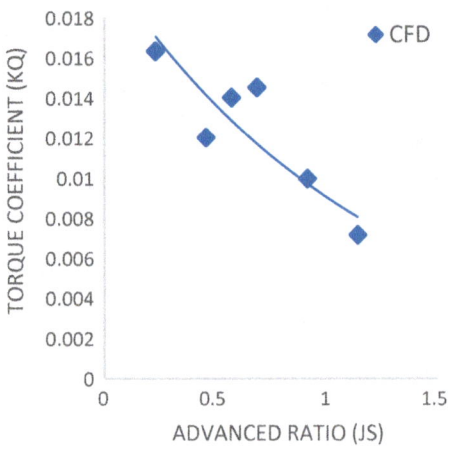

the propeller blade. The water at is static or at no velocity therefore, the blade needs more torque to accelerate the water.

3.3 Efficiency

Figure 5 shows the relationship of propeller efficiency versus advance ratio. From the graph below, the peak clearly occurs at an efficiency of 61.5% with 0.87 advance ratio. At this stage, the propeller efficiency was predicted to occur at an advance ratio of 0.8.

Fig. 5 Graph of efficiency versus advanced ratio

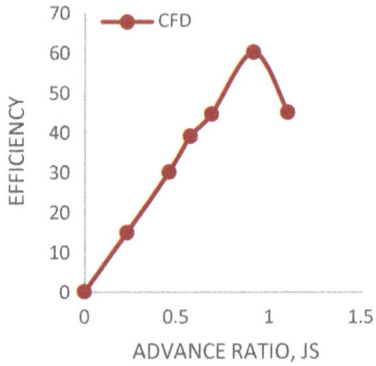

4 Conclusion

The result of CFD shows that the thruster can produce the required power during operation. A 3D scanner can develop the real model of the propeller and increases the accuracy of simulation. The CFD simulation successfully derived the characteristics of the MSI300 propeller in Bollard pull condition. From the CFD simulation, the behavior of the propeller during the operation has been found. The pressure of water hits the surface of the blade where it proves the hydrodynamic acting concept. This numerical study will give advanced understanding of the dynamic behavior and helps to identify the true interference of the AUV and the water in every condition. With these alternative studies, the AUV operation in the water will be more reliable.

References

1. Alam, K., Ray, T., Anavatti, S.G.: Design and construction of an autonomous underwater vehicle. Neurocomputing **142**, 16–29 (2014)
2. Zhang, H., Hao, L., Wang, Y., Wu, Z., Wang, S., Shao, S.: System. No. January 2012, pp. 4–8 (2013)
3. De Barros, E.A., Dantas, J.L.D.: Effect of a propeller duct on AUV maneuverability. Ocean Eng. **42**, 61–70 (2012)
4. Rauber, J.G., Dos Santos, C.H.F., Chiella, A.C.B., Motta, L.R.H.: A strategy for thruster fault-tolerant control applied to an AUV. In: 17th International Conference on Methods Model. Autom Robotic MMAR 2012, pp. 184–189 (2012)
5. Yi, R., Hu, Z., Lin, Y., Gu, H., Wang, C.: Maneuverability Design and Analysis of an Autonomous Underwater Vehicle for Deep-sea Hydrothermal Plume Survey (2013)
6. Husaini, M., Samad, Z., Arshad, M.R.: Autonomous underwater vehicle propeller simulation using computational fluid dynamic. In: Minin, I.V. (ed.) Computational Fluid Dynamics Technologies and Applications, p. 293 (2010)

Mathematical Model of the Manoeuvring Motion of a Ship

Yassen Adnan Ahmed

Abstract There are many kinds of mathematical models that have been proposed to simulate the ship manoeuvring characteristics. The total force model or the simple rudder to yaw response model are useful to do real-time simulation and control. However, each of these models has some limitations. The total force model combined the hull, propeller and rudder forces together. Therefore, if the propeller or rudder is changed in the model, the whole force model needs to be modified. Moreover, the hydrodynamic derivatives for the force terms have less physical meaning to compare. On the other hand, using the rudder to yaw response model, the change in ship velocity or speed drop during manoeuvring cannot be predicted. Considering these facts, this paper describes a widely used mathematical model known as the manoeuvring mathematical group (MMG) or modular model. This model considers not only the hull, rudder and propeller forces separately but also the interactions among them. Each term used in this model has a physical meaning and is constructed as simple as possible. The given model is also verified by the turning tests and speed tests results.

Keywords Mathematical models · Ship manoeuvring · Manoeuvring mathematical group

1 Introduction

Recent developments in hydrodynamic research of ship manoeuvrability have greatly been done by the experimental means on captive models. However, the method of these tests and the expression of the results of analyses are so wide in

Y. A. Ahmed (✉)
Malaysian Institute of Marine Engineering Technology,
Universiti Kuala Lumpur, Dataran Teknologi Kejuruteraan Marin,
Bandar Teknologi Maritim, Jalan Pantai Remis, 32200 Lumut,
Perak, Malaysia
e-mail: yaseen.ahmed@unikl.edu.my

variety that there are almost no unity to compare, except for the comparison made in the form of integral expressions for the predicted motions. If we use different polynomial expressions for fitting the hydrodynamic forces and moments obtained from captive model tests, the common derivatives will carry different meaning in accordance with the difference in the selected slope of the representative terms of the forces and moments, together with the differences in experimental condition. Moreover, the derivative terms may not have a definite physical meaning and the comparison with the theoretical values is difficult. This tendency, as a matter of fact causes problems among the captive model test and a major task is to select better models and derivatives for a better prediction of ship manoeuvring motions.

Incidentally, we are now familiar with another type of mathematical model, known as the 'rudder to yaw response model' which describes macroscopically the ship's rate-of-turn response to the rudder actions by way of manoeuvring indices such as the ones defined by Nomoto. Such a model is good for treating weak motion with negligible variation in the advance speed; however for hard turning with speed loss, stopping or reversing, the model will be significantly diminished. Therefore, it is important to think about possible rationalizations of the mathematical model to use in a wider range with physical meaning in each term. In consideration of the above mentioned objective, a work group named MMG was specially organised in the Manoeuvrability Subcommittee of JTTC (Japan Towing Tank Conference). This paper will explain the model proposed as MMG in view of the betterment of the said frustrating situations as a result of discussions of the MMG group together with brief comments on the theoretical and experimental backgrounds [1]. More details of it can be found in [2–4].

2 Fundamental Requirement for the Mathematical Model

The concept is described by Fig. 1. This paper describes the model for a single screw ship and in deep water condition. However, the model for twin screw and shallow water can be developed with some modifications. The mathematical model needs to fulfil the following fundamental requirements:

1. The mathematical model should be based on the individual open water characteristics of the hull, the propeller, and the rudder.
2. A simplified expressions of the interactions among the hull, the propeller, and the rudder should be needed.
3. The representation of the hydrodynamic force acting on the hull should be as reasonable as possible.

Fig. 1 Schematic diagram for MMG model

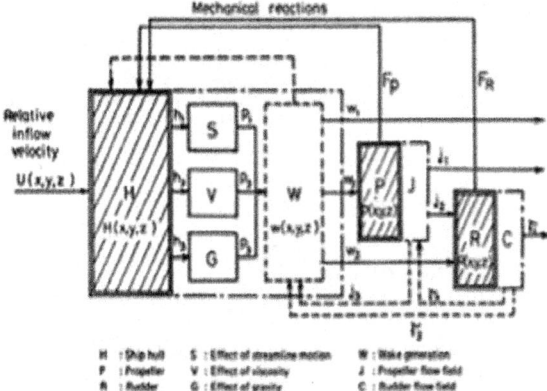

3 Construction of Mathematical Model

The main objective to construct this model is to make it handy for practical usage. The coordinate system used to construct this model is given in Fig. 2.

3.1 Subject Ship

In order to construct the MMG model, 'Esso Osaka' 3-m tanker model is chosen as the subject ship. The model is scaled as 1:108.33. Its details are given in Table 1.

Fig. 2 Coordinate system

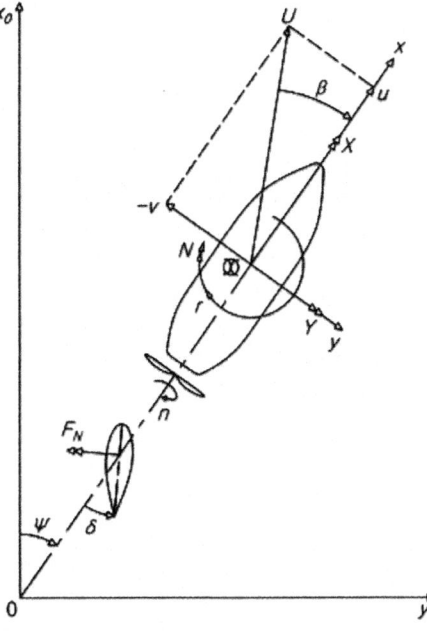

Table 1 Principal particulars and parameters of model ship

Hull		Propeller		Rudder	
L (m)	3	Dp (m)	0.084	b (m)	0.083
B (m)	0.48	P (m)	0.06	h (m)	0.1279
D (m)	0.2	Pitch ratio	0.7151	A_R (m^2)	0.0106
C_b	0.831	Z	5	Λ	1.539

3.2 Equations of Model

The equations of motion are described on the hull fixed coordinate system. Considering the origin at the centre of gravity of the ship as shown in Fig. 1, the equations of motion for surge sway and yaw are given as following:

$$(m+m_x)\dot{u} - (m+m_y)vr = X$$
$$(m+m_y)\dot{v} + (m+m_x)ur = Y \quad (1)$$
$$(I_{ZZ}+J_{ZZ})\dot{r} = N$$

where, m is the mass of the ship, m_x and m_y are the added mass in x and y direction, I_{zz} is the moment of inertia, J_{zz} is the polar moment of inertia, u is the surge velocity, v is the sway velocity, r is the yaw rate and the right-hand side includes the total hydrodynamic forces and moment term due to hull, propeller, rudder. The force terms in Eq. (1) are expressed as follows:

$$X = X_H + X_P + X_R + X_{dis}$$
$$Y = Y_H + Y_P + Y_R + Y_{dis} \quad (2)$$
$$N = N_H + N_P + N_R + N_{dis}$$

where, X_H, Y_H, N_H are the hydrodynamic forces and moments acting on a hull, X_R, Y_R, N_R are the hydrodynamic forces and moments due to rudder, X_P, Y_P, N_P are the hydrodynamic forces and moments due to propeller and X_{dis}, Y_{dis}, N_{dis} are the disturbing forces and moments due to environmental disturbances like wind or current.

The hydrodynamic forces and moments acting on the hull during manoeuvring are usually expressed as a combination of linear and non-linear terms. The hydrodynamic forces and moments, considering advance motion can be described by the following equations.

$$X_H = \frac{1}{2}\rho L d U^2 (X'_{uu}(0) + X'_{bb}b'^2 + X'_{br}b'r' + X'_{rr}r'^2 + X'_{bbbb}b'^4)$$

$$Y_H = \frac{1}{2}\rho L d U^2 \{Y'_b b' + Y'_r r' + Y'_{bb}b'|b'| + Y'_{rr}r'|r'| + (Y'_{bbr}b' + Y'_{brr}r')b'r'\} \quad (3)$$

$$N_H = \frac{1}{2}\rho L d U^2 \{N'_b b' + N'_r r' + N'_{bb}b'|b'| + N'_{rr}r'|r'| + (N'_{bbr}b' + N'_{brr}r')b'r'\}$$

where, the prime in each term denotes non-dimensional values, $X_{uu}(0)$ is the hull resistance in straight motion, $Y_b \sim Y_r$, and $N_b \sim N_r$ are the linear hydrodynamic derivatives, $X_{bb} \sim X_{bbbb}$, $Y_{bb} \sim Y_{brr}$ and $N_{bb} \sim N_{brr}$ are the non-linear hydrodynamic derivate terms. Here, 'b' denotes the drift angle and is considered instead of sway velocity 'v' term.

The hydrodynamic coefficients used in the expressions for the calculations of forces and moment acting on the hull during manoeuvring are determined by curve fitting through the experiment values for forces and moment for different drift angle and yaw rate which are collected from Hirano's [5] paper. *Surge force*: The curve fitted for non-dimensional values is available in the case of the surge force, for 2.5 m model ship. So, the data are modified for a 3 m model by considering the relevant skin friction. The resulting representation after correction is given in Fig. 3. The derivatives for the surge force are calculated for these curve fitted experimental data. *Sway force*: Experimental data are collected for the sway force as well but for 4 m model ship. The hydrodynamic coefficients regarding the sway force and yaw moment are considered with no scale effect because the author believes that using

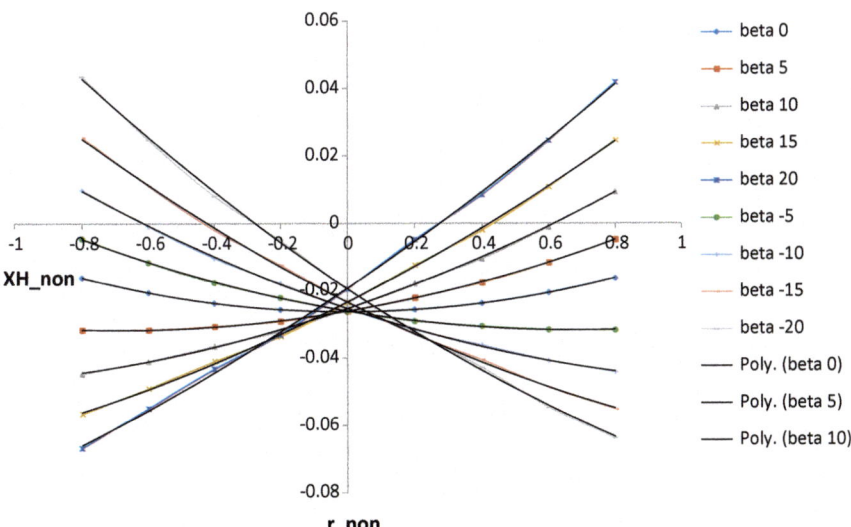

Fig. 3 Corrected curve fitted data for non-dimensional surge force

Fig. 4 Curve fitted data for non-dimensional sway force

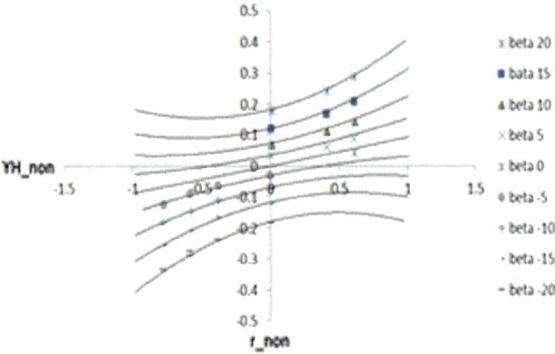

experiment data for a 4 m model ship is still accurate enough to predict the hydrodynamic behaviour of 3 m. Figure 4 shows the data points from experiments together with curves fitted though such points that are used to get the derivatives for the sway force. *Yaw moment*: Same as for the sway force, experimental data for the yaw moment are available for the 4 m model ship. But due to the same reason as explained in the case of the sway force, the author uses these data for curve fitting and to predict the hydrodynamic behaviour of 3 m model ship. Figure 5 shows the data points from experiments together with curves fitted though such points.

The propeller thrust can be described by the longitudinal force of a propeller. The following expression is used to calculate the propeller thrust.

$$X_P = \rho D_p^4 n^2 (1-t) K_T$$
$$Y_P = 0 \qquad (4)$$
$$N_P = 0$$

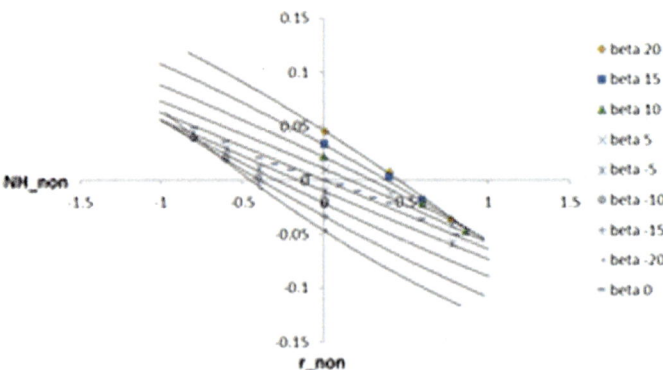

Fig. 5 Curve fitted data for non-dimensional yaw moment

where, K_T is the thrust coefficient and calculated as follows:

$$K_T = C_1 + C_2 J + C_3 J^2$$
$$J = u_p/nD_p$$
$$u_p = u(1 - w_p) \tag{5}$$
$$(1 - w_p) = (1 - w_{p0}) + \tau |v' + x'_p r'| + C'_p(v' + x'_p r')^2$$

Considering the above equations, J is the advance coefficient, $C_1 \sim C_3$ are the constant values from the K_T-J relationship, D_p is the propeller diameter, n is the speed of propeller revolution, u_p is the effective relative inflow velocity in axial direction to propeller, $(1 - w_{p0})$ represents the effect of wake at straight ahead motion, X'_p denotes the x-coordinate of the propeller position, τ and C_p are determined experimentally, t is the thrust deduction factor (approximated by the value for straight ahead motion). Experimental data are available for $(1 - w_{p0})$ versus advance speed, Js. A polynomial curve is fitted by the author for the experimental values and the corresponding value for $(1 - w_{p0})$ is calculated for respective Js. Figure 6 shows the experimental data with curve fitting.

X_R, Y_R, N_R are expressed as the following formulas taking into account the interactions between the hull and rudder as shown in Fig. 7.

$$X_R = -(1 - t_R)F_N \sin \delta$$
$$Y_R = -(1 + a_H)F_N \cos \delta \tag{6}$$
$$N_R = -(x_R + a_H x_H)F_N \cos \delta$$

where, δ is the rudder angle, x_R represents the location of the rudder (=−L/2), and t_R, a_H and x_H are the interactive force coefficients between the hull and rudder. F_N is the rudder normal force and can be described as follows:

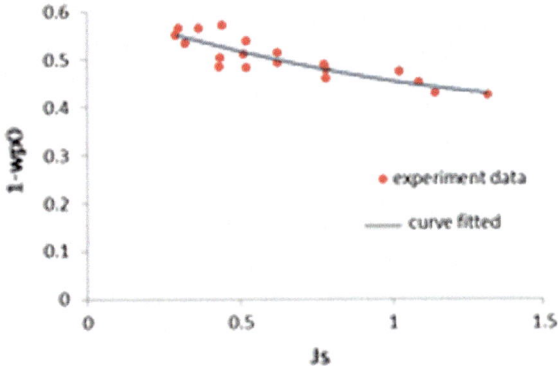

Fig. 6 Curve fitting for experiment data to calculate w_{p0}

Fig. 7 Schematic diagram of rudder force and hull-rudder interaction

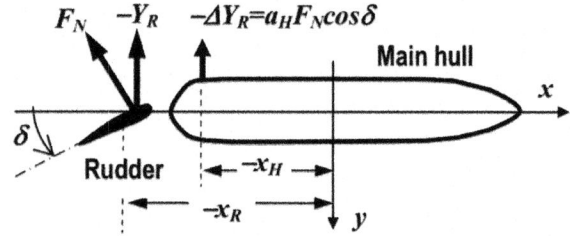

$$F_N = \frac{\rho}{2} A_R f_\alpha U_R^2 \sin \alpha_R \tag{7}$$

where, A_R is the rudder area, f_α is the gradient of the lift coefficient of rudder, and can be approximated as the function of rudder aspect ratio Λ. The following is the well known Fujii's prediction formula (Fig. 8).

$$f_\alpha = 6.13\Lambda/(2.25+\Lambda) \tag{8}$$

As the rudder normal force is highly affected by the propeller stream especially when the rudder is located just behind the propeller. An example of measured rudder normal force for three different propellers thrust is shown in Fig. 9, where it can be seen that the stronger propeller thrust makes the larger rudder normal force. In the MMG model, the effect of propeller stream is included by the longitudinal inflow velocity of the rudder u_R. It can be described as follows.

Fig. 8 Gradient of the lift coefficient of rudder, observed and estimated [6]

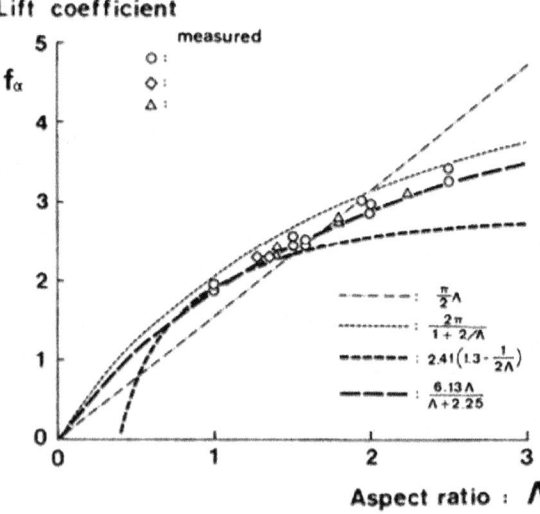

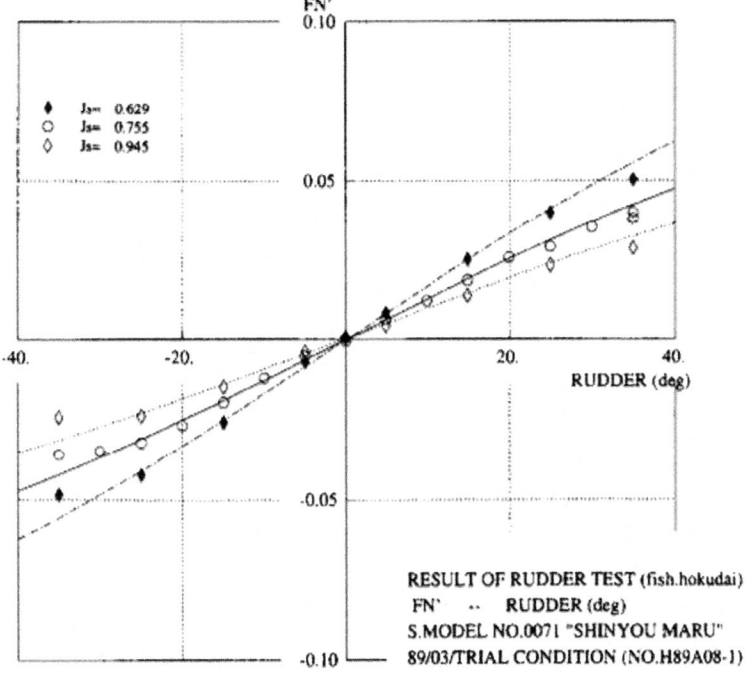

Fig. 9 Rudder normal force for various propeller thrust [7]

$$u_R^2 = (1 - w_R)^2 \{1 + Cg(s)\}$$
$$g(s) = \frac{\eta K(2 - (2 - K)s)s}{(1 - s)^2}$$
$$\eta = D_P/h_R$$
$$K = 0.6(1 - w_p)/(1 - w_R) \qquad (9)$$
$$S = 1.0 - (1 - w_R)U\cos\beta/nP$$
$$w_R = w_{R0}w_p/w_{p0}$$
$$\alpha_R = \delta - \gamma\beta'_R$$
$$\beta'_R = \beta - 2x'_R r'$$

where, γ represents the flow straightening factor of the ship hull. Since the lateral inflow angle is reduced by the ship motion v and r, the rudder normal force also depends on v and r. This effect is known as the course-stabilising factor of the rudder. The definition of γ is illustrated in Fig. 10. Figure 11 shows the schematic diagram of the longitudinal inflow velocity of the rudder, which also defines several parameters used in Eq. (9). The hydrodynamic derivatives and coefficients regarding hull, propeller and rudder are given in Table 2.

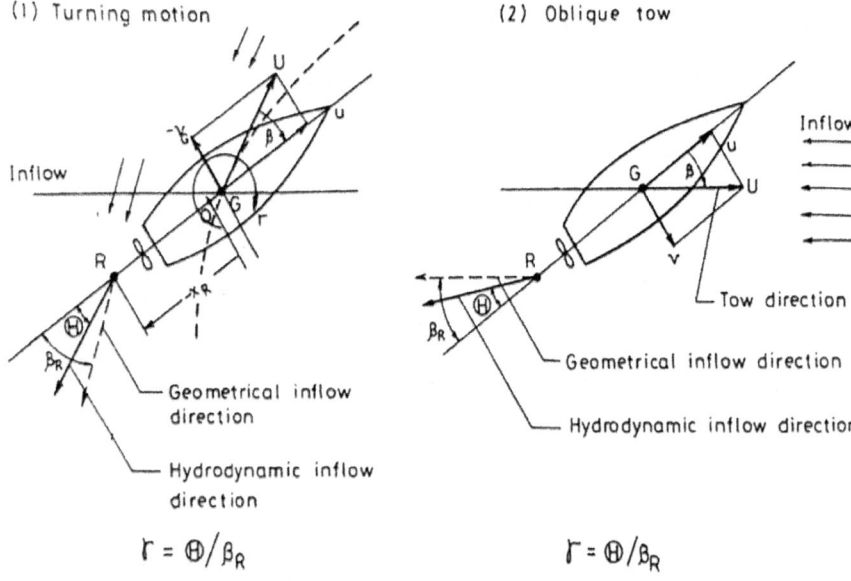

Fig. 10 Definition of flow straightening factor

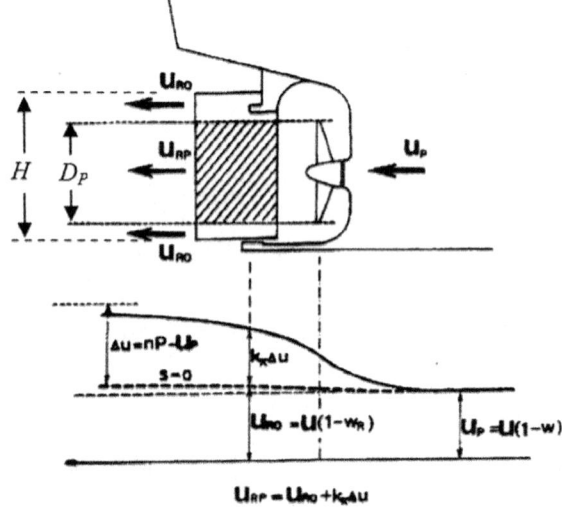

Fig. 11 Schematic diagram of longitudinal inflow velocity of rudder [6]

Table 2 Hydrodynamic derivatives and coefficients

m'	0.2709	m'_x	0.02	m'_y	0.2224	I'_{zz}	0.0172
J'_{zz}	0.00821	X'_{uu}	−0.02639	X'_{br}	0.191559	X'_{bbbb}	0.2751225
X'_{rr}	0.012856	X'_{bb}	0.022823	Y'_b	0.3039	Y'_r	0.0908104
Y'_{bb}	0.5454883	Y'_{bbr}	0.214706	Y'_{rr}	−0.000143	Y'_{brr}	0.332125
N'_b	0.112252	N'_r	−0.063663	N'_{bb}	0.051978	N'_{bbr}	−0.27805
N'_{rr}	0.0027571	N'_{brr}	−0.02597	t	0.2	w_{P0}	0.4710
x'_P	−0.5	C_1	0.32	C_2	−0.2466	C_3	−0.2668
τ	1.45	C'_P	−0.359	t_R	0.2173	a_H	0.398
A_r/Ld	1/59.1	λ	1.599	x_R	−0.82	x_H	−0.442
h_R	0.1278	w_{R0}	0.1792	γ_{port}	0.19	γ_{stbd}	0.23

4 Simulation Results and Comparison with that of Experiments

Using the above mentioned MMG model and the hydrodynamic derivatives mentioned in Table 2, simulations are done to predict the ship manoeuvring characteristics. Then the characteristics are validated using the free running experiment data. The details of the experiment model used is mentioned in Table 1.

4.1 Steady Surge Velocity During Turning

The steady surge velocity during turning is non-dimensionalised with respect to the ship's initial velocity to compare with the non-dimensional experiment value. Such comparison is carried out for different rudder angle turnings for both port and starboard. Figure 12 illustrates such a comparison.

4.2 Speed Test

The speed test is carried out using the mentioned MMG for different propeller revolutions and compared with the experimental values, which are collected from Ueda and Ueno's [7] paper. Figure 13 shows such a comparison for the 'Esso Osaka' 3-m tanker.

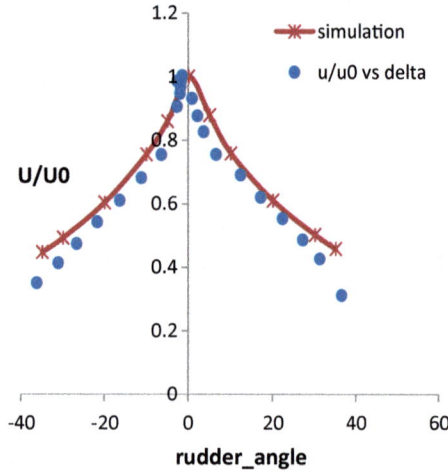

Fig. 12 Comparison of non-dimensional steady surge velocity for different delta

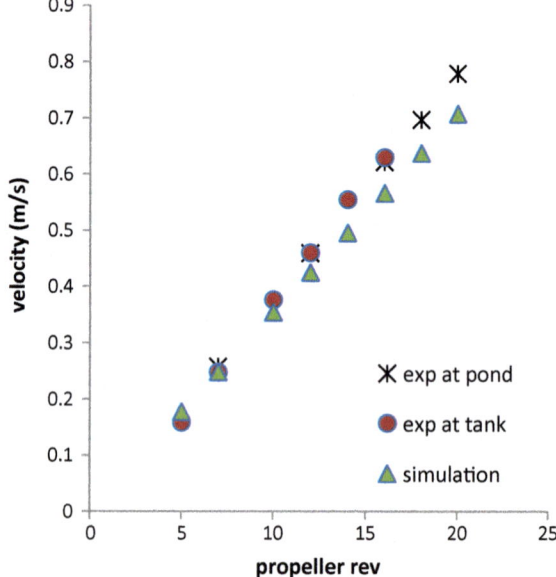

Fig. 13 Comparison of surge velocity for different propeller revolution

4.3 Turning Test

Figures 14, 15, 16 and 17 show the turning test results as compared with the simulation one. These tests are carried out for half-ahead speed. Here, each comparison contains not only the turning trajectory, but also the non-dimensional surge velocity and yaw rate for comparing the initial transition to a steady state value. Considering Figs. 14 and 15, each of the turning circles show a quite promising

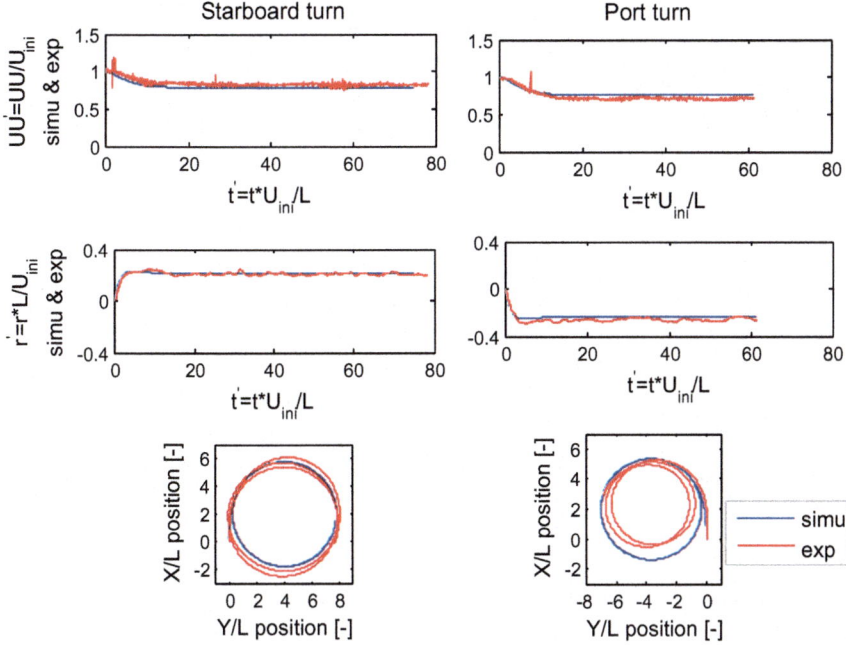

Fig. 14 Turning circle comparison for ±10°

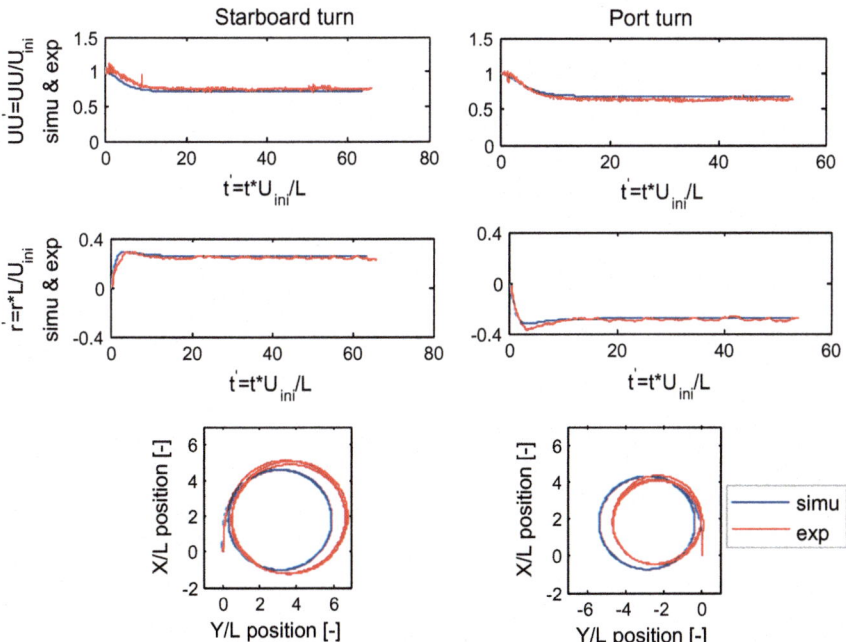

Fig. 15 Turning circle comparison for ±15°

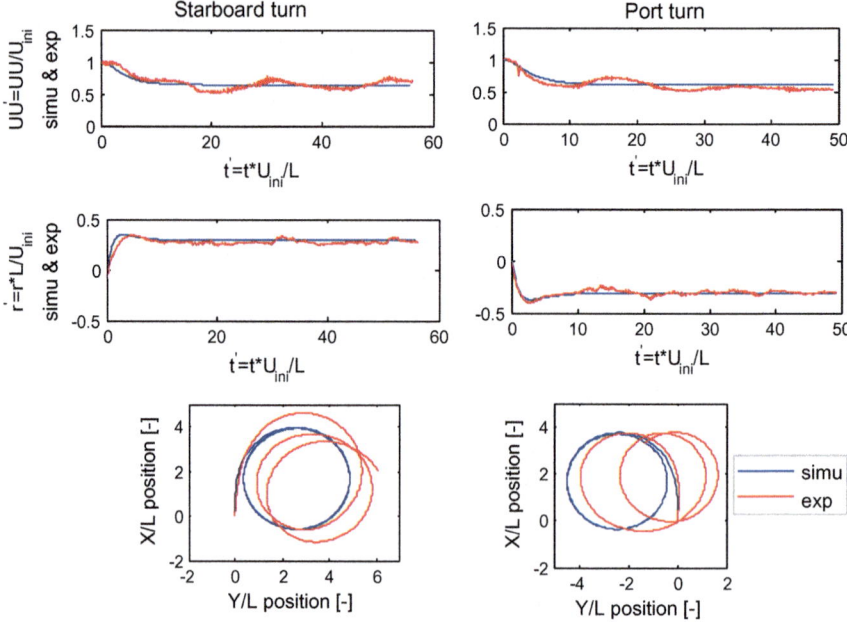

Fig. 16 Turning circle comparison for ±20°

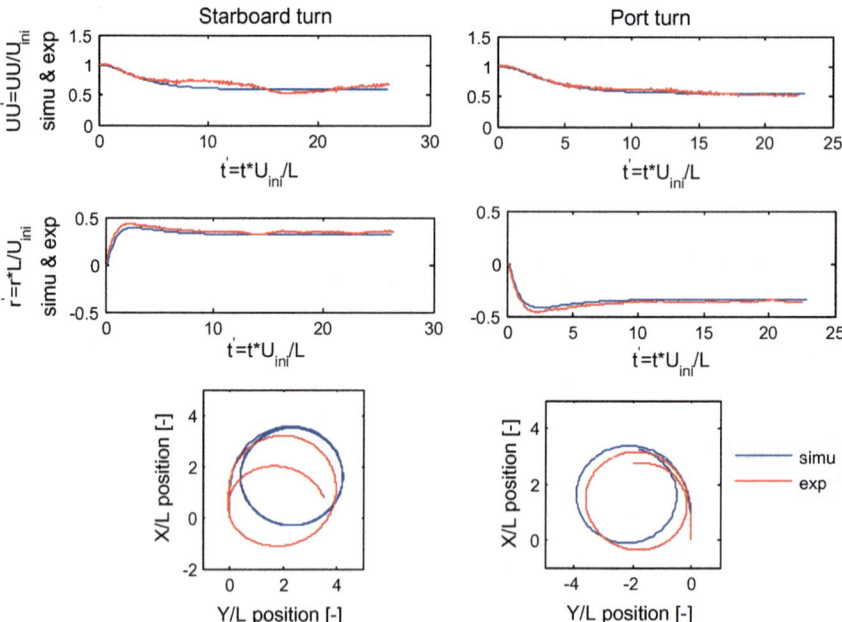

Fig. 17 Turning circle comparison for ±25°

result as compared with experimental one. However, due to the existence of wind during the experiments shown in Figs. 16 and 17, the ship started to drift and the resulting circular trajectories shifted towards the direct of wind. In such cases, comparison of the tactical diameters proves the predictability of manoeuvring motion while using the mentioned MMG model. Although slight discrepancies exist, they are well within the acceptable limit.

5 MMG Including Disturbance Model

The above mentioned MMG is simulated for calm and deep water conditions. However, by considering the terms X_{dis}, Y_{dis} and N_{dis} i.e. forces and moment due to any environmental disturbances like wind or current, the model can be used to predict the realistic ship trajectories. Different types of disturbance models are available, like the Fujiwara wind model [9]. Gust wind can also be taken into account using the Davenport spectrum [10]. Both uniform and non-uniform currents can be considered to simulate the ship manoeuvring characteristics.

6 Conclusions

The largest inconvenience in the field of manoeuvring research is the lack of a standard mathematical model. Hydrodynamic data are available from captive tests at various testing facilities and most of these have a lack of reasonable physical meaning. Therefore, comparison of the experimental results with corresponding theoretical calculations is difficult. This paper explains a well-developed concept proposed by the MMG group to develop a mathematical model, which not only takes hull, propeller and rudder effect separately but also their interactions. Most of the terms used in the model have a physical meaning, therefore for any design modification, one can clearly explain which coefficient needs to be modified instead of doing captive tests right from the beginning. Currently, the MMG model is already developed for twin screw MPVs, special types of ships like fishing trawlers etc. The model is also available considering shallow water effect and low speed manoeuvre.

References

1. Ogawa, A., Koyama, T., Kijima, K.: MMG report-I, on the mathematical model of ship manoeuvring (in Japanese). Bull Soc. Naval Archit. Jpn. **575**, 22–28 (1977)
2. Ogawa, A., Kasai, H.: On the mathematical method of manoeuvring motion of ships. Int. Shipbuild. Prog. **25**(292), 306–319 (1978)

3. Matsumoto, K., Suemitsu, K.: The prediction of manoeuvring performances by captive model tests (in Japanese). J. Kansai Soc. Naval Archit. Jpn. **176**, 11–22 (1980)
4. Inoue, S., et al.: A practical calculation method of ship manoeuvring motion. Int. Shipbuild. Prog. **28**(325), 207–222 (1981)
5. Hirano, M.: Prediction of manoeuvring in shallow water (in Japanese). J. Soc. Naval Archit. Jpn. **69**, (1985)
6. Kose, K., Yumuro, A., Yoshimura, Y.:. Concrete of Mathematical Model for Ship Manoeuvrability (in Japanese). In: 3rd Symposium on Ship Manoeuvrability, SNAJ, pp. 27–80 (1981)
7. Yoshimura, Y., et al.: Coasting manoeuvrability of single CPP equipped ship and the application of a new CPP controller. In: Proceedings of MARSIM'93, pp. 651–660 (1993)
8. Ueda, N., Ueno, M.A.: A comparative study on experimental results for manoeuvring hydrodynamic coefficients (in Japanese). Bachelor Thesis paper, Osaka University (1982)
9. Fujiwara, T., et al.: Estimation of wind forces and moment acting on ships. J. Soc. Naval Archit. Jpn. **183**, 77–90 (1998)
10. Davenport, A.G.: The dependence wind loads on meteorological parameters. In: Proceedings of Conference on Wind Effects on Buildings and Structures (1967)

A Trim Tank Control System for an Autonomous Underwater Vehicle (AUV)

Md Salim Kamil, Noorazlina Mohamid Salih, Atzroulnizam Abu, Muhammad Muzhafar Abdullah, Norshakila Abd Rasid and Mohd Shahrizan Mohd Said

Abstract The current technological developments allow autonomous underwater vehicles (AUV) to operate in deeper water with multiple functions including activities related to marine sciences, scientific researches, naval strategic intelligent and information gathering, oil and gas exploration, offshore platform underwater maintenance and surveillance, sea bed topological surveys and mappings and etcetera. Their capabilities to operate autonomously require a high-tech built-in computerized control system. One of the many important aspects in AUV is the necessity to have a trim tank control system to assist the AUV to be able to dive and surface or manoeuvre during the assigned operations. In this research the AUV is installed with two trim tanks located suitably aft and forward of the hull to provide the maximum trimming moment effects. The tanks are connected by piping to each other and fitted with control valves, a water transfer pump and an accelerometer sensor. Arduino Mega Board (AMB) open technology was used for the circuitries and the source codes design of the trim tank control system. Functional tests on the control system were carried out and it is proven that the main objectives of the study were achieved.

M. S. Kamil · N. Mohamid Salih (✉) · A. Abu · M. M. Abdullah · N. Abd Rasid
Malaysian Institute of Marine Engineering Technology, Universiti Kuala Lumpur, Jalan Pantai Remis, 32200 Lumut, Perak, Malaysia
e-mail: noorazlinams@unikl.edu.my

M. S. Kamil
e-mail: mdsalim@unikl.edu.my

A. Abu
e-mail: atzroulnizam@unikl.edu.my

M. M. Abdullah
e-mail: muzhafar.abdullah@gmail.com

N. Abd Rasid
e-mail: norshakila.abdrasid@unikl.edu.my

M. S. Mohd Said
Universiti Teknologi MARA, 56100 Seri Iskandar, Perak, Malaysia
e-mail: mohds665@perak.uitm.edu.my

© Springer International Publishing AG 2018
A. Öchsner (ed.), *Engineering Applications for New Materials and Technologies*, Advanced Structured Materials 85, https://doi.org/10.1007/978-3-319-72697-7_46

Keywords Trim tank · AUV · Arduino Mega Board · Accelerometer sensor Water transfer pump

1 Introduction

The aim of this paper is to put forward and share the idea to design a trim tank control system for a modular autonomous underwater vehicle (AUV). The AUV is designed to operate in a water of maximum depth of 2 m. The basic design principles of the trim tanks for the AUV are based on similar trim tanks used in a submarine. The sketches and drawings for the trim tank system were done using the SolidWorks software.

Riedel et al. [1], proposed a design and development of a low cost variable buoyancy system for the soft grounding of autonomous underwater vehicles. They discovered the effects of the centre of gravity and calculation for the centre of gravity in a low cost variable buoyancy system.

Gerl [2] with his research project on the development of an autonomous underwater vehicle in an interdisciplinary context found that most commercial AUV using trimming system by changing the overall density of the AUV. He also learned that the AUV pitched downwards by the rudders and its propeller and submerging can also be forced by sets of the propeller. In 2009 Tang et al. [3] researched on modelling for athwartships trim of ship balance and actuator by transferring liquid between tanks. A ship dynamics model for athwartships trim was built based on the ship static knowledge. Alam [4] optimizing design using computer aided.

In 2011, Ji-Qing and Lei [5] studied the heel and trim adjustment of a manned underwater vehicle based on the variable universe Fuzzy S surface control. They noticed that manned underwater vehicles need to have a good manoeuvrability especially the manoeuvrability in trim and heel. Moreover, a ballast water system is a simple and effective way for the adjustment of trim and heel. Listak [6] describing buoyancy control for AUV.

With reference to Fig. 1, the weight of the water in the forward trim tank is p with the longitudinal distance x_1 from the amidships or the centre of moment. The longitudinal distance of the aft trim tank is x_2 aft of the amidships. The difference in distance is $x_2 - x_1$. The centre of gravity moves from point G to point G_1, likewise the centre of buoyancy moves from B to B_1. Initially the AUV floats at waterline

Fig. 1 Control of trim

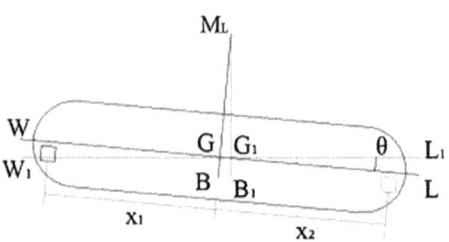

WL and as the water is transferred from the forward to aft trim tanks, the AUV trims by the stern and floats in the new horizontal waterplane W_1L_1. Wang [7] considered the important of AUV application.

Based on the principle of the equilibrium of moments as expressed in Eq. (1), the angle of trim θ in relation to the displacement Δ and the longitudinal metacentric height GML of the AUV and the trimming moment due to the transfer of the water between the trim tanks could be determined using Eq. (2).

$$\Delta GM_L \tan \theta = p(x_2 - x_1) \quad (1)$$

$$\theta = \frac{\tan^{-1} p(x_2 - x_1)}{\Delta GM_L} \quad (2)$$

2 Objectives

The objectives of the research are to design, test and validate the functionality of the trim tank system of an AUV. The design software or source codes were written in the Arduino Mega Board (AMB) and the open technology of the AMB will be verified for its versatility and adaptability.

3 Activities Flow Chart

A design of the trim tank system for the modular AUV is referred to the steps following a manual as shown in Fig. 2. The container is printed by using a 3D printer (plastic filament) to produce the working prototype of the trim tank. The size and dimension is calculated to complete the design. Li et al. [8] presented numerical studies in AUV system.

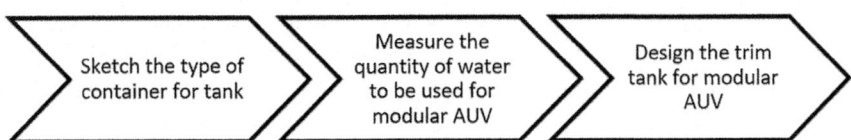

Fig. 2 AUV design process

4 Principal Dimension of the AUV, Control Surfaces and Trim Tanks

Length overall	236.24 cm
Length of cylindrical hull	201.00 cm
Length of aft cone	20.00 cm
Length of bow cap	12.25 cm
Hull diameter	30.48 cm
Rudder area (one rudder)	24.26 cm^2
Hydroplane area (one hydroplane)	22.40 cm^2
Hull volume	159,000 cm^3
The trim tank size	350 mm × 98 mm × 195 mm
Trim tank volume	0.524 m^3

5 Trim Tank Design Consideration

Various aspects had been taken into consideration when designing the trim tanks and the control system for the AUV. This is very important in the sense that the AUV is to be able to operate within the desired trim conditions while performing the duties. The effectiveness of the trim tanks depends very much on the size of the tanks and the right positioning, at which designers must take full advantage from aspects. The trim tank design adopts the basic principle of having two trim tanks, one located at extreme aft and the other tank is located at extreme forward as much as possible. Lillemoen [9] considered software to design the tank.

For the purpose of longitudinal balance of trim, the tanks operation is managed with both are partially filled and emptied, maintaining a constant weight with changing trimming moments. One or both tanks would be partially filled or emptied to allow trim correction to be made and keeping the trim moment constant. In the case of trim correction which affects weight and trimming moment, this could be achieved by filling and emptying the trim tanks simultaneously as appropriate.

Filling up the trim tanks could be done by taking in sea water and venting. However, a special facility is required to discharge out or to get rid of the water in the trim tanks. The AUV is subjected to hydrostatic pressure while underwater which depends on the depth of submergence. The hydrostatic pressure increases with increasing depth. Therefore, the usual way to discharge out the water in the trim tanks is to use high pressure compressed air available in the AUV or to have a pumping system for both tanks that fulfils the full operational scenarios of the AUV.

The design of the trim tanks also considers the variation of the operational loading conditions of the AUV. This would include at fully loaded condition at the lowest possible of the water density and in the lightest condition at the highest water density. The trim tanks total volume could be estimated based on the formulation given in Eq. (3) dependable on the AUV hull volume, range of water densities in percentage plus the specific volume of the variable weights and multiplied by a coefficient called a utility factor. This utility factor is to give due consideration of the internal tank structures and water left out that could not be totally discharged out of the AUV.

$$V_T = [(HV \times \rho_R) + (VW/\rho_W)] \times U_f \qquad (3)$$

where VT is the volume of the trim tank, HV is the hull volume, VW is the variable weight, ρ_R is the density of variable weights, and U_f is the utility factor.

The trim tanks are integrals of the overall tanks system in the AUV which takes into account of compensating tankages. The overall analysis of the design of the trim tanks will eventually give results of the weights in, weights out, buoyancy changes, trimming moments and trim polygon. The trim polygon defines the region or boundary of the trimming moments that covers the full operation of the trim tanks while the AUV is underwater. A typical trim polygon of an under water vehicle is shown in Fig. 3. Figure 4 shows the general longitudinal plan view of the AUV and the positions of the trim tanks at the aft and forward sections are shown in Fig. 5.

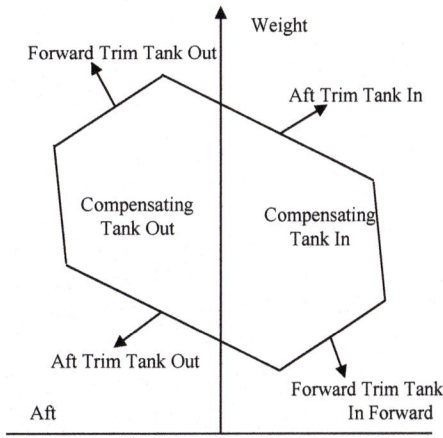

Fig. 3 Trim polygon

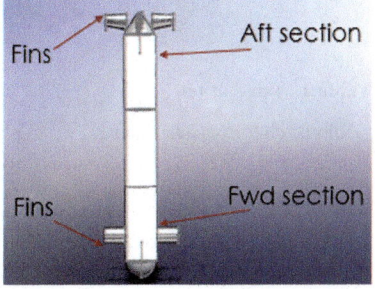

Fig. 4 AUV plan view

Fig. 5 Position of the trim tanks

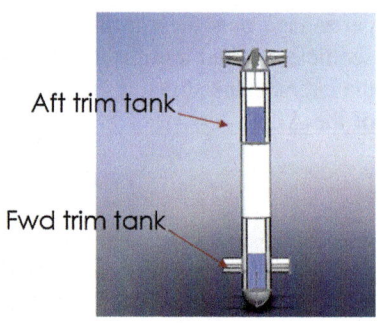

Figure 6 shows the exploded views of the aft and forward trim tanks, in which the amount of volume of water transferred was between 10 ml until 50 ml at a specific angle of trim. The trim tank system contributes to the balancing control of the AUV by adjusting the pitch without changing the weight or buoyancy balance.

Table 1 shows the data recorded from the experiment conducted for variation in times taken which were in the range from 10 s until 40 s. The volume of water transferred was measured from 10 ml until 40 ml. It also shows the change of angles of trim of the AUV in degrees ranging from 5° to 20°.

Figure 7 shows the prototype model of the trim tanks installed with a digital display to show the amount of water transferred between the trim tanks during the conduct of the functional tests. The AMB is embedded to the prototype AUV which

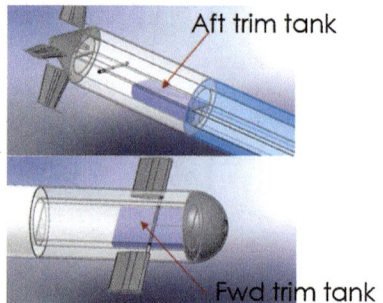

Fig. 6 Exploded view of the trim tanks

Table 1 Table of test results

Time taken to transfer water (s)	Volume of water transferred (ml)	Trim angle (°)
10	10	5
20	20	10
30	30	15
40	40	20

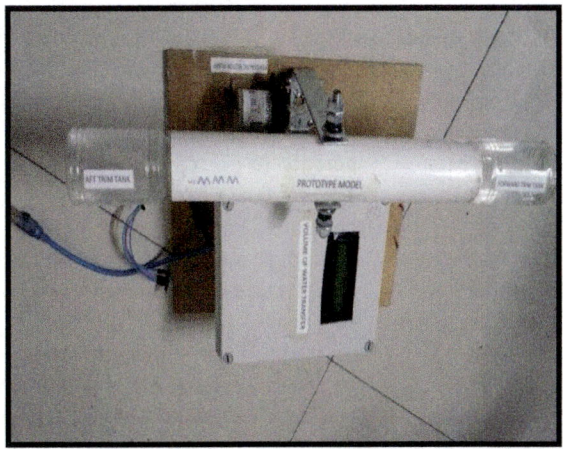

Fig. 7 Prototype design of trim tank control system and the functional test

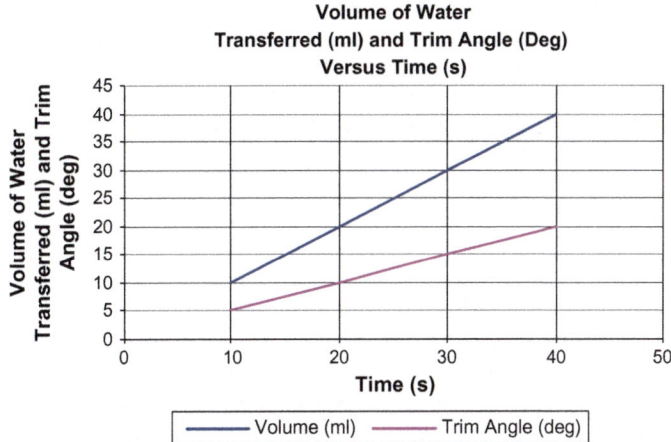

Fig. 8 Graphs of volume of water transferred and trim angle versus time

can control the adjustment of balancing of the AUV and pitch angle for buoyancy balance. Figure 8 shows the plots of the volume of water being transferred and the trim angle versus the times taken. As the time increases the volume of water transferred from between the trim tanks and the trim angle increases. Specifically at 40 s, the trim angle reaches 20° with 40 ml of water being transferred.

```
// MPU control/status vars
bool dmpReady = false;  // set true if DMP init was successful
uint8_t mpuIntStatus;   // holds actual interrupt status byte
uint8_t devStatus;      // return status after each device
                           operation (0 = success, !0 = error)
uint16_t packetSize;    // expected DMP packet size
                           (default is 42 bytes)
uint16_t fifoCount;     // count of all bytes currently in FIFO
uint8_t fifoBuffer[64]; // FIFO storage buffer
// orientation/motion vars
Quaternion q;           // [w, x, y, z]     quaternion container
VectorInt16 aa;         // [x, y, z]        accel sensor measurements
VectorInt16 aaReal;     // [x, y, z]        gravity-free accel sensor
                                            measurements
VectorInt16 aaWorld;    // [x, y, z]        world-frame accel sensor
                                            measurements
VectorFloat gravity;    // [x, y, z]        gravity vector
float euler[3];         // [psi, theta, phi]   Euler angle container
float ypr[3];           // [yaw, pitch, roll]  yaw/pitch/roll container
                           and gravity vector
// packet structure for InvenSense teapot demo
uint8_t teapotPacket[14] = { 'S', 0x02, 0,0, 0,0, 0,0, 0,0, 0x00,
0x00, '\r', '\n' };
```

Fig. 9 Initial coding

Initial coding was written in the AMB to set the trim angle and time taken needed to provide monitoring on the total volume of water transferred. Figure 9 shows the initial coding list for the servo motor to monitor the movement of the AUV. Figures 10, 11 and 12 are the coding for Servo Motor Part 1, Part 2 and Part 3 respectively for the control of the balancing of the AUV system.

```
INITIAL SETUP                ===
void setup() {
lcd.begin(16,2);        // 16,2 for LCD
lcd.clear();            // start with a blank screen
lcd.setCursor(0,0);
  motor.setSpeed(250);//set speed for motor 1
  motor2.setSpeed(250);//set speed for motor 2
  motor3.setSpeed(250);//set speed for motor 3
   // join I2C bus (I2Cdev library doesn't do this automatically)
  #if I2CDEV_IMPLEMENTATION == I2CDEV_ARDUINO_WIRE
    Wire.begin();
    TWBR = 24; // 400kHz I2C clock (200kHz if CPU is 8MHz)
  #elif I2CDEV_IMPLEMENTATION == I2CDEV_BUILTIN_FASTWIRE
    Fastwire::setup(400, true);
  #endif
```

Fig. 10 Coding servo motor part 1

6 Discussion and Conclusion

The trim tanks sizing were performed quite rapidly utilizing the given formulation in Eq. (3) taking into consideration of the hull volume HV, variable weight VW, density of variable weights ρR and the utility factor Uf. The effects on the changes of trim angles due to water transfers between the trim tanks of the AUV were studied and analysed based on the expression given in Eq. (2). The trim polygon covering the operational profiles of the AUV could be generated by analysing the weight changes of the filling and discharging the water to and from the trim tanks and evaluating the resulting trimming moments.

```
// initialize serial communication
// (115200 chosen because it is required for Teapot Demo output
  Serial.begin(115200);
  while (!Serial); // wait for Leonardo enumeration
// initialize device
  Serial.println(F("Initializing I2C devices..."));
  mpu.initialize();
// verify connection
  Serial.println(F("Testing device connections..."));
  Serial.println(mpu.testConnection()
// wait for ready
  Serial.println(F);
  while (Serial.available() && Serial.read()); // empty buffer
  while (!Serial.available());          // wait for data
  while (Serial.available() && Serial.read()); // empty buffer again*/
```

Fig. 11 Coding servo motor part 2

The overall circuitries for the trim tank control system were designed and printed on the circuit board on the Arduino Mega Board (AMB) and the source codes for the control of the servo motors that operate the water transfer pump were developed and easily written or burnt into the AMB. The design of the circuitries, the source codes and the performance of the trim tanks control system were verified in the functional tests conducted which allow for corrections, fine tuning and improvements.

```
// load and configure the DMP
   Serial.println(F("Initializing DMP..."));
   devStatus = mpu.dmpInitialize();
// supply your own gyro offsets here, scaled for min sensitivity
   mpu.setXGyroOffset(220);
   mpu.setYGyroOffset(76);
   mpu.setZGyroOffset(-85);
   mpu.setZAccelOffset(1788); // 1688 factory default for my test chip
// make sure it worked (returns 0 if so)
   if (devStatus == 0) {
      // turn on the DMP, now that it's ready
      Serial.println(F("Enabling DMP..."));
      mpu.setDMPEnabled(true);
   // enable Arduino interrupt detection
      Serial.println(F);
      attachInterrupt(2, dmpDataReady, RISING);
      mpuIntStatus = mpu.getIntStatus();
```

Fig. 12 Coding servo motor part 3

Finally, it can be concluded that the main objectives to design the trim tank control system for the AUV were successfully accomplished. The functionality of the trim tanks had been tested and worked well. It has been proven that Arduino Mega Board open technology is very versatile and can be easily adaptable to various applications.

References

1. Riedel, J.S., Healey, A.J., Marco, D.B., Beyazayn, B.: A design and development of low cost variable buoyancy system for the soft, grounding of autonomous underwater vehicles (2014)
2. Gerl, B.: Development of an autonomous underwater vehicle in an interdisciplinary context. Diploma Dissertation, Department of Electrical, Electronic and Computer Engineering (2006)
3. Tang, M., Guo, J., Nong, W.S.: Modelling for Athwartships Trim of Ship Balance (2009)

4. Alam, K., Ray, T., Anavatti, S.G.: Design and construction of an autonomous underwater vehicle. Neurocomputing **142**, 16–29 (2014)
5. Ji-qing, L., Lei, W.: The heel and trim adjustment of manned underwater vehicle based on variable universe fuzzy S surface control. In: Electronics, Communications and Control (ICECC), International Conference on IEEE (2011)
6. Listak, M., Kruusmaa, M.: Buoyancy Control of a Semiautonomous Underwater Vehicle for Environmental Monitoring in Baltic Sea. Department of Computer Engineering, Tallin University of Technology (Ehitajate) & Institute of technology, Tartu University (Vanemuise) (2003)
7. Wang, W.H., Engelaar, R.C., Chen, X.Q., Chase, J.G.: The State-of-Art of Underwater Vehicles-Theories and Applications (2009)
8. Li, J.H., Lee, P.M.: Design of an adaptive nonlinear controller for depth control of an autonomous underwater vehicle. Ocean Eng. **32**(17–18), 2165–2181 (2005)
9. Lillemoen, N.F.: Development of Software Tool for Identification of Ballast Errors in Autonomous Underwater Vehicles p. 9 (2014)

The Use of Backscatter Classification and Bathymetry Derivatives from Multibeam Data for Seabed Sediment Characterization

Razak Zakariya, Mohd Azhafiz Abdullah, Rozaimi Che Hasan and Idham Khalil

Abstract The multibeam data provides backscatter and bathymetry accurate measurement of water depth information this valuable is related to the topography and composition of the seafloor. From multibeam dataset, the backscatter information and comprising bathymetry derivatives were interpreted using the ArcGIS software. The sediment samples from ground survey were collected using Ponar Grab and classified the size of sediment using Particle Size Analysis (PSA). The sieving method was also applied for coarse sediment samples. Ground survey data were used on verification and determination of sediment map accuracy produced by GIS analysis. In this study, maps of backscatter, bathymetry, slope and rugosity were developed and applied for developing seabed characterization classes. The result shows the seabed sediment from the study area was successfully developed with overall accuracy of 88.2% and kappa coefficient of 0.82. The study provided the opportunity on increasing the accuracy on determining the seabed sediment characteristic within the multibeam coverage area.

Keywords Multibeam echo sounder · Bathymetry · Backscatter
Sediment

R. Zakariya · M. A. Abdullah (✉) · I. Khalil
School of Marine Science and Environment, University Malaysia
Terengganu, 21300 Kuala Terengganu, Terengganu, Malaysia
e-mail: azhafiz88@gmail.com

R. Zakariya
e-mail: ajak@umt.edu.my

I. Khalil
e-mail: idham@umt.edu.my

R. C. Hasan
UTM Razak School of Engineering and Advanced Technology, Universiti Teknologi
Malaysia, Jalan Semarak, 54100 Kuala Lumpur, Malaysia
e-mail: rozaimi.kl@utm.my

© Springer International Publishing AG 2018
A. Öchsner (ed.), *Engineering Applications for New Materials and Technologies*,
Advanced Structured Materials 85, https://doi.org/10.1007/978-3-319-72697-7_47

1 Introduction

A multibeam echo sounder system (MBES) is a system that sends out a broad swath of sound pulse, bouncing over a wide area of seafloor beneath the survey. By calculating the two-way travel time of the sound pulses, the system determines the depth of the water beneath the vessel. Two types of data collected from the multibeam echo sounder system are backscatter data and bathymetry data. Backscatter data allows the user to determine the type of seafloor by analyzing the strength of the returned sound pulse [7]. Hard sediment such as boulder or gravel will sent strong return signals to the system while mud and sand will send weak return [3, 8]. Bathymetry data is returned by calculating the two way travel times, which display the depths of a wide area beneath the survey vessel [13]. Both data can be derived simultaneously in on single transect. The main purpose of the study is to utilize the output of multibeam data on developing seabed sediment characteristic with better accuracy.

2 Study Area

The study area covers part of West of the Pulau Bidong. The coverage for this survey area is approximately 406,183 m^2. The area is selected as the study area due to the availability of bathymetry data and backscatter data to classify the seafloor terrain characteristics. The height range of this area is 5 m up to 33 m. Figure 1 shows the location of the study area.

3 Methodology

The methodology of this study is organized into three main stages (i) data acquisition, (ii) data processing and (iii) data analysis. Figure 2 shows the flowchart of the methodology adopted for this study.

For this study, Multibeam Model R2 sonic 2020 system was used for data acquisition to provide bathymetry data and backscatter data. Table 1 shows the list of the survey equipment used.

Backscatter and bathymetry was processed using the FMGeocoder Toolbox (FMGT) [6] and Qimera software. The backscatter processing using FMGT as part of this research intended to improve the quality of the backscatter mosaic for sediment classification using supervised classification in ArcGIS 10.1 using sediment sample data [2, 9]. Figure 3 shows the backscatter mosaic map of the Pulau Bidong.

The Qimera software is a processing tool to produce a bathymetry grid. First the processing tide, sound velocity profile (SVP) and patch test (roll, pitch and heading)

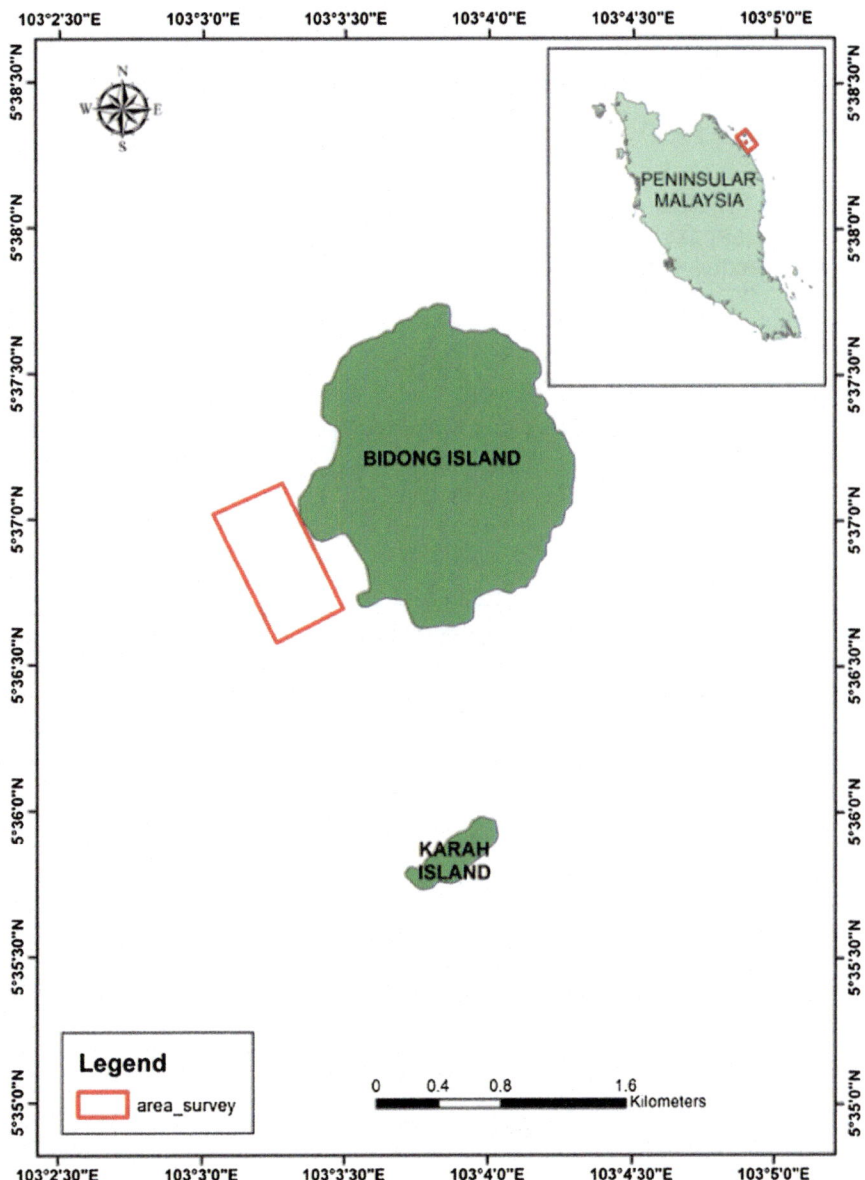

Fig. 1 Location of the study area

is done first. Removal of spikes or noise and other issues related to relatively poor data quality in survey area is required. The bathymetry grid data is exported to .geotiff format so further processing is done in ArcGIS 10.1 using the Benthic

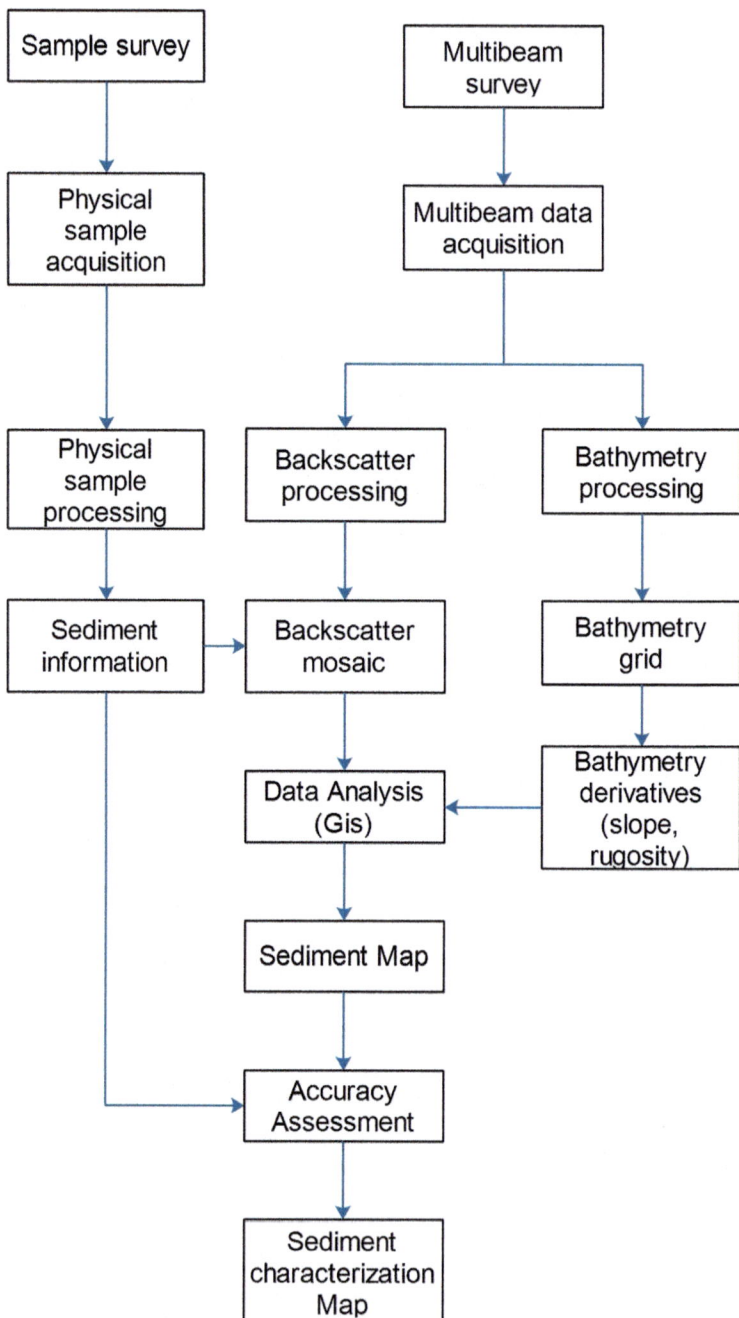

Fig. 2 Flowchart of the research methodology

Table 1 List of the survey equipment used

Equipment	Model
Multibeam echo sounder	R2 Sonic 2020
Motion sensor (MRU/IMU)	TSS DMS 10
Sound velocity profiler	CastAwayCTD
Sound velocity sensor	Veleport Mini SVS
Positioning	Trimble 461
Heading antennas	Trimble 461
Software	Qinsy

Terrain Modeler (BTM) [15]. Figure 4 shows the bathymetry map of the Pulau Bidong.

The sample sediment was collected after the multibeam survey. Total sample was collect 9 stations using a ponar grab. The physical sample processing using a sieve was conducted on all samples. The sample that has been processed can determine the type of the sediment [1, 10].

Further processing involved the bathymetry derivatives [3] and backscatter classification with sediment information (sample survey). Backscatter classification and bathymetry derivatives can be done with the Arcgis 10.3 software using the image classification toolbox [11]. Maximum likelihood classification was used by the supervised classification with the sample sediment to classify all the study area. This method was chosen to classify the pixels. Data from Particle Size Analysis as a guideline to classify certain pixels to identify the distribution of the sediment based on pixels characteristics where it is correlated with different type of sediment. Data variable between backscatter, bathymetry derivatives (slope, rugosity) and ground truth data to know how reliable based on the error matrix. Kappa coefficient will determine the accuracy assessment to evaluate the reliability of the data.

The bathymetry derivatives were generated using Benthic Terrain Modeler (BTM) [15] in the ArcGIS 10.3 software to generate the slope and rugosity [5, 14]. Slope is an identifier of the gradient or rate of maximum change in z-value from each cell of a raster surface. The slope calculation must be correct by casting a z factor when the unit z is applied to units different from the ground x, y.

The rugosity measures the terrain ruggedness as the variation in three-dimensional orientation of grid cells within a neighborhood. Vector analysis is used to calculate the dispersion of the vectors normal (orthogonal) to grid cells within the specified neighborhood. This method effectively captures the variability in slope and aspect into a single measure. Ruggedness values in the output raster can range from 0 (no terrain variation) to 1 (complete terrain variation). Typical values for natural terrains range between 0 and about 0.4.

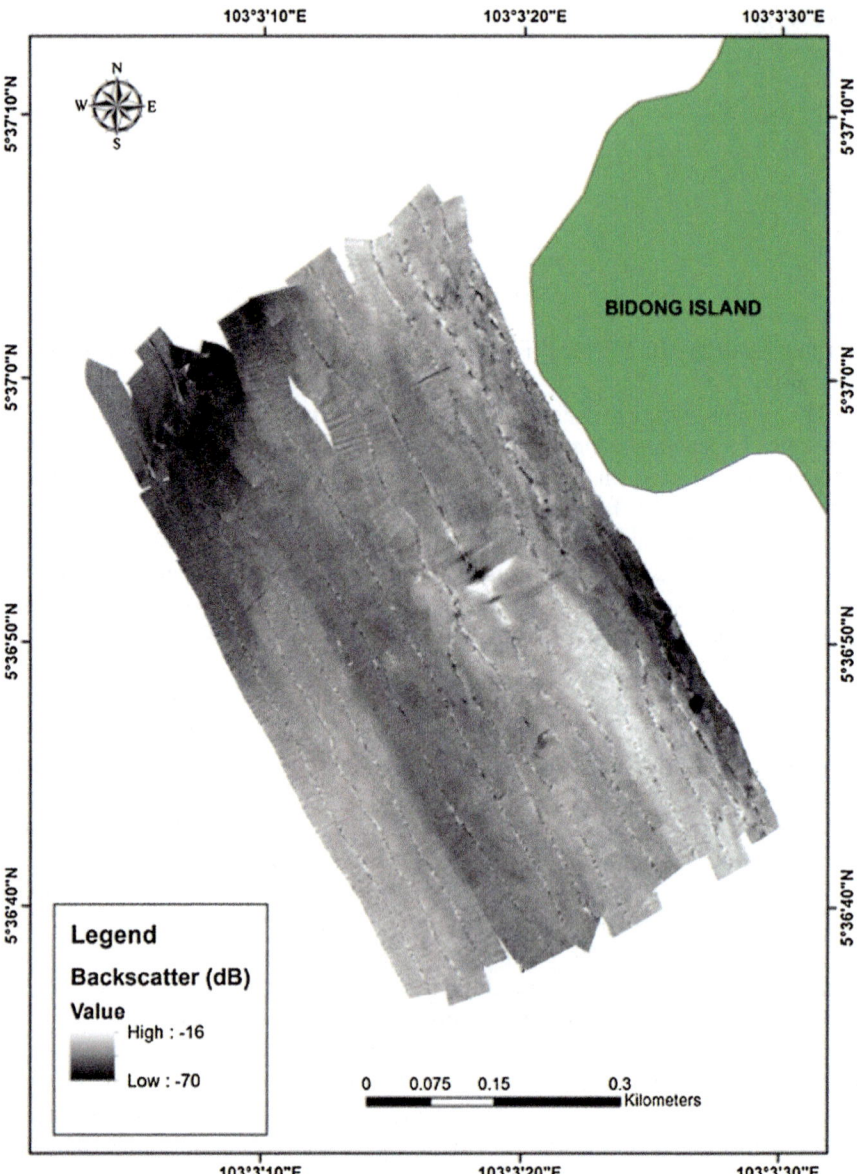

Fig. 3 Backscatter mosaic map of the study area

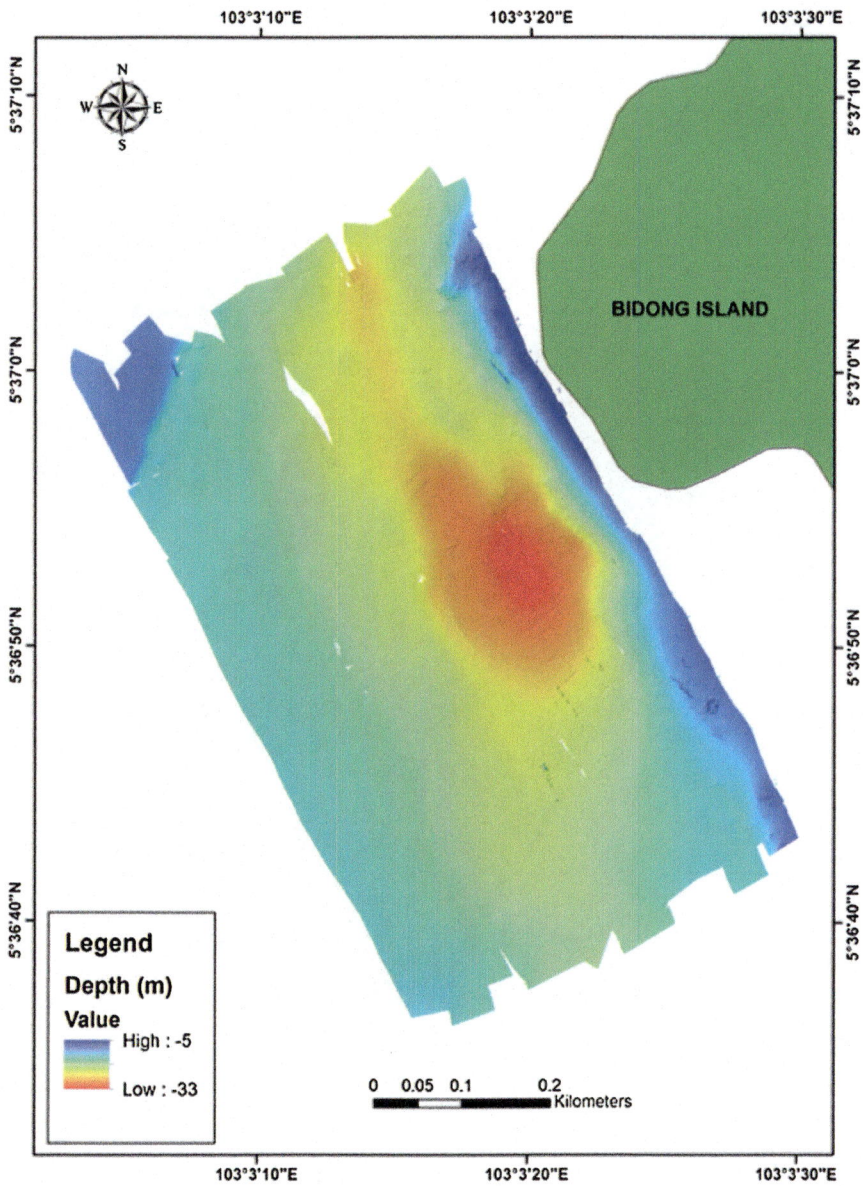

Fig. 4 Bathymetry map of the study area

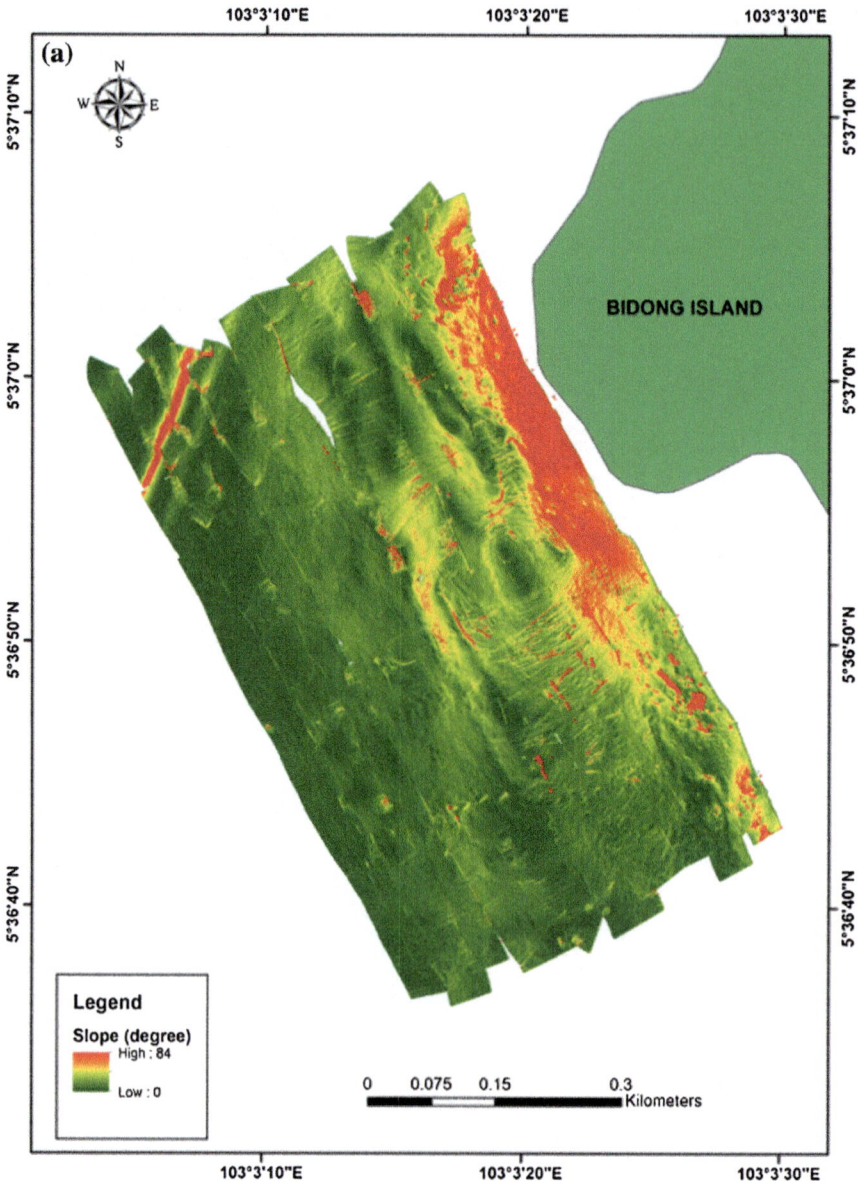

Fig. 5 a Bathymetry derivatives—slope, b Bathymetry derivatives—rugosity

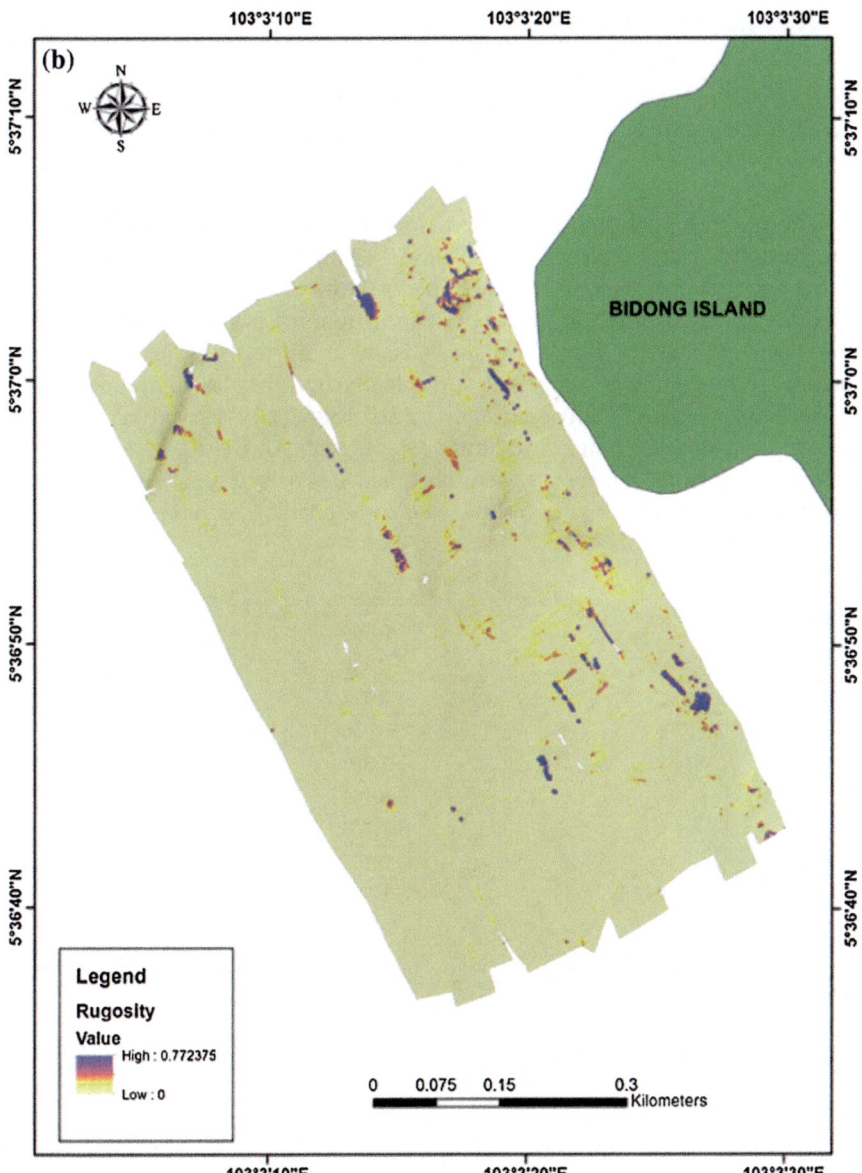

Fig. 5 (continued)

4 Results and Discussion

The sediment information from the sample sediment shown in Table 2. The result was obtained from sediment processing using the sieve method. The results for stations 1–4 show very coarse sediment type sand while the stations 5, 6 and 9 reveal very fine gravel. Station 6 and 8 belong to coarse sand.

Referring to Fig. 5a that the highest slope from the bathymetry derivatives is more than 84° and the lowest slope is below 0°. Most of the areas in the study area have a slope ranging from 0° to 20°. Table 3 shows the class by slope map.

From Fig. 5b, the highest rugosity obtained from the bathymetry derivatives is more than 0.77 and the lowest rugosity is below 0. Most of the area in the study area has a slope ranging from 0 to 0.2. Table 4 shows the class by rugosity map.

The classification of the backscatter data and bathymetry derivatives (slope and rugosity) was done by using a sediment sample with ArcGIS 10.3 software. The backscatter classification results were classified into very coarse sand, very fine gravel and coarse sand. Figure 6 shows the corresponding sediment map.

Table 2 Sediment information from sediment sample

Station	Mean	Type of sediment
1	−0.563	Very coarse sand
2	−0.830	Very coarse sand
3	−0.311	Very coarse sand
4	−0.956	Very coarse sand
5	−1.882	Very fine gravel
6	0.182	Coarse sand
7	−1.76	Very fine gravel
8	0.153	Coarse sand
9	−1.847	Very fine gravel

Table 3 Class by slope map

Class	Total area (m^2)
0–20	401,397.5
21–40	4004
41–60	513.5
61–80	256.75
80>	11.25

Table 4 Class by rugosity map

Class	Total area (m^2)
0–0.2	403,279.5
0.21–0.4	104.75
0.41–0.6	37.5
0.6–0.8	2.25

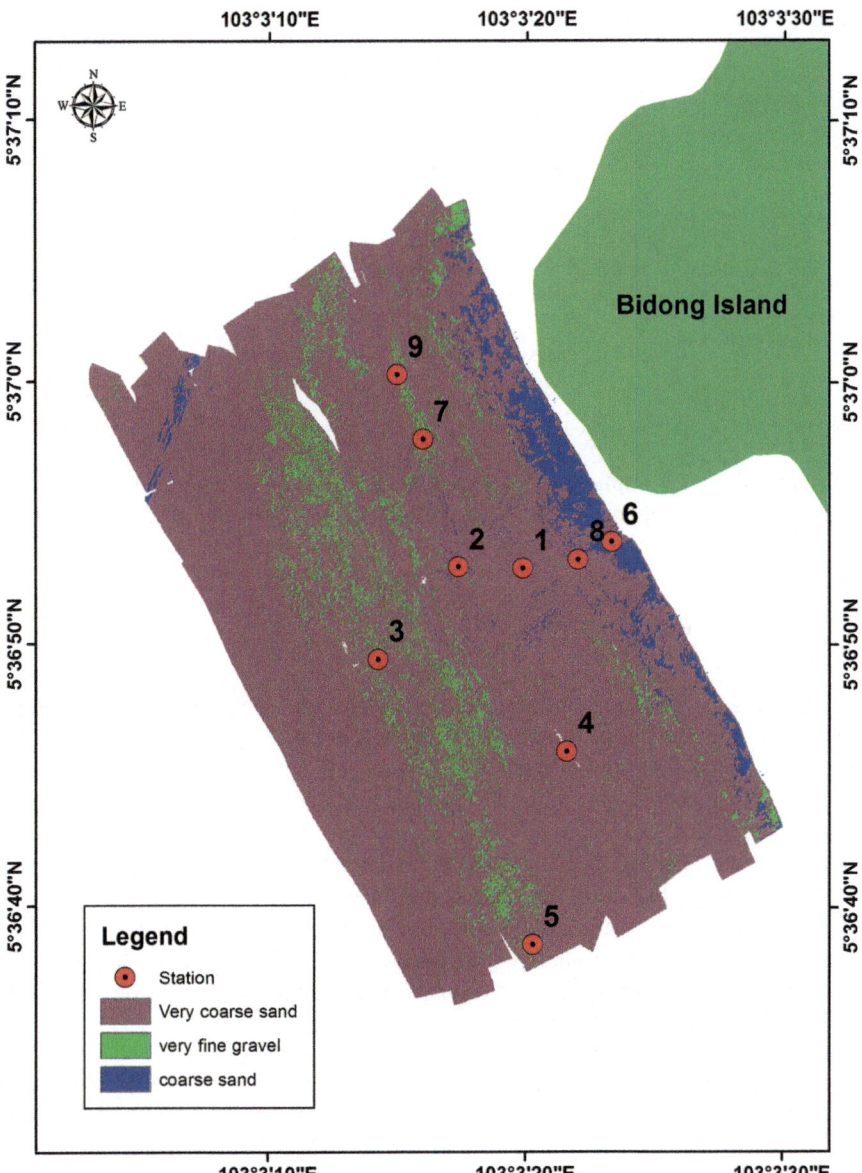

Fig. 6 Seabed sediment map of the study area produced from backscatter and bathymetry derivatives of multibeam echo sounder

Tables 5 and 6 shows the accuracy assessment for the sediment classification. The error matrix was used for the classification because the possible error occurred when classifying the map based on the selected pixels which did not match with the ground truth data. Accuracy assessment is 88.2% and Kappa coefficient is 0.82.

Table 5 Error matrix of accuracy assessment for maximum likelihood classification

Class	Very coarse sand	Very fine gravel	Coarse sand	Total
Very coarse sand	4	0	1	5
Very fine gravel	0	3	0	3
Coarse sand	0	0	1	1
Total	4	3	2	9

Table 6 Accuracy assessment for maximum classification method

Method	Overall accuracy	Kappa coefficient
Maximum likelihood	88.2	0.82

A kappa value of 1 implies complete agreement, while a value less than 0 implies more disagreement than a random assignment of classes to one of the maps [12].

5 Conclusion

Multibeam surveys collected bathymetric data and backscatter data at a very high resolution. Using this type of dataset, large areas can be examined at various scales to get a regional overview of the seafloor characteristic, as well as detailed studies of particular features. The availability within this study of both multibeam bathymetry and ground true sample sediment provides an interesting opportunity to investigate the relationship between seafloor morphology and geology structures.

The results of this study, i.e. backscatter classification and bathymetry derivatives from multibeam sonar can be used to study another application in the guiding the location and extent of infrastructure developments. For fisheries management, a combination of the bathymetry and rugosity layers can provide a surrogate for reef localities and therefore potential reef fish distribution.

Acknowledgements The authors thanked the University Malaysia Terengganu (UMT) and Universiti Teknologi Malaysia (UTM) Razak School of Engineering and Advanced Technology, for research facilities and financial assistance provided for this project.

References

1. Brown, C.J., Cooper, K.M., Meadows, W.J., Limpenny, D.S., Rees, H.L.: Small-scale mapping of sea-bed assemblages in the eastern English channel using sidescan sonar and remote sampling techniques. Estuar. Coast. Shelf Sci. **54**, 263–278 (2002)
2. Burrough, P.A., McDonnell, R.A.: Principles of Geographic Information Systems. Oxford University Press, Oxford (1998)

3. Dolan, M.F.J., Grehan, A.J., Guinan, J.C., Brown, C.: Modelling the local distribution of cold-water corals in relation to bathymetric variables: adding spatial context to deep-sea video data. Deep-Sea Res. Part I **55**, 1564–1579 (2008)
4. Edwards, B.D., Dartnell, P., Chezar, H.: Characterizing benthic substrates of Santa Monica Bay with seafloor photography and multibeam sonar imagery. Mar. Environ. Res. **56**, 47–66 (2003)
5. Guinan, J., Grehan, A.J., Dolan, M.F.J., Brown, C.: Quantifying relationships between video observations of cold-water coral cover and seafloor features in Rockall Trough, west of Ireland. Mar. Ecol. Prog. Ser. **375**, 125–138 (2009)
6. IVS3D Fledermaus: Fledermaus reference manual. Version 7 (2011)
7. Lamarche, G., Lurton, X., Verdier, A.-L., Augustin, J.-M.: Quantitative characterisation of seafloor substrate and bedforms using advanced processing of multibeam backscatter—application to cook strait, New Zealand. Cont. Shelf Res. **31**, S93–S109 (2011)
8. Li, J., Siwabessy, J., Tran, M., Huang, Z., Heap, A.: Predicting seabed hardness using random forest in R. In: Zhao, Y., Cen, Y. (eds.) Data Mining Applications with R. Elsevier (2013, in press)
9. Mcgonigle, C., Brown, C.J., Quinn, R.: Operational parameters, data density and benthic ecology: considerations for image-based classification of multibeam backscatter. Mar. Geodesy **33**, 16–38 (2010)
10. Mcgonigle, C., Collier, J.S.: Interlinking backscatter, grain size and benthic community structure. Estuar. Coast. Shelf Sci. **147**, 123–136 (2014)
11. Micallef, A., Le Bas, T.P., Huvenne, V.A.I., Blondel, P., Hühnerbach, V., Deidun, A.: A multi-method approach for benthic habitat mapping of shallow coastal areas with high-resolution multibeam data. Cont. Shelf Res. **39**(40), 14–26 (2012)
12. Rositter, D.G.: Statistical methods for accuracy assesment of classified thematic maps. Department of Earth System Analysis (2014)
13. Siwabessy, P.J.W., Daniell, J., Li, J., Huang, Z., Heap, A.D., Nichol S., Anderson, T.J., Tran, M.: Methodologies for seabed substrate characterisation using multibeam bathymetry, backscatter and video data: a case study from the carbonate banks of the Timor Sea, Northern Australia. Record 2013/11. Geoscience Australia, Canberra (2013)
14. Tong, R.: Influence of seabed topography on cold-water coral distribution and habitat suitability. Ph.D. Thesis, Jacobs University, p. 119 (2012)
15. Wright, D.J., Pendleton, M., Boulware, J., Walbridge, S., Gerlt, B., Eslinger, D., Sampson, D., Huntley, E.: ArcGIS Benthic Terrain Modeler (BTM), v. 3.0. Environmental Systems Research Institute, NOAA Coastal Services Center, Massachusetts Office of Coastal Zone Management. Available online at http://esriurl.com/5754 (2012)

A Review of Piezoelectric Design in MEMS Scanner

Nur Azirah Abdul Rahim, Ishak Abdul Azid and Loke Kean Koay

Abstract The main objective of this paper is to present a review of the piezoelectric design in MEMS scanner that has been proposed by various researchers. A piezoelectric actuator is a device used for actuating small displacements within the range of 10 pm (1 pm = 10^{-12} m) to 100 μm. Piezoelectric actuators are widely used in scanning technology due to their high precision, especially in micro/nano technology. There are many limitations such as vibration, creep, and hysteresis that can reduce the accuracy and stability of a piezoelectric actuator. Many efforts have been made to overcome the limitations in terms of mathematical modelling and control approaches. This review includes the piezoelectric actuator design for optical scanners and the development of miniaturize scanners. The types of piezo actuators in scanning technology are categorized by the different configurations such as stack actuator, thin film and tube actuator. Besides, the advantages, limitations and the applications of piezo actuator are included. This paper explains the existing piezo actuator design constraints and applications.

Keywords Actuator · Piezoelectric · MEMS · Scanner

1 Introduction

Figure 1 shows the overview of the paper. This paper reviews the configurations of piezoelectric actuators and the working principles for each type of piezoelectric actuator. The generic form of a piezo actuator is the piezo plate. The actuator

N. A. Abdul Rahim · I. Abdul Azid · L. K. Koay (✉)
Malaysian Spanish Institute (MSI), Universiti Kuala Lumpur, Kulim Hi-Tech Park, 09000 Kulim, Kedah, Malaysia
e-mail: lkkoay@unikl.edu.my

N. A. Abdul Rahim
e-mail: nur.azirah@kdupg.edu.my

I. Abdul Azid
e-mail: ishak.abdulazid@unikl.edu.my

displacement is improved when the piezo plates are stacked together in an array [7]. There is another design, the so-called tube actuator, which is applied to small positioning applications. The explanations of the different types of configurations will be revealed in later sections.

Furthermore, the parameters, advantages, limitations and applications are also stated. Besides, the design and control algorithm of the piezoelectric actuator in micro electro-mechanical (MEMS) applications are revealed. Lastly, the drawback of the actuators such as creep, hysteresis and vibration also being discussed.

Piezo actuators can be equipped with amplifiers to measure the displacement and are then suitable for repeatable positioning [1]. Piezoelectric actuators are usually employed in high precision displacement control or high generative force devices [1]. Recent research made by Matsushita and Kanno, 2011, revealed that MEMS piezoelectric thin films provide fast actuation of 23.7 kHz at small drive voltages of 10 Vpp (peak to peak), meanwhile they also provides accurate displacements control [2].

Generally, piezoelectric actuators utilize the reverse effect of piezoelectric materials to generate displacement and mechanical force [1]. The piezoelectric material will be mechanically strained by placing it into the electric field or by applying voltages [3].

According to Troiler and Muralt, ceramic piezoelectric is applied in MEMS technology due to its high resonant frequency structures while maintaining good

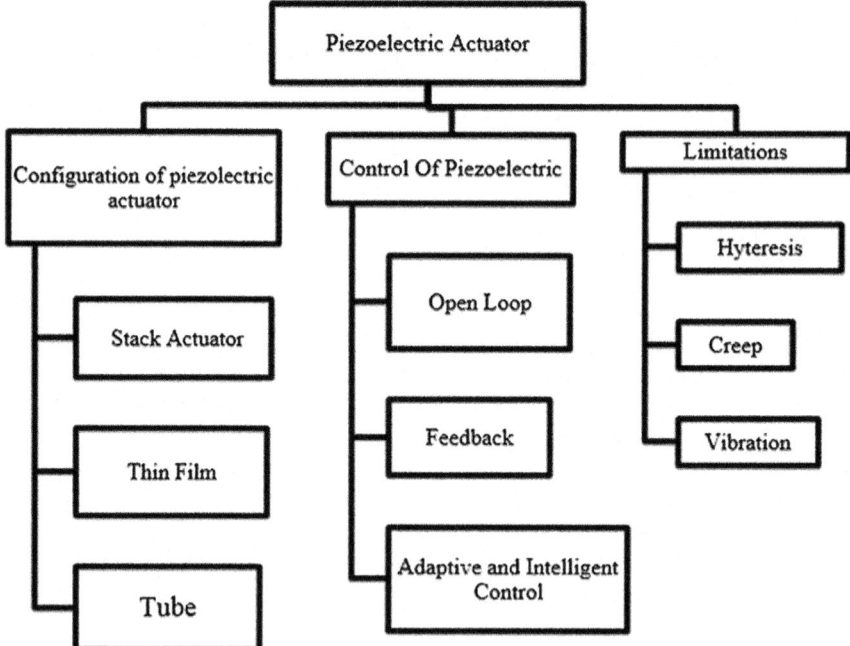

Fig. 1 Overview of piezoelectric actuator configuration

temperature stability [4]. Furthermore, piezoelectric has the ability to perform large amplitude actuation with lower drive voltages and low hysteresis [5].

However, according to Abdalla [6], the limitation of piezoelectric actuators is affected by the mechanical impedances of the piezoelectric stack and load, driven by high force or low displacement characteristics of the piezoelectric. Mechanical impedance is the extent of a structure to resist motion when subjected to harmonic forces. This may affect the stability of the actuator and reduce the accuracy of the system [6].

2 Configuration of Piezoelectric

2.1 Stack Actuator/Transducer

A piezoelectric stack actuator is built using multiple piezoelectric discs wired in parallel and stack mechanically in series as shown in Fig. 2 [7]. In a piezoelectric, 100 multilayers of piezoelectric ceramic sheets are stacked together. The piezoelectric produces a high generative force of 1000 N, fast response within 10 μs and high stability with 100 V power consumption [1]. With those advantages, piezoelectric stack actuators have been applied in the fields of micro/nano manipulation, nanopositioning stage and micro vibrational control [1]. Figure 2 shows the schematic diagram of a piezo stack actuator. A driving voltage is applied to a piezoelectric stack, expansion or contraction along the axial direction (as shown in Fig. 2) of the stack will be induced [8].

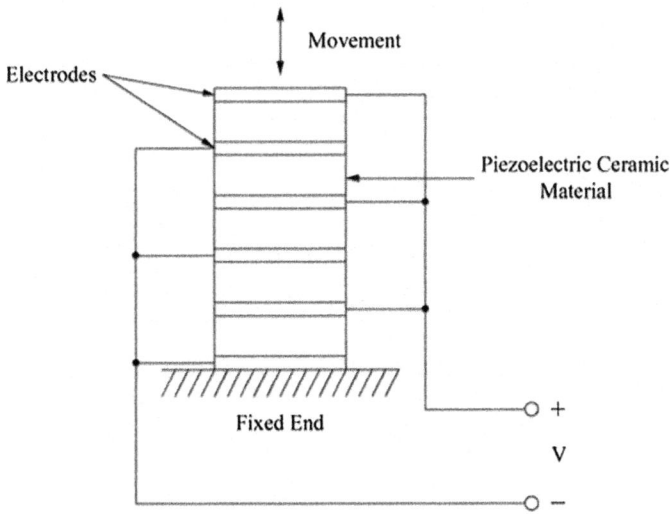

Fig. 2 Stack actuator construction

Fu presented the experimental setup of displacement measurement without external mechanical load as shown in Fig. 3. The experiment is used to identify the voltage and displacement coefficient of the piezoelectric stack actuator [9]. The signal generated by the function generator is amplified by the voltage amplifier to activate the piezoelectric stack actuator as shown in Fig. 3. The actuator under the activation voltage will generate oscillation, which is measured by the laser displacement sensor as shown in Fig. 3. The output signals of the function generator and laser sensor are collected by the data acquisition system. The data shows that the displacement increases as the voltage increases [9].

A design by Wang and Yang [8] shows that a n-shaped piezoelectric stack actuator can generate bending moments. The moments acts on the host structure through the pedestal working as the mechanical lever which can exert a strong actuating power of piezoelectric stack as shown in Fig. 4. The actuator can help the piezoelectric stack and host structure for vibration control applications.

The driving voltage is applied to the piezoelectric stack, and expands along its axial direction [8]. The two ends of the stack are hold by the n-shaped pedestal with a screw as shown in Fig. 4. If the base of the pedestal is bonded to the host structure, actuating moments will be exerted on the structure, thus the pedestal will produce a controlled vibration [8].

However, there are also limitations that affect the system of piezoelectric stack actuators. Since piezoelectric is a dielectric, a PZT stack actuator can exhibit capacitive behavior. Hysteresis is one example of the disadvantages in piezoelectric actuators. This is due to nonlinear movement of the actuator that will affect the accuracy of motion [10]. A research by Yi and Veillette [7] presented the feedback controller to reduce the hysteresis effect on the actuator. The implementation uses an inverting op-amp configuration instead of the non-inverting configuration.

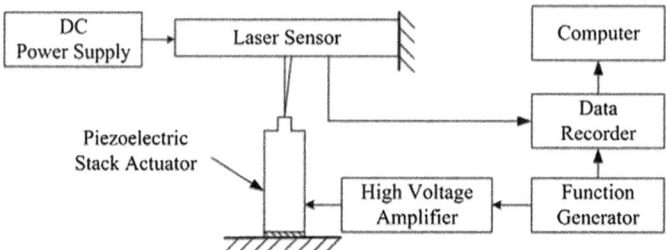

Fig. 3 Schematic configuration of experimental setup for the dynamic characteristics measurement of the actuator without mechanical load

Fig. 4 Detailed design of piezoelectric stack actuator

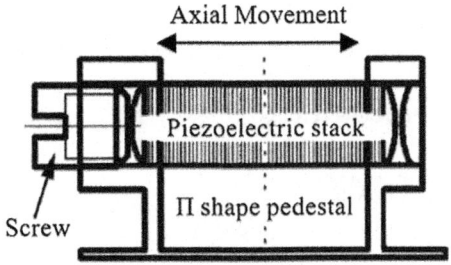

2.2 Thin Film Piezoelectric

Piezoelectric thin films actuators are widely used in MEMS development. Piezoelectric thin films provide large linear forces with fast actuation of 3 kHz at small drive voltages of 5 V. Piezoelectric thin films will provide accurate displacements of 1.2 μm [5]. Thin films of non-ferroelectric aluminium nitride (AlN) and ferroelectric PZT based materials are widely studied for fabrication of piezoelectric actuators and sensors [12]. In the fabrication, a thin film actuator composes of double-layer PZT films, each of the layers have polarities, so when voltage is applied, the actuation of thin film will be displaced transversely [11].

The conventional piezoelectric micro-cantilever was integrated with a piezoelectric film layer sandwiched between two metal electrodes as a sensor or as an actuator. Figure 5a shows a bimorph cantilever that consists of two active layers, while Fig. 5b shows a unimorph cantilever that consists of one active layer and one inactive layer. A researched by Wei and Hong proved that bimorph PZT micro-cantilevers have the force sensitivity (high reaction force) higher than the force of the unimorph PZT micro-cantilevers [12]. This is due to the fact that the two active layers in a bimorph actuator have more energy efficiency compared to unimorph actuators [12] (Fig. 6).

Energy harvesting is a process where electrical energy is derived from external forces where the energy is collected and stored for devices [13]. In a research of piezoelectric energy harvesting (EH) power for MEMS, waste mechanical energy is recycled into electric energy by piezoelectric cantilevers [14]. The EH power

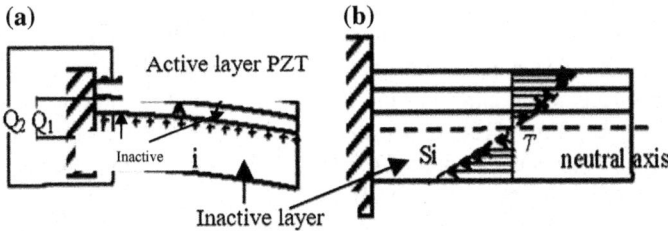

Fig. 5 The sensing performance measurement of the biomorph piezoelectric microcantilevers [12]

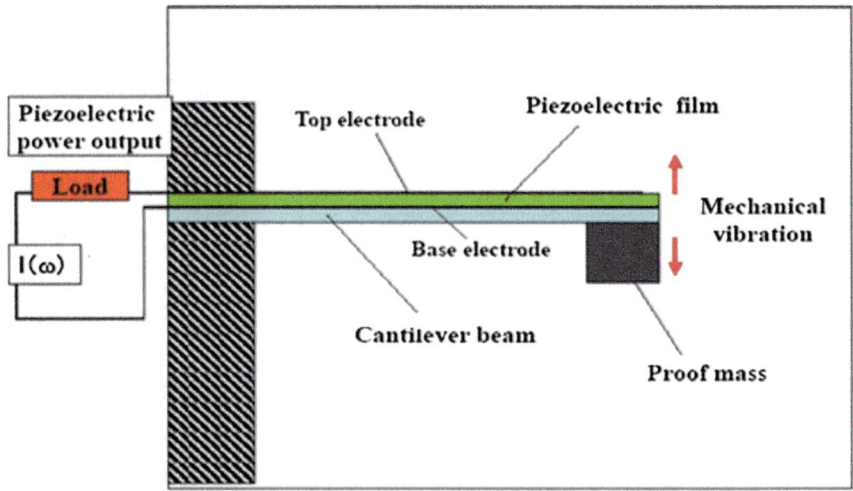

Fig. 6 Typical construction of piezoelectric thin-film energy harvesting MEMS [14]

MEMS consists of a mechanical vibration and the piezoelectric power generating device. Thin-film piezoelectric unimorph cantilevers are used for the fabrication of compact EH power MEMS. Figure 5 shows that the cantilevers compose a narrow substrate beam, piezoelectric thin films deposited on the cantilever substrate beam with top and base electrodes, and proof mass. The mechanical energy is converted into electrical energy by the transverse piezoelectric properties [14].

There are many applications for thin films piezoelectric actuators. They are used for gyrosensors and ink jet printer heads because of their piezoelectric properties, such as fast actuation and positioning accuracy [14]. Optical MEMS also applied the thin film piezoelectric concept to expand the research on optical networks such as tunable lasers, optical switches, and variable optical attenuators [15].

2.3 Piezoelectric Tube Actuator

Piezoelectric tubes are the most widely used positioning actuators in micro and nano scale positioning applications [16]. The accurate displacement control is the main reason that piezoelectric tube actuators are chosen for various applications [17]. The displacement control has the advantages of high resolution, fast response and high output force [18]. They also offer sub-nanometer resolution in all 3-translational degrees of freedom (DOF) together with a compact design which is compatible in various applications [19].

A piezoelectric tube actuator works whenever a differential voltage is applied to the external electrodes, it bends to the corresponding directions [18]. The typical setup of a piezoelectric tube actuator is shown in Fig. 7, the lower end of the tube is

Fig. 7 A piezoelectric tube scanner developing a lateral tip displacement 'd' as a response to an anti-symmetric driving voltage, V [19]

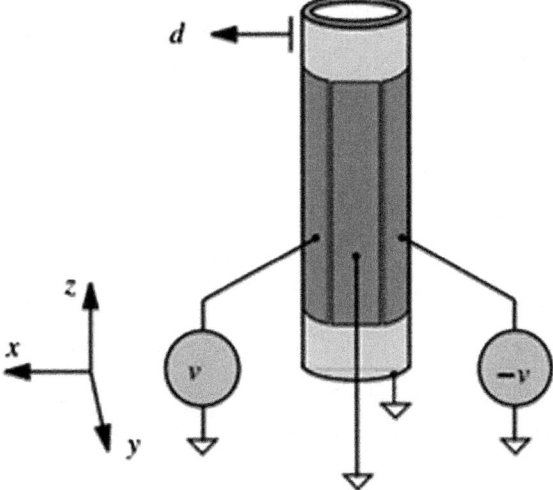

clamped, while the upper one, denoted as tip, are allowed to move in all three directions. The outer surface of the tube is separated into quartered electrodes. Both electrodes of a pair are driven with equal but opposite voltages [20].

The limitation of a piezoelectric tube actuator that reduces the positioning performance is the vibration that occurs on the devices structure [20]. Another limitation is the dynamics-coupling that causes errors [21], nonlinear hysteresis effects in relatively long-range scans and drift caused by creep effects in slow operation modes [22]. The design considerations are the deflection sensitivity and maximum deflection; both of these characteristics are optimized by a large ratio of length to diameter. An effect of designing tubes with large length/diameter ratios is the low mechanical resonance frequency [20]. There are many approaches by researchers to overcome these issues which includes calibration of the scanner, post imaging software image correction, and real-time correction via feedback control [22].

A typical application of this technology is the micro displacement of samples in the X and Y directions with a scanner in an atomic force microscope (AFM) as shown in Fig. 8. This application has contributed to the development of nanotechnology such as imaging atoms [23]. For instance, scanning tunneling microscopes (STMs) and atomic force microscopes (AFMs) are used extensively in diverse areas of science such as crystallography and cell biology [24].

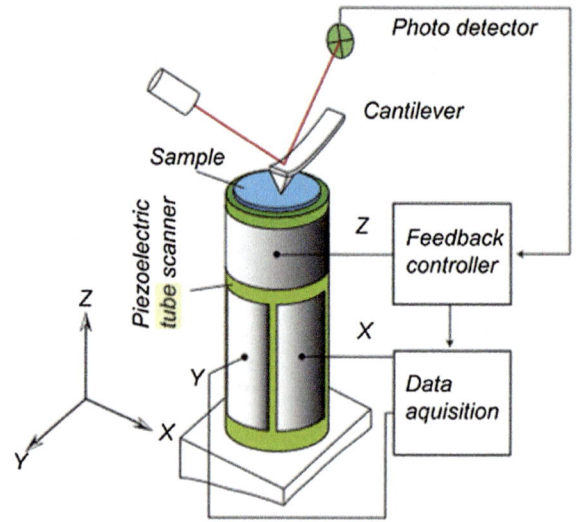

Fig. 8 Schematic of an atomic force microscope with a piezoelectric tube scanner for the sample positioning [47]

3 Limitations of Piezoelectric Actuator

Piezo actuators are typically operated in the linear range to avoid positioning errors caused by nonlinear hysteresis effects [25]. Another cause of loss in precision is due to the creep effects, which become more significant when positioning is required over extended periods of time. For instance, during the slow operation of an AFM. Loss of precision also occurs due to vibrations when the positioning bandwidth is increased relative to the first resonant frequency of the piezo actuator. This vibration-caused loss in positioning precision is typically less at low operating speeds [25].

3.1 Hysteresis

The disadvantage of a piezoelectric actuator is the hysteresis effect. It is a phenomenon that is nonlinear, which makes it difficult to control the displacement precisely [26]. Hysteresis effect is the displacement differences at the same voltage when the voltage is applied in ascending or descending manner [29]. Figure 9 shows the maximum displacement difference in hysteresis [27]. The higher the nonlinear hysteresis contributes to more complication in the control of piezoelectric actuators for high-precision applications. The maximum hysteretic error is typically about 15% in static positioning applications [28].

In piezoelectric actuators, hysteresis exists in both the electric field (voltage)-polarization relationship and the electric field (voltage)-strain relationship [29]. Furthermore, hysteresis in piezoelectric actuators occurs in both slow operating

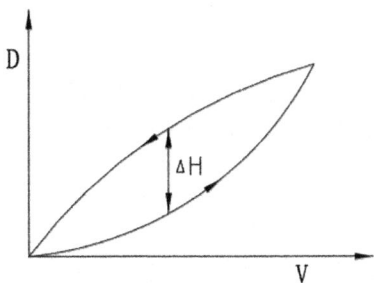

Fig. 9 Hysteresis of piezoelectric materials

conditions [29]. According to Adriaens, the relationship is linear among the actuating force and the resulting displacement of the piezoelectric actuator [3].

However, there are many researches to overcome hysteresis and improve the positioning accuracy. For instance, many researchers applied a mathematical model which describes the piezoelectric hysteresis and error compensation [30]. According to Clayton and Tien, the amount of hysteresis error tends to decrease with operating range. Therefore, a typical hysteresis reduction strategy is to operate the piezo positioner at a small displacement (e.g., nanometers as opposed to microns) [31].

Fleming and Moheimani suggested two techniques to overcome and reduce hysteresis. The first method is to impose a direct current (dc) charge amplifier, thus hysteresis is reduced by avoiding characteristic drift voltage. The second method is to introduce piezoelectric shunt damping [20]. Within the first technique as shown in Fig. 10, the voltage amplifier is replaced by a charge amplifier to reduce the hysteresis effect. Meanwhile, for the second technique, the connection of an electrical impedance to the terminals of one electrode is employed. Generally the piezoelectric shunt damping results in a damped electrical resonance, which is capable of significantly reducing the magnitude of hysteresis [20].

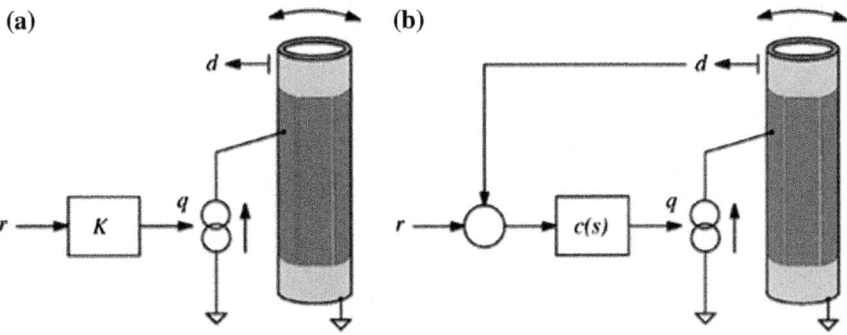

Fig. 10 Charge driven tube scanner. **a** Open-loop with signal pre compensation. **b** Closed-loop with displacement feedback [20]

3.2 Creep

Creep in a piezoelectric system can result in precision loss, especially when operating for a long period at an offset position. Creeping caused the decrease in the amplitude for piezoelectric actuator although the input voltage is maintained. Creep is the expression of the slow realignment of the crystal domains in a constant input voltage over time [30]. Creep causes instability to the system because of the inconsistence actuation [32]. To isolate and model the effects of creep, the positioning range is kept small to minimize hysteresis effects. Similarly, the outputs were measured over extended periods of time (in the five minutes interval) so that the effect of induced vibrations can be neglected [25].

In order to improve the actuation accuracy, the creep effects should be compensated. Similar to hysteresis, creep is related to the effect of the applied voltage on the polarization of the piezoelectric ceramics [30].

In order to remove the effects of the creep and to increase the accuracy of piezo actuators, closed-loop techniques (feedback) have been used. A research by R. Micky presents the block diagram of the creep model from Eq. (1) and its compensator for one degree of freedom piezo actuators. Figure 11 shows that the transfer function of the compensator is inverse multiplicative [33].

$$y(s) = K \cdot D(s) \cdot U(s) + C(s) \cdot U(s) \qquad (1)$$

3.3 Vibrations

Vibrations in piezoelectric actuators are related to mechanical properties of the actuator such as mass, stiffness and damping [27]. At relatively low scan rates, the vibrational effects are negligible and the main cause of image distortions are dully from creep and hysteresis effects [25]. When the scan rate is increased to 30 Hz, ripples appeared in the images obtained without inverse compensation. Vibrations

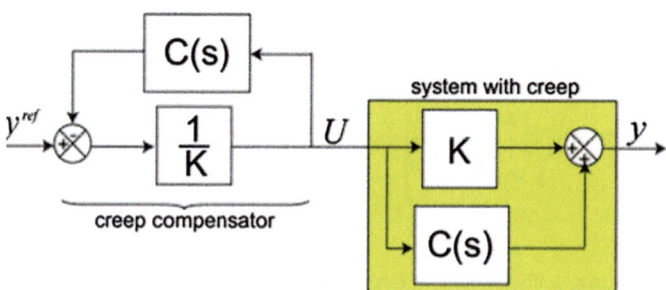

Fig. 11 Mono variable creep modeling and compensation [33]

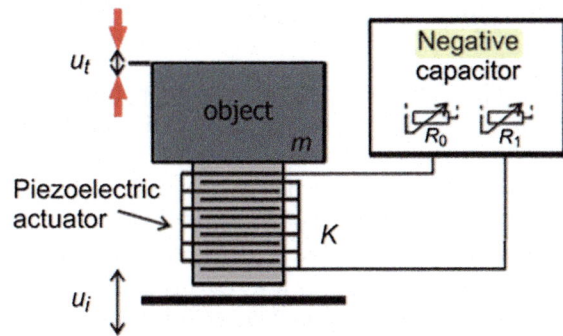

Fig. 12 Vibration isolation system, where the piezoelectric actuator is connected to the negative capacitor [34]

lead to severe distortions, the distortions or ripples caused by vibrations correspond to the first mode natural frequency of the scanner [25].

According to a research conducted by Kodejška, Václavík, the elasticity of an element can be tuned when the transmission of vibrations are reduced through a soft element. The same result can be obtained by connecting a negative capacitor to the piezoelectric actuator [36]. So, by merging these two facts, a new vibration control system is developed. Figure 12 shows the vibration isolation system where the frequency of the incident vibrations coincides with the capacitance matching frequency. The system is able to achieve the suppression of vibration with several harmonic components or with a white noise signal [34].

A research made by Tani and Qiu showed that a combination of active control and disturbance cancellation methods is used in the vibration control of a piezoelectric cylindrical shell. The hybrid control method was employed to control the forced vibration. The obtained results were compared with those merely the active control or the disturbance cancellation methods. The simulation and experiment results show that the hybrid control method is more effective than each individual control method which are active control or disturbance cancellation [35].

4 Control of Piezoelectric Actuator

In order to increase the precision of the piezoelectric actuators, control strategies and models have been studied mainly to characterize the hysteresis nonlinearities [36]. The current main control schemes can be divided into feedforward control, feedback control, and compound control [37]. This subtopic will discuss in detail about these types of controllers. For instance, fuzzy control, and sliding mode based robust control has been used to deal with hysteresis in piezo-actuators [37].

4.1 Open Loop Control

Figure 13 shows a typical open-loop control scheme for piezoelectric actuators, where the effect of hysteresis, creep and vibration are compensated by the inverse transfer function model of piezo to achieve better tracking accuracy [29]. In no load condition the open-loop control schemes have been shown to be highly effective in their applicable frequency ranges [25].

Ru and Wang suggested to use a least-mean-square (LMS) algorithm to identify the main hysteresis loop [38]. However, one major disadvantage of the open-loop control schemes is that their positioning performances are highly sensitive to model errors, external loads, and changes in the dynamics of the piezoelectric actuator [41].

In a research made by Zareinejad and Mohammad Razi [39] the inverse hysteresis model is updated adaptively to account for the effects. However, the influence of the effects on the piezoelectric actuator output, though largely suppressed, can still be clearly observed in the experimental results. Thereby, there are a lot of researches carried out to improve the method in future design.

4.2 Feedback Method

Feedback control is a typical type of closed-loop control as shown in Fig. 14. The hysteresis and creep nonlinearities of piezo actuator are open-loop characteristics, so their influences can be effectively reduced using the closed-loop method [40]. Feedback control schemes as shown in Fig. 14 lead to several of the effects, including model errors, external loads, and changes in the dynamics of the piezo actuator on the position control performances, hence they are widely used in actuators [41].

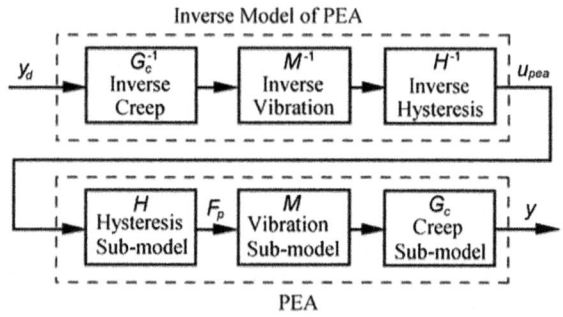

Fig. 13 Open-loop control scheme for a piezoelectric actuator (PEA) [29]

Fig. 14 A feedback control scheme for piezoelectric actuator [41]

Ge presented a proportional integral derivative (PID) feedback controller to control the nonlinearity of piezoelectric actuators [42]. The feedback controller consisted of a gain tuning neural work and a variable gain PID controller; the neural network is to trace the optimal gains of the PID controller. Besides, mentioned controller with nonlinear vector regression in feed-forward loop and hybrid controller are carried out. The related experimental results indicated that the proposed controller is effective in precision control method [43].

4.3 Adaptive Control and Intelligent Control

One of the effective ways to solve the inaccuracy issue due to hysteresis is to use model reference adaptive control [44]. An adaptive displacement tracking control using only displacement feedback is proposed for a piezo-positioning mechanism. In order to develop a robust model to represent the overall system dynamics of the controlled piezo-positioning mechanism, a specific function is proposed by Shieh and Lin [45]. Figure 15 shows a block diagram of the piezo positioning mechanism. Typically, the main type of intelligent control that have been widely used are fuzzy logic controller and neutral network controller [40]. These two controllers have a great potential in the identification and control of nonlinear especially in active vibration control. This is due to, the characteristics of fuzzy control proves that it can use subjective experience and intuition [46]. Furthermore, this controller has learning and adaptive abilities which are able to conduct a large number of operations in a short period [46].

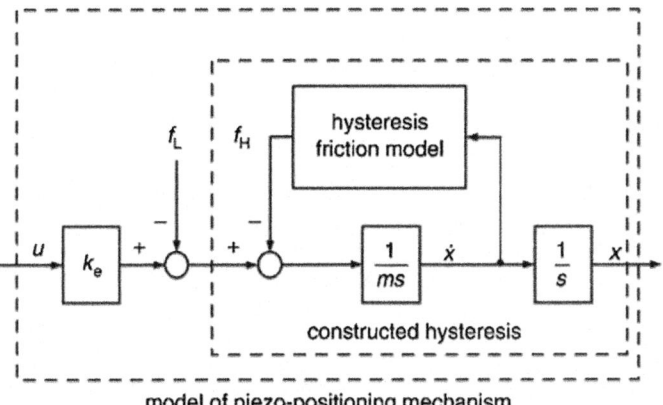

Fig. 15 Block diagram of the proposed model of the piezo positioning mechanism [45]

5 Conclusion

The paper provided a review of piezoelectric actuators. The main aim is to explain the recent technologies of piezoelectric actuators, their limitations and the way to overcome the instability such as hysteresis, vibration and creep. There are many researches that studied the disadvantages of piezoelectric and the problem solving methods. Furthermore, the application of piezoelectric differs for every design and configuration of the actuator depending of its displacement range, direction and size. This review also explained in detail the configuration types of piezo actuators. Finally, the control methods have been reviewed.

References

1. Uchino, K.: Introduction to piezoelectric actuators and transducers Kenji Uchino. International Center for Actuators and Transducers, Penn State University, no. 5, p. 40 (2003)
2. Matsushita, S., Kanno, I., Yokokawa, R., Kotera, H.: Metal-based piezoelectric MEMS scanner mirrors composed of PZT thin films on titanium substrates. In: 2011 16th International Solid-State Sensors, Actuators Microsystems Conference on TRANSDUCERS'11, pp. 574–577 (2011)
3. Adriaens, H.J.M.T.A., de Koning, W.L., Banning, R.: Modeling piezoelectric actuators. IEEE/ASME Trans Mechatron. **5**(4), 331–341 (2000)
4. Trolier-Mckinstry, S., Muralt, P.: Thin film piezoelectrics for MEMS. J. Electroceram. **12**(1–2), 7–17 (2004)
5. Eom, C.-B., Trolier-McKinstry, S.: Thin-film piezoelectric MEMS. MRS Bull. **37**(11), 1007–1017 (2012)
6. Abdalla, M., Frecker, M., Gürdal, Z., Johnson, T., Lindner, D.K.: Design of a piezoelectric actuator and compliant mechanism combination for maximum energy efficiency. Smart Mater. Struct. **14**(6), 1421–1430 (2005)
7. Yi, K.A., Veillette, R.J.: A charge controller for linear operation of a piezoelectric stack actuator. IEEE Trans. Control Syst. Technol. **13**(4), pp. 517–526 (2005)
8. Wang, W., Yang, Z.: A compact piezoelectric stack actuator and its simulation in vibration control. Tsinghua Sci. Technol. **14**, 43–48 (2009)
9. Li, P., Fu, J., Wang, Y., Xing, Z., Yu, M.: Dynamic model and parameters identification of piezoelectric stack actuators. In: 26th Chinese Control Decision Conference (2014 CCDC), no. 2, pp. 1918–1923, May 2014
10. Goldfarb, M., Celanovic, N.: Modeling piezoelectric stack actuators for control of micromanipulation. IEEE Control Syst. Mag. **17**(3), 69–79 (1997)
11. Kuwajima, H., Uchiyama, H., Ogawa, Y., Kita, H., Matsuoka, K.: Manufacturing process of piezoelectric thin-film dual-stage actuator and its reliability for HDD. IEEE Trans. Magn. **38**(5), 2156–2158 (2002)
12. Mengwei Liu, L.W., Cui, T.: Fabrication and Characterization of the Piezoelectric Microcantilever Integrated with PZT Thin Film Microforce Sensor and Actuator, pp. 2255–2258 (2007)
13. Knight, R.R., Frederick, A.A., Clark, W.W.: Fabrication and energy vibrating MEMS piezoelectric. **133**, 1–7 (2011)
14. Wasa, K., Matsushima, T., Adachi, H., Kanno, I., Kotera, H.: Thin-film piezoelectric materials for a better energy harvesting MEMS. J. Microelectromech. Syst. **21**(2), 451–457 (2012)

15. Koh, K.H., Leea, C., Kobayashi, T.: A 3-D MEMS VOA using translational attenuation mechanism based on piezoelectric PZT thin film actuators. Procedia Eng. **5**, 613–616 (2010)
16. Mohammadzaheri, M., Grainger, S., Bazghaleh, M., Yaghmaee, P.: Intelligent modeling of a piezoelectric tube actuator. In: 2012 International Symposium Innovations in Intelligent Systems and Application, pp. 1–6 (2012)
17. Mohammadzaheri, M., Tafreshi, R.: Evaluation of the Induced Voltage in Driven Electrodes of Piezoelectric Tube Actuators for Sensorless Nanopositioning, pp. 1–4 (2015)
18. Lu, H., Fang, Y., Ren, X., Zhang, X.: An improved direct inverse modeling approach for high-speed feedforward tracking control of a piezoelectric tube actuator, October, pp. 27–31 (2014)
19. Maess, J., Fleming, A.J., Allgower, F.: Simulation of piezoelectric tube actuators by reduced finite element models for controller design. In: 2007 American Control Conference, pp. 4221–4226 (2007)
20. Fleming, A.J., Moheimani, S.O.R.: Sensorless vibration suppression and scan compensation for piezoelectric tube nanopositioners. IEEE Trans. Control Syst. Technol. **14**(1), 33–44 (2006)
21. Tien, S., Zou, Q., Devasia, S.: Iterative control of dynamics-coupling-caused errors in piezoscanners during high-speed AFM operation. IEEE Trans. Control Syst. Technol. **13**(6), 921–931 (2005)
22. Tamer, N., Dahleh, M.: Feedback control of piezoelectric tube scanners. In: Proceedings of 33rd IEEE Conference on Decision Control, 1994, vol. 2, pp. 1826–1831, Dec 1994
23. Jong-kyu, P., Won-kyu, M.: Development of XY scanner with minimized coupling motions for high-speed atomic force microscope. J. Cent. South Univ. Technol., 697–703 (2011)
24. Bhikkaji, B., Ratnam, M., Moheimani, S.O.R.: PVPF control of piezoelectric tube scanners. Sens. Actuators A Phys. **135**(2), 700–712 (2007)
25. Croft, D., Shedd, G., Devasia, S.: Creep, hysteresis, and vibration compensation for piezoactuators: atomic force microscopy application. In: Proceedings of the American Control Conference on 2000, vol. 3, pp. 2123–2128, June 2000
26. Lizhi, P., Min, Z., Wendong, W.: Feedforward Controller Based-on Piezoelectric Actuator's Hysteresis Model and Its Performance Simulation of an XY Plane Motion Stage, pp. 1073–1077 (2009)
27. Zhan-hui, L., Yun-xin, W., Xi-shu, D.: Modeling and Compensating of Piezoelectric Actuator Hysteresis in Photolithography G (x) G (x) G (x), pp. 6–8 (2007)
28. Ang, W.T., Khosla, P.K., Riviere, C.N.: Feedforward controller with inverse rate-dependent model for piezoelectric actuators in trajectory-tracking applications. IEEE/ASME Trans. Mechatron. **12**(2), 134–142 (2007)
29. Peng, J., Chen, X.: A Survey of Modeling and Control of Piezoelectric Actuators, pp. 1–20, February 2013
30. Rul, C., Chen, L., Sun, L.: Tracking control method of piezoelectric actuator for compensating hysteresis and creep. In: 2007 2nd IEEE International Conference on Nano/Micro Engineering and Molecular Systems, pp. 186–190, 2007
31. Clayton, G., Tien, S., Devasia, S., Fleming, A.J., Moheimani, S.O.R.: Compensation in Piezoelectric Actuators By Integrating Charge, pp. 812–818 (2006)
32. Muhammad, B.B., Rahman, H., Ridzuan, M.: Modelling and Simulation of Piezoelectric Actuator for Vibration Control (2015)
33. Rakotondrabe, M.: Modeling and compensation of multivariable creep in multi-DOF piezoelectric actuators. In: Proceedings of the IEEE International Conference on Robotic Automation, pp. 4577–4581 (2012)
34. Kodejška, M., Václavík, J., Mokrý, P.: A System for the Vibration Suppression in the Broad Frequency Range Using a Single Piezoelectric Actuator Shunted by a Negative Capacitor. In: Proceedings of 2010 IEEE International Symposium Applications of Ferroelectrics ISAF 2010, Co-located with 10th European Conference on the Applications of Polar Dielectrics ECAPD 2010, p. 4, 2004

35. Qiu, J.: Vibration control of a cylindrical shell using distributed piezoelectric sensors and actuators. J. Intell. Mater. Syst. Struct., 980 (1995)
36. Gao, X., Ren, X., Zhu, C., Zhang, C.: Discrete Composite Control for Piezoelectric Actuator Systems, pp. 4489–4493 (2015)
37. Matthews, C., Dickinson, P., Shenton, A. T.: Fuzzy control of piezoelectric actuator with inverse compensation. System **83**, 27–31 (2014)
38. Ru, C., Wang, K., Ye, X., Yang, Y.: An open-loop operation method of piezoelectric actuators based on a new hysteresis model. Proc. World Congr. Intell. Control Autom. **2**, 8366–8369 (2006)
39. Zareinejad, M., Razi, K., Seifabadi, R.: Hysteresis compensation of piezoelectric actuators under dynamic load condition. In: 2007 IEEE/RSJ International Conference on Intelligent Robotic System, pp. 1166–1171, 2007
40. Chi, Z., Xu, Q.: Recent advances in the control of piezoelectric actuators. Int. J. Adv. Robot. Syst., 1 (2014)
41. Devasia, S., Eleftheriou, E., Moheimani, S.O.R.: A survey of control issues in nanopositioning. IEEE Trans. Control Syst. Technol. SPEC. ISS. **15**(5), 802–823 (2007)
42. Ge, P., Jouaneh, M.: Tracking control of a piezoceramic actuator. IEEE Trans. Control Syst. Technol. **4**(3), 209–216 (1996)
43. Ji, H.-W., Yang, S.-X., Wu, Z.-T., Yan, G.-B.: Precision control of piezoelectric actuator using support vector regression nonlinear model and neural networks. In: Proceedings of the 2005 International Conference on Machine Learning Cybernetics, 2005, vol. 2, pp. 1186–1191, Aug 2005
44. Nealis, J.: An adaptive control method for magnetoristrictive transducers with hysteresis. In: Conference on Decision Control, pp. 4260–4265, Dec 2001
45. Charef, A.: Adaptive tracking control solely using displacement feedback for a piezo-positioning mechanism H.-J. Control Theo. Appl. IEE Proc. **153**(6), 714–720 (2006)
46. Baras, J.S.: A robust control framework for smart actuators. In: Proceedings of 2003 American Control Conference, 2003, vol. 6, pp. 4645–4650, 2003
47. Kuiper, S., Schitter, G.: Active damping of a piezoelectric tube scanner using self-sensing piezo actuation. Mechatronics **20**(6), 656–665 (2010)

Review of the Control System for an Unmanned Underwater Remotely Operated Vehicle

Ahmad Makarimi Abdullah, Nursyahida Izzati Zakaria,
Khairul Arief Abu Jalil, Norhafizah Othman,
Wardiah Mohd Dahalan, Hisham Hamid, Muhamad Fadli Ghani,
Ashman Yussof Iqmal bin Mohd Haidir
and Mohd Iqram Mohd Kamro

Abstract This paper presents a review of control system algorithms for unmanned underwater remotely operated vehicles (ROV), focusing on the proportional integral derivative (PID) controller and fuzzy logic lontroller (FLC). This review covered the field area of the unmanned underwater remotely operated vehicles including the remotely operated vehicles and autonomous underwater vehicles (AUV). The review paper covers the recently published documents for a control system of an unmanned underwater vehicle (UUV). It also describes the control system algorithm from different field areas which has inspired the authors to develop the same control algorithm for the unmanned underwater vehicle area. A comparison of the control system designs from the existing development of the ROV was also included. There has been a lot of developments of ROV which depend on their purposes or applications either for the exploration, low-cost vehicle, shallow water or educational purposes. The review of the control system for ROV will provide insights for the reader to design new control systems for ROVs.

A. M. Abdullah (✉) · N. I. Zakaria · K. A. A. Jalil · N. Othman
W. M. Dahalan · M. F. Ghani
Malaysian Institute of Marine Engineering Technology (MIMET), Universiti Kuala Lumpur,
Jalan Pantai Remis, 32200 Lumut, Perak, Malaysia
e-mail: makarimi@unikl.edu.my

N. I. Zakaria
e-mail: syahida_izzati92@gmail.com

K. A. A. Jalil
e-mail: rockstarfellas@gmail.com

N. Othman
e-mail: norhafizahothman6@gmail.com

W. M. Dahalan
e-mail: wardiah@unikl.edu.my

M. F. Ghani
e-mail: muhamadfadli@unikl.edu.my

© Springer International Publishing AG 2018
A. Öchsner (ed.), *Engineering Applications for New Materials and Technologies*,
Advanced Structured Materials 85, https://doi.org/10.1007/978-3-319-72697-7_49

Keywords Remotely operated vehicle · Autonomous underwater vehicle
Graphical user interface · Unmanned underwater vehicle · Manned underwater
vehicle

1 Introduction

In the last decade, there has been an increased interested in researching and developing technologies for underwater vehicles. In marine industry especially, underwater vehicles are widely used for underwater research, oil and gas exploration and construction. Basically, the underwater vehicle can be divided into two types which are unmanned underwater vehicles (UUVs) and manned underwater vehicles (MUVs). The different between them is that UUV will operate in water without an occupant while the MUV must be operated under human occupant. The UUV that can attract attention in the new era is called as remotely operated vehicles (ROVs) [1]. Remotely operated underwater vehicles are driven by a remote control using a tether cable. The cable is connected to the mother ship to perform several functions by using the tools equipment on the ROV after the ROV reached to underwater area [2]. Furthermore, the ROV is highly maneuverable, unoccupied and is operated by a person at the surface area. So, this review focuses on the control system of the ROV because of its high ability to maneuver the vehicle.

The ROV consists of the mechanical structure that contains thrusters, camera, sensor and additional equipment such as manipulator arm, suction arm and electronic modules to control the ROV. In the ROV system the operator and the control surface platform is connected by using an umbilical cable. The launch system, cable dynamics and also the associated power supply is controlled by the system of the ROV. The ROVs design can be in different size either from smallest vehicles to

---

N. I. Zakaria · N. Othman · A. Y. I. b.M. Haidir · M. I. M. Kamro
British Malaysian Institute (BMI), Universiti Kuala Lumpur, Bt. 8,
Jalan Sungai Pusu, 53100 Gombak, Selangor Darul Ehsan, Malaysia
e-mail: Ash_man93@hotmail.com

M. I. M. Kamro
e-mail: iqramkamro@gmail.com

K. A. A. Jalil · H. Hamid · A. Y. I. b.M. Haidir
France Institute (MFI), Universiti Kuala Lumpur Malaysia, Section 14 Jalan Teras Jernang
Seksyen 14, 43650 Bandar Baru Bangi, Selangor, Malaysia
e-mail: hishamhamid@unikl.edu.my

M. I. M. Kamro
Institute of Product Design and Manufacturing (IPROM), Universiti Kuala Lumpur, No. 9,
Jalan Perdana 7/3 Taman Shamelin Perkasa, 56100 Kuala Lumpur, Selangor, Malaysia

bigger vehicles. The ROV is equipped with a complex work system that includes several dexterous manipulators, video cameras, TV's, sensors and other equipment. Thus, the objective of this paper to conduct some reviews to get an idea to design and control the system of the ROV in order to provide effective control schemes. In addition, to achieve the desired velocity and position of the vehicle, the required relevant signal is needed to perform effective control schemes.

After investigations were carried out in the literature, reviews, and case studies, the major problem considered is in the complex work system of the ROV which contains several manipulators such as robotic and suction arms, TV's, camera, sensors and other equipment. Although a ROV is a powerful tool designed with different purposes, it requires more than one control system to operate.

2 Control System Algorithm of Unmanned Underwater Vehicle (UUV)

The control of an underwater vehicle is not easy and simple because there is the highly nonlinear hydrodynamic effect resulting from the interaction with the environment that cannot be quantified. The problem of the complex control system is due to the lack of accurate models of the UUV hydrodynamics and uncertainty parameters. In addition, the appearance of environmental disturbance such as wave, current and wind also affects the problem in the control system. The most common problem of ROV in the design of the control system is a multiple of nonlinear and parameter uncertainties modelling [3].

According to the Rúa and Vásquez [4], the ROV is driven through a joystick that sends the power commands to a main-board installed in the vehicle. After that, this board will translate the command into input signals for each motor. Figure 1 shows the current open-loop control system that was implemented in the ROV. In order to tune the gains of the PID controller, the heuristic methods were used. The highest proportional gains act rapidly to correct the changes in the references. Due to the problem of the vehicle's dynamic and the interaction with the fluid, a high derivative action is used to present damping to the motion of the vehicle when it reaches to the set point. The small integral action can correct the steady state error of the system. In order to improve the PID control, the gravity compensation and vehicle's kinematics are used. Figure 2 shows the implemented closed loop control system in the ROV to improve the PID control. In order to calculate the output, the controller needs the error and position estimation.

According to the Akkizidis et al. [5], the GARBI underwater robot is used to apply the Fuzzy-like PD controller. The controller is for course-keeping and course-changing for the control performances of both depth and steering control. The Fuzzy-like PD controller is an architecture for this controller design. The structure is designed by combining a fuzzy logic controller and conventional PD controller. Next, by optimising the SF, the Fuzzy-like PD dynamic properties will

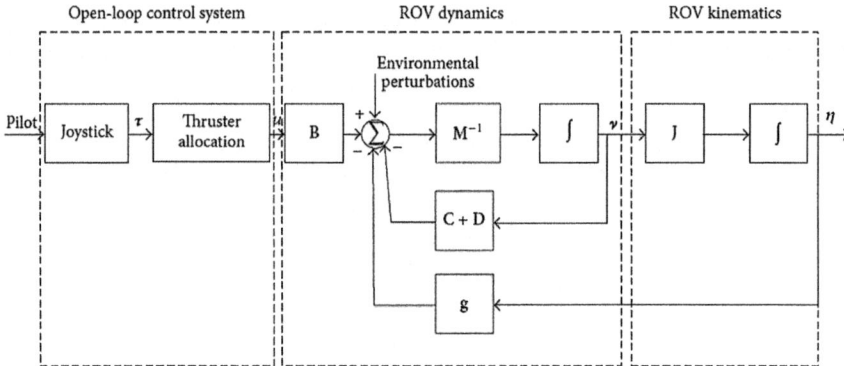

Fig. 1 Open-loop control system in the ROV

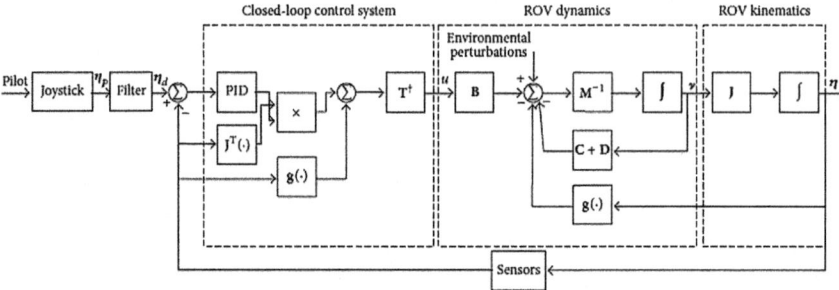

Fig. 2 Closed-loop control system in the ROV

be improved. According to the global effect on the dynamics closed loops system, the gains of input and output of SF is defined. By using the regular method, a Fuzzy-like PD controller has been performed satisfactory since both course-keeping and course-changing performances in minimal error with the desired response.

According to the Henriksen et al. [6], human in the loop (HITL) control refers to the situation where a human is present in the control loop and fulfils one or several control functions. The control loop consists of a controller and a process at which the controller receives the feedback from the process to control the actuator into the desired set point. In the ROV, the human typically takes the place as a controller and uses the video feedback to control the thrusters on the vehicle. For the traditional ROV control system, to appropriate the current position of the ROV, the operator uses a video feedback. Figure 3 shows the human in the loop in traditional ROV control. While, by using an automated control loop, the processing and receiving of the sensor feedback is more faster and precise rather than for a human. Next, due the external disturbance on the ROV, the continuous control actions are needed to keep ROV at the desired position. The proposed position controller enables the operator to change the position and orientation using a joystick.

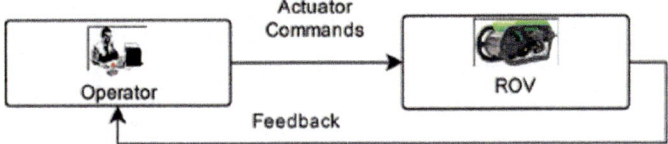

Fig. 3 Human in the loop in traditional ROV control

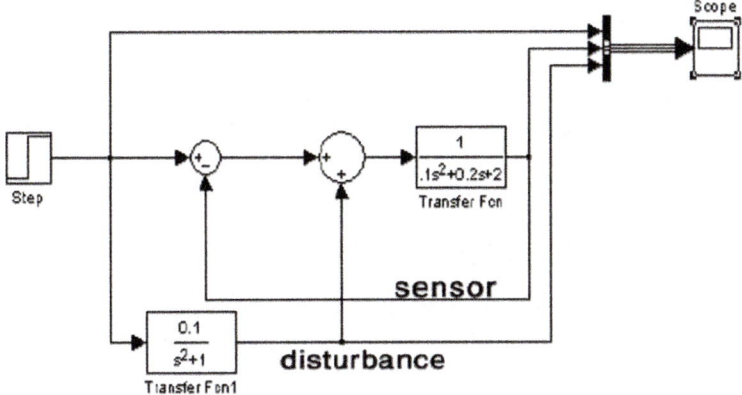

Fig. 4 Closed loop-feedback control system

The position controller applied a nonlinear PID-controller with a feed forward reference. The filter output is a smooth position reference as well as a reference velocity used as a feed-forward to the position controller.

According to Tehrani et al. [7], for the depth regulation of the ROV the authors used two different control schemes and the performances between the two methods were compared. The first method is a unit feedback control system that is developed to keep the amplitude of the overshoot in the limited response. Figure 4 shows the closed loop-feedback control system. Next, the second method of control scheme is a PID controller that shows a more satisfactory performance which is able to reduce the steady state error and the overshoot. In addition, this controller was able to give a robust performance and simple structure in a wide range of operating conditions. Figure 5 shows the PID control system.

According to the Le et al. [8], the hybrid control system is developed by combining the PID and model-based control. The model-based control will provide fast response with reference to signals. While, the PID control is robust with uncertainties of the control and provides small disturbance. So, the combination between these two control systems will provide a good control algorithm due the small steady state error and fast response.

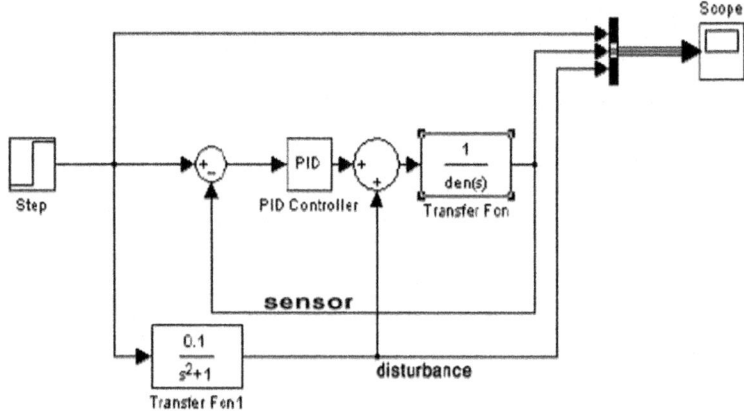

Fig. 5 Control system of PID

3 Control System Algorithm of Proportional Integral Derivative (Pid) Controller and Fuzzy Logic Controller (FLC)

A summary of key papers in control system algorithms focusing on PID and FLC is listed in Table 1.

4 Comparison of Designs of Unmanned Underwater Vehicles (UUV)

In the development of the ROVs, a lot of ROVs have been designed for monitoring, observation, or working purpose either in small or big scale of the ROV. In this review, ten designs have been selected as a comparison to the one designed here to compare the components that need to be controlled by the controller. Every design will provide different control systems depending on their purpose. For the observation purpose, basically the weight of the ROV must be not too heavy and provide a simple work system. While, for the working class ROV, normally the weight is heavy and has a complex work control system. Table 2 shows the comparison of the designs of other existing ROVs.

Table 1 Other relevant works field in control system algorithm using PID and FLC

Author (year)	Title	Important result/findings
Cervantes and Castillo [9]	Type-2 fuzzy logic aggregation of multiple fuzzy controllers for airplane flight control	The author explained the method for combining the multiple controls in the individual controller to achieve the complex plants of control. When there has a complex problem, this kind of method can be used to achieve efficient and accurate total control of the plant. In the control plant, to increase noise in the process, the joystick is included and disturbance is introduced. In order to develop the method, the 2-fuzzy system is implemented and simulation a result is done to shows the desired method is useful for complex problem. It was shown the control error is reduced and improved the overall behaviour of the plants
Shakya et al. [10]	Design and simulation of PD, PID and fuzzy logic controller for industrial application	The authors study the performance between PID controller and fuzzy logic controller for flowing fluids. The performance analysis has been done by using the MATLAB and Simulink. The results prove the fuzzy logic controller has a fast response and small overshoot compared to PID controller. While, the PID controller can attribute the robust performance and simple functionality
Lee and Kim [11]	Digital controller design to control the direct current motor system	The authors designed a digital proportional-integral-derivative (PID) controller in a PLD's device environment to control the speed of the DC motor system using feedback technique. The uses of feedback in a closed loop system will increase the accuracy and reduced the effect of external disturbance and variations. In order to increase the system performance, the sum of three term actions is done which is increase gain margin, stabilize unstable system and rapid control response by proportional control. While, minimize steady state error and derivative control is done by the integral control in order to increase stability system, reducing the overshoot and improve the behaviour of control system during transient

(continued)

Table 1 (continued)

Author (year)	Title	Important result/findings
Yadav and Patil [12]	Fuzzy PID controllers using Matlab GUI based for real time DC motor speed control	The authors designed the control system of DC motor based on Fuzzy logic controller. The parameter of fuzzy controller is improved by using MATLAB-GUI based on FLC and IFLC algorithm. The MATLAB-GUI will reduce the time tuning the controller in other software with less effort. There have three components to the structure of FLC which are fuzzifier that used to measure the input or definition of fuzzy set. Next, the rule base or fuzzy control will provide the necessary decision making logic in the system. Others, the defuzzifier with combine the actions need to be decided and produced a single non-fuzzy output. The uses of fuzzy logic controller are to provide the solutions of ill defined and too complex to model in the control system. Compared with PID controller the IFLC will improve the performance of motor in term of minimizing the overshoot and settling time
Upalanchiwar and Sakhare [13]	Design and implementation of the fuzzy PID controller using MATLAB/SIMULINK model	The authors proposed the PID controller with fuzzy technology to control the DC motor. In order to control the complex system and unclear model system, the Fuzzy PID controller is a better method to used due the tuning parameter of conventional PID controller is difficult, poor robustness and difficult to accomplish optimal state under field condition. The advantages of Fuzzy PID controller is can provide simple and effective control system, good dynamic response, play fuzzy control robustness
Hanafi et al. [14]	Simple GUI wireless controller of Quadcopter	The authors designed the GUI by using Visual Basic (VB) 2008 Express as a communication interface between quadcopter system and Proportional, Integral and Derivative (PID) controller. In order to control Quadcopter simulation model, the fuzzy logic control is implemented

(continued)

Table 1 (continued)

Author (year)	Title	Important result/findings
		and designed. The PID controller is designed and then embedded in the Arduino UNO in order to control altitude motion of Quadcopter. The PID control is used to sustain the distance of Quadcopter altitude motion based on input from ultrasonic sensors
Srinivas [15]	Comparative analysis of conventional PID Controller and fuzzy controller with various defuzzification methods in a three tank level control system	The authors make an analysis the efficiency for a control system of three tank level using a conventional PID controller over fuzzy logic. The methods was used is fuzzification & defuzzification. The result shows the FLC gives the best response in centroid defuzzification method and triangular membership function due the PID controller cannot give corrective action and only initiate the control action after an error has developed. The PID controller produces a slow response with peak overshoot for unit step input. So, the FLC will use to remove the overshoot and reduce the rise-time and settling time
Immanuel et al. [16]	Design and development of real time MATLAB-GUI based fuzzy logic controllers for DC motor speed control system	The authors designed the MATLAB-GUI program to act as an interface tool between user and computer and for DC motor speed control. The GUI is for tuning of controllers and effectively running the control system. The MATLAB-GUI makes the system user friendly to study the performance of various real time control systems with less effort

4.1 Design 1—Design and Development of a Remotely Operated Underwater Vehicle (Mini ROV) [17]

The mini ROV is developed in small size to use for underwater applications. This type of vehicle is equipped with powerful underwater probe sensors in order to collect underwater data, a DC brushless motor for controlling the direction as well as real time camera monitoring. The data of the ROV are sent via an underwater cable to the computer base station on the coast. It is also equipped with a USB camera and IMU-GY85. The camera is used to capture images and record video

Table 2 Other relevant design works for unmanned underwater vehicles (UUV)

Parameter	Design 1-2016 (Mini ROV)	Design 2-2015 (Low cost ROV)	Design 3-2015 (Visor3 ROV)	Design 4-2014 (BabyROV)	Design 5-2013 (Kaxan ROV)	Design 6-2013 (RG-III)	Design 7-2012 (LAUV)	Design 8-2011 (KCROV)	Design 9-2011 (Ariana-I ROV)	Design 10-2010 (DENA ROV)
Controller	GUI joystick	Joystick	Joystick GUI	Joystick	Joystick GUI	Joystick	Joystick GUI	Joystick	GUI	2 Joystick GUI
Camera	1 USB camera	1 camera	1 camera	1 camera	1 camera	1 camera	1 digital camera	1 camera	1 video camera	4 camera
Light	–	–	–	1 light	2 light	–	2 high power LEDs	1 light	–	1 light
Sensor	– Sonar – Temperature – IMU – Light	– Sonar – Pressure – Electrical current – ArduIMU V3	– IMU – Depth – Temperature – Humidity – Flooding	– Pressure – Temperature	– Depth – Gyros – Inclinometer – Compass	–	–	– DVL – IMU – GPS – depth sensor	– Pressure transmitter – 3-axis accelerometer – 2-axis inclinometer – Water leakage – Orientation and altitude	– Arm sensor

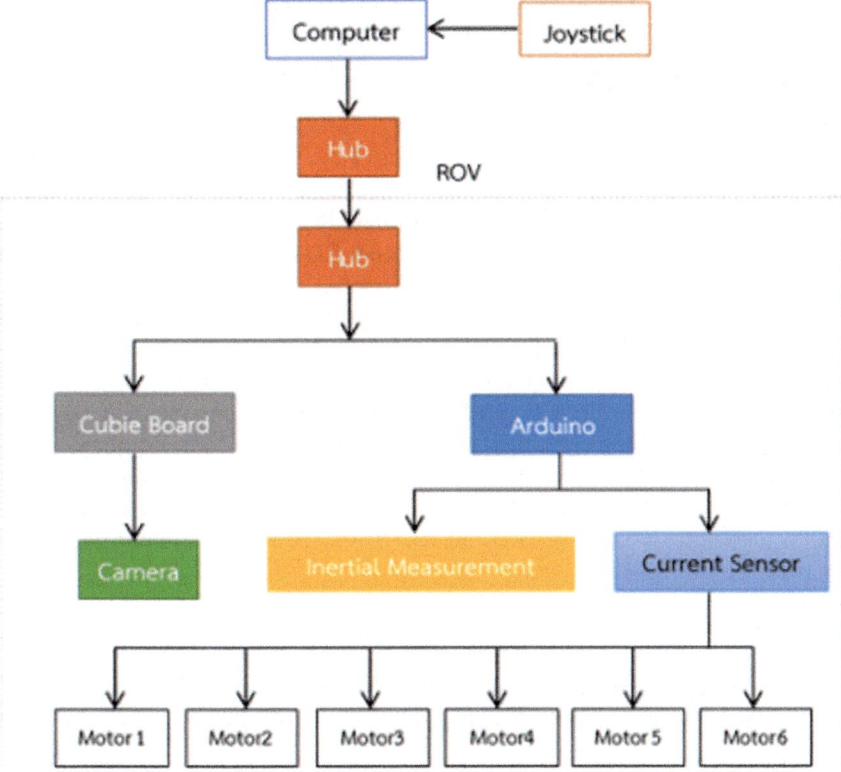

Fig. 6 Control diagram for the mini ROV

streaming. While the IMU-GY85 is the combination of accelerometers, gyroscope and compass in one module to control roll, pitch and yaw angle. Figure 6 shows the control diagram for the ROV. It started from the joystick which will connect to the computer using a processing program. After that, the processing program reads the command from the joystick then sends it to the ROV via internet hub by IP address. Next, the Arduino microcontroller will receive the data to control six motors by following the command from the user. The data from the IMU sensor is read for automatic closed loop drive. The graphical user interface (GUI) is designed using the processing program. It is an open source program and can be freely downloaded. The processing program reads the joystick and then sends the command from the user to the microcontroller. The microcontroller sends the sensor data which is the rotation angle, speed, video and compass to the GUI to monitor the behaviour of the robot.

4.2 Design 2—Development and Control of a Low-Cost, Three-Thruster, Remotely Operated Underwater Vehicle (Low Cost ROV) [8]

The low cost ROV is designed which consists of three thrusters and an open source hardware. The hardware parts of the ROV include the actuators, sensor and control structure. The control structure of three-thrusters ROV contain of three main parts which are the control station (on-board system), ROV controller (offshore system) and joystick controller. Figure 7 shows the control structure of the three-thrusters ROV. In the ROV controller, the microcontroller Arduino Mega 2560 is used to connect with the peripheral sensors. All the information of the sensor data is sent to the control station using serial communication RS-485 at the baud rate of 115,200 bps. The joystick controller generated the control signals in the station computer to send it back to the microcontroller via transmission cable. The control signals are received by the offshore microcontroller to execute them. The advantages of this control structure is that it flexible because it is easier to upgrade and modify the control algorithm in the station computer. Next, one can re-program the microcontroller in the ROV.

4.3 Design 3—Design of an Open Source-Based Control Platform for an Underwater Remotely Operated Vehicle (Visor3 ROV) [18]

The Visor3 ROV is designed to acquire visual information for the maintenance of ships hulls and also the underwater structure in oceanography for research tasks. The new original hardware architecture of Visor3 is divided in two parts which is a surface control station and the ROV vehicle. The surface control station is

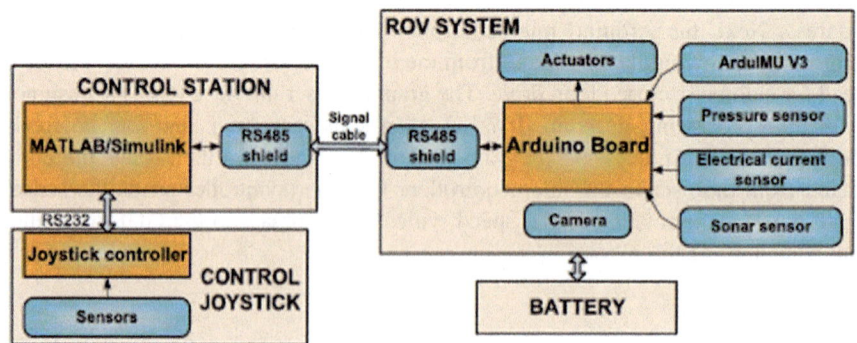

Fig. 7 Control structure of three-thrusters ROV

comprised of an industrial joystick as a command input and computer used as a Human-Machine-Interface (HMI) while for ROV vehicles it comprises an IP camera, an embedded processor, thrusters driver, DC power units, fiber optic communications and sensors. Figure 8 shows the control system architecture of the Visor3 ROV. For the development of the modular software framework, programming languages such as C++, JavaScript and LabVIEW have proved to be adequate tools for robotic devices. In addition, the software development is using Python, a high level programming language that allows one to implement control algorithms that require complicated matrix computations.

4.4 Design 4—Design of a New Low Cost ROV Vehicle (BabyROV) [19]

The BabyROV is designed for a new low cost ROV. The ROV is controlled by a joystick controller through a network cable and able to submerge up to 20 m into the water to perform several tasks. The control system of the BabyROV is depending on the communication between the underwater and land control circuits. The communication between the underwater and land control used the UART protocol. The land control circuits used SKPS and PS2 controller. Figure 9 shows the flowcharts of control system for the BabyROV. For the software part, the program is written for the ROV PIC microcontroller with the C language through Microchip MPLAB IDE v8.10 software.

4.5 Design 5—Modelling, Design and Robust Control of a Remotely Operated Underwater Vehicle (Kaxan ROV) [20]

The Kaxan ROV was designed for shallow water areas. The main objective of this Kaxan ROV is to develop a basic visual inspection of oil platforms and hydroelectric dams. The electronic architecture of the ROV consists of two computers which are at the surface and electronic container. The image taken by the underwater camera or sensors can be monitored by the user using a surface PC. A joystick is used by the user to control the ROV. The electronic architecture was monitored and controlled by the GUI that was developed using LabVIEW from NI. The GUI is able to display the values of the sensors, image of the camera, container humidity level, system for voltage and current and power electronic temperature. Figure 10 shows the electronic control system architecture for the Kaxan ROV.

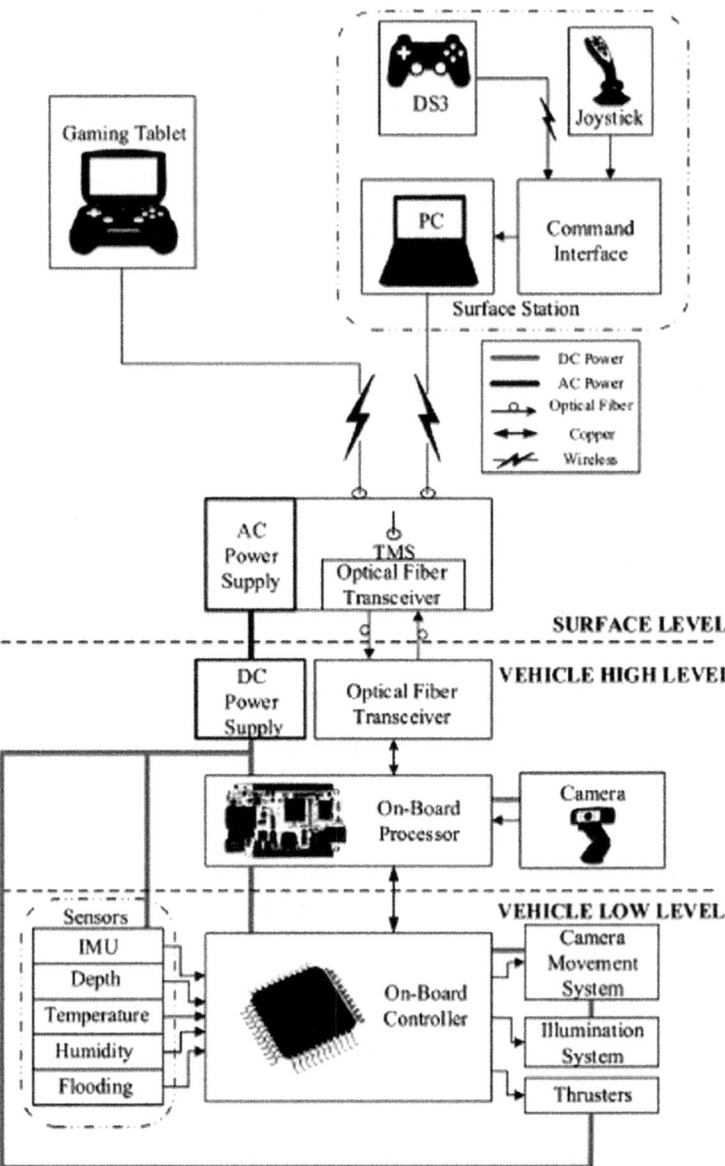

Fig. 8 Control system architecture of Visor3 ROV

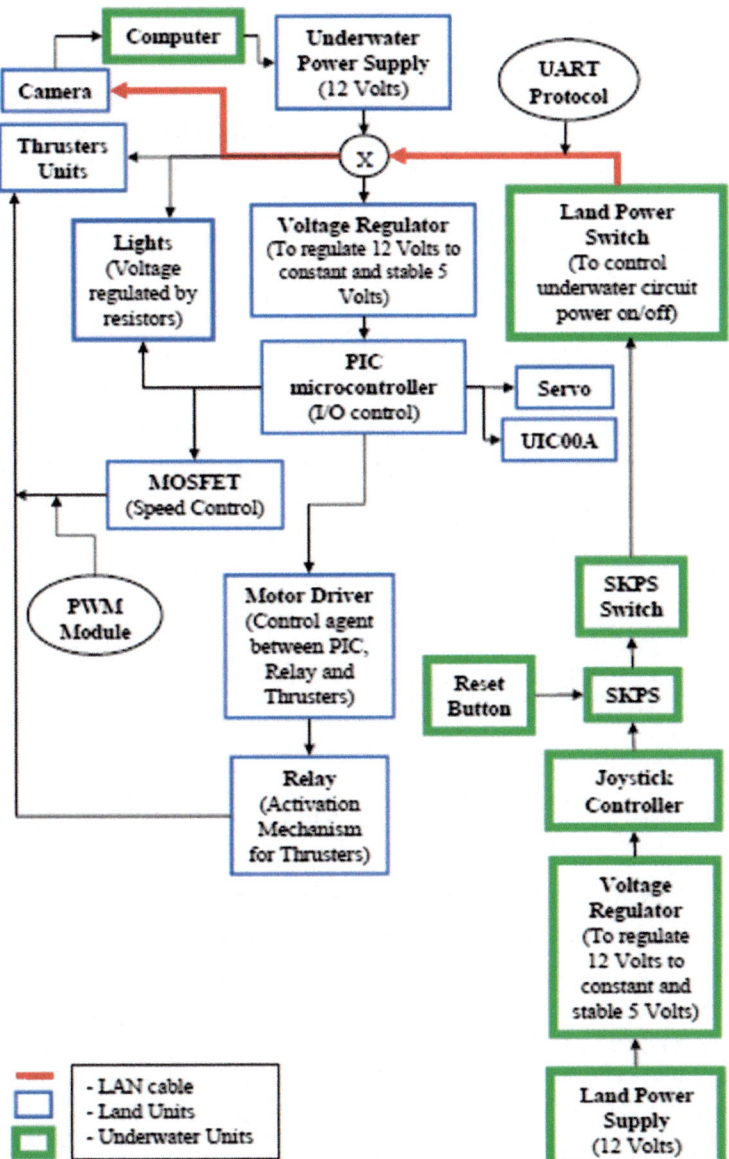

Fig. 9 Flowcharts of control system for baby ROV

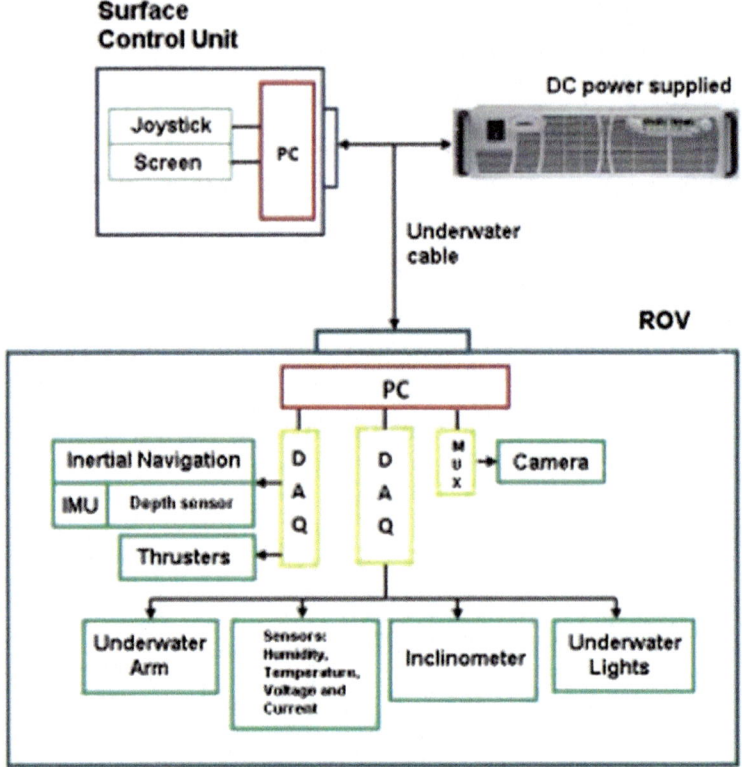

Fig. 10 Electronic control system architecture for Kaxan ROV

4.6 Design 6—Hybrid Robot Crawler/Flyer for Use in Underwater Archaeology (RG-III) [21]

The RG-III is designed to be hybrid vehicle to operate in depths down over 100 m. The RG-III is a remotely operated crawler that allows direct contact with the seafloor and at the same time it is able to "fly" when necessary. The control system of this hybrid vehicle is split into two categories which are the analog and digital part. The digital microcontroller is used as the primary control and the analog microcontroller is used for testing and backup the control system. On the surface, all controller inputs are gathered by a microcontroller via USB and then transmitted to a microcontroller on board the crawler. This kind of method will arrange the communication between surface controls and the RG-III which is handled between two microcontrollers. For the surface controller, the Arduino Mega 256 is used while for the underwater controller the Arduino UNO is used. The RS-485 protocol is used to communicate between the two Arduino microcontrollers. For the software parts, it was written in the native Arduino integrated design environment (IDE) for both surface and ROV controller. The signed byte 9600 Baud rate is transmitted to the ROV.

4.7 Design 7—LAUV: The Man—Portable Autonomous Underwater Vehicle [22]

LAUV was designed to be a lightweight vehicle that can be easily operated, launched and can recovered with minimal operation. This vehicle is developed to be a cost effective vehicle for oceanographic, environmental and security surveys. LAUV was controlled by the Control Command Unit (CCU) which was developed by the Universidade of Porto. It will support networked operations of inter-operated unmanned vehicles in a mixed initiative environment. A standard laptop or desktop computer is used to develop the Graphical User Interface (GUI) and to be operating system independently, as shown in Fig. 11. Next, it will provide the necessary support for different stages of a mission life-cycle which is planning, simulation, execution, data analysis and monitoring. The GUI console layout can be edited by removing, adding or placing individual console widgets. In addition, by using the mission review and analysis interface, all the collected data can be analysed once the mission is executed. In addition, to avoid possible obstacles during launching and recovery area, the operator may control the vehicle directly using a simple joystick. It was also equipped with a digital camera and two high power LEDs. The digital camera will provide the ability to acquire high quality images even in scenarios with poor optical conditions. While, two high power LEDs are used to improve the lighting condition of the digital video camera acquisition.

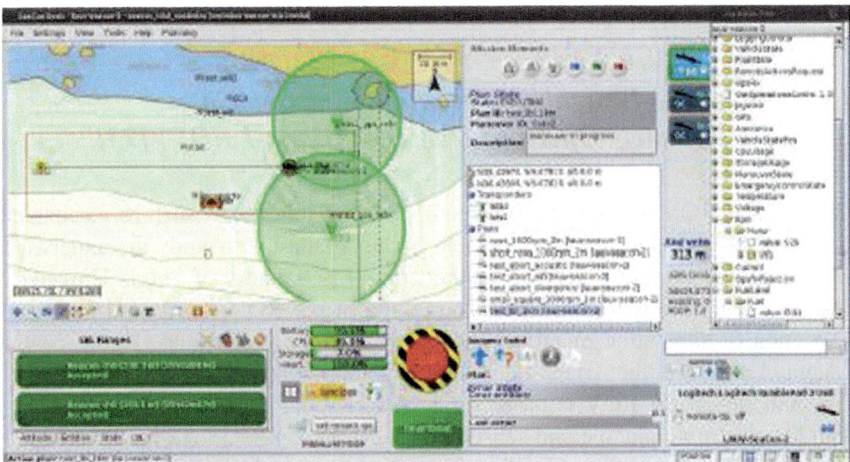

Fig. 11 GUI control command unit (CCU)

4.8 Design 8—Design and Control of a Convertible ROV (KCROV) [2]

The KCROV is designed for various purposes either to be used like the ROV or the AUV. The open architecture is designed to be function like the ROV. The ROV has a manipulator and is remotely operated by the power supply via a tether cable. While to be operated like an AUV, it needs to remove the manipulator, tether cable and ROV frames. In the control systems for the ROV and the AUV, the KCROV has developed a different architecture but the systems still use the same processors. Next, in both modes ROV and AUV the main microprocessor which is TMS320c28335 is connected to the data processing ATxmega128A1. In addition, the sensors are also connected to the microprocessor via a simple RS-435 communication link. In both modes ROV and AUV, similar sensors are used such as IMU, DVL, USBL and depth sensors. There are two different modes in the control algorithm for the KCROV which are for the AUV vehicle navigation is controlled by using an autonomous control algorithm and for the ROV vehicle motion is controlled by the operator orders. The architecture of the ROV contains a control system, power system and RS485 communication. The control system of the ROV is to control the thrusters by using the signal transfer from the operator via an optical fiber bus. While, to approximate the vehicle's current position, the sensor system which is USBL, IMU, DVL and depth sensors is installed. In the AUV mode, when the AUV rises to the surface, the wireless communication system is required for the surface communication. The navigation sensor data is transmitted in communication parts into the control system of the ROV. For the RS485 communication, the data is converted to the TTL level between the control board and thrusters. Next, the sensor data is used to the autonomous navigation algorithm of the AUV by using state estimation and sensor fusion filtering. Figure 12 shows the overall control system for the KCROV.

4.9 Design 9—Design, Construction and Control of a Remotely Operated Vehicle (Ariana-I ROV) [23]

The Ariana-I ROV is designed to be equipped with other tools such as mechanical arms. The objective behind development of the Ariana-I is to study the underwater navigation and advanced control for the Control Laboratory of Shiraz University. The next objective is to construct a prototype of the ROV to be used for visual inspection of submarine pipelines and oil platforms power cables. In the control system of the ROV, the Ariana-I can operate autonomously using the onboard pilot. The autopilot of the ROV is based on an Atmel AVR microcontroller interfacing with the electronic module and sensors in the ROV. Figure 13 shows the autopilot control system of the Ariana-I ROV. The interfacing to the sensors in the ROV is using the microcontroller ADUC8041. The signal data is transfered via a RS232

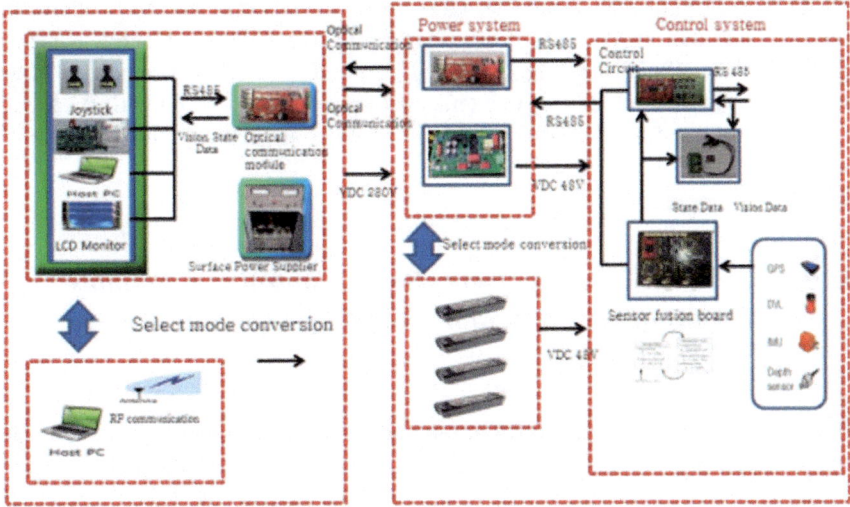

Fig. 12 Overall control system of KCROV

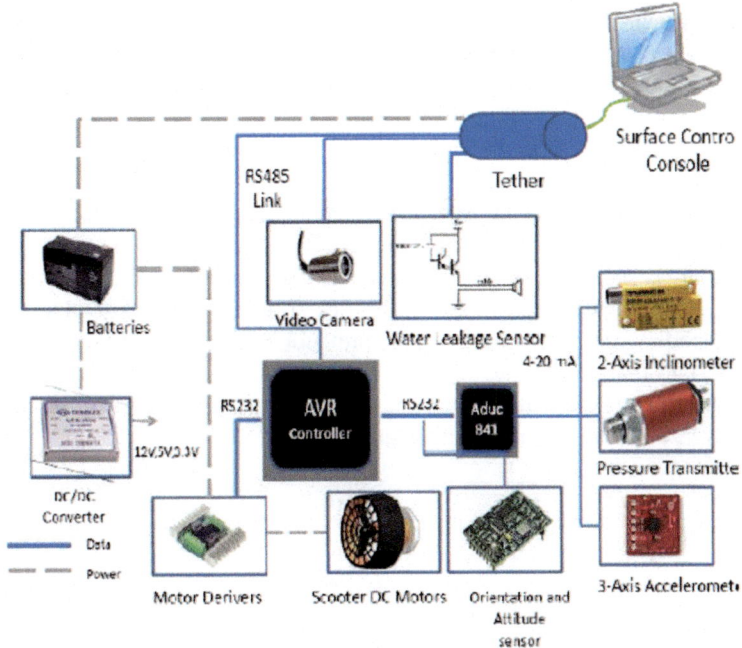

Fig. 13 Autopilot control system of Ariana-I ROV

link to the autopilot main controller (AVR microcontroller). The AVR microcontroller sends the signal data from sensors to the Surface Control Console (SCC) over the RS485 bus. The Graphical User Interface (GUI) is designed in the Surface Control Console (SCC) to interact between operators with the ROV. For all six degrees of freedom, the different control schemes are applied onboard or on the Surface Control Console (SCC).

4.10 Design 10—Development, Depth Control and Stability Analysis of an Underwater Remotely Operated Vehicle (DENA ROV) [7]

The DENA ROV is developed for an upgradeable platform and low cost vehicle. The development was focused on the design of a controller to reduce the overshoot limitation in the system response. When the ROV operates in a dangerous environment, the overshoot system response is important. So, the PID depth controller is designed to overcome the problem. The control system of DENA ROV has five electronic boards which are power, control, arm-sensor and two thrusters board. Figure 14 shows the control system of the DENA ROV. In the control board, the joystick is attached to the computer to be controlled by the pilot. The command is received by the joystick to translate into specific code words by using a tether cable RS232. The board consists of two AVR Atmega16 microcontrollers which are the "Main Micro" and the "Equipment Micro". The "Main Micro" is used for the moving command of the ROV while the "Equipment Micro" is used to handle the operations tools on the ROV. The advantages of using two microcontrollers are easy to change the ROV program for different missions. Next, the program of the ROV is written by using Visual Basic.NET for the main control program. Then, the program is executed on the computer at the area of surface to receive the instruction either from the keyboard or joystick. The instruction is send to the ROV in proper code words based on a defined protocol between the surface unit and the ROV. Besides that, to provide the real time information regarding the status of the ROV, the programs consist of three layers which is a Graphical User Interface (GUI) layer, software layer and the interface layer.

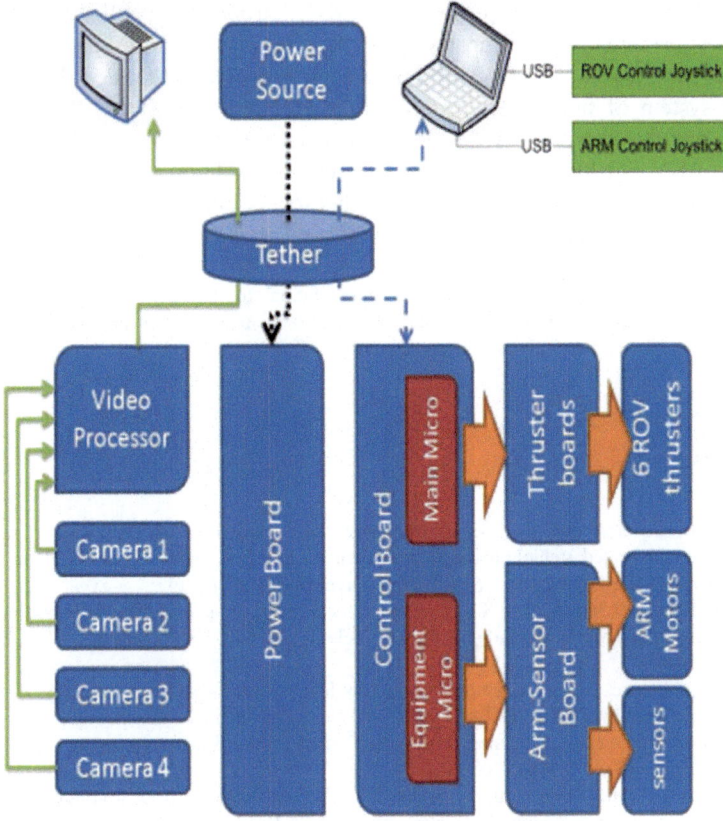

Fig. 14 Control system of the DENA ROV

5 Conclusion

This paper provides an explanation of the control system algorithm for Unmanned Underwater Remotely Operated Vehicle (ROV) focusing on the use of the Proportional Integral Derivative (PID) controller and the Fuzzy Logic Controller (FLC) for the control system of the ROV. Based on the review, there are advantages and disadvantages between these two controllers. For the FLC, it will provide fast response and reduced small overshoot. Other than that, the tuning parameter of the fuzzy controller is easy to implement. Otherwise, for the PID controller, it will increase the accuracy and reduce the external disturbance. So, by combining these two controllers which is Fuzzy PID, it will provide a simple and effective control system, good dynamic response and fuzzy control robustness. Based on the journal "Simple GUI wireless Controller of Quadcopter", the fuzzy logic is implement to the control model simulation while the PID controller is embedded and developed in the Arduino UNO to control the altitude motion. So, the outcome from the review paper is to implement the Fuzzy PID controller but applied in the underwater vehicle area. The comparison of the control system design of the same concepts of ROV is also discussed. The review provided the idea designing the control system of the project in terms of tools, microcontroller, communication system and sensors. So, for the project will take several design ideas from the existing ROV but different tools will be used and additional tools will be added. Lastly, the review provided valuable insights for the reader to design a new technique of control system and algorithm that is relevant in their field of research.

Acknowledgements Special gratitude and appreciation to Khairul Arieff Abu Jalil and NorHafizah Othman for providing the financial support from their grant application as well as moral support to complete this project successfully.

References

1. Aras, M.S.M., Abdullah, S.S.: Development and Modelling of Unmanned Underwater Remotely Operated Vehicle using System Identification for Depth Control (2013)
2. Choi, H., et al.: Design and Control of a Convertible ROV (2011)
3. Mohd Shahrieel Mohd Aras, S.S.A., Azis, F.A.: Review on auto-depth control system for an unmanned underwater remotely operated vehicle (ROV) using intelligent controller. **7**(1), 47–56 (2015)
4. Rúa, S., Vásquez, R.E.: Development of a low-level control system for the ROV Visor3. Int. J. Navig. Obs **2016**, 1–12 (2016)
5. Akkizidis, I.S., et al.: Designing a fuzzy-like pd controller for an underwater robot. Control Eng. Pract **11**(4), 471–480 (2003)
6. Henriksen, E.H., Schjølberg, I., Gjersvik, T.B.: Adaptable joystick control system for underwater remotely operated vehicles. **49**(23), 167–172 (2016)
7. Tehrani, N.H., et al.: Development, depth control and stability analysis of an underwater remotely operated vehicle (ROV), pp. 814–819 (2010)
8. Le, K.D., Nguyen, H.D., Ranmuthugala, D.: Development and control of a low-cost, three-thruster, remotely operated underwater vehicle. **9**(1), 67–75 (2015)
9. Cervantes, L., Castillo, O.: Type-2 fuzzy logic aggregation of multiple fuzzy controllers for airplane flight control. Inf. Sci **324**, 247–256 (2015)
10. Shakya, R., et al.: Design and simulation of PD, PID and fuzzy logic controller for industrial application. 2016 **4**(4), 363–368 (2014)
11. Lee, Y.H., Kim, H.S.: Digital controller design to control the direct current motor system. Int. J. Control Autom **7**(9), 283–288 (2014)
12. Yadav, S.: Fuzzy PID controllers using matlab GUI based for real time DC motor speed control. **4**(3) (2014).
13. Upalanchiwar, T., Sakhare, P.A.V.: Design and implementation of the fuzzy PID controller using MATLAB/SIMULINK model. **3**(3), 369–372 (2014)
14. Hanafi, D., et al.: Simple GUI wireless controller of quadcopter. Int. J. Commun. Netw. Syst. Sci **06**(01), 52–59 (2013)
15. Srinivas, P.: Comparative analysis of conventional Pid controller and fuzzy controller with various defuzzification methods in a three tank level control system. Int. J. Inform. Technol. Control Autom **2**(4), 75–86 (2012)
16. Immanuel, J., et al.: Design and development of real time MATLAB-GUI based fuzzy logic controllers for DC motor speed control system. **3**(1), 133–139 (2011)

17. Joachim, C., Phadungthin, R.: Design and development of a remotely operated underwater vehicle (2016)
18. Aristizábal, L.M., et al.: Design of an open source-based control platform for an underwater remotely operated vehicle. Dyna **83**(195), 198–205 (2016)
19. Ahmed, Y.M., Yaakob, O.: Design of a new low cost ROV vehicle (2014)
20. Govinda, L., et al.: Modelling, design and robust control of a remotely operated underwater vehicle. Int. J. Adv. Robot. Syst. 1 (2014)
21. Wood, S., et al.: Hybrid robot crawler/flyer for use in underwater archaeology, pp. 1–11 (2013)
22. Sousa, A., et al.: LAUV: the man-portable autonomous underwater vehicle. IFAC Proc. Vol **45**(5), 268–274 (2012)
23. Marzbanrad, A., et al.: Design, construction and control of a remotely operated vehicle (ROV), pp. 1295–1304 (2011)

Maneuvering and Submerged Control System for a Modular Autonomous Underwater Vehicle

Mohamid Salih Noorazlina, Md Salim Kamil, Mohammad Amiruddin Hashim, Nordiana Jamil and Hanisah Johor

Abstract An autonomous underwater vehicle (AUV) is considered a vehicle that can be submerged and autonomously operate underwater, fitted with various systems that include the electrical power supply system, propulsion system, tanks system, maneuvering and submerged control system. All these systems are controlled by an integrated built-in computer system. AUVs, as in the case of submarines, need control surfaces and the associated control systems for maneuvering while being underwater or on the surface. In the current study, rudder blades and hydroplanes made from carbon fiber were designed and installed suitably at stern and forward positions of the AUV. Arduino Mega board (AMB) module and its open technology are used to design the main automation circuit and written source codes for the maneuvering and submerged controlled system. The advantages of the AMB are its flexibility for the simulation using fritzing to test the performance of the system through computerisation of the control system and programmable to meet specified requirements as well as easy in interfacing the software and the hardware.

Keywords Autonomous underwater vehicle · Rudder blades · Hydroplanes Arduino mega board

---

M. S. Noorazlina (✉) · M. S. Kamil · M. A. Hashim · N. Jamil · H. Johor
Malaysian Institute of Marine Engineering Technology, Universiti Kuala Lumpur, Jalan Pantai Remis, 32200 Lumut, Perak, Malaysia
e-mail: noorazlinams@unikl.edu.my

M. S. Kamil
e-mail: mdsalim@unikl.edu.my

M. A. Hashim
e-mail: mohammadamiruddinhashim@gmail.com

H. Johor
e-mail: hanisah@unikl.edu.my

© Springer International Publishing AG 2018
A. Öchsner (ed.), *Engineering Applications for New Materials and Technologies*,
Advanced Structured Materials 85, https://doi.org/10.1007/978-3-319-72697-7_50

1 Introduction

Malaysia is a maritime and islanded country, therefore the use of AUV is suitable for underwater research activities and subsea cruise. Many enthusiastic roboticists invent or build AUVs as a hobby. AUV is also popularly known as unmanned underwater vehicle, programmable to specific functions, integrated with robotic functions, able to hover skim on surface, move underwater without control by human operators. Some AUVs communicate with operators periodically or consistently through satellite signals or submerged acoustic reference points to allow some level of control by the mother ships.

AUVs are used for underwater survey missions such as searching, detecting and locating submerged wrecks, rocks, underwater topographical mapping and other research activities. AUVs allow researchers to lead different analyses from a surface ship while the vehicle is off collecting data or gathering information elsewhere on the surface or in the deep ocean. Some AUVs can also make decisions on their own, changing their mission profile based on environmental data they receive through sensors while under way. All systems in AUVs use components, sensors and devices that are embedded in the associated systems such as fitted cameras, lights, sonar, etc.

A new design of maneuvering and submerged control system is developed for a modular AUV and capable to operate within the range of 2 m of operational submerged depth. Besides, it can be considered as a new design concept of modular AUV utilizing the AMB technology. The research focuses on designing the maneuvering and submerged control system for a cost effective modular AUV. Drawings of the control surfaces i.e. the rudders and hydroplanes were developed using SOLIDWORKS.

Nickell and Woolsey discovered in their study that if the AUV is slightly buoyant it may be incapable of generating sufficient down-force to maintain depth [1]. In 2001, Miyamoto et al. conducted a research project on a maneuvering control system design of an AUV in which all movements in yaw and pitch due rudders, thrusters and hydroplanes respectively were controlled by an integrated microcontroller [2]. In 1994 and 2005 Fossen studied the control system and quoted that it is responsible for providing the corrective signals to enable the AUV to follow the desired paths [3, 4]. These were achieved based on the signals received for the maneuvering or navigating the AUV. The operational pitch and yaw angles were then calculated and the corresponding forces and moments were applied through the use of the various actuating mechanisms fitted on the AUV. Nickell in 2005 also stated that the AUV dived by reversing the propeller and driving backwards to pull the AUV underwater [4]. The propeller direction of rotation was then reversed to provide the forward thrust to drive the AUV. By adjusting the mass in a rear position of a gravitational moment also helped the AUV to submerge. On the other hand, Ming et al. used a double server, warm-cold backup scheme in the control system design, and a fuzzy-PI motor control method was adopted in order to make the AUV to adapt to the complex underwater environment which cannot be precisely modelled [5]. Besides that, Ismail et al. proposed a novel robust dynamic region based control scheme, which was able not only to track a moving target as a region but also to position itself inside the region [6]. In 1991, Blidberg et al. gave explanation and some understanding about the AUV that is required for the economic potential of the ocean and to increase research and development in the area of underwater robots. Evans and Nahon in 2004, described a dynamics model of an AUV. Huggins and Packwood, with project title The Optimum Dimensions for a Long-Range, Autonomous, Deep-diving, Underwater Vehicle for Oceanographic Research, concluded from their study that a vehicle of displaced volume 3.75–4.4 m^3, travelling at a constant velocity of 2.5 m/s, of a low-drag, laminar-flow hull-design, with the layout suggested, may be capable of a transoceanic traverse [7–9]. In 2009, Jun et al. [10] developed a small AUV named IsiMI, to work as a test-bed AUV for the development and validation of various algorithms and instruments required to enhance the AUV's functions. Moreover, Stevenson et al. [11] studied about combining practical and hydrodynamic considerations to build a better AUV. Phillips et al. stated that steady state CFD has been used to successfully replicate the yawed towing tank and rotating arm experiments for a torpedo style AUV to derive the steady state hydrodynamic derivatives. Abrebekooh and Rad investigated the drag force and drag coefficient in a different flow direction for different Reynolds numbers of a special underwater model [12, 13]. Meanwhile, Kumar et al. [14] stated an approximate analytical solution to the optimal control problem which was derived for a stable underwater vehicles, both for two-dimensional and three-dimensional motions. In 2003, Fernandes et al. in their project reviewed the application of AUVs to fisheries- and plankton-acoustics research [15]. They conclude considering developments that may turn AUVs from objects sometimes perceived as science fiction into instruments used routinely to gather scientific facts.

2 Research Objectives

The objectives of the research are to design an integrated maneuvering and submerged control system for a modular AUV using the Arduino Mega board technology and to test and analyse the performance of the integrated maneuvering and submerged control system.

3 Methodology

Figure 1 shows the flow chart of the research activities and is self-explanatory. The hardware and the software design of the control circuit is described herein. The method involved the step by step processes from designing the hardware to the coding of the microcontroller. This design made use of the SolidWorks 2016 software to design and create the model into drawings and to get the overall views of the configuration of the actual AUV. The activities included the gathering of data and information, determining the dimensions of the rudder and hydroplanes, development of the control system coding, conduct of tests and analysis of the test results. The performance of the control surfaces can be validated by using different types of models and it can be easily modified for any conditions of the AUV.

4 Principle Dimensions of the AUV

The principal dimensions can be stated as follows:

Length overall	−236.24 cm
Length of cylindrical hull	−201.00 cm
Length of aft cone	−20.00 cm
Length of bow cap	−12.25 cm
Hull diameter	−30.48 cm
Rudder area (one rudder)	−24.26 cm^2
Hydroplane area (one hydroplane)	−22.40 cm^2
Hull volume	−159,000 cm^3

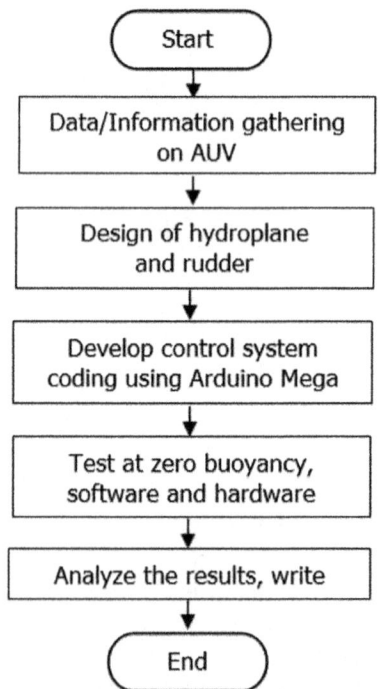

Fig. 1 Flow chart of activities

Figure 2 illustrates the test flow chart which interprets the angle of the hydroplane is either up or down conditions. Motors were used to move the fins in the direction of up and down for the AUV. Figure 3 shows the side view of the AUV which comprises of forward hydroplane that gives the movement of upward and downward deflection. It also shows the rudder at port and starboard deflections. At the aft hydroplane, it moves in the upward and downward directions. Figure 4 shows the plan view of the AUV for forward and aft hydroplane at starboard. At port also located the forward and aft hydroplane. Figure 5 demonstrates the components of the hydroplane and rudder that has been assembled completely. The testing and data gathered were from results observation set in the coding of AMB. Figure 6 shows the circuit developed to show the connections of motors and AMB in the system. The project has been simulated to obtain the simulation movement of the hydroplanes.

5 Arduino Mega Board Coding

The Arduino Mega board coding written for the maneuvering and submerged control system are given in Appendix A.

6 Conduct of Tests and Results

Results were taken to observe the movement of the hydroplane and fin angle to maneuver the AUV. At initial condition, the hydroplane was set at 900 as shown in Fig. 7. Next, the hydroplane was deflected to an angle of 1200. At this condition, the AUV was submerging. After 60 s, the hydroplane was then deflected to 600 and in this condition the AUV was surfacing. In this condition, all angles were set at 900 before it was being tested. At 120 s set in the coding, the movement of hydroplane continued running to perform deeper submerging at an angle of 1200 with a larger pitch angle and trim angle. As expected, the AUV was controlled by adjusting the hydroplane at a particular angle of dive as stated in Table 1.

7 Conclusion

An AUV is useful for use in underwater exploration and research activities. Maneuvering and submerged control system for a modular AUV is a vital system for the operation of the AUV. In conclusion, this research was successful and the objectives to design, test and analyze the performance of an integrated maneuvering and submerged control system for a modular AUV were achieved. It is potentially

Fig. 2 Tests flow chart

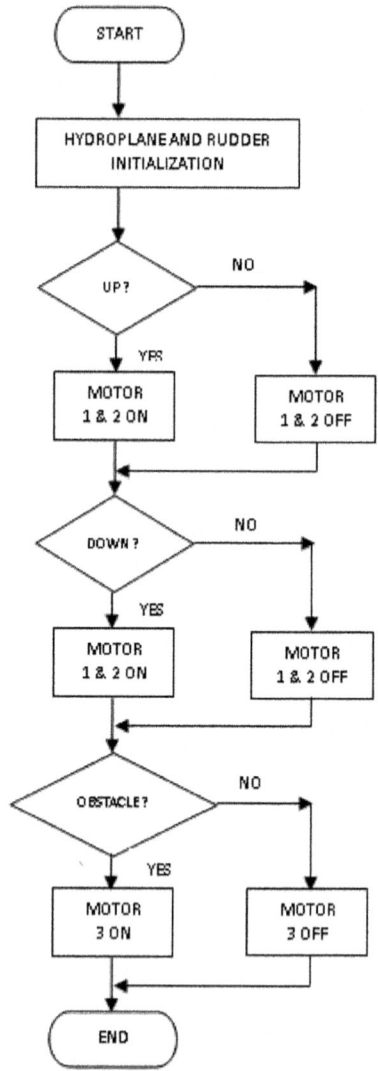

viable to be further developed for other scientific applications and purposes in the future. The research has successfully invented and has proven a typical functional system embedded with advanced system coding based on the Arduino Mega board (AMB) open technology capable of storing useful writeable codes for application in mechatronics control systems. Its main advantage is obvious that the AMB is easily programmable and adaptable to changing hardware for different applications.

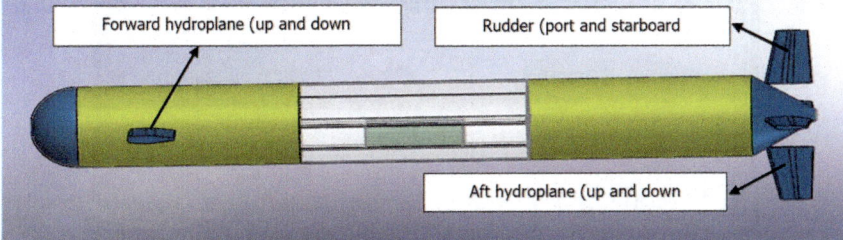

Fig. 3 Side view of the AUV

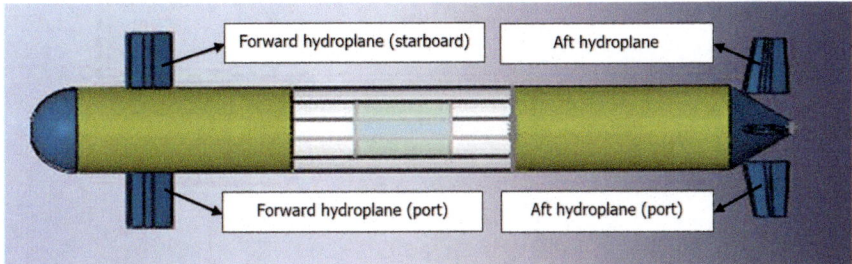

Fig. 4 Plan view of the AUV

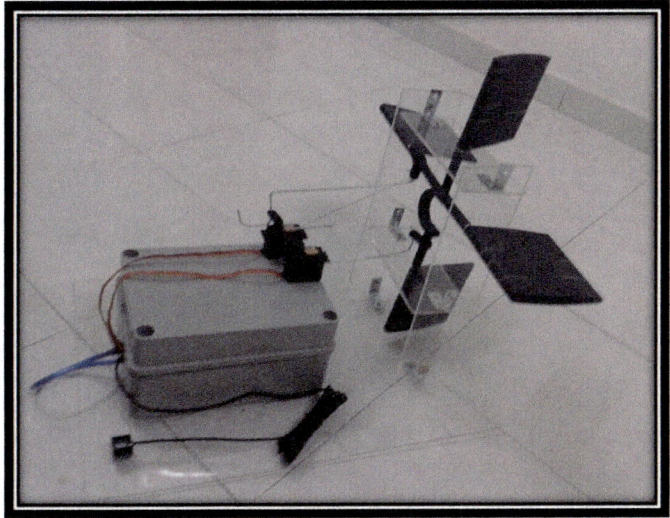

Fig. 5 Components of the hydroplane and rudder

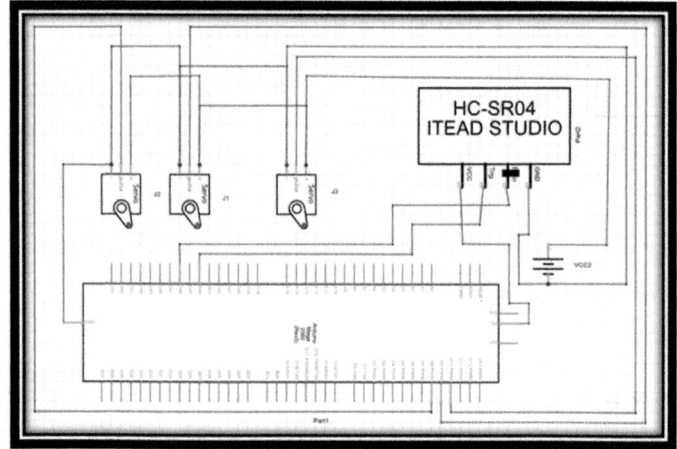

Fig. 6 The schematic circuit

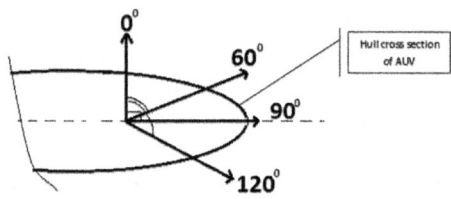

Fig. 7 Movement plane angle

Table 1 Data of fin at 120 s

Angle of hydroplane set (°)	Timer set (s)	Condition observation
60	120	Submerging
90	120	Initial (origin)
120	120	Surface

Appendix A

Arduino Mega Board Coding

```
//56266212266
#include <Servo.h>
#define maximumRange 4500
#define minimumRange 5
Servo myservo;
Servo myservo2;
```

```
Servo myservo3;
const int trigPin = 44;
const int echoPin = 46;
long duration, inches, cm;
unsigned long time;
unsigned int msecs, secs, mins;
int bal, flag, angle, obstacle, detect;
int motion [] = {90,180,90,0};
int motion_pointer = 0;
int a = 0;
int j = 0;
int OutOfRange = 0;
int Get_Sonar()
{
pinMode(trigPin, OUTPUT);
digitalWrite(trigPin, LOW);
delayMicroseconds(2);
digitalWrite(trigPin, HIGH);
delayMicroseconds(10);
digitalWrite(trigPin, LOW);
pinMode(echoPin, INPUT);
duration = pulseIn(echoPin, HIGH);
inches = microsecondsToInches(duration);
cm = microsecondsToCentimeters(duration);
Serial.print("Range = ");
Serial.print(cm);
myservo3.write(90);
myservo2.write(90);
}
void loop()
{
Serial.print("cm");
Serial.println();
if (cm < 100) {
obstacle = 1;
}
else {
obstacle = 0;
}
return obstacle;
}
long microsecondsToInches(long microseconds)
{
return microseconds / 74 / 2;
}
```

```
long microsecondsToCentimeters(long microseconds)
{
return microseconds / 29 / 2;
  }
  void loop()
  {
    Serial.println("————————————————");
    detect = Get_Sonar();
    if (detect == 1) {
     myservo2.write(120);
     //delay (1000);
    }
    else {
     myservo2.write(90);
     //delay (1000);
    }
    time = millis(); //prints time since program started
    bal = time/1000;
    delay(1000); // wait a second so as not to send massive amounts of data
    Serial.print("Seconds = ");
    Serial.println(bal);
    flag = bal%20;
    //Serial.print("Flag = ");
    //Serial.println(flag);
    if (flag == 0) {
      angle = pointer_motion(a);
      Serial.print("Angle = ");
      Serial.println(angle);
      myservo.write(angle);
      myservo3.write(angle);
      //delay(2000);
  }
    //Serial.print("————————————————");
    //Serial.print();
}
int pointer_motion(int x) {
  int motion [] = {60,90,60,90};
  //int j;
  if (j < 3) { //depend on motion[] size-1
   j = j + 1;
   x = motion[j];
   //j ++;
   return x;
  }
  else {
```

```
  j = 0;
  return motion[j];
 }
}
```

References

1. Nickell, C. L.: Modular modification of a buoyant AUV for low-speed operation. http://citeseerx.ist.psu.edu/viewdoc/summary?doi=10.1.1.460.267 (2005)
2. Miyamoto, S., Aoki, T., Maeda, T., Hirokawa, K., Ichikawa, T., Saitou, T., Iwasaki, S.: Maneuvering Control Syst. Des. Auton Underwater Veh. **1**, 482–489 (2001). https://doi.org/10.1109/OCEANS.2001.968771
3. Fossen, T.I.: Guidance and control of ocean vehicles. https://www.ntnu.edu/employees/thor.fossen. ISBN 0-471-94113-1 (1994)
4. Fossen, T.I., Smogeli, Ø.N.: Nonlinear time-domain strip theory formulation for low-speed maneuvering and station-keeping. Model. Ident. Control. **25**(4), (2004)
5. Ming, C., Qiang, Z., Sanlong, C.: Control Syst. Des. Auton. Underwater Veh. 1–6, (2006). http://doi.org/10.1109/RAMECH.2006.252736
6. Ismail, Z.H., Mokhar, M.B.M., Putranti, V.W.E., Dunnigan, M.W.: (A robust dynamic region-based control scheme for an autonomous underwater vehicle. Ocean Eng. **111**, 155–165 (2016). https://doi.org/10.1016/j.oceaneng.2015.10.052
7. Blidberg, D.R., Turner, R.M., Chappell, S.G.: Autonomous underwater vehicles: Current activities and research opportunities. Robot. Auton. Syst. **7**, 139–150 (1991)
8. Evans, J., Nahon, M.: Dynamics modeling and performance evaluation of an autonomous underwater vehicle. Ocean Eng. **31**, 1835–1858 (2004)
9. Huggins, A., Packwood, A.R.: The optimum dimensions for a long-range, autonomous, deep-diving, underwater vehicle for oceanographic research. Ocean Eng. **21**, 45–56 (1994)
10. Jun, B.H., Park, J.Y., Lee, F.Y., Lee, P.M., Lee, C.M., Kim, K., Lim, Y.K., Oh, J.H.: Development of the AUV 'ISiMI' and a free running test in an Ocean engineering basin. Ocean Eng. **36**, 2–14 (2009)
11. Stevenson, P., Furlong, M., Dormer, D.: AUV design—shape, drag and practical issues. Sea Technol. **50**(1), 41–44, 1 Jan (2009)
12. Phillips, A., Furlong, M., Turnock, S. R.: The use of computational fluid dynamics to determine the dynamic stability of an autonomous underwater vehicle. In: 10th Numerical Towing Tank Symposium (NuTTS'07), Hamburg, Germany, pp. 23–25, 2007
13. Abrebekooh, Y.N., Rad, M.: Experimental and numerical investigation of drag force over tubular frustum. Scientia Iranica **18**(5), 1133–1137 (2011)
14. Kumar, R.P., Dasgupta, A., Kumar, C.S.: Real-time optimal motion planning for autonomous underwater vehicles. Ocean Eng. **32**, 1431–1447 (2005)
15. Fernandes, P.G., Stevenson, P., Brierley, A.S., Armstrong, F., Simmonds, E.J.: Autonomous underwater vehicles: future platforms for fisheries acoustics. ICES J. Mar. Sci. **60**, 684–691 (2003)

Development of an Electric Turbo Generator for Automotive Application

Khairul Shahril Shaffee, Muhammad Idham Hassan Azmi,
Shahril Nizam Mohamed Soid, Muhammad Najib Abdul Hamid
and Tajul Adli Abdul Razak

Abstract A turbo charger is a device driven by exhaust gas that increases the engine power by pumping air into the combustion chambers. This project is to develop the electric turbo generator by assembling the turbine from the turbo charger with the alternator. This experiment is about to use the function of the turbo charger to generate the current for further needs of the electricity generation in automotive components. This project is focusing on the field of harvesting energy which manipulates the waste from the exhaust gas to move the turbine to rotate. The advantage of using the turbine is that the input of the turbine is only exhaust gas. So, there is no any extra fuel or electric source needed to operate the turbine. Besides, the waste gas from the engine is manipulating to be meaningful source to generate current. The result from the electric turbo generator average can produce only 9.3 V due to limited material and the consideration of the safety of the engine.

Keywords Turbo charger · Automotive application · Electric turbo generator

---

K. S. Shaffee (✉) · M. I. H. Azmi · S. N. Mohamed Soid
M. N. Abdul Hamid · T. A. Abdul Razak
Mechanical Section, Malaysia Spanish Institute, Universiti Kuala Lumpur, 09000 Kulim Hi-Tech Park, Kulim, Kedah, Malaysia
e-mail: khairuls@unikl.edu.my

M. I. H. Azmi
e-mail: idhamhassanazmi@gmail.com

S. N. Mohamed Soid
e-mail: shahrilnizam@unikl.edu.my

M. N. Abdul Hamid
e-mail: mnajib@unikl.edu.my

T. A. Abdul Razak
e-mail: tajuladli@unikl.edu.my

1 Introduction

An automobile is a self-propelled vehicle, which contains the engine as power generation, the transmission and clutch to connect and disconnect the power to the drive line, and the drive line to drive the vehicle moving on the road of the surface. The automotive generation became larger and trending with its own design, power, performance, and material of the vehicle. Researchers study in various fields to increase the performance of the engine by introducing turbo chargers. This precious component is capable to boost the performance of the engine by forcing extra air into a combustion chamber. Therefore, the efficiency of the engine increases rapidly and the vehicle can move faster with responsive acceleration and the speed can be achieved faster. Nowadays, the era of automotive is introducing intelligent components such as detector sensors, auto-drive management units, balancing sensors and the latest invention is the hybrid/electric vehicle technology. The aim of this study is to develop a turbo generator to produce electricity by using the turbine of the turbo charger. The turbine generator was developed by taking the turbine from the turbo charger and assembling the turbine to the dynamometer by a custom coupling system. Current voltage can be produced and used for the electrical components in the vehicle.

1.1 Problem Statement

Renewable energies and energy efficiency have consistently been need for action issues and they play an important role towards fulfilling climate targets and commitments towards energy security and access to energy. The application of a turbine generator is to generate current by using the part of the turbo charger. The purpose strongly supports the idea to produce current without more load to the engine like what alternators are doing now. Secondly, a higher current might be produced because the speed of the turbine also able to rotate higher and faster than the engine crank. The advantage of using the turbine is that the input of the turbine is the exhaust gas. This means that there is no any extra fuel or electric source needed to operate the turbine. Besides, the waste of engine is manipulated to be a meaningful source to generate current.

1.2 Objective of the Study

The main objectives are to design the turbo generator for automotive application, to fabricate and assemble the turbo generator and to measure the current voltage produced by the generator.

1.3 Overview of the Turbo Generator

In order to reach the main objective of this project, the best turbine should be identified and considered. If the turbine is larger and beyond the pressure of the hot gas of the car, it might tend to turbo lagging. Turbo lagging is the term of under pressure to move the blade of the turbine. Thus, the turbine will fail to function. Therefore, the type of car selected for this experiment also very importance. The space under the bonnet and the maintenance should be seriously taken in order to avoid the redo and excessive budget to use. This turbo generator performs slightly the same as the turbo charger. The difference is only that there is no compressor to force extra air to the combustion chamber. However, the compressor is replaced by the alternator to produce electricity.

2 Turbo Generator Systems

2.1 Turbo Charger Concepts

A turbo charger is made up of two main sections: the turbine and the compressor. The turbine consists of the turbine wheel and the turbine housing. It is the job of the turbine housing to guide the exhaust gas into the turbine wheel. The energy from the exhaust gas turns the turbine wheel, and the gas then exits the turbine housing through an exhaust outlet area. The compressor also consists of two parts: the compressor wheel and the compressor housing. The compressor's mode of action is opposite to that of the turbine. The compressor wheel is attached to the turbine by a forged steel shaft, and as the turbine turns the compressor wheel, the high-velocity spinning draws in air and compresses it. The compressor housing then converts the high-velocity, low-pressure air stream into a high-pressure, low-velocity air stream through a process called diffusion. The compressed air is pushed into the engine, allowing the engine to burn more fuel to produce more power [1]. The components of the turbo generator are shown in Fig. 1.

2.2 Turbo Charger Failure

Turbo charges deal with hot temperature from the exhaust gas and together produce the high boost of pressure. The high temperature can drive the material of the turbo to deform and then the high boost from the compressor is also able to fracture the turbo by experiencing the over stressing through the component. Therefore, the turbo must be equipped to the right engine according the level of horse power in order that the turbo and engine operates efficiently without failure [2]. The example of a turbo failure is shown in Figs. 2 and 3 [3].

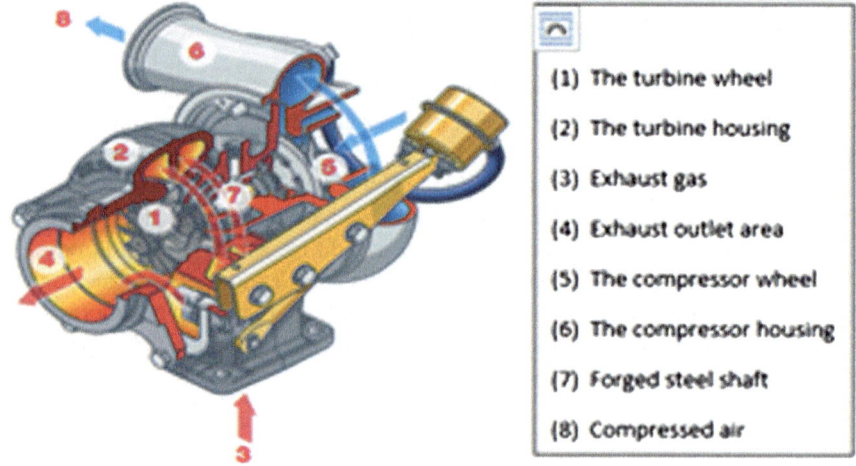

Fig. 1 The components of turbo charger

Fig. 2 Propeller failure

2.3 *Alternator*

An automotive charging system is made up of three major components: the battery, the voltage regulator and an alternator. The alternator works with the battery to generate power for the electrical components of a vehicle, like the interior and exterior lights, and the instrument panel. An alternator gets its name from the term

Fig. 3 Lubricity failure

alternating current (AC). An alternator supplies the power to run the vehicle electronics such a ignition, lights and to charge the battery. Without alternator, the battery soon will have lost all its powers just after starting the vehicle since the battery has to deal with a huge power drain when starting [4] (Fig. 4).

Fig. 4 Typical automotive alternator

2.4 Alternator Voltage

Normally the alternator has to rotate about 1400–2000 rpm at shaft to achieve 50% rated output. But some of the alternators are able to achieve its 100% output at 2500. The efficiency of the alternator is higher when more current voltage is produced with low rpm. The desire of alternator output in automotive application is around (12 V, 40 amp) or the output must not be less than 12 V and 30 amp. Figure 5 presents the output of the alternator at three stages of rpm [5].

2.5 Type of Coupling

In order to transmit the torque between two shafts that either tend to lie in the same line or slightly misaligned, a coupling is used. Based on the area of applications there are various types of couplings available. These are generally categorized in the following varieties [6].

1. Rigid Couplings
2. Flexible or Compensating Couplings
3. Gear couplings

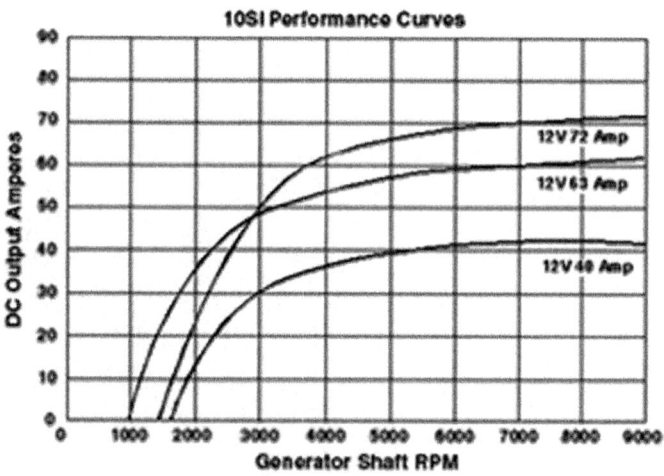

Fig. 5 The graph of alternator output versus the speed of rpm

2.6 Universal Joint

A joint or coupling that allows parts of a machine not in line with each other for limited freedom of movement in any direction while transmitting rotary motion is also called a universal coupling. This amazing joint allows the user to connect a shaft even at an uneven surface. Then, the vibration cause of the not straight connection can be avoided [7].

2.7 Car Selection

The car selection is a Perodua Myvi EZi 1.3. The decision to choose this car is because the power from the engine is able to operate the type of turbo charger that was already provided in this project and also the part availability.

2.8 Turbo Charger Selection

The turbo charger that used in this project is a Nissan Skyline Garret M24 AR42 with the working range from 250 to 360 HP, as shown in Fig. 5. The recommended engine sizes for the turbocharger are 1800–3000 cc for twin turbo while the minimum is 1300 cc for the single turbo application [8] (Fig. 6).

Fig. 6 Garret m24 AR42 turbo charger

3 Methodology

3.1 Overall Methodology

Figure 7 explains about the whole planning about the step and flow for the development of the electric turbo generator. The initial stage of this project is idea generation and brainstorming in regards the electric turbine generator, and then project planning, identification of all the components needed for the development, choosing the type of automotive engine to be used and representing the overall look of the project in the Catia software as the presented model for the project that will be fabricate in the next stage. The fabrication process is to build the electric turbine generator by assembling the turbine to the alternator. After equipping the turbine

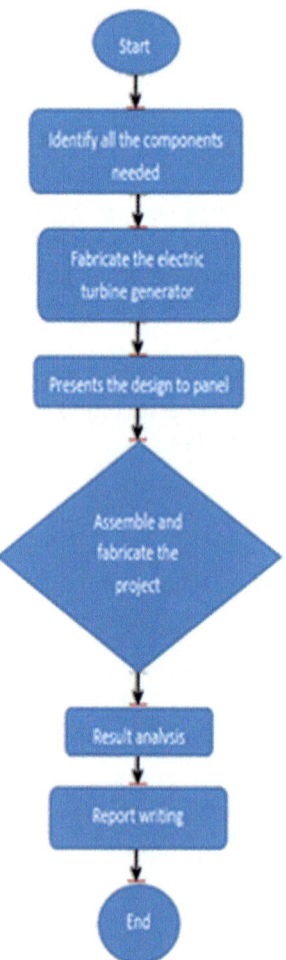

Fig. 7 Flowchart of the overall project

with the engine and running the project to measure the output which means how much current and voltage it can produce.

3.2 Experiment Setup

Figure 8 shows the flow chart for the experimental setup. The flow chart shows the important step by step of the process before obtaining a result. The experimental setup consists of two stages of experiments. Firstly, ensuring all the components are

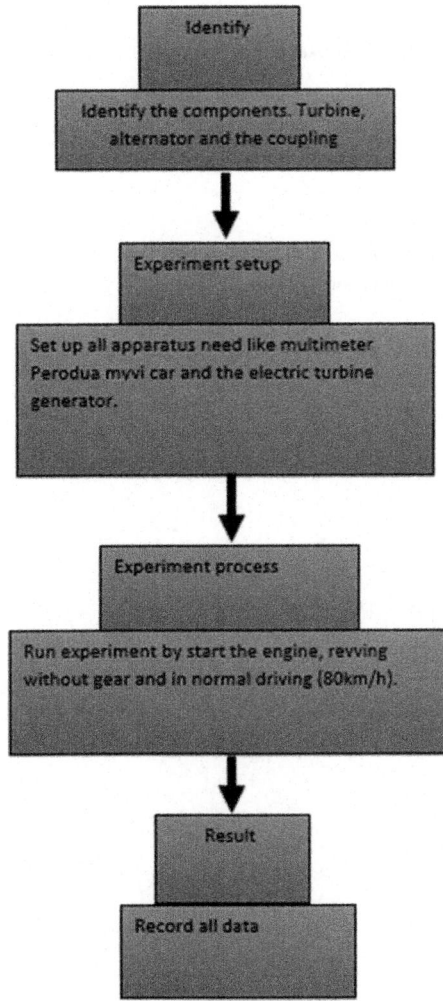

Fig. 8 Flow chart of experiment process

in good condition for performance very well. Secondly, assemble all the components to become an electric turbo generator.

3.3 Experimental Setup for Component

Refer to the Fig. 9, which shows the correct flow how to assemble all the components to develop the electric turbine generator and then run the experiment for result measurement.

3.4 Turbine Preparation

The turbine is separated from the compressor and replaced by the alternator, as shown in Fig. 10.

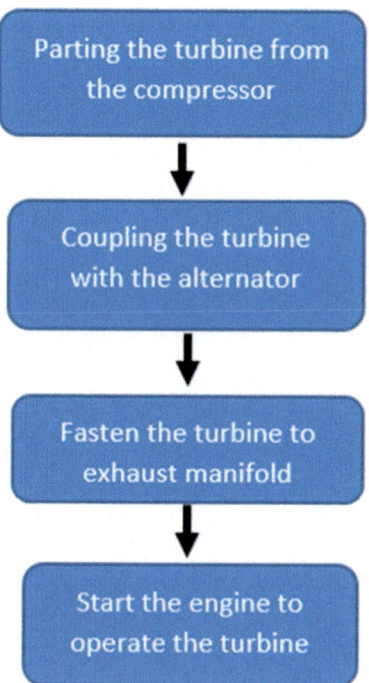

Fig. 9 Flow chart on setup the component for experiment

Fig. 10 Turbine separation

3.5 CATIA Design

Figure 11 presents the overall look of the electric turbine generator and the specification of the components.

3.6 CATIA Design for New Assembly

Figure 12 represents the new assembly of the electric turbo generator. When comparing the old design of Fig. 11, it shows that the compressor remains fitted while the alternator is connected to the turbine shaft by using the steering joint replacing the coupling. The changing reason is the limited space at the engine area, where it is impossible to fit the alternator at the compressor site. If the design would follow the old design, then a lot modification would be needed such cutting, welding and removing material of the engine.

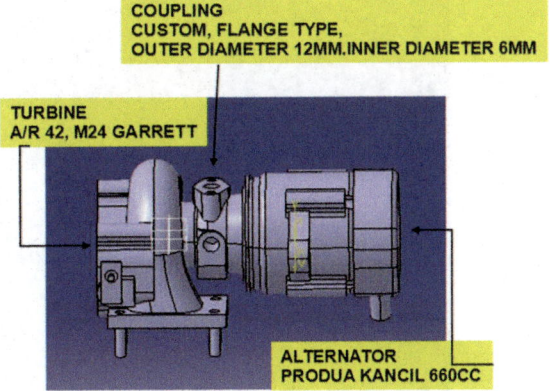

Fig. 11 Assembly modeling in Catia software

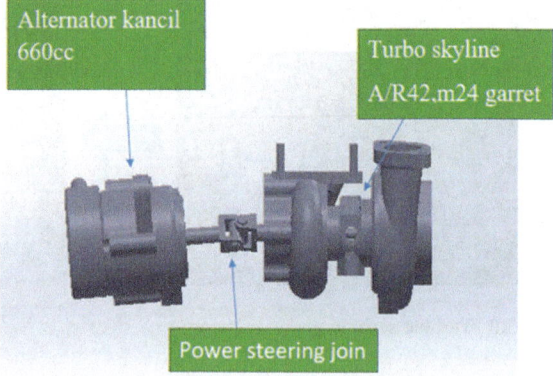

Fig. 12 New assembly of electric turbo generator

3.7 Assembly Process

The turbo is mounted to the custom manifold which is a special design for the Perodua Myvi's engine as shown in Fig. 13.

3.8 Modification of Steering Joint

The process in modification and assembling of the steering joint is to merge with both the shaft alternator and the turbine shaft including welding, grinding and cutting processes, as shown in Fig. 14. Refer to Fig. 15, the complete electric turbo

Fig. 13 Turbo and manifold assembly process

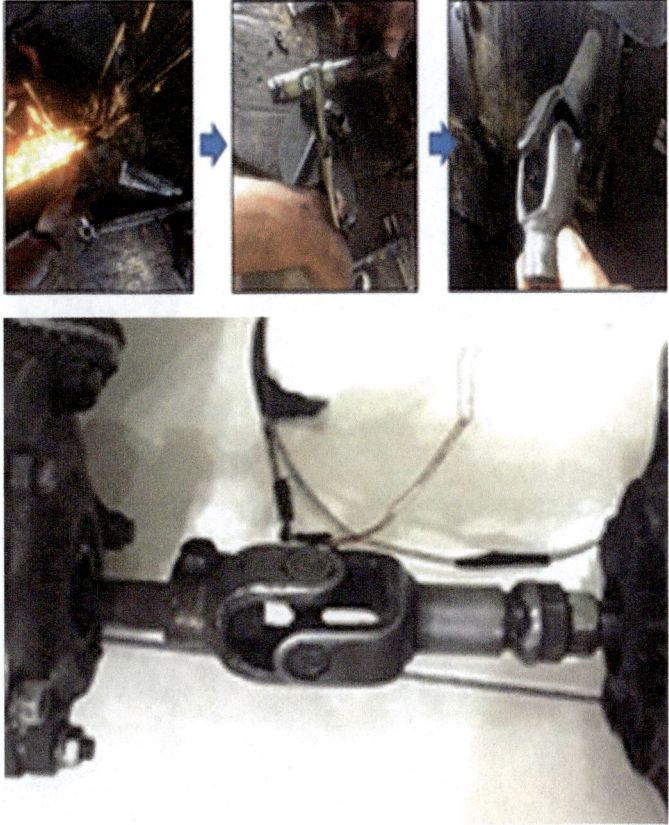

Fig. 14 Preparation of shaft coupling

generator consists of the compressor part. Even the assembly is more a hybrid configuration which means the turbo can compress air and generate electricity, but the aim of this study is to measure the current generated. Figures 16 and 17 show that the turbo generator is well fitted to the engine and completes the assembly process of the electric turbo generator.

3.9 Electric Turbine Generator Electric Circuit Diagram

From the diagram as shown in Fig. 18, the bulb initially is ON. When the engine is started and revving, the hot gas from the exhaust starts forcing to the turbine

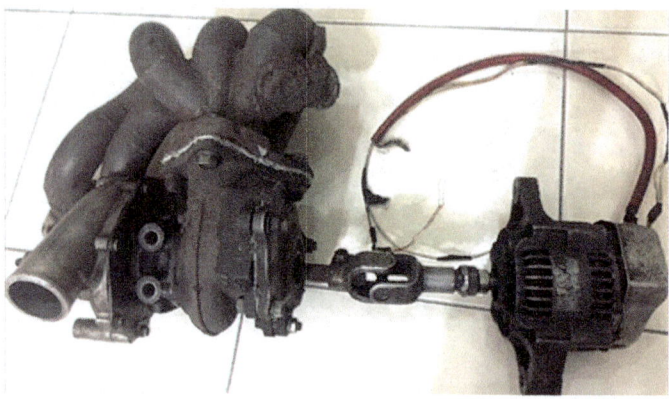

Fig. 15 Complete assembly of electric turbo generator

Fig. 16 Turbo is fitted to the engine

impeller thus, rotating the alternator shaft. The alternator starts to produces voltage and recharges the battery. When this occurs, the bulb will be OFF to indicate the charging process. If the light is still ON when the shaft is rotating, then it mean a failure of the recharging process.

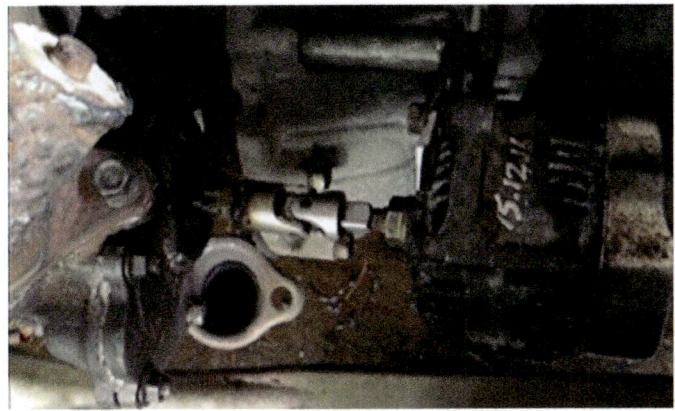

Fig. 17 Complete assembly process of the electric turbine generator

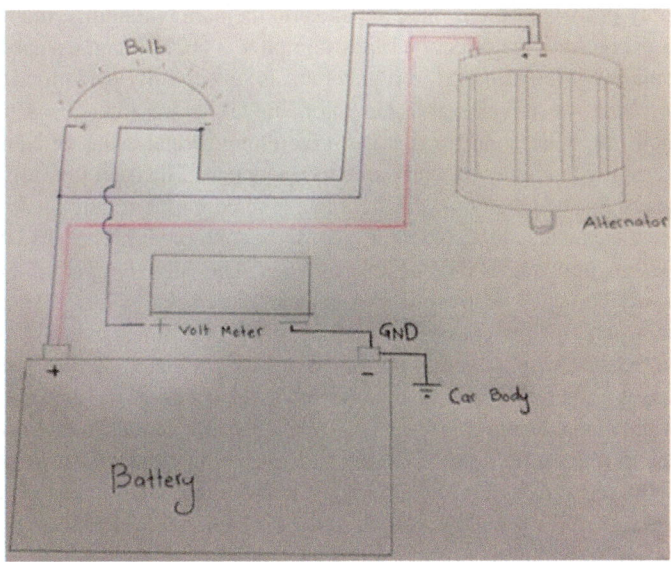

Fig. 18 Electric circuit diagram

4 Results and Discussion

4.1 Results

From the results the voltage produced by the alternator is 9.4 V for experiment 1, 9.3 V for experiment 2, and 9.2 V for experiment 3. The maximum output a normal alternator can reach is 12 V. For this special brand, it can reach up to 14 V.

Fig. 19 3 tests conducted

The result from the electric turbo generator average can produce only 9.3 V due to the limited material and the consideration of the safety of the engine (Fig. 19).

4.2 Discussion

The assembly process of the project depends on the part availability and the budget constraints. Therefore, the selection of the component is important to make sure that the development of the electric turbo generator is successfully performed. The used parts were chosen to develop the generator, but there are some risks to be considered about the ware and tare that can give effect, because the efficiency of the part to function will a bit lower compared to a new part. The used steering joining is also too heavy for the generator. As the load is increased the rotation also will slow down. The experiments are also unable to push beyond the limit because there is no cooling system applied to the turbine bearing. The high temperature from the generator will damage the overall system and the engine's life. The maximum voltage output that the generator can produce is still questioned and a higher output still can be reached with a cooling system and the right materials. From the overall assembly design, there are several changes in the design as shown in Catia design 1 in Fig. 11 and Catia design 2 in Fig. 12. This is because of the limited space of the engine area. Furthermore, if the assembly process follows exactly the design as first plan, heavy modifications to the engine part like cutting, welding a new mounting housing and a lot of cost will be needed to do so. Then, the new assembly design is the best design to avoid all of those disadvantages.

5 Conclusions and Recommendation

From the results the overall conclusion shows the project has achieved the main objective with the voltage produced by the electric turbo generator. The selection of material is also very important to avoid any factor that can affect the final result. The assembly work has been conduct properly according to the plans. Therefore, the electric turbo generator still can be improved by fitting the right materials and a cooling system. To explore more about the output of the electric turbo generator and the intend to get a proper result at every stage of engine rpm there are some recommendations from the experiments, which need to be considered for future testing:

1. Selection of the turbine—The best is to consider a smaller turbine because it can produce more pressure depending on the type of the engine used.
2. Type of coupling—Rigid coupling might reduce the load and vibration to the connecting shaft and provide more speed to the shaft to rotate.
3. Cooling system—The most important matter to make sure that the overall experiment is safer because of the extreme temperature.

References

1. History of the exhaust gas driven turbocharger. 1 Nov 2016. Retrieved from (http://en.turbolader.net/Technology/History.aspx)
2. Batzel, T.D., Swanson, D.C., Defenbaugh, J.F.: Predictive diagnostics for the main field winding and rotating rectifier assembly in the brushless synchronous generator. In: Proceedings of 2003 IEEE symposium on diagnostics for electric machines, power electronics and drives Conference, pp. 349–354
3. How Turbochargers Work. 2 Nov 2016. Retrieved from (http://auto.howstuffworks.com/turbo3.html)
4. Fuchs, E.F., Rosenberg, L.T.: Analysis of an alternator with two displaced stator winding. IEEE Trans. Power Appar. Syst. **93**, 1776–1786 (1974)
5. Comnac, V., Cernat, M., Mailat, A., Vittek, J., Rabinovici, R.: New 42 V automotive supply system. Based on conventional 14 V alternator". In: Proceedings OPTIM 2008, Brasov, Romania, vol. II, pp. 271–276, 22–24 May 2008
6. Darpe, A.K., Gupta, K., Chawla, A.: Coupled bending, longitudinal and torsional vibrations of a cracked rotor. J. Sound Vibr. **269**, 33–60 (2004)
7. Hummel, S.R., Chassis, C.: Configuration design and optimization of universal joint. Mech. Mach. Theory **33**(5), 479–490 (1998)
8. Gale Banks—Why Turbos Are the Best. 3 Nov 2016. Retrieved from http://turbochargerspecs.blogspot.my/

Printed by Printforce, the Netherlands